D0238122

Green Development, 2nd edition

Since the first edition of *Green Development* was published, sustainable development has become a vital part of government policy and a key issue in corporate boardrooms around the world. This new edition has been completely rewritten, but it retains the clear and powerful argument that so characterised the original. It gives a valuable analysis of the theory and practice of sustainable development and suggests that at the start of the new millennium we should think radically about the challenge of sustainability. This fully revised edition discusses:

- the roots of sustainable development thinking and its evolution in the last three decades of the twentieth century;
- the dominant ideas within mainstream sustainable development (ecological modernisation, market environmentalism and environmental economics);
- the nature and diversity of alternative ideas about sustainability (for example ecosocialism, ecofeminism and deep ecology);
- the problems of environmental degradation and the environmental impacts of development;
- ideas about political ecology and risk society, applied to deforestation, conservation and pollution;
- development project planning and environmental assessment in theory and practice;
- strategies for building sustainability in development from above and below.

Green Development is unique in offering a synthesis of theoretical ideas on sustainability North and South, and of practical policy development in the Third World. It gives a clear discussion of theory and extensive practical insights drawn from Africa, Latin America and Asia. It now has further reading and chapter outlines and summaries for the student reader.

W.M. Adams has worked for over twenty years researching the problems of environment and development in Africa. He has written books on wildlife conservation, water resources management and dryland agriculture. He is Reader in Conservation and Development in the Department of Geography at the University of Cambridge.

Green Development, 2nd edition

Environment and sustainability in the Third World

W.M. Adams

Routledge
Taylor & Francis Group

LONDON AND NEW YORK

First published 1990 by Routledge
2 Park Square, Milton Park, Abingdon, Oxon OX14 4RN

Simultaneously published in the USA and Canada
by Routledge
270 Madison Avenue, New York, NY 10016

Second edition 2001

Reprinted 2003, 2004, 2006, 2007

Routledge is an imprint of the Taylor & Francis Group, an informa business

Typeset in Galliard by
Florence Production Ltd, Stoodleigh, Devon
Printed and bound in Great Britain by
Biddles Ltd, King's Lynn, Norfolk

British Library Cataloguing in Publication Data
A catalogue record for this book is available from the British Library

Library of Congress Cataloging in Publication Data
Adams, W. M. (William Mark)
  Green development : environment and sustainability in the Third World /
  W.M. Adams.—2nd ed.
    p. cm.
  Includes bibliographical references and index.
  1. Sustainable development—Developing countries.   2. Green movement—
  Developing countries.   I. Title.
  HC59.7 .A714 2001
  338.9′27′091724—dc21
  2001016595

ISBN 978–0–415–14765–1 (hbk)
ISBN 978–0–415–14766–8 (pbk)

To my father H.C. Adams, 1917–1987

Contents

Plates

Figures

Tables

Preface

The first edition of this book was conceived in 1984 and written in the second half of the 1980s. In retrospect, the question of environment and development at that time was still a backwater, navigable reasonably simply in a single volume. Since then it has become a turbulent intellectual and popular torrent. Massive changes have taken place in the ways in which those active in development view the environment, and the ways in which environmentalists understand development. Academic writing in the field of environment and development has flowered and seeded wildly. Internationally, the political road-show of the Rio Conference transformed the terms within which debates about environment and development were held, and gave them hugely inflated profiles. There has been an astonishing growth in apparently 'green' ideas and statements of intent from development agencies, up to and including the World Bank, and governments. In the early 1990s, what had been a minority concern about the course and costs of 'development' suddenly became conventional wisdom. Environmentalist radicalism became mainstream environmental planning. Environmentalists came in from the cold to talk at boardroom tables; indeed, the environment became a significant growth industry in itself, a vital part of the corporate portfolio of the eager executive in rapidly globalising companies.

For an author, this explosion of activity is in a sense a delight – but it has to be admitted that it also presents problems. When *Green Development* was conceived, the debates it explored were not only marginal to development planning, but marginal academically. A decade on, at the start of a new century, nothing could be more different. In the 1980s it became accepted by those working in development that the environment was important: remarkably, in the 1990s, it became fashionable among social scientists as a whole, the centre of a vibrant debate and the fertile sparring ground of new talent and ideas.

Academics are as susceptible to the allure of a bull market as any other group of entrepreneurs, and government policy concern has fed rapidly through into funding opportunities and in turn into theoretical development. Environmental economics, ecological economics, political ecology, ecofeminism and deep ecology have all developed extensive, demanding and exciting literatures, and debates about the social construction of nature have extended the once unfashionable area of 'environmental issues' by opening challenging links with wider developments in social science associated with the 'postmodern turn'. A sophisticated and vigorous theoretical debate about sustainability grew to parallel policy debate about environment and development. The sheer weight of paper devoted to the subject of sustainability should give all those of us who have contributed to it pause for reflection.

My own understanding of both sustainability and development has continued to grow and evolve in the past ten years. I have been distressed by the way in which the radical potential of debates about poverty and environment has been dissipated, and the ease with which key words and phrases have been taken up and incorporated as a greenwash over 'business as usual'. Truly, the path to sustainable development is paved with good intentions, but the rhetorical vagueness of that master-phrase 'sustainable development' has made it too easy for hard questions to be ignored, stifled in a quilt of smoothly crafted and well-meaning platitudes. My own ideas continue to be informed by both a sometimes tense mixture of humanistic concern for human needs, and a biocentric awareness of human demands on nature. These concerns stem from different strands of moral argument, and can conflict in certain places and at certain times. Both are important: non-human nature has intrinsic value but also underpins economy and society by its capacity to provide resources to meet human needs. It seems to me that if the 'sustainable development' debate is to have any value, it must address the challenge of relationships between people in their use of nature, and between humans and the rest of the biosphere. In this second edition of *Green Development*, as in the first, I have continually been forced to challenge my own beliefs.

The second edition of *Green Development* differs from the first in several ways. First, it must perforce be less encyclopedic in its coverage of the literature. The diversity and sheer volume of writing on environment and development precludes the kind of panoptic treatment in which I indulged in the first edition. Partly because of the enterprise of the many distant strangers who have put finger to keyboard, the book still remains a record of a relentlessly steepening learning curve. It has been exhilarating to write, although daunting. The second edition is also somewhat different in presentation. It became clear that although the first edition was not written primarily as a textbook, it was mostly being read as one. Many people, both those who inflicted it on their students and those who endured its pages, pointed out that it was not very user friendly: there were no illustrations, no guides for further reading, precious few tables. Much of it was less than easily readable. I have therefore tried to overcome some of these shortcomings in this edition. Equally, the first edition sought to be a book that made an argument – a book with a clear beginning, middle and end, rather than simply being a smorgasbord of ideas and examples from which busy teachers and students could pick a tasty plateful. I hope that this aspect of the first edition has survived. I hope, too, that while this is obviously an academic's book, it is not an overly academic book, in the sense that I have been successful in making it readable.

A word is needed on the use of language. First, the phrase 'Third World', used in the book's subtitle. This is just one of the many euphemistic and disparaging labels for less-industrialised countries (the 'South', 'developing' or 'underdeveloped' countries). In this edition I have used 'South' and 'Third World' interchangeably. Neither is satisfactory. The collapse of the Iron Curtain in Europe made the original distinction of the label 'Third World', chosen by those non-aligned countries lying between capitalist 'First' and socialist 'Second' worlds, outdated. However, as will become clear, none of the words in the book's title is uncontested: the idea of development is shot through with contradictions, and the concept of sustainability is likewise confusingly complex in its meaning. Anyway, as will also become clear, this book is not a handbook for making something called 'sustainable development' happen faster or in particular places. The shortcomings of the labels 'the South' or 'the Third World' are obvious but unavoidable. I would argue that, paradoxically, their manifest inadequacy is their strength.

Second, I have chosen not to litter the text with '*sic*' where I have quoted authors from previous decades (or the names of current organisations) who choose to talk of 'man and nature' or 'man and the environment' rather than using less sexist language; times change – in this small way, at least, for the better – but past times used past language. I can but apologise if my usage offends.

Green Development tries to bridge two important gulfs. The first is between environmentalism and development, and the second between armchair theory and practice. To create the first bridge I have devoted most of the first six chapters of the book to a discussion of the theory of sustainable development. These chapters argue that environmentalists still have much to learn from radical views of political economy, although (regrettably, but perhaps sensibly) I do not pretend that they offer a simple short cut to an adequate radical Green theory.

The second gulf that the book tries to stretch across is that between the ivory tower (in which it will be clear that this has been written) and the real world of development decisions. To tackle this, later chapters explore environmental problems at smaller scales, focusing on environmental aspects of development projects and programmes, in drylands, forests and water resources and in urban and industrialised contexts. The environmentalist challenge to development has been mounted on the basis of the critique of development practice, and calls for the reform of environmental planning procedures. I have discussed both the nature of the environmental impacts of development, and the attempt to reform the development process by imposing 'sustainable development from above' through ecological modernisation, as well as the potential for achieving 'sustainable development from below'.

This book argues that the 'greening' of development calls for a quite fundamental reassessment of the concept of development itself. The mainstream environmentalist critique of development sees a need for reform, changes of planning procedures, better valuation of nature, and better techniques for trading off economic growth, poverty and environmental degradation. This kind of thinking about sustainable development has dominated the 'greening' of development that has taken place. It has been the core of the key documents on sustainable development from the *World Conservation Strategy* and the Brundtland Report to *Agenda 21*, and the dominant arguments about sustainability by environmentalist pressure groups, businesses and governments.

However, behind this pragmatic (and within limits effective) 'mainstream' view lie deeper and more subversive visions of the nature and scale of the human impact on non-human nature, and the success of the capitalist world economy in providing for human needs in a way that endures. There are important questions of power, over both environment and people. At present this power is wielded by banks, aid donors and governments, and increasingly by businesses that have a global reach. Reforming the way that power is used such that it promotes equity and sustained opportunities for resource-based wealth creation is a reasonable goal: indeed, it is one whose widespread acceptance seemed most unlikely in the 1980s.

However, achieving sustainable development according to this paradigm is only part of the Green challenge to development. The world is not a machine, to be run by privileged super-mechanics, however skilled in environmental housekeeping. Rather than simply contributing to the enhanced efficiency of centralised bureaucratic and technocratic power, 'green' development must also address the capacity of individuals and groups to plan and run their own lives, and control their own environments.

If there is a single conclusion to be drawn from what I have written, it is simply this: it is no good talking about how the environment is developed, or managed, unless this is seen as a political process. Any understanding of environment and development must come back to this issue. The radical potential of 'greenness' in development planning, therefore, is to be sought not in its concern with ecology or environment *per se*, but in how it addresses questions of control, power and self-determination in the social engagement with nature.

Acknowledgements

Many people have helped in the writing of this book. I have learned a great deal from colleagues with whom I have worked or travelled, particularly Martin Adams, Tim Bayliss Smith, Alan Bird, Lisa Campbell, Pat Dugan, Polly Gillingham, Jack Griffith, Dick Grove, David Hulme, Mark Infield, Uwem Ite, Kevin Kimmage, Mike Mason, Charles Megaw, Mike Mortimore, Marshall Murphree, John Murton, Ted Scudder, David Stoddart, David Thomas, Bhaskar Vira and Liz Watson. I remain enormously grateful for the opportunity to have worked with Ted Hollis, whose energy, passion and intelligence were an inspiration.

I have also gained greatly from the chance to teach the many enthusiastic and often frighteningly well-informed students in the Department of Geography at Cambridge and the Faculty of Environmental Design in Calgary. I am grateful to Chris Park for inviting me to write the first edition of *Green Development*, to Ian Agnew for drawing the diagrams for this one, to those who have commented on draft manuscripts and chapters, including Glenn Banks, Aaron Padilla and Simon Batterbury and to referees.

Above all, I would like to thank Franc for teaching me to stretch. Emily and Thomas have also helped in their ways, and perhaps most significantly as a reminder that people and nature matter.

Copyright acknowledgements

The author and publisher would like to thank the following for granting permission to reproduce material in this work:

Arnold, London, for figure 3.3, p. 70, from *Environmental and Social Impact Assessment: an introduction,* by Barrow, C.J., Arnold, 1997 (Figure 11.1).

Blackwell Science Inc., for figure 2, p. 40, from 'Natural capital and sustainable development', by Costanza, R. and Daly, H.E., in *Conservation Biology* 6: 37–46, 1992 (Figure 5.11).

Cambridge University Press, for figure 1.1 from 'The urgent need for rapid transition to global sustainability', by Goodland, R.J., Daly, H.E. and El Serafy, S., in *Environmental Conservation* 20: 297–309, 1993 (Figure 5.6).

The Copyright Clearance Center Inc., for figures 1.1 and 1.2 from *Environmental Economics and Sustainable Development,* by Munasinghe, M., in World Bank Environmental Paper No. 3, World Bank, Washington, 1993 (Figures 5.8 and 5.9).

Edward Elgar Publishing, Cheltenham and Edward Barbier, for figure 16.1, p. 347, from 'Valuing environmental functions: tropical wetlands', in Barbier, E.B., *The Economics of Environment and Development: selected essays,* Edward Elgar, 1998 (Figure 5.4).

Elsevier Science, for a figure from 'Tropical deforestation: balancing regional development demands and global environmental concerns', by Wood, W.B., in *Global Environmental Change,* 1: 23–41, 1990 (Figure 9.4).

Greenpeace, for Plate 4.1.

IUCN, for two text diagrams entitled 'The need to integrate conservation with development' and 'Depletion of living resources' from *The World Conservation Strategy,* by the International Union for Conservation of Nature and Natural Resources, United Nations Environment Programme and the World Wildlife Fund, IUCN, 1980 (Figures 3.1 and 3.2).

International Institute for Environment and Development, London, for figure 1.3 by R. H. Behnke and I. Scoones, in *Rethinking Range Ecology*, ODI/IIED, 1991 (Figure 7.5).

Island Press, California, for figure 1.1, p. 9, figure 1.2, p. 10, figure 1.3, p. 10 and figure 8.2, p. 133, all from *Investing in Natural Capital: the ecological economics approach to sustainability*, by Jansson, A., Hammer, M., Folke, C. and Costanza, R. (eds) Island Press, 1994 (Figures 5.1, 5.2, 5.5 and 5.7).

New Scientist for the figure 'Pesticide use', p. 5, from 'Japanese aid may upset Cambodia's harvests', by Hadfield, P., in *New Scientist* 1864, 1993 (Figure 10.3).

Oxford University Press Inc., New York, for figures 1.4 and 1.6, from *World Development Report 1992*, by The International Bank for Reconstruction and Development/The World Bank, 1992 (Figures 5.3 and 5.10).

Oxford University Press, Oxford, for figure 3, from 'Drought, agriculture and environment: a case study from the Gambia, West Africa', by Baker, K.M., in *African Affairs* 94: 67–86, 1995 (Figure 7.1).

Oxford University Press, Oxford, for figure 5.6, p. 96, from 'Climate change within the period of meteorological records', by Hulme, M., in *The Physical Geography of Africa*, Adams, W.M., Goudie, A.S. and Orme, A.R. (eds), 1996 (Figure 7.3).

SCOPE, for figure 1.8, in SCOPE 36, *Acidification in Tropical Countries*, by Rodhe, H. and Herrera, R. (eds), J. Wiley and Sons, Chichester, 1988 (Figure 10.2).

Springer-Verlag, for the figure 'Livestock/herbivore biomass against rainfall', in 'Biomass and production of large African herbivores in relation to rainfall and primary production', by Coe, M. J. Cummings, D. H. and Phillipson, J., in *Oecologia* 22: 341–54, 1976 (Figure 7.4).

Taylor & Francis Publishers, for figures 3.2 and 3.3, from 'Towards a green political theory', by Carter, A., in *The Politics of Nature*, Dobson, A. and Lucardie, P. (eds), Routledge, 1993 (Figures 6.3 and 6.4).

Taylor & Francis Publishers, London, for figure 1, p. 36, from *Ecology and Equity*, Gadgil, M. and Guha, R., Routledge, 1995 (Figure 10.1).

United Nations Environment Programme, Nairobi, for a figure based on the original map 'Desertification in the world's susceptible drylands', in *World Atlas of Desertification* (2nd edition), by Middleton, T. and Thomas, D.S.G. (eds), 1997 (Figure 7.2).

United Nations Publications, Department of Public Information, New York, for the figure 'World Population Projections to 2150', by United Nations Population Division, 1998 (Figure 6.1).

World Conservation Monitoring Centre, for figure 20.1, from *Global Biodiversity: status of the earth's living resources*, 1992 (Figure 9.3).

Every effort has been made to contact copyright holders for their permission to reprint material in this book. The publishers would be grateful to hear from any copyright holder who is not here acknowledged and will undertake to rectify any errors or omissions in future editions of this book.

Abbreviations and acronyms

ACIWLP	American Committee for International Wildlife Protection
BCSD	Business Council for Sustainable Development
CAMPFIRE	Communal Areas Management Programme for Indigenous Resources (of Zimbabwe)
CBA	cost–benefit analysis
CDC	Conservation for Development Centre (of IUCN)
CFC	chlorofluorocarbon
CGIAR	Consultative Group on International Agricultural Research
CIDA	Canadian International Development Agency
CILSS	Comité Inter-États de la Lutte contre la Sécheresse Sahélienne
CITES	Convention on International Trade in Endangered Species
COP	Conference of the Parties
CSA	Conseil Scientifique pour l'Afrique
CSD	Commission on Sustainable Development
DFID	(UK) Department for International Development
DPCSD	(UN) Department for Policy Coordination and Sustainable Development
EA	environmental assessment
EC	European Community (now the European Union)
ECOSOC	United Nations Economic and Social Council
EDP	environmentally adjusted net domestic product
EIA	environmental impact assessment
EIC	environmental impact coefficient
EM	ecological modernisation
ENSO	El Niño/Southern Oscillation
EU	European Union
FAD	food availability decline
FAO	Food and Agriculture Organisation (of the United Nations)
FSC	Forest Stewardship Council
GDP	gross domestic product
GEC	global environmental change
GEF	Global Environment Facility

GIS	geographical information system
GNP	gross national product
HCFC	hydrochlorofluorocarbon
HDI	Human Development Index (of the United Nations Development Programme)
IARC	International Agricultural Research Centre (e.g. International Institute for Tropical Agriculture or the International Rice Research Institute)
IBP	International Biological Programme
ICBP	International Council for Bird Protection (forerunner of BirdLife International)
ICDP	integrated conservation–development programme
ICID	International Commission in Irrigation and Drainage
ICOLD	International Commission on Large Dams
ICRAF	International Centre for Research on Agroforestry
ICSU	International Council of Scientific Unions
IFAD	International Fund for Agricultural Development
IIED	International Institute for Environment and Development
IITA	International Institute for Tropical Agriculture (Ibadan, Nigeria)
INCRA	National Institute for Colonisation and Agrarian Reform (Brazil)
INEAC	Institut National pour l'Étude Agronomique du Congo Belge
IOPN	International Office for the Protection of Nature (forerunner of the International Union for the Conservation of Nature and Natural Resources)
IPAL	Integrated Project on Arid Lands
IPCC	Intergovernmental Panel on Climate Change
IPM	integrated pest management
IRRI	International Rice Research Institute
IRSAC	Institut pour la Recherche Scientifique en Afrique Centrale
ISEW	Index of Sustainable Economic Welfare
ITDG	Intermediate Technology Development Group
ITTO	International Tropical Timber Organisation
IUBS	International Union of Biological Sciences
IUCN	International Union for the Conservation of Nature and Natural Resources (World Conservation Union)
IUPN	International Union for the Protection of Nature (forerunner of the International Union for the Conservation of Nature and Natural Resources)
LIRDP	Lwangwa Integrated Resource Development Project
LPI	Living Planet Index (of the World Wide Fund for Nature)
MAB	Man and the Biosphere Programme
MSD	mainstream sustainable development
NEAP	National Environmental Action Plan

NEPA	(US) National Environmental Policy Act
NGO	non-governmental organisation
NIE	new institutional economics
NOAA	(US) National Oceanic and Atmospheric Administration
ODA	(UK) Overseas Development Administration (now the Department for International Development)
OECD	Organisation for Economic Cooperation and Development
OEHA	(World Bank) Office of Environmental and Health Affairs
OMS	(World Bank) Operation Manual Statement
OMVS	Organisation pour la Mise en Valeur du Fleuve Sénégal
OPEC	Organisation of Petroleum Exporting Countries
ORSTOM	Office du Recherche Scientifique et Technique d'Outre Mer
PACD	Plan of Action to Combat Desertification
PCB	polychlorinated biphenyl
POP	persistent organic pollutant
RBDA	River Basin Development Authority
SCOPE	Scientific Committee for the Protection of the Environment
SD	sustainable development
SEA	strategic environmental assessment
SEAA	System of Economic Accounting
SIA	social impact assessment
SNA	System of National Accounts
SPFE	Society for the Protection of the Fauna of the Empire
SWIFT	Solomon Western Islands Fair Trade
TFAP	Tropical Forests Action Plan
TNC	transnational corporation
TRIPS	Trade Related Intellectual Property Rights (agreement)
TVA	Tennessee Valley Authority
UN	United Nations
UNCED	United Nations Conference on Environment and Development (Rio Conference)
UNCOD	United Nations Conference on Desertification
UNCTAD	United Nations Conference on Trade and Development
UNDP	United Nations Development Programme
UNEP	United Nations Environment Programme
UNESCO	United Nations Educational, Scientific and Cultural Organisation
UNGASS	United Nations General Assembly Special Session on Sustainable Development
UNSCCUR	United Nations Scientific Conference on the Conservation and Utilisation of Resources (Lake Success Conference)
USAID	United States Agency for International Development
WCD	World Commission on Dams
WHO	World Health Organisation
WID	Women in Development

WMO	World Meteorological Organisation
WRI	World Resources Institute
WWF	World Wide Fund for Nature (formerly World Wildlife Fund)

1 The dilemma of sustainability

'The question is,' said Alice, 'whether you *can* make a word mean so many different things.' 'The question is,' said Humpty Dumpty, 'which is to be master – that's all.'

(Lewis Carroll 1872)

Sustainable development

It is a familiar cliché that actions speak louder than words, but this denies the power of words, and the ideas of those who make and spread them. As Jonathan Crush (1995) notes in *Power of Development*, the words written and spoken about development, the 'discourse of development', have enormous power. Development action is driven forwards by texts ranging from humanitarian tracts to national development plans. These portray the world in particular ways, often in crisis of some kind, and almost always as requiring management and intervention by the development planner (ibid.). These texts also determine who has the authority to act and establish the basis of knowledge that frames such action. The words we use to talk about development, and the way our arguments construct the world, are usually seen as 'self-evident and unworthy of attention' (ibid., p. 3), but they are not: in development nothing is self-evident, even if many choices or options remain hidden from view.

Poverty, hunger, disease and debt have been familiar words within the lexicon of development ever since formal development planning began, following the Second World War. In the past decade they have been joined by another, sustainability. 'Sustainable development' has become one of the most prominent phrases in development discourse – indeed, Lélé suggested it was 'poised to become the development paradigm of the 1990s' (1991, p. 607), and in many ways it did. The capacity of the phrase to restructure development discourse and to reorganise development practice, a sure reflection of its power, will be discussed below.

Where did the new phrase come from? Its usage grew from small and at first unpromising roots, as will be explored in Chapter 3. Suffice it to say here that the concept began to be widely adopted following the 1972 United Nations Conference on the Human Environment in Stockholm (see Chapter 3).

Subsequently, under the label 'ecodevelopment', the concept was taken up by a number of authors (e.g. Riddell 1981, Sachs 1979, 1980, Glaeser 1984b). Sustainable development became the central concept in the *World Conservation Strategy* published in 1980 (IUCN 1980), and the foundation of the report of the World Commission on Environment and Development seven years later (Brundtland 1987). When it was launched in April 1988, the World Commission on Environment and Development claimed that its report set out a 'global agenda for change'. It was an agenda that now began to command attention in the core of the development universe: in a major shift of culture and policy, the President of the World Bank spoke in May 1988 of the links between ecology and sound economics in a major statement of the Bank's policy on the environment (Hopper 1988). Such 'greening' of development thinking was a characteristic feature of the 1980s (e.g. Harrison 1987, Conroy and Litvinoff 1988).

Sustainable development's place in the discourse of development was assured in the early 1990s when it became the driving concept behind the United Nations Conference on Environment and Development in Rio in 1992 (UNCED, or the 'Earth Summit'). This was attended by representatives of over 170 governments, most of whom made some kind of public proclamation of support for environmentally sensitive economic development. UNCED was also a forum for a vast range of non-governmental organisations, many of them from the First World, that strove both to capture media headlines and to influence intergovernmental debate through the parallel Global Forum (Holmberg *et al.* 1993, Chatterjee and Finger 1994). A vast media circus danced attendance, and the conference was thus promoted as a global event, although there were as many column inches decrying the glaring contradiction of privileged delegates and the urban poor of Rio de Janeiro's *favelas* as there were discussing sustainability, or analysing the political economy of international environmental diplomacy. The media had built up hopes that UNCED would bring about a new environmental world order, and once the razzmatazz had died down, many commentators reported that the chance had been blown. But what was going on behind the endless dry diplomatic debates about texts in conference rooms and the windy rhetoric of politicians pursuing the evanescent Green vote, and how realistic were hopes that the conference would bring about a change in business as usual? What was this 'sustainable development' that was on the table at Rio? What kind of environmentalist critique of development did ideas of sustainability represent?

Concern about environment and development in the Third World has been an important feature of debate about development studies since the late 1970s, and even then awareness of the environmental aspects of development was not new, whether among scholars, practitioners or participants in development. What was new in the last decade of the twentieth century was the scope and sophistication of critiques of the environmental dimensions of development in practice, and the high profile being given to the environment in the context of social and economic change (McCormick 1992).

To a large extent, credit for the infusion of environmental concerns into development discourse must lie with Northern environmentalists. Goodland *et al.* capture the passion of the environmentalist vision: 'the transition to sustainability becomes urgent because global life support systems – the environment – have a time-limit' (1993, p. 297). The loss of species and natural habitat caused by development projects had been a potent focus for the extension of environmental pressure group politics familiar in the industrialised world since the 1970s. 'Save the rainforest' campaigns followed logically enough from concerns about pollution, whales or First World countryside.

Concern for Third World environments reflects in part simply the growing integration of the global village, and environmentalist pressure can be seen as an extension of traditional concerns about environmental quality in that village's new countryside, in the Third World. First World environmentalism, however, has done more than simply broaden its field of concern (McCormick 1992). There has been a self-conscious effort to move beyond environmental protection and transform conservation thinking by appropriating ideas and concepts from the field of development. In extending their focus from hedgerows to rainforests, environmentalists found (or claimed to have found) much common ground with Third World peoples' groups opposing development projects that threaten breakdown in indigenous and subsistence ways of life. In environmental opposition by environmental groups to investment in large projects such as dams, the threats they represent to the rights and interests of indigenous peoples are likely to be at least as prominently expressed as threats to biodiversity. Indeed, the links between the two are likely to be drawn explicitly and prominently (e.g. Pearce 1992).

The display of development agencies and environmental groups dancing to the same 'sustainable development' tune in the 1990s is remarkable, but not entirely accidental. It reflects in part the success of environmentalist pressure on aid donors through the 1980s, backed by willing and effective media coverage (e.g. Harrison 1987), combined with renewed concern about the global environment, particularly the 'ozone hole' and the 'greenhouse effect'. The response of development agencies to environmental issues also reflects the more general 'greening' of politics in Western industrial countries in the 1980s, epitomised for UK observers by Margaret Thatcher's famous speech to the Royal Society in 1988.

The question remains, however, how deep the apparent revolution in development thinking goes. Has there really been a 'greening' of development? Has there, for example, been a revolution in ideology in any way analogous to Charles Reich's celebrated account of new thinking in the USA in the 1960s, *The Greening of America* (Reich 1970)? Commentators agree that the environmentalism of the 1960s and 1970s was a new social movement of profound significance (e.g. Cotgrove and Duff 1980, Hays 1987), but to what extent did this embrace thinking about the Third World, let alone thinking *within* the Third World? Was the 'greening of development' evidence of a paradigm shift in development thought, or simply an exercise in relabelling? At a time when there is such visible enthusiasm about new perspectives and new alliances over the environment, this is a hard and unpopular question, but it is an important one.

The answer turns on the extent to which 'sustainable development' or 'green' development or 'ecodevelopment' are words backed up by logical theoretical concepts rather than simply convenient rhetorical flags under which ships of very different kinds can sail. Are those phrases synonyms for some more general concept? If not, are the differences between them consistent and important in terms of the values they draw upon or the policy responses they reflect and demand? The diversity of ideas about sustainable development will be discussed later (in Chapters 5 and 6); for now, it is most important to ask what the phrase 'sustainable development' itself means, and whence it derives its power to attract such a large and disparate following.

Power without meaning?

The phrase 'sustainable development' is now widely employed, in the fields of policy and political debate as well as research. It seems to contain the potential to unlock the doors separating disciplines, and to break down the barriers between academic knowledge and action. It does this partly because the term is at the same time superficially simple and yet capable of carrying a wide range of meanings and supporting sometimes divergent interpretations. Both radical environmentalists and conventional development policy pragmatists have seized the phrase and used it to express and explain their ideas about development and environment. In the process they have created a powerful new term in the lexicon of development studies, and a theoretical maze of remarkable complexity (Dixon and Fallon 1989, Daly 1990, Lélé 1991).

Sustainable development has many definitions (see, for example, Pearce *et al.* 1989). Some definitions have a strong element of social justice; Eckholm (1982), for example, calls for 'economic progress that is ecologically sustainable and satisfies the essential needs of the underclass' (p. 8). The dominant definition, however, has undoubtedly been that of the Brundtland Report, in *Our Common Future*: 'development that meets the needs of the present without compromising the ability of future generations to meet their own needs' (Brundtland 1987, p. 43). Notwithstanding the rhetorical and slightly vague character of this definition (attributes it shares with many others – see, for example, the discussion in Lélé 1991), it has proved to be popular and compelling for those concerned about poverty, and inter- and intragenerational equity in human access to nature and natural resources. It has commended itself to those concerned to speak for nature itself, arguing for the preservation of habitats and species. The appealing, moralistic but slightly vague form of words of the Brundtland Report allowed sustainable development to become, in Conroy's term, the 'new jargon phrase in the development business' (1988, p. xi). It became also a vital element in the discourse of researchers trying to explain the relations between economy, society and environment, and to influence these (e.g. Redclift 1984, 1987, 1996, Clark and Munn 1986, Redclift and Benton 1994).

However, the Brundtland definition of sustainable development is a better slogan than it is a basis for theory. Words about sustainable development, whether in academic journals or the soundbites of politicians, very often prove to have no coherent theoretical core. The literature is strewn with the terms 'sustainability' and 'sustainable development', and before them with related terms like 'ecodevelopment', but too seldom are any of them given a clear and consistent meaning. With some reason, Redclift comments that sustainable development 'seems assured of a place in the litany of development truisms' (Redclift 1987, p. 3).

Such phrases are versatile, allowing users to make high-sounding statements with very little meaning at all. Their flexibility and their 'beguiling simplicity' (O'Riordan 1988, p. 29) only add to their attraction. On the one hand, environmentalists speak of 'sustainable development' in trying to demonstrate the relevance to development planners of their ideas about proper management of natural ecosystems. The conviction behind works such as the *World Conservation Strategy* is that sustainable development is a concept that truly integrates environmental issues into development planning. In using terminology of this sort, environmentalists have attempted to capture some of the vision and rhetoric of development debates. Sadly, they often have no understanding of their context or complexity. Environmentalist prescriptions for development, shorn of any explicit treatment of political economy, can have a disturbing naïveté.

On the other hand, the phrase 'sustainable development' is attractive to development agencies and theorists looking for new labels for liberal and participatory approaches to development planning. Development bureaucrats and politicians have undoubtedly welcomed the opportunity to fasten onto a phrase that suggests radical reform without actually either specifying what needs to change or requiring specific action. Thus the UK response to the Brundtland Report emphasised that there was continuity between its concept of sustainable development and existing British government policy, albeit in 'strands which have previously had separate currency' (Department of the Environment 1988): Brundtland's sustainable development was apparently acceptable to the British government precisely because it did not demand radical change of policy direction, just some linking up of existing policies and procedures.

Since the 1980s, sustainability has become an increasingly plausible element of government policy in the UK as elsewhere. Much of the thinking and rhetoric about sustainability has been focused on the domestic environment and economy, but the same language is used of intentions for development aid. The 1997 UK government White Paper on international development made a specific commitment to the elimination of poverty in poorer countries through sustainable development; specific objectives include the promotion of 'sustainable livelihoods', a focus on the poor, and protection and better management of the environment (Carney 1998a, DFID 2000). Arguably the rushed application of green camouflage paint to existing policies that characterised the late 1980s has been replaced with more carefully constructed thinking and policies; at the very least, even the most hard-bitten cynic will admit that the quality of the paint-work has improved. As a way of talking about development, sustainability is recognised and used everywhere.

The discourse of development

One reason for the overlapping meaning of sustainable development is the highly confused question of what development itself means. This is a semantic, political and indeed moral minefield (Goulet 1971). Seers (1977) argued that if poverty, inequality and unemployment were decreasing without a loss of self-reliance (for example through foreign ownership of manufacturing plants), then development was taking place. Versions of this formula have formed the standard basis for development discourses, but development itself nonetheless remains an ambiguous and elusive concept, prey to prejudice and preconception. It is 'a Trojan Horse of a word' (Frank 1987, p. 231), a term which is sufficiently empty that it can be filled at will by different users to hold their own meanings and intentions. The word 'development' is used both descriptively (to describe what happens in the world as societies, environments and economies change) and normatively (to set out what *should* happen; Goulet 1995).

Sachs speaks of development as 'a perception which models reality, a myth which comforts societies, and a fantasy which unleashes passions' (1992a, p. 1). It is what Howard (1978) described as a 'slippery value word' (p. 18), used by 'noisy persuaders' such as politicians 'to herd people in the direction they want them to go' (p. 17). Advocates for particular ends in development, or means to achieve those ends, make explicit use of the slipperiness of the word, and the confusion created by its heavy ethical burden.

Such value-laden words become political battlegrounds. To return to Crush (1995), the discourse of development promotes and justifies very real interventions and practices, and is inextricably linked to sets of material relationships, to certain kinds of specific activities and to the exercise of power. Arturo Escobar argues that reality has been so 'colonised by the development discourse' that 'those who were dissatisfied with this state of affairs had to struggle for bits and pieces of freedom within it, in the hope that in the process a different reality could be constructed' (1995, p. 5).

Concepts of development have a complex pedigree and etymology. There is a long history of 'development thinking before development began' (Brookfield 1975, p. 2). The word 'development' came into the English language in the eighteenth century and soon acquired an association with 'organicism' and ideas of growth (Watts 1995). Cowen and Shenton (1995) explore its evolution, and the origins of ideas of *under*development, in nineteenth-century European thought. By the start of the nineteenth century, development had become a linear theory of progress, bound up with capitalism and Western cultural hegemony, and advanced through mercantilism and colonial imperialism. Tracing what Watts (1995) calls the 'genealogy' of development reveals the complexity of its meanings over time. In the light of this, it is somewhat surprising that there has been such a uniformity to development thinking in the period since the Second World War.

This capitalist and Eurocentric 'developmentalism' (Aseniero 1985) presents 'development' as the process that recreates the industrial world: industrialised, urbanised, democratic and capitalist. Development has been depicted as a 'crucible' through which successful societies emerge purified, both modern and affluent (Goulet 1971). Developmentalism suggested that countries developed through different stages, on 'a linear path towards modernisation' (Chilcote 1984, p. 10), and that progress down that path could be measured in terms of the growth of the economy, or some economic abstraction such as per capita gross domestic product. The word 'development' then came to mean the projects and policies, the infrastructure, flows of capital and transfers of technology which were supposed to make that imitation possible. Development thus involved the imposition of the established world order on the newly independent periphery. Illich commented:

> There is a normal course for those who make development policies, whether they live in North or South America, in Russia or Israel. It is to define development and set its goals in ways with which they are familiar, which they are accustomed to use in order to satisfy their own needs, and which permit them to work through the institutions over which they have power or control.

He concluded, harshly, that this formula 'has failed and must fail' (1973, p. 368).

In practice, 'orthodox development thinking' (Oman and Wignarajah 1991, p. 5) sought to follow the success of the Marshall Plan by applying the same approach (injecting foreign aid for capital for investment in infrastructure) to the non-industrialised world: 'it was assumed that rapid industrialisation and generalised improvement in material conditions of life could be won quickly by following the formula that had worked in reconstructing war-damaged Europe' (Goulet 1992, p. 468). With that aid went the hegemony of values. The modernisation paradigm was built on the conceptual separation of 'modern' and 'traditional' (or 'Western' and 'non-Western') societies. Such concepts, which welded seamlessly into ideas of development, came from the same roots in Western Enlightenment rationality, and built on profoundly encoded Western preconceptions about civilisation and improvement versus barbarism (Slater 1993).

In *The Development Dictionary*, Wolfgang Sachs dates the start of the 'age of development' to the inaugural speech by US President Harry Truman in January 1949, in which he referred to the southern hemisphere's 'underdeveloped areas' (Sachs 1992a, p. 2). From this exercise in labelling grew the exercise of economic and cultural power that has become development practice. Development discourse is built on this definition of the non-'developed' and non-Western 'other' as a fitting, needy and legitimate target for action. Much of Edward Said's account of the power of Orientalism could be applied to development (Crush 1995; cf. Said 1979). The manner of that representation, and the material actions that flowed from it, have been highlighted and challenged in the extensive writings of 'post-colonialism' (e.g. Spivak 1990).

Esteva (1992) claims that the notion of 'underdevelopment' also began with Truman's speech (although he was not the first to coin the word), suggesting that from that day 2 billion people became underdeveloped. 'Development', its meaning soon narrowed to economic growth, thenceforth was defined as the escape from that sorry condition. Of course, escape proved impossible for most countries and most people, even in these narrowly defined terms. Sachs describes the project of development as 'a blunder of planetary proportions' (1992a, p. 3). For him, development is obsolete, standing 'like a ruin in the intellectual landscape' (ibid., p. 1).

By the time of the United Nations First Development Decade (1960–70), the certainties of developmentalism had begun to falter. Social and economic conditions for the majority of the population in many of the countries of the capitalist periphery steadily worsened in the immediate post-war years (Frobel *et al.* 1985). Commentators from a wide range of persuasions began to admit (and theorise about) the glaring gap between bland and simplistic expectation and reality. Debate about the nature and causes of the apparent failure to 'develop' has created the burgeoning disputes of development studies, and the proliferation of development theory (Chilcote 1984).

The 1980s saw the rise to authority of a 'counter-revolution' in development theory and practice, one that was opposed both to the established neo-Keynesian approach to planning, and to structuralist and Marxist theories of development (Toye 1987, 1993, M. Robinson 1993). The counter-revolution emphasised the benefits of free markets and the minimisation of the activities of the state. The conversion of key Western governments (and hence of the World Bank) to the doctrine of economic liberalisation (as in 'Reaganomics' and 'Thatcherism') for a while carried all before it. Thus the world financial institutions, spearheaded by the implacable economists of the World Bank and the International Monetary Fund, imposed structural adjustment to counteract the 'longstanding weaknesses in every economy and in international arrangements' revealed by the recession of the early 1980s, in pursuit of recovery and 'sustained and rapid growth of the kind the world enjoyed for twenty-five years after World War Two' (World Bank 1984b, p. 1). In particular, the counter-revolution demanded that governments slim down. As the *World Development Report* commented (in the related context of the transformation of the economies of Eastern Europe and the former Soviet Union), 'the state has to move from doing many things badly to doing its fewer core tasks well' (World Bank 1996, p. 110).

Debate in development studies has reflected changing ideas about the meaning of development, and the policies necessary to achieve it. With the rise of conservative economic policy in the industrialised world, and the collapse of the Iron Curtain in Europe, old certainties broke down and old enemies wavered and became confused. In radical development theory, there was extensive but unresolved debate within and about Marxism and post-Marxism (Corbridge 1993), while the rise of postmodernism, cultural theory and post-colonialism undermined established certainties. It was widely seen that there was an impasse in development studies (Schuurman 1993). Academics, being enthusiastic arguers, mapped and remapped ways out of that impasse, proposing a renewed dependence on the redemptive powers of neo-populism, 'new social movements' or a renewed and radically modernist post-Marxism (Corbridge 1993, Crush 1995, Escobar 1995). The crisis in development studies released a torrent of words. Meanwhile, the problem of global poverty persisted and deepened, the account of human misery growing almost unchecked.

The challenge of poverty

Whatever the state of development theory, there is no doubt of the ethical imperative of tackling human poverty (Corbridge 1993, Goulet 1995). As Goodland *et al.* (1993) succinctly put it, 'poverty is a massive global outrage' (p. 297). The perception of dramatic and unsolvable problems in the countries of the 'South' is common to politicians, aid agencies, academic analysts and the media. Indeed, such perceptions have long made crisis the commonplace motif of development writing (e.g. Brandt 1983, Frank 1981, Frobel *et al.* 1985), particularly in Africa (Timberlake 1985, Morgan and Solarz 1994). The 1984 drought in Sahelian Africa intensified this perception, leading both to the astonishing individual generosity of those unkindly (but perhaps accurately) described as 'the fat and happy in rich countries' (Harrell-Bond 1985, p. 13) and to the stereotyping of the continent and its peoples as locked, helpless, in a rictus of crisis (Watts 1989).

The 'crisis' of development, or the lack of it, embraces the problems of debt, falling commodity prices, falling per capita food production, growing poverty and socio-economic differentials both within Third World states and between countries. This idea of the Third World locked in crisis tends to favour 'fire-fighting' approaches to development as against discussions of deeper ills and the treatment of symptoms, not causes. This urgency leaves little time for lateral thinking. Thus Julius Nyerere commented, 'African starvation is topical, but the relations between rich and poor countries which underlie Africa's vulnerability to natural disasters have been relegated to the sidelines of world discussion' (Nyerere 1985).

The dimensions of the problems facing the Third World are substantial and real enough, and the litany of statistics on poverty soon acquires a grimly familiar ring. The World Bank's *World Development Reports* show that the notion of a world where all countries are experiencing economic growth and gains in quality of life (let alone all people in those countries) is an illusion. The kinds of aggregate statistics used to discuss such questions are deeply unsatisfying, but inasmuch as they are accurate they have the merit of relative consistency. They show that of a world gross national product (GNP) of US$28,862.2 trillion, only 6.4 per cent stems from low-income countries, this figure falling to only 1.7 per cent if India and China are excluded (World Bank 2000). Public debt was over 93 per cent of GNP in the world's poorest countries in 1997 (UNDP 1999).

Gross national product per capita in 1998 was US$520 in low-income countries, and $380 per capita if China and India are excluded. By contrast, average incomes were $21,400 in the UK, $29,340 in the USA and $40,080 in Switzerland (World Bank 2000). Moreover, not only are average incomes far higher in these and other industrialised economies, but wealth is more equitably distributed: if the rich and super-rich elite of poor countries were removed from calculations, the plight of the mass of the poor in poor countries would stand out even more starkly. Even without this, it is clear that some countries are well adrift from any model of equitable world development. Average annual income was less than $300 per head in a whole swath of countries, including Burundi, Cambodia, Chad, the Democratic Republic of Congo, the Central African Republic, Eritrea, Ethiopia, Malawi, Mali, Nepal, Niger, Nigeria, Rwanda, Sierra Leone, Tanzania and the Yemen Republic. Many, although not all, of these poorest countries are in Africa, and many are also suffering the destruction brought by civil or international war (e.g. Rwanda, Sierra Leone, Eritrea, Ethiopia). The share of the global economy enjoyed by the world's poorest countries, and by the world's poorest people in all countries, is catastrophically low and falling.

However, economic measures are a crude way to seek to measure development; as the *Human Development Report* notes, 'human development is the end – economic growth is a means' (UNDP 1996, p. 1). Since 1990 UNDP has calculated a Human Development Index (HDI) on the basis of longevity (life expectancy at birth), educational attainment (adult literacy and primary, secondary and tertiary education enrolment ratios, and standard of living (measured as gross domestic product (GDP) per capita). The HDI ranking has been calculated for 174 countries. In 1999, forty-five countries had high HDI ranks (HDI > 0.80), ninety-four had medium (HDI 0.5–0.79) and thirty-five had low ranks (HDI < 0.5). The average HDI value for the world in 1999 was 0.71. For industrialised countries it was 0.92. For the least developed countries it was 0.43, and for sub-Saharan Africa 0.46 (UNDP 1999). This index usefully throws attention on issues of quality of life, but even so, it sanitises crude geographical differences in life quality and life chances. It translates into life expectancies at birth in 1999 of 39 years in Malawi, 58 years in Bangladesh, or 53 in the Lao People's Democratic Republic, compared to 80 years in Japan or 79 in Canada (UNDP 1999). In many poor countries, particularly in Africa, investments in healthcare are being overwhelmed by AIDS, and life expectancies are plunging, while attempts to invest are disabled by the death of key people at every level from families to governments.

In 1999 the average proportion of the population in least developed countries with access to safe water was 41 per cent. In Bangladesh, out of a population of 125.6 million, 5 per cent had access to safe water, 26 per cent to health services, 57 per cent to sanitation; 56 per cent of children are underweight. Across all developing countries 2.5 billion people lacked access to safe water. Partly for this reason, infant mortality rates for those countries with the lowest HDI scores are high, 92 per 1,000 live births in 1996. This is a marked improvement on 1960, when the figure was 168 per 1,000 births, but the infant mortality rate is still greater than 1 per cent in twenty-seven countries, many of them at war (e.g. Sierra Leone, 165 per thousand, and Afghanistan, 163 per thousand).

These aggregate statistics numb the brain, and hide as much as they reveal. In particular, they need to be broken down by gender. In almost every country, women are systematically in a worse position than men in their exposure to poverty and its effects, and in their access to clean water, health services and education. Average life expectancies at birth in least developed countries were 53 years for men and 51 for women (UNDP 1999). Average adult literacy in least developed countries was 38 per cent for women and 59 per cent for men, and as low as 7 per cent for women in Niger and 10 per cent in Bhutan. The massive burden of household reproduction, and in particular the care of children, falls predominantly to women (Momsen 1991). Children's development and welfare suffers in tandem with the immiseration of women.

Obviously the accuracy and usefulness of aggregate statistics of this sort are limited. Nonetheless, the overall picture of the human dimensions of the challenge in the Third World is clear. Globally, decades of development investment have not driven the problem away. Indeed, the disparities between the world's rich and poor have increased; the gap between the economic power of industrialised countries and the levels of consumption of the majority of their people, and the economic weakness and grinding poverty of the least industrialised countries, has grown. Furthermore, the globalisation of economies and the inter-visibility provided by technology make these inequalities more and more glaring. The magnitude of the continuing problem of poverty is the chief evidence for the failure of the practical project of 'development' (i.e. the failure of development bureaucracies to solve obvious problems). Behind this failure lies the failure of both conventional economic thought and its new-right and reconstructed-left critique, and the discomforting impasse of development studies. This gloomy scene has provided fertile ground for ideas about sustainable development to flourish in the 1980s and 1990s. Perhaps sustainable development could provide an alternative paradigm, certainly a new start. Out with the old (Keynesianism, Marxism, dependency theory, even harsher versions of the new 'market' orthodoxy), in with the new: as the *Human Development Report 1996* had it, by the middle of the 1990s the concept of development had deepened and broadened to include dimensions of empowerment, cooperation, equity, sustainability and security (UNDP 1996). The idea of sustainable development was welcomed by development thinkers and practitioners because it seemed to provide a way out of the impasse and away from past failure, a means of re-routing the lumbering juggernaut of development practice without endangering belief in the rightness and feasibility of its continued forward movement.

The challenge of environmental change

Ideas about sustainable development draw on critiques of the development process, for example from populist writings (including failures of distribution and the plight of the poorest), from radical ideas (such as dependency theory), and from more pragmatic critiques of development project appraisal and implementation. However, in sustainable development these have been wedded to rather different concerns about the *environmental* impacts of development, both the costs in terms of lost ecosystems and species, and (latterly) the impacts of development action on natural resources for human use. Above all, environmentalist critiques of development have presented a picture of the Third World environmental crisis. Indeed, the notion of *global* crisis was an important element of environmentalism in the 1960s and 1970s, and became a central element in debates about sustainable development.

The problems of the environmental impacts of development will be explored in detail in later chapters. I will simply note here that the literature on the global environment of the past twenty years has portrayed a second crisis, paralleling that of poverty, a crisis of environmental degradation. In the last decades of the twentieth century, academics and journalists identified many heads to this particular monster, most notably desertification (Grainger 1982), fuelwood shortage (Munslow *et al.* 1988) and the logging of tropical rainforest (Caufield 1982, Myers 1984). These problems and others were widely reviewed, for example by Myers (1985). From 1972 the United Nations Environment Programme (UNEP) published annual 'state-of-the-environment' reports on particular issues, and carried out a review of 'world environmental trends' ranging from atmospheric carbon dioxide and desertification to the quality of drinking water (Holdgate *et al.* 1982).

Many accounts of the 'state of the world's environment' have been completed since, all of them relentlessly negative in their account of rapid declines in forest cover, rapidly rising global levels of energy use and carbon dioxide production (particularly in Asia), overexploitation of fisheries, depletion of soil resources, and shortages of food not met by international trade and aid flows (e.g. Groombridge 1992, Holdgate 1996, World Resources Institute *et al.* 1996, UNEP 2000). Such statistics suffer from the same problems of quality and completeness as data on development, or the lack of it (discussed above). Despite more than three decades of satellite remote sensing data, and the application of increasingly sophisticated computers to the analysis, storage and retrieval of data, information on global or regional environmental change is still patchy and in some instances (in spite of the work of international organisations such as the World Resources Institute in the USA and the UNEP–World Conservation Monitoring Centre in the UK) of limited reliability (Groombridge 1992). Perhaps the urgency and melodramatic style of environmental groups desperate for enough media attention to win a hearing from politicians and government decision-makers has sometimes further muddied the waters.

Debates about global environmental crisis have often descended into slanging matches between environmentalist Cassandras crying disaster and conservative (often corporate) sceptics claiming that they exaggerate. The debate is dogged not only by lack of data, but also by the lack of a clear 'headline' statistic, for example something that might parallel the debates about poverty in the development field (although that is by no means easy to define, as the vast literature debating its measurement shows). One attempt to derive such a 'headline' is the World Wide Fund for Nature's 'Living Planet Index' (LPI; Loh *et al.* 1999). This was first calculated in 1998, as an attempt to answer the simple question 'How fast is nature disappearing from the Earth?' (ibid., p. 1). The LPI includes three indicators of the state of natural ecosystems; global natural forest cover (excluding plantations), populations of 102 freshwater vertebrate species (birds, mammals, amphibians and fish) and populations of 102 vertebrate marine species (birds, reptiles, mammals and fish). The LPI fell by 30 per cent between 1970 and 1995, at around 1 per cent per year. Natural forest cover declined steadily since the 1960s, with about 10 per cent lost between 1970 and 1995, although losses of biodiversity may be greater owing to changes in forest quality, particularly on temperate forests (Loh *et al.* 1999). Freshwater species showed a 45 per cent decline and marine species a 35 per cent decline between 1970 and 1995 (ibid.).

Whatever the limitations of indices of this sort and of the data on which they are based, there can be no doubt that the human impact on the biosphere is very extensive, and accelerated rapidly during the twentieth century. Global mapping suggests that three-quarters of the habitable surface of the earth has been disturbed by human activity (Hannah *et al.* 1994). The UNEP *Global Biodiversity Assessment* (UNEP 1995) suggested that between 5 per cent and 20 per cent of the perhaps 14 million plant and animal species on earth are threatened with extinction. Rates of species extinction are hard to estimate with any accuracy, but Edward Wilson suggests that human activities have increased previous 'background' extinction rates by at least between 100 and 10,000 times. He comments, 'we are in the midst of one of the great extinction spasms of geological history' (E.O. Wilson 1992, p. 268). This conclusion about rapid loss of species is widely held (e.g. Prance 1991), and in particular is tentatively supported by analysis of known data on extinctions in the IUCN 'Red Lists' (Smith *et al.* 1993).

The fundamental dynamic of environmentalist concern about development in its broad sense – the expansion of industrial capacity, and the urbanisation and socio-cultural changes that accompany it – is the scale of human demands on the biosphere. The conventional environmentalist target (particularly in the harsh neo-Malthusianism of the 1960s and 1970s) has been population growth. The more serious problem is consumption. Vitousek *et al.* (1986) calculated that 40 per cent of potential terrestrial net primary production was used directly by human activities, co-opted or forgone as a result of those activities. This consumption embraces food and other products directly consumed (e.g. crops, fish, wood, etc.), and that consumed by livestock, as well as production consumed less directly, for example in fires or human-induced soil erosion. As the authors of this study recognised, such figures are bound to be full of errors of all kinds; nonetheless they suggested quite reasonably that 'an equivalent concentration of resources into one species and its satellites has probably not occurred since land plants first diversified' (ibid., p. 372).

In the 1990s, climate change came to share with biodiversity loss a key place in environmentalist discourse about human impacts on the biosphere. The World Climate Conference in Geneva in 1978 called attention to the problem of greenhouse gases and anthropogenic climate change, and the work of the Intergovernmental Panel on Climate Change (IPCC) from 1988 established a strong global scientific consensus that human action was indeed affecting global climatic patterns. Human impacts on climate are superimposed on natural variation, and the global (and even more the regional) ocean–atmosphere system is notoriously hard to model satisfactorily. However, the IPCC consensus has held, so that, for example the second report in 1995 concluded that the global mean temperature of the twentieth century was at least as warm as any since 1400 (which is as far back as data allow comparison); the mid-range climate change scenario predicted an increase in global mean surface air temperature of 2°C between 1990 and 2100, which would give a sea level rise of 50 cm by 2100 due to expansion of the oceans as they warm up, and ice melting (Houghton *et al.* 1995). The impacts of this, globally and regionally, are complex (warmer, wetter winters in northern polar regions, for example, and a reduced North Atlantic thermohaline circulation). The implications for environment and society are, however, potentially very great, with what is rather blandly described as 'an enhanced global mean hydrological cycle', meaning more droughts and floods and storms.

Anthropogenic climate change has become more than some environmentalist bogey; it is now accepted scientific fact. The Framework Convention on Climate Change, signed at Rio, reflects that acceptance, although views about who should take what action and when vary a great deal (see Chapter 4). To environmentalists, the evidence for human impacts on climate has offered clear evidence of the unacceptably large scale of human demands on the biosphere. It represents a significant challenge to developmentalism and its conventional strategies of industrialisation and economic expansion.

Clearly, just as the idea and practices of development are an unprecedented human enterprise of the past century or two, there is a novelty to that enterprise's demands on the natural systems of the earth. As environmentalists, from the 1960s onwards, have said repeatedly, humans have not been here before: there are no road-maps for the future. How is life (human and non-human) to be sustained? It was to answer such questions that the discourse of sustainable development was created.

Environment and development: one problem, two cultures

The threat of multidimensional global crisis has therefore been a key theme within debates about sustainability: a crisis of development, of environmental quality and of threats to the material benefits supported by natural biogeochemical processes and sinks. The 1992 *World Development Report* opened with the assertion that 'the achievement of sustained and equitable development remains the greatest challenge facing the human race' (World Bank 1992, p. 1). By the 1990s the point that there are close links between the problems of development and environment had sunk in, although, despite significant advances (Blaikie 1985, Blaikie and Brookfield 1987, Redclift 1984, 1987, Redclift and Benton 1994, Elliott 1999), theoretical understanding of the links still tends to lag behind practical and rhetorical recognition of the problem (see Chapters 4 and 5).

It is recognised that tight and complex links exist between development, environment and poverty (e.g. Broad 1994, Reardon and Vosti 1995, Blaikie 1995). The poor often endure degraded environments, and in some instances contribute to their further degradation. Urban air and water pollution are both rising rapidly, even in those countries in which economic growth is taking place, and the degradation of agricultural, forest and wetland resources is extending the depth and breadth of deprivation in many rural areas. Enduring problems, for example the lack of clean drinking water, are getting more and not less serious: 2 million children die of intestinal diseases due to unclean water each year (World Bank 1992).

The reciprocal and synergistic links between poverty and environmental degradation force what Blaikie describes as the 'desperate ecocide' of the poor (1985, p. 138). Access to and control over cultivable land, fuelwood or other usable attributes of nature are uneven. Blaikie emphasises the political dimensions of rights over resources, stressing the need for those seeking to understand environment–development problems to explore the links between environment, economy and society that he calls 'political ecology'. Blaikie and Brookfield argue that 'land degradation can undermine and frustrate economic development, while low levels of economic development can in turn have a strong causal impact on the incidence of land degradation' (1987, p. 13). Poverty and environmental degradation, driven by the development process, interact to form a world of risk and hazard for both urban and rural communities. Understanding the reality of this world, and the environmental, economic and political factors that create it, lies at the heart of the widespread contemporary concern for sustainable development.

However, even in the decade of sustainable development that followed publication of the Brundtland Report in 1987, the fields of developmental and environmental studies were far from unified. The one language of sustainability has hidden the separation of two cultures, which have often remained remote from each other both conceptually and practically.

Despite the rise of careers in 'environment and development', and of massive sources of funding such as the Global Environmental Facility that fuel them, development planning and environmental planning have remained separate fields. Sociologically, they still have their own separate cadres and culture, their own self-contained arenas of education and theory formation, their own technical language and research agendas, and – above all – their own literature. So-called experts (in the sense used by Chambers 1983) rarely claim expertise in both, and seldom understand the theoretical linkages between them. Development and environment work are still the fruit of distinct cultures. Although the two overlap a great deal, and indeed make confident inroads onto each other's territory with scant regard for the exact meaning or purpose of terminology, there is rarely if ever any integration.

The need for effective interdisciplinarity to make sense of the problems of environment and development is blindingly obvious. As Piers Blaikie comments, 'environmental issues are by definition also social ones, and therefore our understanding must rest on a broader interdisciplinary perspective that transcends institutional and professional barriers' (1995, p. 1). In practice, however, both academics and practitioners are reluctant to cross disciplinary boundaries. Our individual 'disciplinary bias' is deeply coded by our training, and is a severe constraint on innovative thinking (Chambers 1983). This problem is not confined to the Third World. Thus it is recognised that research on global environmental change must be pursued through collaboration between the natural and social sciences; however, such work is by no means easy to achieve successfully (Miller 1994). Unrealistic expectation, problems of data and measurement, and problems with the ways in which research questions are framed all represent challenges to interdisciplinarity. In particular, the differing perspectives of ecologists and economists provide a difficult terrain for effective engagement on issues of sustainability (Tisdell 1988).

A further problem is that there is an unavoidable and hence critical ideological component in understanding problems of environmental resource use and environmental degradation. The importance of ideology in understanding 'environmental' phenomena is most clearly analysed in the context of soil erosion and erosion control projects planned and enforced by the state, particularly in colonial territories (see Blaikie 1985, Blaikie and Brookfield 1987, Anderson 1984), and in their successors, anti-desertification projects (Swift 1996). These are discussed in detail in Chapters 7 and 9. They are a particular example of the power of environmental narratives to condition and constrain even 'impartial' scientific research on the environment (Leach and Mearns 1996). Environmental scientists, and environmentalists, persistently fail to recognise the ideological burden of ideas and policies. People trained in natural science disciplines in particular find it difficult to transcend the notion of the impartiality and 'truth' of science, and hence to agree a common approach to understanding (let alone tackling) field problems of poverty and environment (Seeley and Adams 1987).

Development crises and environmental crises exist side by side in the literature, and together on the ground, yet explanations often fail to intersect. Environmentalists and social scientists speak different languages. Very often theirs is a dialogue of the deaf, carried on at cross purposes and frequently at high volume. The complex and multidisciplinary nature of the links between development, poverty and environment makes them difficult to identify and define. They often go unnoticed, fall down the cracks between disciplines, or get ignored because they fit so awkwardly into the structures of academic analysis or discourse. Nonetheless, in the real world these links are real enough. They explain why development policy often causes rather than cures environmental problems. Development and environmental degradation often form a deadly trap for the poor.

Chambers (1983) argued that it is the plight of the poor that should set the agenda for development action, and in his approach to sustainability he directed attention to the concept of sustainable rural livelihoods, defined as the secure access to sufficient stocks and flows of food and cash to meet basic needs. He suggested that there are both moral and practical imperatives for making sustainable livelihood security the focus for development action. This is the principle underlying the arguments in this book: the touchstone for debate about environment and development is the human needs of the poor, both environmental and more conventionally developmental. Among other things, it makes debate about the environment in the Third World, like that about development, inherently political; but, as Redclift argues, it is an illusion to believe that environmental objectives are 'other than political, or other than distributive' (Redclift 1984, p. 130).

Outline of the book

This book is not another attempt to find the winning formula, the mix of sticks and carrots, rhetoric, capital flows and environmental knowledge that will achieve 'real' development. Rather, it is about the peculiar difficulty of talking sensibly about the environmental dimension to development in the Third World. Its aims are, first, to discuss the nature and extent of the 'greening' of development theory. It does this by examining the key concept of sustainable development. It looks at the origins and evolution of these ideas, and offers a critique of their articulation in the *World Conservation Strategy*, the Brundtland Report and the documents of the Rio Conference. It is argued that the ideology of sustainable development is eclectic and often confused. Sustainable development is essentially reformist, calling for a modification of development practice, and owes little to radical ideas, whether claiming a Green or a Marxist heritage. Second, the book attempts to draw a link between theory and practice by discussing the nature of the environmental degradation and the impacts of development. In doing so it attempts to address the question of the limitations of reformist approaches. It argues that, ultimately, 'green' development has to be about political economy, about the distribution of power, and not about environmental quality.

The first part of the book is largely concerned with ideas and theories about environment and development, and focuses in particular on the global scale.

Chapter 2 discusses the origins and growth of sustainable development ideas, looking in particular at their roots in nature preservation, colonial science, and the internationalisation of scientific concerns in the 1960s and 1970s. This is an account of institutions and organisations as well as ideas, and takes the form of a historical account. Attention is focused on the 1970s and the Conference on the Human Environment in Stockholm in 1972, at which sustainable development became a specific and identified area of concern. The *World Conservation Strategy* was a direct development of the thinking at that time.

Chapters 3 and 4 discuss the evolution of 'mainstream sustainable development', arguing that a coherent set of ideas has persisted through the 1980s and 1990s, in the *World Conservation Strategy* and the Brundtland Report, in *Caring for the Earth*, and (in Chapter 4) in the work of the Rio Conference. These chapters explore these ideas in some detail, analysing both what they say and the nature of their ideologies. They draw on both technocentrist and ecocentrist worldviews, the first being rationalist and technocratic, and leading to approaches to the environment involving management, regulation and 'rational utilisation', and the latter being romantic and transcendentalist, embracing ideas of bioethics and the intrinsic values of non-human nature (O'Riordan 1981, O'Riordan and Turner 1983, Worster 1985). Turner (1988b) suggests that a coalition may be possible between less extreme examples of these divergent areas of thought, 'accommodating technocentrism' (a conservationist position of sustainable growth) and 'communalist ecocentrism' (a preservationist position emphasising macroenvironmental constraints on growth, and decentralisation). However the field is classified, the history of ideas goes some way to explaining why current visions of sustainable development are rather messy: enthusiastic, positive and committed without, in general, being overtly political.

The sustainable development mainstream is essentially reformist, a broadly neo-populist vision of the world being allied with a call for more technically sophisticated environmental management. The theoretical dimensions of this mainstream view are explored in Chapter 5, which discusses market environmentalism, ecological modernisation and the role of environmental economics in delivering sustainable development. However, there are more radical ideas about world development. These will be discussed in Chapter 6, where eco-socialism, ecoanarchism, deep ecology and ecofeminism will be analysed, and their relations to both the conventional reformism of mainstream sustainable development thinking and the mainstream of radical thought will be discussed.

The second half of the book will provide a commentary on the theoretical ideas in the first half by discussing environment and development in practice, moving down the scale continuum to focus attention on development projects. In Chapter 7 the links between sustainability and environmental degradation will be explored, particularly climate change, desertification and overgrazing, as will the scientific thinking that underlies them. Chapter 8 will consider the environmental costs of development, looking at water resources and the impacts of dams. Chapter 9 will discuss the political ecology of sustainability in the context of tropical forests, conservation, and projects and famine. In Chapter 10 the environmental problems of urbanisation, industrialisation and pollution will be considered within the framework of ideas about 'risk society', and Chapter 11 will examine the prospects for improved technical planning using established techniques of environmental appraisal, and the 'greening' of aid. Chapter 12 will explore the potential of ideas of achieving development not 'from above' through better planning, but 'from below', through the participation of local people. The final chapter of the book (Chapter 13) will move beyond these various 'reformist' approaches to sustainability to consider more radical strategies.

This book does not offer a synthesis, in the sense of a shopping list of environmental desiderata, nor an attempt to set out a blueprint for 'sustainable development'. There is no comfortable celebration of the achievements of the new green bankers and aid bureaucrats. Instead, it reveals the tensions in the heart of the environmental critique of development practice which challenge the calculated reformism of mainstream sustainable development. Important elements in green critiques of development are radical and not reformist, and they are both awkward and inconvenient. Mainstream sustainable development is bureaucratically and politically acceptable, because it seeks to reprogram the juggernaut of development through reformist thinking, involving better measurement of social and environmental impacts, better assessment of costs and benefits, better 'clean' technologies and efficient planning procedures. Alternative countercurrents within sustainable development offer politically far more risky waters, for they challenge the global status quo and raise painful and radical questions. Between these two broad views there is tension. The ethics of sustainability demand rather more than merely reform of the development process. The 'greening' of development demands a more radical analysis, and a more transforming response.

In 1970 Charles Reich wrote with passion and hope in *The Greening of America* of the new consciousness abroad in the USA, 'arising from the wasteland of the corporate state like flowers pushing up through the concrete pavement' (Reich 1970, p. 328). Three decades later, a generation for many and a lifetime for many in the South, the confident exuberance of that time is gone. However, many of the seeds sown in the 1960s have taken root, and environmentalism is one of them. Despite their flaws, the growth of sustainable development ideologies has had an impact on the consciousness that informs development thought and action. There is a new environmental awareness in development which is perhaps evidence of a 'greening' of a kind. To date it has been largely superficial, a thin green layer painted onto existing policies and programmes. The challenge of more profound change still lies ahead.

Summary

- Sustainable development has become a central concept in development studies, building on environmental, social and political critiques of development theory and practice.

- There is no simple single meaning of 'sustainable development': a wide range of different meanings are attached to the term. Far from making the phrase useless, it is precisely because of its ability to host divergent ideas that sustainable development has proved so useful, and has become so dominant.
- One reason for the complexity of concepts of sustainable development is the confused and contested meaning of development itself. The idea of sustainable development has gained currency in the 1990s at a time when development thought is widely held to have reached an impasse.
- The use of the term 'sustainable development' reflects in particular the prominence at the end of the twentieth century and the beginning of the twenty-first about the problem of acute global poverty and global environmental degradation. Although it is now acknowledged that these crises are linked, problems of environment and development are often addressed independently. They have to be tackled in an integrated way; the challenge of doing so is inevitably political. There are choices to be made between reformist and radical ideas about sustainability and development.

Further reading

Corbridge, S. (1995) *Development Studies: a reader*, Arnold, London.
Crush, J.C. (1995) *Power of Development*, Routledge, London.
Elliott, J.A. (1999) *An Introduction to Sustainable Development*, Routledge, London (2nd edition).
Holdgate, M.W. (1996) *From Care to Action: making a sustainable world*, Earthscan, London.
McCormick, J.S. (1992) *The Global Environmental Movement: reclaiming paradise*, Belhaven, London.
Middleton, N. (1999) *The Global Casino: an introduction to environmental issues*, Arnold, London (2nd edition).
Redclift, M. (1996) *Wasted: counting the cost of global consumption*, Earthscan, London.
Redclift, M. (2000) *Sustainability: life chances and livelihoods*, Routledge, London.
Sachs, W. (ed.) (1992) *The Development Dictionary: a guide to knowledge as power*, Zed Books, London.
UNEP (2000) *Global Environment Outlook 2000*, Earthscan, London, for the United Nations Environment Programme.

Web sources

<http://www.un.org/esa/sustdev/>   United Nations home page on sustainable development (including information on *Agenda 21*, the United Nations Commission on Sustainable Development, and the particular problems of 'Small Island Developing States').
<http://www.undp.org/>   The website of the United Nations Development Programme, with data from the UNDP Human Development Reports, and on UNDP's work on democratic governance, pro-poor policies, and crisis prevention and recovery.
<http://www.fao.org/waicent/faoinfo/sustdev/>   Information   from   the   Sustainable

Development Department (SD), Food and Agriculture Organisation of the United Nations (FAO), for example on sustainable food security, land tenure and people's participation.

<*http://www.wri.org/*> The World Resources Institute website, with up-to-date information on environment, resources and sustainable development, including biodiversity, forests, oceans and coasts, water and health; also information.

<*http://www.unep-wcmc.org/*> The UNEP-World Conservation Monitoring Centre website has data and maps on conservation and sustainable use of the world's living resources, including the status of species, freshwaters, forests and marine environments.

<*http://www.worldbank.org/poverty/*> The World Bank on poverty; the first place to look for material and ideas on poverty and its alleviation.

2 The origins of sustainable development

Telling me, a harried public official who must answer to 48 million restless, hungry and thirsty people, to 'Ensure development is sustainable and humane' is like warning me 'Operate, but don't inflict new wounds'. I know that. What I don't know is how to do it.

(Kader Asmal, chair [World Commission on Dams 2000, p. iv])

Environmentalism and the emergence of sustainable development

The phrase 'sustainable development' has become the focus of debate about environment and development. It is not only the best-known and most commonly cited idea linking environment and development, it is also the best documented, in a series of publications, beginning with the *World Conservation Strategy* and the Brundtland Report, *Our Common Future*, and leading to the documents arising out of the Rio Conference. These mainstream documents are the subject of the next two chapters, which discuss their arguments and assess the nature of the ideology that shapes their ideas about development. However, the concept of sustainable development cannot be understood in a historical vacuum. It has many antecedents, and over time has taken on board many accretions and influences. These are the subject of this chapter.

The history of thinking about sustainable development is closely linked to the history of environmental concern and about the conservation of nature in Western Europe and North America. An understanding of the evolution of sustainable development thinking must embrace the way essentially metropolitan ideas about nature and its conservation were expressed on the periphery in the twentieth century, initially on the colonial periphery and latterly within the countries of the independent Third World. This focuses attention in particular on the rise of international environmentalism (Boardman 1981, McCormick 1992, Holdgate 1999). The phenomenon of that emergence is well described elsewhere (e.g. O'Riordan 1976a, Sandbach 1980, Cotgrove 1982, Lowe and Goyder 1983, Hays 1987, McCormick 1992). Its intellectual roots lie a great deal further back and are beyond the scope of this book to unravel. They have been explored, for example, by Thomas (1983), Merchant (1980), Pepper (1984), Grove (1990a, 1992) and Grove *et al.* (1998).

As McIntosh (1985) points out in his history of ecological science, *The Background of Ecology*, stories of the development of ideas divorced from social and intellectual context are of little value. It is also unhelpful to look for clear and simple 'roots' to ideas which in fact relate to each other through time in a complex and fluid way, and which at any given time are held and articulated in diverse ways by different people. Ideas about non-human nature, and particularly about ways people or society should treat or manage nature, are both subtle and intractable. They reflect changing ideas about society itself, and are slippery, and hard to trace through space and time. An account of the evolution of different strands of thought about what we have come to call 'sustainable development' is therefore not entirely straightforward.

However, if the nature of the thinking about sustainable development that emerged in the 1960s is to be understood, it is necessary to tease out some of the strands in the complex fabric of its past. This chapter does this by selecting eight themes, overlapping in time. The first is that of the importance of the tropical regions in the development of environmental concern. The second is the development of nature preservation, and later conservation, in tropical countries. The third theme is the rise of tropical ecological science, particularly in Africa; the fourth is the importance of ecological ideas about the 'balance of nature'; and the fifth is the growth of ecological managerialism. The sixth theme is the growth of concern about the ecological impacts of development, and the seventh is the rise of perceptions of global environmental crisis, and particularly the perceived threat of human population growth. The final theme is the increasingly international organisation of scientific concern about the environment, particularly in the form of the UNESCO Man and the Biosphere Programme.

Tropical environmentalism

The attempt to write about the history of sustainable development is made difficult by the abundance and Eurocentric and Americocentric focus of the literature on environmentalism. The 'global environmental movement' whose growth is charted by McCormick (1992) has been, for most of its history, an almost exclusively northern hemisphere phenomenon. Redclift (1984) argues that its ethnocentric nature should make us wary of international comparisons that are in fact based on European or North American experience. Southern environmental NGOs (non-governmental organisations) began to appear from the 1970s onwards, but even in recent decades the rapid growth in the number, size and influence of international environmental NGOs has been driven by the power of those based in the First World (Princen and Finger 1994).

This implies Northern roots of concern about human relationships with the environment. Such roots certainly exist, and run deep (Thomas 1983). The destructive power of human activities was widely appreciated in North America and Europe, particularly in the second half of the nineteenth century, most famously in George Perkins Marsh's *Man and Nature* (1864). Marsh observed, 'man is everywhere the disturbing agent. Wherever he plants his foot, the harmonies of nature are turned to discords. The proportions and accommodations which ensured the stability of existing arrangements are overthrown' (Marsh [1864] 1965, p. 36). Marsh's classical education, his boyhood in Vermont and his sojourns in Europe created a truly modernist critique of industrialisation and the environmental demands of economic growth. Nature, he explained and demonstrated, 'avenges herself upon the intruder, by letting loose on her defaced provinces destructive energies hitherto kept in check by organic forces destined to be his best auxiliaries, but which he has unwisely dispersed and driven from the field of action' (ibid., p. 42). Similar perceptions, if less scientifically expressed, had surfaced elsewhere in the industrialising world, most significantly perhaps in the Romantic movement in Britain (Bate 1991, Veldman 1994). Concern for the conservation of nature in the USA and in Europe tapped these concerns in a direct way (Adams 1996).

Twentieth-century environmentalism, with its rather belated concern for the unindustrialised, tropical and colonial parts of the world, therefore has immediate precursors in the industrialising world in the nineteenth century. However, this is far from a complete picture. For example, a series of studies by Richard Grove (1990a, 1995) maps a very different evolution of environmental concern as neither a local response to the conditions of Western industrialisation, nor something derived exclusively from Northern attitudes. He describes the way in which European ideas of nature were transformed from the fifteenth century onwards by global trade and travel. These changes developed further with the rise of colonial trade and conquest. It was new ideas fed back into Europe from imperial possessions that informed changing philosophical ideas about nature, and about God.

Such ideas greatly influenced the development of natural science. In her book *Imperial Eyes*, Mary Pratt focuses on the work of the Swedish taxonomist Carl Linnaeus in the eighteenth century. She argues that the Linnaean system of classifying organisms not only drew upon biological collections from explorers and colonists, but 'epitomised the continental, transnational aspirations of European science' (Pratt 1992, p. 25). Indeed, she suggests that the scientific task being carried out in Northern Europe of naming and classifying the complex unknown organisms revealed by exploration 'created a new kind of Eurocentred planetary consciousness' (ibid., p. 39).

While science appropriated the vast biological complexity of the tropics, Grove argues that the newly described worlds were appropriated spiritually also, becoming associated with ideas of paradise or Eden. The tropics (and especially tropical islands) were 'increasingly used as the symbolic location for the idealised landscapes and aspirations of the Western imagination' (Grove 1990a, p. 11). What Grove describes as the 'full flowering' of 'the Edenic island discourse', in the mid-seventeenth century, coincided with realisation that economic demands on islands threatened to destroy their natural beauty and bounty, as for example on the Canary Islands and Madeira. Grove (1990a, 1992, 1995) suggests that this led to an awareness of the ecological impacts of emergent capitalism and colonial rule. In due course this led to an awareness of environmental limits and the need for conservation, as experience of ecological change on islands was translated into more general fears of environmental destruction on a global scale (Grove 1992, 1995). Grove discusses the work of the French 'Physiocrats' on the island of Mauritius from 1768 to 1810, the development of forest protection in British Caribbean territories from 1764, and the interest in desiccation in the Royal Society of Arts. Scientists employed by the trade companies as surgeons and botanists (the Dutch and English East India Companies and the French Compagnie des Indes) developed and disseminated ideas about environmental limits. The Indian Forest Service, in particular, was a critical arena within which environmentalist ideas were developed (McCormick 1992, Grove *et al.* 1998, Rajan 1998). Above all, this wider appreciation of pre-twentieth-century environmentalism demonstrates that it was in the colonial periphery, where full-throated capitalist and imperialist expansion met tropic societies and ecosystems for the first time, that environmentalist ideas developed. They are thus, in a sense, not narrowly 'European' at all.

Nature preservation and the emergence of sustainable development

In many ways, wildlife or nature conservation has been the most deep-seated root of sustainable development thinking. Indeed, sustainable development was put forward as a concept partly as a means of promoting nature preservation and conservation. The history of nature conservation in countries of the industrial metropole, for example in Britain (Sheail 1976, Allen 1976, Evans 1992) or the USA (Nash 1973, Worster 1985, Hays 1959, 1987), is well established. Although the intellectual roots of a concern for nature (either for its own sake, or for fear of repercussions of its misuse for people) lay deeper and further back (Thomas 1983, Pepper 1984, Grove 1995), the foundation of formal institutions to carry out and promote conservation began in the nineteenth century. Thus in Britain the second half of that century saw legislation for the protection of seabirds and the establishment of a number of conservation organisations such as the Commons Open Spaces and Footpaths Preservation Society (1865), the Royal Society for the Protection of Birds (founded initially in 1893), the National Trust for Places of Historic Interest and Natural Beauty (1894), and the Society for the Promotion of Nature Reserves (1912). Elsewhere in Europe, International Ornithological Congresses were held in Vienna in 1884 and Budapest in 1895, leading eventually to a treaty signed in 1902 to protect bird species 'useful in agriculture' (Boardman 1981, p. 28). The early years of the twentieth century saw the foundation in 1909 of the Swiss League for the Protection of Nature (primarily to raise funds for a national park, achieved in 1914), and of the Swedish Society for the Protection of

Nature. There were parallel developments in Germany (Conwentz 1914). In the USA the Yellowstone National Park was established in 1872, the Boone and Crocket Club formed in 1887, and the Sierra Club in 1892. Activities were not confined to the industrial and colonial metropole. National parks were established in the 1880s and 1890s in Canada, South Australia and New Zealand (Fitter and Scott 1978). In Britain debate about the need for national parks ran through the early decades of the twentieth century, before they and government nature reserves were eventually made possible by Act of Parliament in 1949 (Sheail 1976, 1984, 1996).

These developments were primarily aimed at promoting the protection of nature within the industrialised nations themselves. However, from an early date there was also concern about conservation on a wider geographical scale, in imperial or colonial possessions. Pratt suggests that in the eighteenth century, 'nature' had come to be defined in terms of the absence of human impact, specifically 'regions and ecosystems which were not dominated by "Europeans"' (Pratt 1992, p. 38).

Conservation action began before the end of the nineteenth century in over-seas territories (Grove 1987, McCormick 1992). In Africa, for example, concern about the depletion of forests in the Cape Colony developed in the early nineteenth century. This, with pressure for government money for the botanic garden (*inter alia* from the Royal Botanic Gardens at Kew, itself taken over by the government in 1820; Worthington 1983) led to the appointment of a Colonial Botanist in 1858. Legislation to preserve open areas close to Cape Town was passed in 1846, and further Acts for the preservation of forests (1959) and game (1886) followed (Grove 1987, J.M. MacKenzie 1987). Grove argues that by about 1880, a pattern of conservation (derived from a mix of Indian and Cape Colony philosophies) was established in Southern Africa. The utilitarian basis of these ideas brought them into conflict with settler interests. Holistic conservation ideas could not be reconciled with 'the driving interests of local European capital' (Grove 1987, p. 36). In India these forces were resisted, but not in Africa, where conservation moved away to an obsession with big game hunting, parks to protect game from 'poaching', and associated land alienation. In India, for example, 30 per cent of non-agricultural land in some provinces had been brought under the control of the Forest Department (Grove 1990b).

Various motives for conservation can be identified within colonial states. First, conservation allowed resources to be appropriated, both for the use of private capital and as a source of revenue for the state itself; second, conservation was a response to concern about environmental degradation and climatic change; third, conservation reflected idealist (and orientalist, cf. Said 1979) ideas of tropical 'nature' as 'Eden' and the need to protect it from rash humanity (Grove 1990b). This last concern provided a direct colonial parallel to the institutional development of nature or wildlife conservation in the colonial metropole.

The charismatic megafauna of the plains of Africa has probably been the most influential context for the development of Western ideas about conservation outside Europe and North America, certainly until the sudden advent of enthusiasm for rainforests in the 1970s. Here conservation has persistently been seen as a response by remorseful hunters to destructive hunting practices. J.M. MacKenzie (1987) identifies three phases in the extension of European hunting in Africa. The first was commercial hunting for ivory and skins: it was through the contacts with white hunters in the 1850s that African rulers first began 'riding the tiger of European advance' (p. 43). This grew into a second phase, of hunting as a subsidy for European advance: meat for railway construction workers, or to feed and finance trade and missionary activity. In time this gave way to a third phase, a ritualised and idealised 'hunting', with its obsession with trophies, sportsmanship and other ideals of British boys' education (MacKenzie 1989). As rifles improved and the number of hunters rose, astonishing numbers of game animals were killed, their deaths lovingly chronicled and recorded. As Graham picturesquely comments, 'The swirling torrents of bloodlust that were gratified are beyond our powers of measurement. Sport hunting of real wild animals will never see its like again' (1973, p. 54). By the last decades of the nineteenth century, substantial areas of Southern Africa were more or less emptied of game, certainly near white settlements, railways and wagon trails.

One aspect of the phenomenon of 'the hunt' was the growth in interest in natural history and 'pseudo-science' (J.M. MacKenzie 1987) such as specimen and trophy hunting and taxonomy. This led to the emergence of ideas of controlling hunting, and hence eventually ideas of conservation. Such a move parallels the evolution in Britain from the collection of natural history specimens to conservation (Allen 1976). Theodore Roosevelt perhaps personifies this best. An avid hunter and collector of trophies, he was by the early years of the twentieth century, as President, the centre of the policy debate in the USA about conservation, taking the utilitarian line of the forester Gifford Pinchot against the more romantic notions of John Muir (Hays 1959). He subsequently became one of the 'pioneer statesmen of the movement' (Fitter and Scott 1978, p. 16).

The most obvious aspect of the conservation based on this hunting ethos was the complete denial of hunting to Africans. White men hunted; Africans poached. This denial was achieved through controls on firearms, and latterly by the establishment of game reserves. The Cape Act for the Preservation of Game of 1886 was extended to the British South African Territories in 1891 (J.M. MacKenzie 1987). In 1892 the Sabie Game Reserve was established (to become the Kruger National Park in 1926), and in 1899 the Ukamba Game Reserve was created in Kenya, including land in what became the Amboseli National Park (Lindsay 1987). In 1900 the Kenyan Game Ordinance was passed, effectively banning all hunting except by licence (Graham 1973).

The work of game departments in Africa varied both between territories and over time, but the 'poaching problem' was a regularly repeated motif (Graham 1973, Steinhart 1989, Beinart and Coates 1995). The classic argument about the destructiveness of hunting by Africans (as opposed to whites) tended to run as follows:

It is commonly thought that the visiting sportsman is responsible for the decline of the African fauna. That is not so. The sportsman does not obliterate wild life. True, he kills. But seldom is the killing wholesale or indiscriminate. What the sportsman wants is a good trophy, almost invariably a male trophy, and the getting of that usually satisfies him. . . . The position is not the same with the native hunter. He cares nothing about species or trophies or sex, nor does he hunt for the fun of the thing. What the native wants is as many animals as possible for the purpose either of meat or barter.

(Hingston 1931, p. 404)

Plate 2.1 Elephant in the floodplain of the lower Zambezi, Zimbabwe. Hunting for ivory in the second half of the nineteenth century took a staggering toll of elephant populations. John Mackenzie reports that 40,000 lb of ivory was traded on the Zambezi in 1876, implying the killing of 850 elephants; one hunter, Henry Hartley, shot 1,000–1,200 elephants in his career (MacKenzie 1987). Elephant populations declined drastically, contributing to the wider perception by 'penitent butchers' of the need for conservation. Through the twentieth century, elephants have continued to hold a prominent if controversial place in debates about African conservation, from the IUCN African Special project of the 1960s through to the burning of stockpiled ivory in Kenya in the 1990s, and the ongoing argument about the moral rights and wrongs of safari hunting, and the merits of banning or legalising international trade in ivory (e.g. Barbier *et al.* 1990a).

This particular formulation appeared at a later period, between the two world wars, but such views were typical of hunter-conservationists in the early twentieth century. They remained an important element in thinking about wildlife in developing countries. The importance of such ideas in the establishment of reserves for wildlife conservation is considerable (see Chapter 9). The paternalism and ideological blindness of these views (or perhaps more accurately the pretence at a 'scientific' view which is beyond ideology) as well as the sheer political unacceptability of imposed rules for environmental management resurface in debates about environmentalism and development.

Hunting in Africa also led to significant institutional developments. A conference of African colonial powers (Britain, Germany, France, Portugal, Spain, Italy and the Belgian Congo) met in London in 1900 and signed a Convention for the Preservation of Animals, Birds and Fish in Africa (Fitter and Scott 1978, McCormick 1992). Three years later the possibility that a game reserve north of the River Sobat in the Sudan would be de-gazetted led to the establishment of the Society for the Preservation of the Wild Fauna of the Empire, with a powerful list of ordinary and honorary members (including Kitchener of Khartoum, President Theodore Roosevelt and the British Secretary of State of the Colonies; Fitter and Scott 1978). The 'penitent butchers', as they became nicknamed, helped place preservation on the agenda of colonial management. The 1900 Convention was never set in operation, but in the early decades of the century there were developments in a number of other colonial nations. For example, in the mid-1920s a French Permanent Committee for the Protection of Colonial Fauna was established, and in 1925 King Albert created the gorilla sanctuary that became the Parc National Albert (now the Virunga National Park), the first African national park (Fitter and Scott 1978, Boardman 1981).

More interesting in many ways was the pressure growing in industrialised countries, primarily in Western Europe, for an international organisation to promote conservation (Boardman 1981). The idea was mooted at an International Congress for the Preservation of Nature held in Paris in 1909, and in fact an Act of Foundation of a Consultative Commission for the International Protection of Nature was signed at Berne in 1913 by delegates from seventeen European countries. However, war prevented any substantive action, and despite the recommendation of another congress at Paris in 1923, it was not until the assembly of the International Union of the Biological Sciences in 1928 that an Office International de Documentation et de Corrélation pour la Protection de la Nature was established (Boardman 1981). This was consolidated into the International Office for the Protection of Nature (IOPN) in 1934 (Holdgate 1999). The IOPN joined the International Committee for Bird Preservation (ICBP, now BirdLife International), formed in 1922 to promote nature protection on a global scale.

Concern for nature in colonial territories, particularly Africa, persisted, and in the 1930s the IOPN published a series of reports on African colonial territories. In this they were particularly supported by American interest. Following President Roosevelt's lead, American concern about Africa, 'that region where scientists and naturalists felt the greatest threats to the world's fauna and flora existed' (Boardman 1981), grew, although it remained fragmentary until after the Second World War. In 1930 the Boone and Crockett Club set up the American Committee for International Wildlife Protection (ACIWLP). The ACIWLP assisted the IOPN financially and also itself promoted nature protection and carried out research, particularly during the Second World War when the work of the IOPN (based in Amsterdam) was severely disrupted.

Africa also remained the target of pre-war concern by the Society for the Protection of the Fauna of the Empire (SPFE; the word 'wild' was dropped from their title after the First World War). For example, Hingston's visit to Africa was at the request of the SPFE, and his paper to the Royal Geographical Society on the need for national parks was plain and persuasive: 'It is as certain as night follows day that unless vigorous and adequate precautions be taken several of the largest mammals of Africa will within the next two or three decades become totally extinct' (Hingston 1931, p. 402). Following the International Congress for the Protection of Nature in Paris in 1931, a new intergovernmental conference was called for 1933 (Boardman 1981).

After the war, efforts to strengthen international nature protection were renewed through both the IOPN and the ICBP, and the Swiss League for the Protection of Nature. The idea of a new organisation was mooted at a conference in Basle in 1946, and taken up by the newly established UNESCO (the United Nations Educational, Scientific and Cultural Organisation) in the person of its first Director, Julian Huxley (Huxley 1977). Huxley had been chairman of the government committee which in 1947 had recommended government involvement in nature conservation in Britain and the establishment of a Biological Service (later substantially implemented in the creation of the Nature Conservancy; Sheail 1976, 1984, Adams 1995). Huxley also had long-standing interests in Africa, for example as a member of the African Survey Research Committee formed in Britain in the early 1930s (Huxley 1930, Worthington 1983).

Given this experience, it is perhaps not surprising that the UNESCO General Conferences in 1946 and 1947 were persuaded to take nature conservation on board. A Provisional Union for the Protection of Nature was set up at a further meeting the same year. Political difficulties abounded, but a constitution of the International Union for the Protection of Nature (IUPN) was adopted at a conference at Fontainebleau in 1948 (Holdgate 1999). It comprised an unusual blend of governmental and non-governmental organisations, and its purposes were to promote the preservation of wildlife and the natural environment, public knowledge of the issues, education, research and legislation (ibid.). UNESCO granted financial support a month later.

In 1949 IUPN ran an 'International Technical Conference on the Protection of Nature' concurrently with the United Nations Scientific Conference on the Conservation and Utilisation of Resources (UNSCCUR) at Lake Success in New York State. McCormick believes that UNSCCUR has been underrated by environmental historians, and sees it as 'the first major landmark in the rise of the international environmental movement' (1992, p. 37). The parallel IUPN meeting was attended by representatives from thirty-two countries and seven international organisations. One of its resolutions suggested that IUPN should, with development agencies, consider carrying out surveys of the ecological impact of development projects (McCormick 1992). This notion, which lay unimplemented for another twenty years, demonstrates the stirring of a specifically conservationist concern for environmental aspects of development.

Despite constraints of funding, the IUPN survived and prospered. In 1956 it changed its name to the International Union for Conservation of Nature and Natural Resources (IUCN). From the first, its focus was firmly placed wider than the industrialised countries of the West. Data collection on endangered species began, and was strengthened by extra funding after 1955 which put a biologist into the field in Africa, South Asia and the Middle East to report on threatened mammals. The Red Data Books, as they appeared from 1966, contained many species from developing countries. IUCN began several projects in the early 1960s which had specific reference to the emerging 'developing world'. A Commission on Ecology was set up in 1954, and Project MAR (with an ecosystem focus on wetlands) was launched in 1961 (McCormick 1992). A Convention on Wetlands of International Importance, Especially as Waterfowl Habitat was signed in 1971 at Ramsar in Iran (Holdgate 1999).

There was also interest in the tropics, and particularly in Africa, on the part of other organisations and key individuals. These included the Frankfurt Zoological Society through the work of Bernard Grzimek in Kenya and Tanzania, and the Conservation Foundation (set up in 1948 by the New York Zoological Society under Fairfield Osborn) through Frank Fraser Darling. Concern about African wildlife was a feature of both the Fontainebleau meeting in 1948 and the Lake Success conference the following year. National parks were declared in several colonial territories following the Second World War, for example the Nairobi National Park in Kenya in 1946 and that at Tsavo two years later, Wankie in Southern Rhodesia, Serengeti in Tanganyika in 1951, and Murchison Falls and Queen Elizabeth National Parks in Uganda in 1952 (Fitter and Scott 1978). A special conference on African conservation problems convened at Bukavu in the Belgian Congo in 1953. In what McCormick describes as 'a tangential departure from previous thinking' (1992, p. 43), this suggested broadening the concerns of conservation from simply the preservation of fauna and flora to the wider human environment, and suggested a new convention to address the whole natural environment and focus on the needs of Africans (McCormick 1992).

By 1960 Africa had become for IUCN 'the central problem overshadowing all else' (Boardman 1981, p. 148). African countries were becoming independent, and political control was shifting away from the metropole; the poachers were turning gamekeepers, and might not follow the same policies. In 1961 IUCN therefore joined with the FAO to launch the 'African Special Project' to influence African leaders (Holdgate 1999).

The centrepiece of the African Special Project was the Pan African Symposium on the Conservation of Nature and Natural Resources in Modern African States, held in Arusha in Tanzania in 1961, the 'Arusha Conference'. Much was made of the Arusha Declaration on Conservation, the nominal author of which was Julius Nyerere (and not to be confused with the Arusha Declaration of 1967 when Nyerere launched Tanzania into 'African Socialism'). This represented a strong personal statement of commitment to wildlife conservation, but also stressed the project's focus on wider concerns of resource development. It stated:

> The survival of our wildlife is a matter of grave concern to all of us in Africa. These wild creatures amid the wild places they inhabit are not only important as a source of wonder and inspiration but are an integral part of our natural resources and of our future livelihood and well-being.
>
> (Worthington 1983, p. 154)

IUCN followed up the African Special Project with missions to seventeen African countries (Fitter and Scott 1978). However, although Africa was important, it was not unique in its problems. Conferences to try to reproduce the spirit of Arusha were also held around the world, in Bangkok in 1965 and at San Carlos de Bariloche in Argentina in 1968 (ibid.). Elsewhere, IUCN was involved in Ecuador in the establishment of the Charles Darwin Foundation for the Galapagos Islands in 1959, and the Fauna Preservation Society in 'Operation Oryx', to reintroduce the Arabian oryx in the Middle East, in 1961–2. IUCN also made a series of approaches to governments in the early years about the status of particular rare species, for example Indonesia (orangutang and elephant), and Nepal and India (rhinoceros). However, these had limited success, often being greeted by 'polite rebuffs, disguised either as bland encouragement or the forwarding of notes to other departments' (Boardman 1981, p. 76).

One reason for this was IUCN's Eurocentric and 'First World' image, which continued to limit its credibility in the Third World. This was if anything made worse by the foundation of the World Wildlife Fund (WWF; now the World Wide Fund for Nature) in London in 1961. This was conceived of initially to raise funds for conservation, working with IUCN and providing capacity to undertake projects (Holdgate 1999). However, WWF from the first had a different culture from IUCN (business and not science), and as money began to come in from national appeals in Europe and North America, WWF developed its own agenda; between 1962 and 1967, less than 13 per cent of WWF expenditure went to IUCN (McCormick 1992).

IUCN sought to spearhead conservation in the Third World, its work being dominated in the 1960s by Africa, yet, as Boardman comments, 'Western Europe and North America were the habitat of the conservationist' (1981, p. 114). Thus although by 1976 twenty African countries were represented by government agencies or non-governmental organisations as members of IUCN, they were swamped by the number of First World members: 57 of the 244 NGO members came from the USA. This relative domination by First World organisations has led to periodic complaints (for example by the Kenyan delegation at the United Nation, General Assembly in New Delhi in 1969; Boardman 1981).

However, the New Delhi meeting also adopted a definition of conservation as 'the management . . . of air, water, soil, minerals, and living species including man, so as to achieve the highest possible quality of life' (McCormick 1992, p. 46). This reflected the growing acceptance within IUCN that conservation had to embrace resources in a wider sense than simply wildlife, and had to have regard for their long-term management (Holdgate 1999). This was a response in particular to influence from the USA, where conservation as 'wise use' of resources had been part of public life since the progressive conservation movement associated with Gifford Pinchot at the start of the twentieth century (Hays 1959). This change in focus was reflected in IUCN's title in 1956 (also the result of American pressure; McCormick 1992). This change, to include the words 'conservation' and 'natural resources', followed discussion of population and resource issues at the 1963 IUCN General Assembly in Nairobi. It was said to have

> symbolised the conviction reached over the previous eight years that 'nature', the fauna and flora of the living world, is essentially part of the living resources of the planet; it also implied that social and economic considerations must enter into the problem of conservation.
>
> (Munro 1978, p. 14)

The African Convention on the Conservation of Nature and Natural Resources (eventually adopted by the Organisation of African Unity in Addis Ababa in 1968, after considerable political haggling between IUCN and the FAO; Boardman 1981) also broadened the definition of conservation. The convention embraced not only the established preservationist concerns of fauna and flora, but also the conservation of the more immediately obviously 'economic' resources of soil and water. It suggested that these were all resources to be managed 'in accordance with scientific principles' and 'with due regard for the best interests of the people' (McCormick 1992, p. 46).

Within IUCN there was increasing recognition of the need to make conservation more 'relevant' to the needs of the emerging Third World in the 1960s, and it progressively repackaged its message to embrace development. IUCN took part in a conference at Washington University in Virginia in 1968 on ecology and international development (subsequently published; Farvar and Milton 1973), and went on to cooperate with the Conservation Foundation in producing a 'guidebook' for development planners (McCormick 1992, p. 155), *Ecological Principles for Economic Development* (Dasmann *et al.* 1973). This is discussed on p. 44.

This thinking was a direct forerunner of the idea of 'sustainable development' contained in the *World Conservation Strategy* in 1980. The emergence of this document is discussed in the next chapter. Before that, it is necessary to backtrack slightly to examine other themes that fed into concepts of sustainable development. The African Convention referred to 'scientific principles'; where were these to come from, and what was the role of science and scientific thinking in views of tropical environments?

Tropical ecology and sustainable development

The science of ecology developed at the end of the nineteenth century in Europe and the USA (Lowe 1976, Worster 1985, McIntosh 1985, Sheail 1987). There were close links from an early date between the new science and the preservation or conservation movement, particularly in the UK (Sheail 1976, 1987). Ecology was also closely associated with the rise of environmentalism in the 1970s. At this period (and to a lesser extent today) the word 'ecology' was used widely even where ideas owed little or nothing to scientific ideas or method (Enzensberger 1974, Lowe and Warboys 1975, 1976), but even then it was true that many of the prominent figures of the environmental movement were trained in ecology (Chisholm 1972).

Among many other attributes, ecology has at different times seemed to offer new, value-free and apolitical ways of not only understanding but also managing the environment. Robin (1997) has described ecology as a 'science of empire', and pointed out (in the context of Australia) some of the strategic links between science and politics, ecology and empire. This ecological 'managerialism' was particularly attractive in places such as Africa at the end of the Second World War. The natural environment seemed relatively little affected by people, and there were few effective models for the burgeoning task of development planning. Thus the acquisition of scientific understanding of tropical environments was followed in time by a concern for the application of that knowledge to development. This utility of ecological science paralleled the rising demand for conservation and formed a second strand in the growth of sustainable development thinking.

The tropics have been an important hearth of innovation in biological science, and particularly ecology. Indeed, Goodland (1975) argued that the tropics have played a pivotal role in the development of ecology, choosing as a hook on which to hang his argument the occasion of Eugen Warming's jubilee. Certainly the attention of the practitioners of 'self-conscious ecology' (McIntosh 1985) in Europe and North America was soon extended to the study of the diversity of nature overseas. This engagement was simply a carrying forwards of the close and reciprocal links which had existed between science and exploration in the tropics through the eighteenth and nineteenth centuries (Stoddart 1986; see also Crosby 1986). Grove (1990b) argued that science directed and legitimised the extension of colonial state power over non-agricultural lands, and the people within them.

In the UK, the British Ecological Society was founded in 1914 (Salisbury 1964, Allen 1976). The first volume of its new *Journal of Ecology* in that year contained reviews of publications on the forests of British Guiana (Anderson 1912) and studies of vegetation in Natal and the eastern Himalayas. The second volume reviewed works on the vegetation of Aden and the highlands of north Borneo and Sikkim, and the following year added work on the dipterocarp forests of the Philippines and montane rainforest in Jamaica.

Interest in the ecology, and particularly the vegetation, outside Britain was maintained in subsequent years, and in volume 12 (1924) a specific section was instituted, quaintly entitled 'Notices of publications on foreign vegetation'. The majority of publications reviewed concerned the USA, Australia and South Africa, but there were others, for example Cuba (volume 15, 1927). The journal also in time carried substantive papers on the vegetation of the Empire. The first full paper concerned the vegetation of South Africa (Bews 1916), and this was followed by a steady trickle of papers on various parts of the world, both inside and outside the Empire, for example work on the rainforest of south Brazil (McLean 1919) and the forests of the Garhwal Himalaya (Osmaston 1922). Nor was such work confined to British ecologists. Studies on the vegetation and soils of Africa were begun in 1918 at the instigation of the American Commission to Negotiate Peace, involving a journey from the Cape to Cairo from 1919 to 1920 and a monograph on African vegetation, soils and land classification (Shantz and Marbut 1923).

The Imperial Botanical Conference, held at Imperial College in London in 1924, stressed the need for 'a complete botanical survey of the different parts of the Empire' (Brooks 1925, p. 156). This conference was the successor to the International Botanical Congresses in Paris (1900), Vienna (1905) and Brussels (1910), and was deferred until the British Empire Exhibition took place. The argument being made was that science should serve Imperial commerce: 'in such a Survey the Imperial Government has good reason to take a prominent part, as it depends so much on the overseas portions of the Empire for the supply of raw materials for manufacture, and of foodstuffs' (Davy 1925, p. 215). Throughout, the tone of the conference was strongly economic and utilitarian:

> I submit that it is our duty as botanists to enlighten the world of commerce, as far as may lie in our power, with regard to plants in their relation to man and their relation to conditions of soil and climate.
>
> (Hill 1925, p. 198)

The Imperial Botanical Conference passed a resolution establishing the British Empire Vegetation Committee, under the chairmanship of A.G. Tansley (first editor of the *Journal of Ecology*, and already a leading scientific figure in conservation in the UK). The committee published its first monograph, *Aims and Methods in the Study of Vegetation*, in 1926, although a further twelve years passed before the first in a planned series of regional monographs appeared, *The Vegetation of South Africa* (Adamson 1938).

The later 1920s saw British ecologists working on the ground in a number of tropical locations. For example, there were expeditions from Oxford to British Guiana in 1929 (Hingston 1930, Davis and Richards 1933) and to Sarawak in

1932 (Harrison 1933), two expeditions to the East African Lakes (Worthington 1932, 1983), and a Cambridge botanical expedition to Nigeria in 1935 (Richards 1939, Evans 1939). By that time ecological research on disease in Africa was well established, notably on the tsetse fly (Buxton 1935, Huxley 1930, Ford 1971). Biological science was readily harnessed in the service of colonial development. In 1928, for example, the *Journal of Ecology* carried a paper by the Assistant Economic Botanist in the Bombay Department of Agriculture on the aquatic weeds of irrigation canals in the Deccan (Narayanayaya 1928), and the geographer Dudley Stamp (1925) described an aerial survey of the Irrawaddy Delta forests. Ecologists discovered that their scientific work had considerable bearing on economic problems, and found a ready audience in the powerful but ignorant officers of the colonial state. In reviewing research on the East African Great Lakes many years later, E.B. Worthington commented, 'advice on the existing and potential fisheries was a primary objective, but to provide this it was necessary to elucidate the ecology of each lake' (1983, p. 13). Development needed careful and extensive scientific research.

Ecology, however, was seen to have more to offer than simply the scientific task of describing vegetation. Vegetation analysis and classification could serve 'a most practical purpose', allowing the definition of natural regions 'for use in the development of both agriculture and forestry' (Shantz and Marbut 1923, p. 4). In 1931 Phillips advocated a 'progressive scheme' of ecological investigation in East, Central and Southern Africa with the aim of helping develop resources. He believed that ecology could contribute to agriculture ('the rational use of biotic communities'), grazing (the 'wise utilisation of natural grazing'), forestry (the development of 'progressive forestry policies'), soil conservation (the 'prevention of soil erosion and its concomitant evils'), catchment water conservation and research into tsetse fly (Phillips 1931, p. 474). Phillips was practical enough to include detailed suggestions on staffing, a research programme and administrative arrangements.

Practical application of botanical survey to development began in what was then Northern Rhodesia (contemporary Zambia) in the 1930s, in the vegetation and soils surveys of three scientists, Colin Trapnell, Neil Clothier and William Allan (Trapnell and Clothier 1937, Trapnell 1943). Together they developed concepts of carrying capacity and critical population density. These concepts, and their analysis of the *chitemene* bush-fallow cultivation system, have become controversial (Moore and Vaughan 1994), but they were significant early expressions of ideas that came to prominence half a century later, when they formed one line of thinking about sustainable development (see Chapter 12).

The argument about the usefulness of ecology as an input to development in Africa was set out in the work of the African Survey (Hailey 1938), carried out under the aegis of an African Research Survey Committee, on which the two biologists were Julian Huxley (later Director of UNESCO) and John Orr (later Director of the FAO). In 1935 Sir Malcolm Hailey (later Lord Hailey) drove in two Fords from Cape Town to the Mediterranean. With him for part of the way went Barton Worthington, who wrote a contributory volume *Science in Africa* (Worthington 1938). This report was influential, both on the main volume and elsewhere. It took the utilitarian line that science (particularly ecology) was useful in development (or 'bonification' as Worthington called 'the promotion of human welfare'):

> a key problem was how *Homo sapiens* could himself benefit from this vast ecological complex which was Africa, how he could live and multiply on the income of the natural resources without destroying their capital ... and how he could conserve the values of Africa for future generations, not only the economic values but also the scientific and ethical values.
>
> (Worthington 1983, p. 46)

Science in Africa broke new ground. Not only did it place science in the service of a loosely defined 'development', it also used an explicitly ecological structure in presentation. The report moved from geology and meteorology through botany and zoology to agriculture, fisheries, disease, population and anthropology, and stressed the fact that interrelations between the sciences had 'important practical applications' (Worthington 1938, p. 3).

Following the end of the Second World War, institutions were put in place to promote and support science in Africa. The Conseil Scientifique pour l'Afrique (CSA), established following an Empire Scientific Conference held by the Royal Society in London in 1946 and a further conference in Johannesburg in 1949, developed similar arguments about the importance of science to development (Worthington 1958). Increased interest in scientific research in African territories at this time saw the establishment of the British Colonial Research Council (under Lord Hailey), the French Office de Recherche Scientifique et Technique d'Outre Mer (ORSTOM) and the Belgian Institut pour la Recherche Scientifique en Afrique Centrale (IRSAC) which joined the existing agricultural body, the Institut National pour l'Étude Agronomique du Congo Belge (INEAC) (Worthington 1983). Post-war colonial development drew heavily on a scientific approach to the environment and its management.

Ecology and the balance of nature

Ecology's most obvious contribution to development was its scientific methodologies for the descriptive analysis of the living environment. At first these primarily addressed the problems of vegetation description. However, following Charles Elton's *Animal Ecology* (1927), and his presentation of concepts of food chain, pyramid of numbers and niche, and his emphasis on the dynamics of populations in space and time, studies expanded to consider animal population ecology (for example in Worthington's limnological studies of the African lakes; Worthington 1932). The significance of ecology for the development of sustainable development was, however, also significant at a more profound level. Within ecology, a whole series of concepts had been developed to describe patterns of change in natural systems, and these came to provide a powerful conceptual basis for sustainable development. Chief among them was the concept of the ecosystem and the idea of balance between predator and prey species.

In the first decades of the twentieth century, 'nature' was seen to be essentially static, an array of habitat fragments as natural objects set in a landscape of change. This drew on a powerful organic metaphor of nature, a view of nature balanced and integrated and threatened by change from 'outside', from human action (Botkin 1990, Livingstone 1995). The links between ecology and conservation in industrialised countries were very close. In the UK, for example, Tansley's *Types of British Vegetation* (Tansley 1911) provided a classificatory framework for the first lists of nature reserves in the UK (Sheail 1976, Adams 1995). In *Research Methods in Ecology*, the American ecologist F.E. Clements provided a scientific basis for the identification of vegetation 'types' (McIntosh 1985). He developed ideas about plant succession, the notion of progressive change towards a 'climatic climax', with the vegetation formation like a complex organism 'developing' through time, in a way deliberately analogous to the growth of individual organisms. Clements's ideas of vegetation as organism were challenged by another American ecologist, H.A. Gleason, and also by Tansley. In 1920 Tansley argued against the idea that all aggregations of plants had the properties of organisms, and in 1935 he published a sharp critique of Clementsian thinking about the climatic climax (McIntosh 1985). He proposed a more complex pattern of succession, with soils, physiography and human action all driving change under different conditions, and framed the new concept of the ecosystem (Tansley 1935, 1939, Sheail 1987).

In the 1930s and 1940s, work on animal ecology was strongly quantitative, and focused on the dynamics of populations, whereas studies of plant population dynamics lagged until the 1970s (McIntosh 1985). The 1920s had seen the foundation of theoretical population ecology, built around the logistic curve. The potential for geometric growth of populations had been noted by Thomas Malthus in the eighteenth century, and the logistic curve was formulated by Velhust in 1828 (McIntosh 1985). It was rediscovered by Raymond Pearl in the 1920s, who suggested (controversially) that there was a 'law of population growth', with specific reference to human population. Two mathematicians, Lotka and Volterra, independently developed mathematical equations to describe the fluctuations of two interacting species, and from these, and extensive laboratory experimentation with highly simplified 'ecosystems' (with flour or wheat beetles for example) by Chapman, Pearl and Gause, the discipline of population ecology grew (McIntosh 1985). Within it developed the notion that animal population existed in balance in nature, sustained by density-dependent competition. This idea, developed by the Australian A.J. Nicholson in the 1930s, persisted as the centre of research and disputation between ecologists (McIntosh 1985). It also sank into the wider consciousness of ecologists and environmentalists formulating ideas about sustainable development.

These ideas drew also on experience of marine biology and the scientific management of fisheries. Fisheries science had two roots, one in Victorian marine biology (for example, the voyage of the *Challenger* in the 1870s), and the other in the systematic decline of fish catches as capture became industrialised with the advent of steam-driven boats and other innovations (Cushing 1988). The industrial revolution in Europe stimulated demand and provided the technology to meet it, driving fishing in the North Sea further and further away from port as stocks were fished out. Declining stocks generated government inquiries in the second half of the nineteenth century, and eventually the International Conference for the Exploration of the Sea in 1899 proposed scientific inquiries to promote 'rational exploitation of the seas' (ibid., p. 194). Fisheries biology eventually came to provide some of the leading ideas about population ecology, notably in W.F. Thompson's work on the Pacific halibut in the 1920s.

By the 1930s the idea (and the mathematics) of a maximum sustainable yield had been calculated. Furthermore, international institutions were established to try to use this emerging science to regulate fishing, from the International Council for the Exploration of the Sea established in 1902 in Copenhagen, through a series of other regional institutions, the Overfishing Convention agreed in London in 1946, and the International Whaling Commission. Neither these nor their successors achieved the sustained exploitation of any significant open-water fish stock. The boom–crash cycle of sealing in the nineteenth century and that of whaling in the twentieth are perhaps particularly glaring, but are absolutely typical of humans' management of stocks of their cold-blooded relatives (Small 1971, Cushing 1988).

However, within fisheries management the principle of using science to define what harvest was sustainable was established, and provided a powerful model for application outside fisheries, in other areas of resource development. The language within which analysis of fish populations was expressed suggests a considerable influence of economics: stocks were renewed or depleted, and calculations included estimates of catch per unit effort. In turn, economics reflected evolving understanding of the dynamics of resources, particularly in the distinction between renewable (flow) and non-renewable (stock) resources (Ciriacy-Wantrup 1952). In this form, too, ideas about the dynamics of natural systems and their response to human management fed into thinking about sustainable development.

Ecology's contribution to thought about sustainability reflected more than population biology alone. As theoretical and experimental approaches to ecology developed, the 'organismic' approach gave way to an essentially mechanistic framework of analysis, involving the concept of the ecosystem ecological energetics and systems analysis (Tansley 1935, Lindemann 1942, McIntosh 1985, Botkin 1990). However, even in the 'systems ecology' associated particularly with the work of E.P. Odum, the fundamental notion that ecosystems tended towards equilibrium endured (McIntosh 1985). Donald Worster comments that both Odum's view of nature as 'an automated factory' and its predecessor the 'Clementsian super-organism' implied that nature 'tended toward order' (1993, p. 160).

Ideas of equilibrium and stability, the 'classical paradigm' or 'equilibrium paradigm' (Steward *et al.* 1992), dominated ecology until the 1970s. It suggested that ecological systems were closed, and that ecosystems were self-regulating so that if disturbed they would tend to return towards an equilibrium state. This paradigm in turn fed ideas in the wider environmental movement that there was a 'balance of nature', easily upset by inappropriate human action. Both systems ecology and population ecology emphasised equilibrium and stability. Nature was portrayed as a homeostatic machine (Pahl-Wostl 1995), and ecosystems were analysed as if they were 'nineteenth-century machines, full of gears and wheels, for which our managerial goal, like that of any traditional engineer, is steady-state operation' (Botkin 1990, p. 12). Nature was a system whose state was maintained by processes of internal feedback, but it was also susceptible to external control. Human action could upset the machine, but fortunately the ecologist could predict how and where this upset might occur, and diagnose how to put the balance right. Ecological science could therefore be used to generate technocratic recipes for managing nature. Words and concepts drawn from thermodynamics and engineering (such as energetics, equilibrium and control), were used blithely by ecologists to describe nature. Conservationists, schooled in ecology, presented themselves as 'ideal scientific "managers" of the environment, the engineers of nature' (Livingstone 1995, p. 368). In the decades that saw the rise of formal 'development', between the end of the Second World War and the rising wave of the environmental revolution in the 1960s, ecology provided a powerful script for the emerging dialogue between environmental protection and economic development.

Ecological managerialism

This policy role for ecology is particularly clear in post-war Africa. As the idea of development planning began to take root, ecology was seen to provide not only valuable data, but *a model for the practice of development itself.* Thus in a remarkable polemic in support of a managerial role for environmental science, Culwick (1943) suggested that development in Tanganyika was essentially a problem of 'controlling the environment and exploiting it in such a way that not only this but succeeding generations may be able to attain their full potentialities, both physical and mental'. This task of development was 'primarily a scientific problem in which all branches of science, physical and social, have an essential part to play' (ibid., p. 1).

Furthermore, development not only had to involve scientists, but arguably needed to be conceived of within the paradigms of ecology. Culwick called for reform of the institutions of government after the war, so that life could be 'planned as an oecological whole'. In the past, scientists had been marginal to the essentially political business of government, regarded as 'appendages to the main administrative body, called in every now and again for consultation when a scientific question cropped up'. However, the notion of development (at this time of course still only loosely conceived; see Chapter 1) demanded a new approach. Government should be regarded 'primarily as a scientific affair', and administration should therefore become 'merely the mechanism for putting the big scientific plan into action' (Culwick 1943, p. 5). Science, and particularly ecology and environmental science, could not only inform development, but direct it.

Development in the post-Second World War period was conceived as an organised and coherent attempt to overcome constraints on economic growth, and often explicitly aimed at overcoming environmental constraints on that growth. Croll and Parkin suggest that development may be conceived of 'as a form of self-conscious or planned construction, mapping and charting both landscapes and mindscapes' (1992, p. 31). That planning and action was increasingly driven by science, technology, a capitalist market economy and formal organisation. In his book *Rationality and Nature*, Raymond Murphy (1994) describes these as dimensions of 'rationalisation', following the work of Max Weber (1922). Rationalisation is the process by which human reason rationally dominates both nature and society. The attraction of the science of ecology in development planning was its power to rationalise both comprehension of development 'problems' and action. Development planning can be seen as the imposition of rationalisation, where science provides knowledge that can be used to control environment, economy and society in such a way that change can be directed in desired directions.

Murphy identifies a number of dimensions of rationalisation. The first is the development of science and technology: 'the calculated, systematic expansion of the means to understand and manipulate nature', and the scientific worldview 'belief in the mastery of nature and of humans through increased scientific and technical knowledge' (Murphy 1994, p. 28). The second dimension of rationalisation is the expansion of the capitalist economy (with its rationally organised and in turn organising market); the third dimension is formal hierarchical organisation (the creation of executive government, translating social action into rationally organised action). The fourth is the elaboration of a formal legal system (to manage social conflict and promote the predictability and calculability of the consequences of social action). All these things were features of the colonial state, in Africa as elsewhere. Ecology provided ample contributions to the first of these dimensions, as a science able to produce rational stories in the face of novel environmental complexity (for example the 'useful purpose' served by surveys that 'properly analysed and classified vegetation'; Shantz and Marbut 1923, p. 4). Ecology also provided a highly relevant model of the wider relevance of the ratonalising and ordering power of science for planning and structuring action.

In the colonial world of the tropics, the hegemony of ecology was at best only partial. Even in Africa, where colonial states were slender and recently rooted, and the environmental unknowns were glaring, the contribution of ecology in practice was limited. The colonial administration in Tanganyika, for example, never reduced itself to Culwick's 'mere mechanism' for applying science. However, what Low and Lonsdale (1976) have called a 'second colonial occupation' began in East Africa after the Second World War, and scientists acquired new and important roles in some fields, most notably perhaps agriculture. The engagement with the development problems of rural Africans inevitably demanded an engagement with the ecology of their production systems and the wider landscapes of which they were part. The importance of the biology of agriculture, grazing, forestry and disease, and the physical geography of erosion and water supply, were recognised, and scientific expertise in each field won a place within what had been predominantly legal, political and fiscal institutions of colonial government.

This attention to indigenous production systems sometimes generated positive official understandings, as in West Africa (e.g. Faulkner and Mackie 1933, Stamp 1938), or in Northern Rhodesia, where William Allan's book *The African Husbandman* (published long afterwards, in 1965) demonstrated the logic of indigenous agricultural systems. More generally, the 'second colonial occupation' generated a conventional scientific wisdom that human use of resources led almost inevitably to degradation, as for example in the work of Pole-Evans and Phillips on range management in South Africa (Scoones 1996; see also Chapter 7). An editorial in the scientific journal *Nature* commented in 1952, 'if the human race continues to multiply at the present rate, catastrophe will only be avoided if methods of cultivation are used which, far from destroying the fertility of the soil, maintain or even increase it' (*Nature* 1952, pp. 677–8).

Whether favourable or hostile in its portrayal of indigenous resource users, ecological managerialism penetrated the development planning process in the colonial world following the end of the Second World War. Ecological ideas were important in the First Ugandan Development Plan, although the new Governor in 1946 still thought it 'much too departmental' (Worthington 1983, p. 92). In this it was not unusual; *Nature* complained in 1948 of one plan which was 'an agglomeration of departmental suggestions masquerading as a development plan' (*Nature* 1948). The model being offered in Uganda, which depended to such a large extent on living resources, was explicitly ecological: 'fundamentally, the problem of development was one of human ecology, the diverse people reacting with their varied environments' (Worthington 1983, p. 97). This approach to economic development of human ecology was not unique (a famous example being Frank Fraser Darling's *West Highland Survey* (1955)), but it represents a significant element in the development of environmentalist thinking about the development process.

Ecology and economic development

Understanding of the adverse ecological impacts of development led to attempts to identify specific formulae for avoiding or minimising these impacts. This essentially technocentrist search for 'environmentally benign' development was developed in particular through the work of the International Union for

Conservation of Nature (IUCN) and the International Biological Programme (IBP) and the Man and the Biosphere Programme (MAB) in the 1960s and early 1970s.

A conference on 'the ecological aspects of international development' was held late in 1968 at Airlie House in Virginia, organised by the Conservation Foundation and Washington University of St Louis. Its proceedings were published as *The Careless Technology: ecology and international development* (Farvar and Milton 1973). These papers presented a readable and authoritatively researched catalogue of environmental problems associated with or caused by economic development, from the downstream impacts of dams to pollution. It set the style for many subsequent accounts of environment and development, and many of the issues raised remain important (and unresolved).

A series of meetings followed this between international conservation, environment and development organisations, including IUCN, the Conservation Foundation, UNDP, UNESCO and FAO. In 1970 there was a meeting at the FAO headquarters in Rome of interested parties, including now USAID and the Canadian aid agency CIDA (the Canadian International Development Agency) to consider further the ecological impacts of development. It was decided that IUCN and the Conservation Foundation should publish guidelines for development planners. These appeared as *Ecological Principles for Economic Development* (Dasmann et al. 1973). Initially the book was to include discussion of 'interrelationships between economic development, conservation and ecology', but in the event it was restricted to an exploration of ecological concepts 'useful in the context of development activities'. Particular emphasis was placed on tropical rainforests and semi-arid grazing lands, the effects of tourism (particularly in fragile environments such as high mountains and coasts) and the development of agriculture and river basins (Dasmann et al. 1973, p. vi).

This book encapsulated post-war thinking by conservationists and ecologists on development, and it formed the basis of subsequent thinking and writing about sustainable development. Its basis was the need to apply concepts and insights from the science of ecology to development activities. The central argument is that ecology has a direct bearing on what can, or should, be done in development. If the 'lessons' of ecology are ignored, 'entirely unexpected consequences can often result from what are intended to be straightforwardly beneficial activities'. Thus, for example, the replacement of tropical rainforest with a palm oil plantation 'sets in motion complex ecological forces', which may involve the loss of equilibrium and pest outbreaks (Dasmann et al. 1973, p. 44).

Dasmann et al. picture ecology as an integrative science, an interdisciplinary way of thinking which could be instilled in the minds of 'forester, agricultural specialist, range manager ... development economist or engineer'. The use of ecology in development planning has the aim of both 'enhancing the goals of development' and 'anticipating the effects of development activities on the natural resources and processes of the larger environment'. Ecology had in the past been confined to assessing the potential productivity of a resource; now the adverse impacts of 'certain kinds of development and management technology' on local and global environments demanded that these impacts be 'anticipated in the planning process, and to the extent possible, evaluated', so that decisions about developments are made 'in full knowledge of possible consequences'. A decision to develop despite environmental impacts might be justified by counterbalancing benefits, but 'it should never be taken blindly' (Dasmann *et al.* 1973, pp. 21–2).

Ecological Principles for Economic Development was a pragmatic attempt to express the views of environmentalists in a way that those responsible for decisions in development would understand and take account of. Adverse environmental impacts of development needed to be understood and dealt with during development planning if major problems were to be avoided. Ecology was therefore extremely useful. Its application would help planners 'to make sure of success' (Dasmann *et al.* 1973, p. 21).

This pragmatic approach to development on the part of environmental organisations such as the Conservation Foundation and IUCN was matched by interest from the technical disciplines they hoped to influence. This is shown most clearly in the response to the question of the environmental impacts of dams. The apparent 'conversion' of engineers to wider environmental concerns needs to be seen in the context of continuing suspicion, hostility and pressure from environmentalists (e.g. Morgan 1971, McCully 1996). By the 1960s, large dams were sufficiently common for it to be worth the construction industry compiling a global register, and by the 1970s engineers were starting to acknowledge and review environmental effects (Turner 1971). By the second half of the 1970s, such reviews were commonplace in journals and magazines read by engineers and hydrologists (e.g. Biswas and Biswas 1976, writing in *Water Power and Dam Construction*), and it was possible to synthesise more than a decade of ecological research (Baxter 1977). The Man and Biosphere Programme designated Project 10 on the effects of major engineering works on humans and their environment (UNESCO 1976). In 1977 the UN Water Conference received a report called *Large Dams and the Environment: recommendations for development planning* (Freeman 1977). The International Commission on Large Dams (ICOLD) appointed a committee on damming and the environment in 1972, which in 1981 published *Dam Projects and Environmental Success*, a paper 'intended to illustrate the concern and knowledge of dam engineers related to environmental matters' (ICOLD 1981, p. 8). This paper, with its review of impacts and its stress on the fact that environmental effects could be beneficial as well as adverse, and that 'dam engineers are very aware of the importance of environmental requirements and give full and early consideration to possible difficulties of this type' (ibid., p. 7), neatly combined professional training and public image creation. It also demonstrates the success, at least on a rhetorical level, of attempts to imbue development with ecological awareness.

Such attempts were further developed through the formulation of principles of environmental impact assessment (EIA) in the late 1960s. EIA became an important part of public policy in the industrialised world with the US National

Environmental Policy Act (NEPA) of 1969. A number of industrialised countries followed the USA to a greater or lesser degree down the path of institutionalising the assessment of environmental impacts (Barrow 1997). One of the international applications of EIA procedures was again by ICOLD, which designed and tested a version of the Leopold Matrix (Leopold *et al.* 1971) to take account of the impacts of large dam construction (ICOLD 1980). The procedure of EIA was taken up in the 1970s by the Scientific Committee for Problems of the Environment (SCOPE). In 1975 SCOPE published a volume reviewing techniques in environmental impact assessment, *Environmental Impact Assessment: principles and procedures.* This was subsequently revised at a meeting in 1977 and republished (Munn 1979).

Environmentalism, population and global crisis

The rise of environmentalism in the industrialised world in the 1960s had enormous significance for debates about the role of ecology and conservation in development. Myers and Myers (1982) argue that even in the 1980s, perception of environmental issues rarely transcended national boundaries or national interests. This may be true, as regards the extent to which environmentalism reflected self-interest as opposed to some more egalitarian global consciousness. What is clear, however, is that the perception that there *were* environmental issues of global significance was a distinctive and novel feature of the 'new environmentalism' (Cotgrove 1982) that arose in North America and Western Europe in the 1960s and 1970s. The image of the earth as a blue ball spinning in the darkness of space ('Spaceship Earth', a term used by Boulding (1966)) became environmentalism's icon, the threat to its perfection portrayed in the global 'doomsday syndrome' about which Maddox (1972) complained.

The global vision of environmentalism in the 1960s and 1970s is nicely captured in the title of the book written by Barbara Ward and René Dubos for the 1972 UN Conference on the Human Environment in Stockholm, *Only One Earth* (Ward and Dubos 1972), and by that of Max Nicholson's book published two years before, *The Environmental Revolution: a guide for the new masters of the world* (Nicholson 1970). Partly through the Stockholm Conference, awareness of environmental problems (particularly pollution; Dahlberg *et al.* 1985) on a global scale became a key theme in the environmental revolution of Europe and North America in the 1970s. Parallel with it grew an apocalyptic vision of neo-Malthusian crisis. It is within this context that ideas about sustainable development emerged.

Such globalism in environmental concern was not new. In 1954 Fairfield Osborn wrote, 'man is becoming aware of the limits of his earth. The isolation of a nation, or even a tribe, is a condition of an age gone by' (1954, p. 11). One fruit of this realisation was the rise of concern about global population. Ecologists such as Raymond Pearl discussed the phenomenon of human population growth in the 1920s (Pearl 1927), and the spectre of population growth was raised by commentators such as Carr-Saunders (1922, 1936). In

1948 Osborn commented, 'the tide of the earth's population is rising, the reservoir of the earth's living resources is falling' (1948, p. 68).

Under Julian Huxley, UNESCO became involved in the 1950s in the application of scientific evidence in the debate about population and development. For example, UNESCO ran the World Population Conference with the FAO in Rome in 1954, thereby linking development issues to one of the central concerns of environmentalism. The 'population problem', as it became widely known, was commented on extensively in the 1950s, for example by Boyd-Orr (1953), Stamp (1953), Russell (1954) and in the famous conference volume *Man's Role in Changing the Face of the Earth* (Thomas 1956). Neo-Malthusian arguments were a prominent feature of environmentalism in the 1960s and 1970s, most notably in the work of Ehrlich and Ehrlich (1970), in the apocalyptically titled *The Population Bomb* (Ehrlich 1972), and in Garret Hardin's paper in the journal *Science* in 1968, 'The tragedy of the commons'. This drew attention to issues of common-pool resource mangement, subsequently the focus of the discipline of new institutional economics (e.g. Ostrom 1990, North 1990, Cleaver 2000). However, its starting point was the debate about global population growth, and the observation that 'a finite world can support only a finite population; therefore population growth must eventually equal zero' (Hardin 1968, p. 1243).

The scope of neo-Malthusian thinking has been remarkably persistent. In 1970 the world population was 3.5 billion, and 'Spaceship Earth' was said to be 'filled to capacity and beyond and is running out of food' (Ehrlich and Ehrlich 1970, p. 3). In July 1987 global population had risen, according to estimates by the United Nations Fund of Population Affairs, to 5 billion people; the threat it seemed to represent was unchanged: 'whichever way one looks at the population problem – whether as a biologist, sociologist, theologian, medical doctor, industrialist, administrator or politician – it is obvious that it presents the greatest menace to the future of the biosphere' (Worthington 1982a, p. 98). To some environmentalists, population growth still offered a dire and global threat, and the rhetoric of neo-Malthusianism was still shockingly prominent: 'the remedy is left to Nature's ways of shortage and deprivation, famine or pestilence, and to Mankind's own way of increasing violence and slaughter' (Polunin 1984, p. 296). (It is worth noting that with a global population at the end of the twentieth century at over 6 billion and rising, views about this issue continue to range from the sanguine to the panic-stricken.)

'Catastrophist' environmentalist thinking about pollution and population growth in the 1960s and 1970s was matched by a parallel and closely related debate about economic growth. Over the 1960s and 1970s, economic growth in industrialised countries gave way to recession and inflation. Most analysts followed Kahn and Wiener (1967), who offered a fairly optimistic range of scenarios of the future. Within a few years the tone of forecasts had changed. In 1969 Mishan looked at the costs of economic growth (coining the delightful term 'growthmania'), and concluded rather cautiously that 'the continued pursuit of economic growth by Western societies is more likely on balance to reduce rather than increase social welfare'.

This still cautious view was overtaken by the results of attempts to produce global computer models of the 'world system', notably Forrester's *World Dynamics* (1971). This approach was developed by a team from the Massachusetts Institute of Technology for the 'Club of Rome', an international group set up with the backing of European multinational companies (Golub and Townsend 1977). The work was published as *The Limits to Growth* (Meadows *et al.* 1972), and with the British utopian polemic *Blueprint for Survival* published in the same year (Goldsmith *et al.* 1972), it forms one of the two most commonly quoted (although perhaps less commonly read) treatises of 1970s environmentalism.

The idea of zero growth and a steady-state economy has attracted sober support (e.g. Daly 1973, 1977, Mishan 1977), but has also generated a body of fairly vituperative criticism. *The Limits to Growth* has been the more marked target. Beckerman suggested that the Club of Rome report was 'guilty of various kinds of flagrant errors of fact, logic and scientific method' (1974, p. 242), while Simon dismisses it as 'a fascinating example of how scientific work can be outrageously bad and yet be very influential' (1981, p. 286). Maddox (1972) said simply, 'the doomsday cause would be more telling if it were more securely grounded in facts' (p. 2).

The issue of zero growth will be discussed in more detail in Chapter 5. The important point to note here is the way the notion of 'global crisis' in the 1960s and 1970s contributed to the internationalisation of both ecology and environmentalism. Although the 'ecological prescriptions for managing the human use of the earth' that were on offer were indeed extremely limited (Stoddart 1970, p. 2), the grandiose claims and global fears of environmentalism were far from ineffective.

Global science and sustainable development

Theoretical and practical links between ecological science and development were fostered in 1964 by the establishment of the International Biological Programme (IBP). The IBP was launched by the International Union of Biological Sciences (IUBS) under the International Council of Scientific Unions (ICSU) in 1964, the model of international scientific cooperation initially coming from the International Geophysical Year 1957–8 (Worthington 1975, 1982b, 1983). Planning began in 1959, with the vision of studying 'the biological basis of man's welfare' (Waddington 1975, p. 5). However, it was only at a meeting at the IUCN headquarters in Morges in 1962 that the IBP planning committee adopted the sevenfold grouping of research that came to form the structure of the programme. Perhaps unsurprisingly given the venue, one group was dedicated to terrestrial conservation (Nicholson 1975).

In the event, extensive work was done by the terrestrial conservation section, in particular the gathering of global data on areas of scientific importance, and an attempt to establish a network of biome research stations. However, Nicholson argues that the biological community 'never fully endorsed in practise the inclusion of conservation' (1975, p. 14). There were also substantial

problems of integration. Boffey (1976), for example, points out that the level
of cooperation between First World scientists and their Third World counterparts was small. The terrestrial conservation section faced particular problems.
It had to be Janus-headed and address two audiences: research biologists,
to encourage them to apply their ideas, and natural resource managers, to
encourage the application of ecological theory (Worthington 1975, p. 86). The
results were mixed.

The IBP, however, was only one element in the expansion of research into
the global environment, and awareness of global environmental problems. There
were a series of meetings called in the post-war period to 'bring together key
actors of the globe' to address environmental issues on a global scale (Dahlberg
et al. 1985). In some ways the most fundamental was a symposium held by
the Wenner-Gren Foundation at Princeton, New Jersey, in 1955, Man's Role
in Changing the Face of the Earth (Thomas 1956), which was chaired by Carl
Sauer, Marston Bates and Lewis Mumford. This symposium claimed roots in
the work of George Perkins Marsh (1864) on the influence of man on nature,
and similar thinkers from the late nineteenth century onwards. Man, 'the ecological dominant on this planet', needed the insights of scholars to understand
his 'impress' on the earth. The symposium was 'a first attempt to provide an
integrated basis for such an insight and to demonstrate the capacity of a great
number of fields of knowledge to add to our understanding' (Thomas 1956,
p. xxxvii). This theme was one which fitted easily into the new internationalism of science in the IBP, and with the rising globalism of environmentalist
thinking in the 1960s.

The global nature of environmental problems was debated at the Third General
Assembly of the IBP in Bulgaria in April 1968, and in 1969 the Scientific
Committee for Problems of the Environment (SCOPE) was established under the
International Council of Scientific Unions (ICSU) (Worthington 1982b, 1983).
SCOPE's work was to focus on ways of understanding specific environmental
problems, particularly at a global scale (Table 2.1). It reported in 1971 and 1973
on global environmental monitoring, in 1975 and 1979 on global geochemical
cycles, and in 1979 on Saharan dust. The proceedings of the SCOPE/UNEP

Table 2.1 The mandate of the Scientific Committee for Problems of the Environment
           (SCOPE)

- To assemble, review and assess the information available on man-made environmental changes and the effects of these changes on man;
- to assess and evaluate the methodologies of measurement of environmental
  parameters;
- to provide an intelligence service on current research;
- by the recruitment of the best available scientific information and constructive
  thinking to establish itself as a corpus of informed advice for the benefit of centres
  of fundamental research and of organisations and agencies operationally engaged
  in studies of the environment.

Source: Munn (1979)

symposium on 'environmental sciences in developing countries' in Nairobi were published in 1994. SCOPE's work also collated material on specific environmental issues, for example environmental impact assessment (Munn 1979).

To an extent, the work of SCOPE was overtaken (although not replaced) by the Man and the Biosphere Programme (MAB) which grew out of the 'Biosphere Conference', the Intergovernmental Conference of Experts on a Scientific Basis for Rational Use and Conservation of the Biosphere held in Paris in 1968. This made explicit the growing engagement by conservationists and environmentalists with the development process (Caldwell 1984). It was a further step in the incorporation of Third World countries into the world of international environmental concern. The initiative stemmed primarily from a realisation of the continuing failure to create a truly international environmentalism. Boardman comments that 'in the last analysis, the one major political cleavage that the issue of nature conservation has failed to bridge adequately is that between industrialised and developing countries' (1981, p. 19).

Like their close contacts in IUCN, conservationists within UNESCO realised increasingly through the 1960s that they could not influence decisions about the use of natural resources in the Third World unless they were prepared at least to talk in the new language of development. UNESCO had adopted key resolutions that explicitly linked conservation and development at its 1962 General Conference. It sponsored the symposium Man's Place in the Island Ecosystem at the Tenth Pacific Science Congress in Honolulu in 1961, which developed the idea of human actions as a functioning part of an ecosystem (Fosberg 1963). UNESCO also joined other UN agencies (the FAO and UNDP), IUCN, the Conservation Foundation and the World Bank in the discussions about ecology and development that followed this conference (Dasmann *et al.* 1973), and had been involved in a review of natural resources in Africa at the request of the Economic Commission for Africa in 1959 (UNESCO 1963).

The Biosphere Conference had a complicated history. The idea of an international conference on endangered species was initially suggested at the IUCN General Assembly in Nairobi in 1968. The broader 'biosphere' approach came from ECOSOC (the United Nations Economic and Social Council) and UNESCO (Boardman 1981). The Biosphere Conference called for the establishment of an interdisciplinary and international programme of research on the rational use of natural resources and 'to deal with global environmental problems' (Gilbert and Christy 1981). This proposal was heavily influenced by those in national delegations with experience of the IBP (Worthington 1983, p. 175). The MAB programme that was launched in 1971 had a strong scientific base, and was in many ways a direct successor of the IBP (Gilbert and Christy 1981, Worthington 1983, Holdgate 1999). There was considerable passage of scientific information between the two programmes (Worthington 1975, 1983).

MAB's function was 'to develop the basis within the natural and social sciences for the rational use and conservation of the resources of the biosphere and for the improvement of the global relationships between man and the environment' (Gilbert and Christy 1981). Consequently, MAB was to be both useful and down-to-earth: 'ivory tower research' was of little use to 'those who have to make management decisions in a world of increasing complexity'. MAB would be different, breaking down 'obsolete barriers' between natural and social scientists and decision-makers, and offering instead 'an interdisciplinary, problem-oriented approach to the management of natural and man-modified ecosystems' (UNESCO not dated). MAB was given an exhaustive and astonishingly open-ended range of specific objectives. These included the study of the 'structure, functioning and dynamics of natural, modified and managed ecosystems' and of the relations between 'natural' ecosystems and 'socio-economic processes', and the identification and assessment of human impacts on the biosphere. There were also aims to promote 'global coherence of environmental research', environmental education and specialist training, and 'global awareness of environmental problems' (Gilbert and Christy 1981, pp. 704–5).

These aims were obviously not all attainable in full, so a series of specific fields were set out for action. These focused either on particular environments of concern to environmentalists (for example rainforests, Project 1, or semi-arid zones, Project 3) or on particular impacts of development (for example major engineering works, urban systems and energy, or pollution, Projects 10, 11 and 14, the latter added in 1974). The main shortcomings of the programme were due quite simply to its ambition. Government response in the Third World was favourable, but overall international action was slow to get off the ground and underfunded (Batisse 1975). Some projects did become operational in the Third World. For example, under MAB Project 1 on the ecological effects of increasing human activities on tropical and subtropical forest ecosystems, the San Carlos de Rio Negro project was begun in 1976 through a newly established International Centre for Tropical Ecology in Caracas. Studies of the structure, composition and production of undisturbed rainforest soils were followed by research on nutrient cycling and the effects of disturbance (Gilbert and Christy 1981). Under MAB Project 3 on the impact of human activities and land use practices on grazing lands, the Integrated Project on Arid Lands (IPAL) was begun in 1976 with UNESCO and UNEP funding, and produced a stream of integrated research studies on semi-arid vegetation and the ecology of pastoralism (e.g. Herlocker 1979). Under MAB Project 7 on island ecosystems, a pilot project was organised in Fiji from 1974 to 1976, supported by the UN Fund for Population Activities (Bayliss-Smith *et al.* 1988), and a second in selected islands of the eastern Caribbean in 1979 (di Castri 1986).

These projects, and others, have a good claim to be the forerunners of 'sustainable development' thinking, linking natural ecosystems and human use in an innovative or wholly research-based structure. However, within MAB, traditional 'nature' conservation also remained important, even if dressed in the new clothes of human ecology. The best example of this is the notion of 'biosphere reserves' (Project 8). These were to be zoned nature reserves, whose aim was to conserve 'natural areas and the genetic information they contain' in core zones, while allowing suitable human activities to continue in outer zones. Existing nature reserves could be reclassified (and sometimes extended) to fit the MAB framework as biosphere reserves. This initiative was an important linkage between 'pure' wildlife conservation and the much broader aims of MAB, and between nature preservation and the idea of conservation of natural resources for human use. Thus Batisse wrote, 'The greatest merit of the "Biosphere Conference" was perhaps the assertion, for the first time in an intergovernmental context, that the conservation of environmental resources could and should be achieved alongside that of their utilisation for human benefit' (1982, p. 101).

Although the MAB programme as a whole failed to live up to the hopes it raised at its inception, it undoubtedly contributed through the 1970s to the growing belief that there was an ecologically sound approach to development which would be 'sustainable' and acceptable, and that this could be discovered for specific environments and circumstances through research done in new, open and interdisciplinary ways.

The biosphere reserves represent this vision at its rosiest tint. Batisse wrote that

> experience already shows that when the populations are fully informed of the objectives of the biosphere reserve, and understand that it is in their own and their children's interests to care for its functioning, the problem of protection becomes largely solved. In this manner, the biosphere reserve becomes fully integrated – not only into the surrounding land-use system, but also into its social, economic, and cultural, reality.
>
> (1982, p. 107)

This vision – of conservation integrated with and serving some rather vaguely defined human (and hence economic) purpose – was central to the MAB programme, and through the 1970s it was fostered by it. The identical notion resurfaces in 1980 in ideas about sustainable development in the *World Conservation Strategy*. This, and the other documents of what I term the 'sustainable development mainstream', are the subject of the next chapter.

Summary

- Ideas about sustainable development that emerged in the 1980s had deep roots. These include the place of the tropics in the development of environmentalism, concern for nature preservation, the science of ecology, and ideas about the balance of nature and the need for scientifically based ecological management, concern about global population growth and the development of global scientific networks.
- The tropics, which were the focus of much of the environmental concern and development action in the late twentieth century, were also important to the development of environmentalism at much earlier periods, particularly in the seventeenth and eighteenth centuries.

- Nature preservation is important to sustainable development both as a source of the impulse to balance human need and human claims on nature, and also because of the role of international conservation organisations (especially the World Conservation Union, and IUCN) in generating the thinking that stimulated the formulation of the concept of sustainable development, and organising the meetings where it was first set out.
- The science of ecology has contributed to development, and development planning, in various ways, for example in the periods following the First and Second World Wars.
- Ecological ideas such as 'the balance of nature', the concept of the ecosystem and maximum sustainable yield provide an essential underpinning of concepts of sustainable development.
- Global scientific collaboration, notably in the International Biological Programme, provided an authoritative, apolitical and effective arena within which ideas of sustainable development could develop and be discussed in the 1970s.
- Concern about limits to growth and global population growth were fundamental to the environmental revolution of the 1970s, and provide the background to the emergence of formal statements of the idea of sustainable development.

Further reading

Boardman, R. (1981) *International Organisations and the Conservation of Nature*, Indiana University Press, Bloomington, IN.
Evans, D. (1992) *A History of Nature Conservation in Britain*, Routledge, London.
Griffiths, T. and Robin, L. (eds) (1997) *Ecology and Empire: environmental history of settler soceities*, Keele University Press, Keele.
Grove, R.H. (1995) *Green Imperialism: colonial expansion, tropical island Edens and the origins of environmentalism, 1600–1800*, Cambridge University Press, Cambridge.
Grove, R.H., Damodaran, V. and Sangwan, S. (eds) (1998) *Nature and the Orient: the environmental history of South and South East Asia*, Oxford University Press, Delhi.
Holdgate, M. (1999) *The Green Web: a union for world conservation*, Earthscan, London.
McCormick, J.S. (1992) *The Global Environmental Movement: reclaiming paradise*, Belhaven, London.
Worster, D. (1985) *Nature's Economy: a history of ecological ideas*, Cambridge University Press, Cambridge.
Worthington, E.B. (1983) *The Ecological Century: a personal appraisal*, Cambridge University Press, Cambridge.

Web sources

<http://www.kenya-wildlife-service.org/contable.htm>   The website of the Kenya Wildlife Service, including an interesting account of the history of conservation in Kenya, and the Kenya Wildlife Service.
<http://www.wing-wbsj.or.jp/birdlife/history.htm>   BirdLife International and its history

(as the International Council for Bird Preservation, founded 1926). Information on current BirdLife programmes and partners is on <*http://www.wing-wsbj.or/birdlife/*>.

<*http://www.unesco.org/mab/*> The UNESCO Man and the Biosphere Programme: offers a complete list of all biosphere reserves (391 sites in ninety-four countries in 2000).

<*http://www.iucn.org/*> The World Conservation Union (IUCN), with details on its programme, publications, meetings and specialist groups.

<*http://www.toursaa.com/krugepark/*> Website for the Kruger National Park in South Africa: Eden in cyberspace, including a virtual safari.

<*http://www.fao.org/fi/default.asp*> The Food and Agriculture Organisation's Fisheries programme – insights into the continuing struggle to make fisheries management sustainable.

<*http://biom3.univ-lyon1.fr/Ecology/Ecology-WWW.html*> A huge list of websites on ecology, ecological research, ecosystems and everything related.

<*http://www.zpg.org/*> Zero population growth is alive and well on the Internet. Another site features the work of Paul and Anne Ehrlich: <*http://dieoff.com/page27.htm*>.

<*http://www/fauna-flora.org*> Fauna and Flora International – whatever happened to the 'penitent butchers' of the Society for the Preservation of the Wild Fauna of the Empire? Information on history, philosophy structure, methodology.

3 The development of sustainable development

Unless the penguin and the poor evoke from us an equal concern, conservation will be a lost cause. There can be no common future for humankind without a better common present. Development which is not equitable is not sustainable in the long term.

(Monkombu Swaminathan, Opening Address to IUCN General Assembly, Perth, 1991 [Holdgate 1999, p. 206])

Before the mainstream: the Stockholm Conference

Sustainable development was codified for the first time in the *World Conservation Strategy* (WCS), a document prepared over a period of several years in the later 1970s by IUCN with finance provided by UNEP and the World Wildlife Fund (IUCN 1980). It was then further developed through the report of the World Commission on Environment and Development, *Our Common Future* (Brundtland 1987), and the follow-up to the WCS, *Caring for the Earth* (IUCN 1991), before its appearance in *Agenda 21* at the Rio Conference in 1992. These documents differ, but have a remarkably consistent core of ideas, a 'mainstream' that has persisted through the two decades between Stockholm and Rio (Adams 1995). At the heart of these documents is a vision of sustainable development strongly influenced by science, by ideas about wildlife conservation, by concerns about multilateral global economic relations, and by an emphasis on the rational management of resources to maximise human welfare. The stock from which all these mainstream documents descended, and the forum at which ideas of sustainable development were first brought onto the international agenda, was the United Nations Conference on the Human Environment held in Stockholm in June 1972.

McCormick (1992) argues that in many ways the Stockholm Conference simply developed ideas already raised at the Biosphere Conference in Paris in 1968 (Chapter 2). However, the Stockholm meeting is usually identified as the key event in the emergence of sustainable development. It was only partly, and belatedly, concerned with the environmental and developmental problems of the emerging Third World. The primary motivation behind the UN's decision to hold such a conference came from industrialised countries, and was the fruit of the classic concerns of First World environmentalism, particularly pollution associated with industrialisation. Sweden itself was particularly concerned about acid rain (as indeed it remained) (McCormick 1986). The Swedish ambassador to the United Nations submitted a proposal for a conference on the human environment to ECOSOC (the United Nations Economic and Social Council) in July 1968, and the resolution was approved in December of that year. Conference planning began in 1968, and a twenty-seven-nation 'Preparatory Committee' began meeting under the chairmanship of Maurice Strong in 1970 (Holdgate 1999).

However, the proposed conference did not command support from all countries. Third World countries in particular believed that environmental problems and development problems had become separated, and the sense of integration and of shared problems between developing and industrialised countries was lost (Russell 1975). Third World governments mistrusted neo-Malthusian ideas, whether of zero growth or lifeboat ethics:

> some 'developing' countries felt that the concept of global resources management was an attempt to take away from them the national control of resources. Furthermore, as industrialised countries used the lion's share of resources and contributed to most of the resulting pollution, the Third World countries did not see much reason to find and pay for the solutions.
>
> (Biswas and Biswas 1984, p. 36)

Faced with urgent short-term problems of poverty, hunger and disease, longer-term environmental problems associated with industrialisation seemed not only remote, but a possible means by which industrialised economies might wriggle off the hook of responsibility for supporting a rapid drive for development.

It seemed possible that controversy over the relative priorities to be accorded environment and development in the Third World might cause the Stockholm Conference to fail. A meeting of a preparatory committee of twenty-seven experts at Founex in Switzerland in June 1971 sought to soothe the concerns of Third World countries, allaying fears about the economic effects of environmental protection policies. It proffered assurance that environmental protection would not go against their interests and would not affect their position in international trade (e.g. by anti-pollution barriers), and that rapid industrialisation could still be pursued, but in such a way that its most adverse impacts were avoided (McCormick 1992).

The Founex meeting was certainly a political success inasmuch as the developing countries duly came to Stockholm. In the preparation for the Stockholm Conference, it was becoming clear that the position of non-industrialised countries was gaining recognition, just as (through their voting power in the UN) they were gaining power. The scope of the conference was expanded in December 1971 to include issues such as soil erosion, desertification, water supply and human settlement. Founex made the case that the environment was relevant to less industrialised countries, indeed that environmental issues were a central issue in successful development (McCormick 1992). The argument was made that the apparent dichotomy of 'environment versus development' was false, and should not be recognised, let alone fostered (Biswas and Biswas 1984).

At the same time, the Founex meeting did not break new conceptual ground. Like Maurice Strong in his meetings with Third World governments, it simply made a statement of faith that development and environment *could* be combined in some way which would optimise ecological and economic systems, without explaining how. It promised that the Stockholm meeting would 'point the way towards the achievement of industrialisation without side-effects', but it did not say *how* this desirable trajectory of change was to be achieved (Clarke and Timberlake 1982, p. 7).

It was the same at the Stockholm Conference itself. Although the phrase 'pollution of poverty' was coined to headline poverty and lack of development in 'environmental' language, there was little discussion of the links between poverty and environmental degradation. Few of the conference's Declaration of Principles did more than recognise the nature of the particular problems of the Third World. Many Third World countries remained sceptical, and a common theme in their leaders' speeches was that 'environmental factors should not be allowed to curb economic growth' (McCormick 1992, p. 99). Even the humane and encyclopedic popular book written for the conference by Barbara Ward and René Dubos, *Only One Earth* (1972), offered relatively little to the 'developing regions'. It recognised the hard inheritances of colonialism and exploitative trade, and discussed the problems of population, possible policies for growth in agriculture and industry, and the question of urban environments. The synthesis, however, was global, and there was little beyond general exhortation in this volume – which became one of the classics of 1970s environmentalism – about how environment and development could be integrated in the Third World. It was clear that this *should* happen, but less clear how it could be done.

The conference itself in June 1972 was attended by 113 nations. The German Democratic Republic was not invited, and in protest the USSR and most of the East European countries did not attend. Five hundred non-governmental organisations participated in a parallel 'fringe' meeting, the Environmental Forum. There were bitter debates, for example about colonialism, Vietnam, whaling and nuclear weapons testing, before compromise was reached in agreement on 26 Principles and 109 recommendations for action (Clarke and Timberlake 1982). The Principles were wide-ranging, from human rights (1) and disarmament (26) through to the need for environmental education and research (19 and 20). There were general exhortations about pollution (6 and 7), the need to 'safeguard' wildlife and natural resources, the need to 'share' non-renewable resources (5), and the need to cooperate over international issues. There was stress on the right of individual nations to determine population and resource policies (16 and 21).

Most importantly, in the light of the influence of the 'spirit of Stockholm', there were some deliberate attempts to address the problems of the Third

World. The fundamental point was that development need not be impaired by environmental protection (Principle 11). This was to be achieved by integrated development planning (13) and rational planning to resolve conflicts between environment and development (14). Furthermore, development was needed to improve the environment (8), and this would require assistance (9), particularly money to pay for environmental safeguards (12), and reasonable prices for exports (10).

Like Founex, the Stockholm meeting itself stated the need to resolve conflicts between environment and development without demonstrating how. The suggested solutions, 'rational planning' or 'integrated development', were words without any detailed substance. They owed much to the technocratic element in ecological and environmentalist thinking developing at that time. In very mild form they reflected the authoritarian idea that environmental harmony should be sought through central control (see Pepper 1984): they were based on the premise that planning was neutral and perfectible, that conflicts could be planned away.

The need to see environment and development as an integrated whole was well argued at Stockholm. However, few of its recommendations addressed the issue: only 8 of the 109 Recommendations for Action referred to development and environment, and they were 'extraordinarily negative' (Clarke and Timberlake 1982, p. 12), concerned chiefly with minimising possible costs of environmental protection. In the ensuing decade there was little progress in this field, the only specific report by the United Nations Conference on Trade and Development (UNCTAD) failing to identify effects of economic policies on trade, and by 1982 the debate was said to be 'largely dead' (ibid., p. 23). Stockholm focused most of its energy on industrial country concerns (Holdgate 1999).

The most conspicuous result of the Stockholm Conference was the creation of the United Nations Environment Programme (UNEP). This was established by resolution of the General Assembly of the United Nations in December 1972 to act as a governing council for environmental programmes; a secretariat would focus environmental action within the whole UN system and an environment fund would finance environment programmes. UNEP was located in Nairobi in Kenya, the first UN body outside the developed world, a symbolic (and politically astute) decision. The conference secretariat had proposed a wholly new intergovernmental body within the UN, but the suggestion of a new full UN agency to deal with environmental problems was strongly opposed by the existing UN agencies (Ferau 1985). The outgoing UN secretary-general in 1971 favoured a 'switchboard' linking separate sectoral organisation (McCormick 1992, p. 93), and this is essentially what was created. UNEP therefore is not a UN agency like UNESCO or the FAO, and these remain responsible for the environmental aspects of their own activities. UNEP seeks to act as a catalyst and think-tank, the 'conscience of the UN system' (Clarke and Timberlake 1982, p. 49).

Sadly, UNEP's small size, poverty, relative weakness within the UN system and peripheral position in Nairobi have limited its effectiveness. It is officially a unit of the UN Secretariat and gets administrative funds from that source, but money for projects comes from the Environment Fund. Contributions to that are voluntary, and have fallen short of needed targets. UNEP's influence on the UN agencies has been relatively small, and they have gone about their business much as before. As Myers and Myers commented, 'we wanted an Environmental Programme of the United Nations. Instead we got an United Nations Environment Programme' (Myers and Myers 1982, p. 201).

UNEP was saddled with an impossibly broad remit and a vague list of priorities from the Stockholm Conference. It went on to develop several notable activities such as the Global Environment Monitoring System (GEMS), begun in 1975 (Gwynne 1982), and the Regional Seas Programme. This was launched in 1974, and had roots directly in the Stockholm Declaration on marine pollution. UNEP acts as a coordinator of intergovernmental action based on an agreed Action Plan, an approach tackling a global problem through regional action. By 1982 the programme covered 10 regions and 120 coastal states (Bliss-Guest and Keckes 1982). UNEP also organised the UN Conference on Desertification in Nairobi in 1977 (United Nations 1977). This was brought about by a resolution of the UN General Assembly in 1974 which had discussed the Sahelian drought of 1972–4. The UN General Assembly endorsed the conference's Plan of Action to Combat Desertification (PACD), and although other UN bodies had significant experience in the field (such as UNESCO in arid zone research), coordination of the PACD was entrusted to UNEP (Karrar 1984). In the event, international, national and local 'anti-desertification' action fell far short of expectations, and the movement spearheaded by UNEP achieved very little. Indeed, by the 1980s and 1990s the whole subject had become beset by controversy (not lessened by the Convention on Desertification that eventually followed the Rio Conference two decades later; Swift 1996). Desertification is discussed in detail in Chapter 7.

Following Stockholm, international debate about sustainable development began to be more extensively influenced by concerns about the need (and potential) for development in the Third World. An international meeting at Geneva in April 1974 flew in the face of the environmentalist rhetoric of the time by eschewing the notion of limits to growth, and emphasising that the key problem was the distribution of natural resources and the benefits that flowed from them, not their global scarcity (McCormick 1992). The advent of OPEC on the international scene had demonstrated the political power of resource-rich states, and in May 1974 the UN General Assembly adopted a declaration calling for a New International Economic Order. This was to: correct inequalities and redress existing injustices, making it possible to eliminate the widening gap between the developed and developing countries and ensure steadily accelerating economic development' (Lummis 1992, p. 44).

The priority of economic growth, which had been such a concern for Third World countries at Stockholm, was now receiving much stronger emphasis. This trend was furthered at a meeting of experts held at Cocoyoc in Mexico in October 1974, which looked at environmental problems from the perspective of the Third World, and particularly the Third World poor. This was attended by Maurice Strong, secretary-general of the Stockholm Conference, and chaired by Barbara Ward, and the resulting 'Cocoyoc Declaration' pointed to the problem of the maldistribution of resources and to the 'inner limits' of human rights as well as the 'outer limits' of global resource depletion. It stressed the priority of basic human needs, and called for a redefinition of development goals and global lifestyles. It called for global resource management, international regimes for the management of common resources, and development policies aimed at the poor (McCormick 1992).

These ideas drew extensively from the debates in development which had emerged through the 'First Development Decade' of the 1960s, and were to emerge again in the global interdependence arguments of the Brandt Report (Brandt 1980). They represented the productive fusion of those debates with those of Western environmentalism. By the time the *World Conservation Strategy* emerged six years later, the two had, superficially at least, merged.

The *World Conservation Strategy*

In the 1960s, thinking within IUCN began to embrace greater concern for economic development (McCormick 1986). The idea of a strategic approach to conservation was considered at the IUCN General Assembly New Delhi in 1969, and conservation and development was the theme of the 1972 General Assembly at Banff, Canada. Work on the strategy for nature conservation began in 1975, when IUCN joined UNEP, UNESCO and the FAO to form the 'Ecosystem Conservation Group', and gradually the notion of a 'world conservation strategy' took shape (Boardman 1981, McCormick 1986, 1992, Holdgate 1999).

In 1977 UNEP commissioned IUCN to draft a document to provide 'a global perspective on the myriad conservation problems that beset the world and a means of identifying the most effective solutions to the priority problems' (Munro 1978). Preliminary drafts were discussed at the IUCN General Assembly in Ashkhabad, in the USSR, in 1978. At that stage the strategy was a textbook of wildlife conservation, about the conservation of species and special areas rather than the integration of conservation and of development (Munro 1978). The focus was subsequently changed substantially to include questions of population, resources and development (Boardman 1981).

The evolution of the *World Conservation Strategy*, and its attempt to imbue development with environmental ideas and principles, was influenced by (and itself in turn contributed to) ideas of 'ecodevelopment' (Sachs 1979, 1980). This term was coined by Maurice Strong, secretary-general of the Stockholm Conference (Boardman 1981). It was subsequently developed and promoted by UNEP (UNEP 1978), and widely discussed internationally, for example at meetings in Belo Horizonte in Brazil in 1978 (the International Workshop in Ecodevelopment and Appropriate Technology), in Berlin in 1979 (the Conference on Ecofarming and Ecodevelopment; Glaeser and Vyasulu 1984), and in

Ottawa in 1986 (the IUCN Conference on Conservation and Development; Svedin 1987). It was also, for example, adopted by the South Pacific Commission and the South Pacific Bureau for Economic Cooperation in a Comprehensive Environmental Management Programme (Dasmann 1980).

Behind the notion of ecodevelopment lies the awareness of the intrinsic complexity and dynamic properties of ecosystems and the ways they respond to human intervention, and the need to ensure the 'environmental soundness' of development projects (*Ambio* 1979, p. 115). The challenge was that of 'improving the economic wellbeing of people without impairment of the ecological systems on which they must depend for the foreseeable future' (Dasmann 1980, p. 1331). These ideas about ecodevelopment are fundamental to the *World Conservation Strategy* (WCS).

The *World Conservation Strategy* was eventually published in 1980 in the name of IUCN, UNEP and the World Wildlife Fund, which had together formed 'the Ecosystem Conservation Group' (IUCN 1980). It had already been presented to the FAO and UNESCO, and publication had been delayed to include their amendments (McCormick 1986).

The WCS was the culmination of more than two decades of thinking by conservationists, particularly those in IUCN, about ways to further nature conservation on a global scale. It was a conservation document that addressed the issues and problems raised by economic development, rather than a document about development and environment as such. The WCS broke new ground in that, in the words of the chairman of the WWF, Sir Peter Scott, it suggested for the first time that development should be seen as 'a major means of achieving conservation, rather than an obstruction to it' (Allen 1980, p. 7). It was also intended to be an outgoing, even evangelistic document, seeking to show the relevance of conservation to the development objectives of others, 'in governments, industry and commerce, organised labour and the professions' (Allen 1980). The WCS was aimed at government policy-makers, conservationists and development practitioners, and seen not only as a conceptual but also as a practical document, offering both 'an intellectual framework and practical guidance' (IUCN 1980, p. i). It was intended 'to stimulate a more focused approach to the management of living resources and to provide policy guidance on how this can be carried out' (ibid., p. vi).

The WCS consisted of twenty sections in three groups, following an introduction to define key terms. The first group (Sections 2–7) described objectives for conservation, their relevance for human survival, and priority requirements for achieving them. The second group (Sections 8–14) set out a strategy for action at national and sub-national levels, and identified obstacles and possible ways to deal with them. The third group (Sections 15–20) outlined the international action required to stimulate and support action at smaller scales.

The WCS identified three objectives for conservation (Table 3.1). First, it called for the maintenance of 'essential ecological processes and life-support systems' (Section 2). These processes were essential for food production, health, and other aspects of human survival and sustainable development' (para. 2.1).

Table 3.1 Objectives of the *World Conservation Strategy*

(a) To maintain essential ecological processes and life-support systems (such as soil regeneration and protection, the recycling of nutrients and the cleansing of waters) on which human survival and development depend;
(b) to preserve genetic diversity (the range of genetic material found in the world's organisms), on which depend the functioning of many of the above processes and life-support systems, the breeding programmes necessary for the protection and improvement of cultivated plants, domestic animals and micro-organisms, as well as much scientific and medical advance, technical innovation, and the security of the many industries that use living resources;
(c) to ensure the sustainable utilisation of species and ecosystems (notably fish and other wildlife, forests and grazing lands), which support millions of rural communities as well as major industries.

Source: The *World Conservation Strategy* (IUCN 1980)

Their maintenance demanded maintenance of the ecosystems that govern, support or moderate them, including agricultural land and soil, forests, and coastal and freshwater ecosystems. Threats to these systems included soil erosion, pesticide resistance in insect pests, deforestation and associated sedimentation, and aquatic and littoral pollution.

The second objective in the WCS was the preservation of genetic diversity, the genetic material both in different varieties of locally adapted crop plants or livestock and in wild species (Section 3). This genetic diversity was both an 'insurance' (for example against crop diseases), and an investment for the future (for example for crop breeding or pharmaceuticals) (para. 3.2). The WCS's third objective was to ensure 'the sustainable utilization of species and ecosystems', particularly fisheries, wild species which are cropped, forests and timber resources, and grazing land (Section 4).

These three main objectives of the WCS were then broken down into a list of priority requirements (Sections 5–7). These were drawn up on the basis of criteria of significance (how important is it?), urgency (how fast is it getting worse?) and irreversibility (Section 5). They are listed in Table 3.2. They ranged from the sublime intention to 'prevent the extinction of species' (Table 3.2, B1) to the detailed requirements of site protection for species conservation (Table 3.2, B3).

The requirements for the conservation of ecological processes basically involved the rational planning and allocation of land use, so that crops could be given priority on the best land (but not on marginal land), and areas such as watersheds and littoral zones were set aside and used appropriately. The conservation of genetic diversity demanded site-based protection of ecosystems (essentially the familiar nature conservation strategies for the protection of the habitats of rare and unique species and typical ecosystems in protected areas) and the timely creation of banks of genetic material.

The third objective, the sustainable utilisation of resources, had no fewer than twelve priority tasks demanding linked research and action to determine

Table 3.2 Priority requirements of the *World Conservation Strategy*

A. Priority requirements: ecological processes and life support systems (Section 5)
1 Reserve good cropland for crops.
2 Manage cropland to high, ecologically sound standards.
3 Ensure that the principal management goal for watershed forests and pastures is protection of the watershed.
4 Ensure that the principal management goal for coastal wetlands is the maintenance of the processes on which the fisheries depend.
5 Control the discharge of pollutants.

B. Priority requirements: genetic diversity (Section 6)
1 Prevent the extinction of species.
2 Preserve as many kinds as possible of crop plants, forage plants, timber trees, livestock, animals for aquaculture, microbes and other domestic organisms and their wild relatives.
3 Ensure on-site preservation programmes protect:
   • the wild relatives of economically valuable and other useful plants and animals and their habitats;
   • the habitats of threatened and unique species;
   • unique ecosystems;
   • representative samples of ecosystem types.
4 Determine the size, distribution and management of protected areas on the basis of the needs of the ecosystems and the plant and animal communities they are intended to protect.
5 Coordinate national and international protected area programmes.

C. Priority requirements: sustainable utilisation (Section 7)
1 Determine the productive capacities of exploited species and ecosystems and ensure that utilisation does not exceed those capacities.
2 Adopt conservation management objectives for the utilisation of species and ecosystems.
3 Ensure that access to a resource does not exceed the resource's capacity to sustain exploitation.
4 Reduce excessive yields to sustainable levels.
5 Reduce incidental take as much as possible.
6 Equip subsistence communities to utilise resources sustainably.
7 Maintain the habitats of resource species.
8 Regulate international trade in wild animals and plants.
9 Allocate timber concessions with care and manage them to high standards.
10 Limit firewood consumption to sustainable levels.
11 Regulate the stocking of grazing lands to maintain the long-term productivity of plants and animals.
12 Utilise indigenous wild herbivores, alone or with livestock, where domestic stock alone would degrade the environment.

Source: IUCN (1980)

and achieve resource utilisation at sustainable levels. These included some which even at the time of publication were well established, for example the regulation of international trade in wildlife products, and others such as the ingenuous suggestion to 'limit firewood consumption to sustainable levels' (Table 3.2,

C10), which must surely defy policy implementation in the real world. The implications of some of this thinking are discussed below.

The WCS then discussed the priorities for national action (Sections 8–14). These proposed the preparation of separate national strategies by governments or NGOs. These strategies should review development objectives in the light of the conservation objectives and establish priority requirements, identify obstacles and propose cost-effective ways of overcoming them, determining priority ecosystems and species for conservation and establishing a practical plan of action. The 'strategic principles' offered sought to integrate conservation and development by doing away with narrow sectoral approaches; to manage ecosystems so as to retain future options on use (this reflecting the poor state of knowledge about tropical ecosystems in particular); to mix cure and prevention; and to tackle causes as well as symptoms.

Two key problems were highlighted: first, the relative weakness of conservation institutions in the context of national policy-making, combined with the sectoral nature of such planning (Section 9), and second, the fact that environmental planning rarely allocates land uses rationally (Section 10). It was proposed that these should be tackled by 'anticipatory and cross-sectoral' environmental policies, more and better evaluation of the capacity of ecosystems to meet human demands, improved prediction of the environmental effects of development, and better procedures for matching capacities and uses of land and water resources. For development policy to be 'ecologically as well as economically and socially sound', conservation objectives needed to be considered at the beginning of the planning process, not factored in at the end once impacts had been caused (Figure 3.1). Other problems identified were the inadequacy of legislation and weak and overlapping natural resource management agencies (Section 11), lack of ecological data, shortcomings in training and education of conservation personnel and in research (Section 12), and lack of support for conservation policies (Section 13).

Thus far, this is a fairly familiar analysis of the inadequate scope of ecological and environmental planning. However, the WCS also highlighted the way in which the environment (and its conservation) was not seen to be relevant to development priorities: there was a 'lack of awareness of the benefits of conservation and of its relevance to everyday concerns' (para. 13.2). To remedy this, planning must not only be better technically, but also involve more public participation and community involvement, and public education. Finally, the WCS also highlighted the lack of conservation-based rural development, 'rural development that combines short term measures to ensure human survival with long term measures to safeguard the resource base and improve the quality of life' (para. 14.5).

In conclusion, the WCS turned to international actions needed to promote conservation (Sections 15–20), recognising that many living resources lie partly or wholly outside national boundaries. These sections covered the obvious areas of international law and conventions, the responsibilities of bilateral and multilateral aid donors (Section 15), the cooperative management of the global

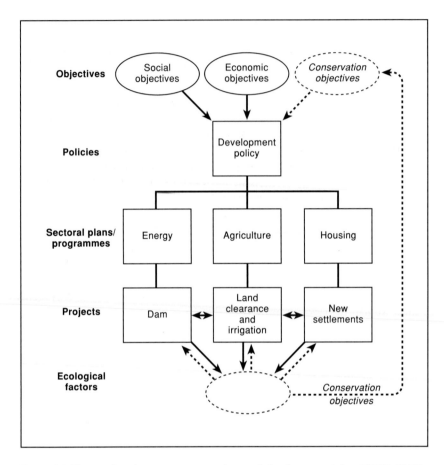

Figure 3.1 The need to integrate conservation and development (after IUCN 1980).

commons of the open ocean, the atmosphere and the Antarctic (Section 18), and international river basins and seas (Section 19). More surprising, perhaps, is a rather unfocused call for 'international action' to conserve tropical forests and to support the Plan of Action on Desertification (Section 16), and a call for a 'global programme for the protection of genetic resource areas' (Section 17). This programme would involve the site-based protection of concentrations of economic or useful varieties (for example wild relatives of cultivars), concentrations of threatened species, ecosystems of 'exceptional diversity', and ecosystems that are poorly represented in existing protected areas. This would require financing internationally, because 'many countries particularly rich in genetic resources are developing ones that can ill-afford to bear alone the burden of their on-site protection' (para. 17.11), or possibly through commercial participation or sponsorship.

Sustainable development in the *World Conservation Strategy*

What impact did the WCS have? The promotion of sustainable development formed one of IUCN's seven Programme Areas for the period 1985–7. IUCN established a 'Conservation for Development Centre' in 1979, marking a move into field programmes with partner organisations (Holdgate 1999). The CDC undertook projects to advise donor agencies how to build conservation into their work, projects to help Southern countries achieve conservation, and projects to promote international conventions in Southern countries. The second category dominated, with a series of projects to support the creation of national conservation strategies. The original plan to revise the WCS every three years gave way to progressive adaptation as national conservation strategies were produced under IUCN guidance (IUCN 1984a, McCormick 1986). A number of countries in both the First and the Third Worlds duly produced national strategies (IUCN 1984b, Nelson 1987, Bass 1988, Holdgate 1999).

At this nominal level, the WCS might therefore be judged a success. Caldwell describes it as 'the nearest approach yet to a comprehensive action-oriented programme for political change' (1984, p. 306). It was successful too in aiding the proliferation of the phrase 'sustainable development' in the media and as part of development terminology. On the other hand, Michael Redclift argues that 'despite its diagnostic value the World Conservation Strategy does not even begin to examine the social and political changes that would be necessary to meet conservation goals' (1984, p. 50). In several ways the WCS is the child of 1970s environmentalism.

First, the *World Conservation Strategy* is clearly neo-Malthusian in its approach. Thus it is argued that every country should have a 'conscious population policy' to achieve 'a balance between numbers and environment' (IUCN 1980, para. 20.2). New approaches to resource management are needed because of the impact of population growth and rising demand, and it is 'the escalating needs of soaring numbers' that have led to 'short-sighted' approaches to natural resources' (ibid., p. i). The WCS graphically illustrated the challenge of growth in population and consumption as shown in Figure 3.2, predicting relative rates of degradation of arable land (the stalk of wheat), reduction in unlogged productive tropical forest (the shrinking tree) and the global population (the human giant).

This neo-Malthusianism is evident in the essentially determinist vision of the WCS. It offers a 'conservation or disaster' scenario not far removed from the classic polemics of the 'ecodoomsters' (Hill 1986). It identifies ecological and environmental limits for human action, and applies ideas drawn from wildlife management (particularly the notion of 'carrying capacity') directly to people without discussion of the political, social, cultural or economic dimensions of resource use. Moreover, it does so on a global scale, seeing people in biological terms: 'Human beings, in their quest for economic development and enjoyment of the riches of nature, must come to terms with the reality of

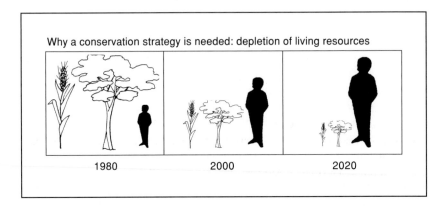

Figure 3.2 Depletion of living resources, as portrayed by the *World Conservation Strategy* (after IUCN 1980). Relative rates of degradation of arable land are shown by the stalk of wheat symbol, reduction in unlogged productive tropical forest is shown by the tree symbol, and the global population growth is shown by the human figure.

resource limitation and the carrying capacities of ecosystems' (IUCN 1980, p. i). The WCS argues not only that ecology should determine human action in the way development is attempted, but also that it sets limits on the scope of human action.

The second way in which the WCS reflects the thinking of 1970s environmentalism is its globalist focus, particularly its attempt to define a global agenda for environmental action. It claims that the 'global interrelatedness of actions, with its corollary of global responsibility', is one of two features that 'characterise our time' (IUCN 1980, p. i). Global responsibility demands global strategy, both for development and for conservation of nature and natural resources, and this the WCS set out to provide. Both IUCN and UNEP were global agencies with high profiles and small budgets. The WCS was of considerable importance in establishing a global remit for national action which centred on their coordinating role.

A third way in which the WCS reflected 1970s environmentalism was in its ethics. It welded together scientific utilitarianism and romantic 'holist' or 'vitalist' thinking into a form of 'bioethics' (O'Riordan 1981, Pepper 1984, Worster 1985). Wild species were to be conserved for two reasons: first, because they were useful to human society and economy, and second, because it was morally right to conserve them. The utilitarian argument for conservation, which adhered to the WCS's conservative principle of keeping options open, was simple: 'we cannot predict what species may become useful to us' (IUCN 1980, para. 3.2). This principle underlay the relatively numerous and well-thought-out proposals concerning the sustainable utilisation of living resources (fisheries management, game farming, regulation of wildlife trade; see Table 3.2). These contrast markedly with the more open-ended and speculative proposals elsewhere in the document. The WCS also picks up the point that not all utilitarian values are monetary, pointing out the symbolic, ritual and cultural importance of wildlife (para. 4.11).

However, the WCS was happy to have the best of both worlds, and not to put all its eggs in the basket of utilitarianism. It did not shrink from stating moral arguments for conservation which make assumptions about the rights of other organisms. Thus there is 'an ethical imperative', rather tritely expressed in the belief that 'we have not inherited the earth from our parents, we have borrowed it from our children' (IUCN 1980, para. 1.5). Because of the power of humans to transform the biosphere (and influence the process of evolution), conservation is a matter of moral principle: 'we are morally obliged – to our descendants and to other creatures – to act prudently' (para. 3.3).

This attempt to argue for conservation along two parallel tracks is by no means accidental. It reflects, of course, old divisions between technological and ecological environmentalism (Pepper 1984), or between technocentrism and ecocentrism (O'Riordan 1981), which create such schizophrenic confusion within the conservation movement and for individual conservationists. In practical terms this dualism is extremely useful (Norton 1991). On the one hand, the utilitarian argument allows conservation to be packaged to be attractive to the anthropocentric materialism which underlies thinking about development. The WCS makes the self-evident case that ecosystems and species must be used sustainably to ensure their continued availability 'almost indefinitely' (IUCN 1980, para. 4.1). On the other hand, moral arguments can be employed where they are more effective, for example among conservationists in industrialised countries. Here economies have substantially been freed from primary dependence on renewable resources, a freedom that developing countries would like to emulate. The rational argument for sustainable utilisation of ecosystems seems less critical to development strategies, but the moral argument can be rapidly marshalled to fill the gap. Although greater economic diversity and flexibility might reduce the need to utilise certain resources sustainably, the WCS argues that there is less excuse not to do so (IUCN 1980, para. 4.1). The WCS's double-barrelled justification of sustainable development is both versatile and robust.

The WCS can thus be seen as conservationist environmentalism refocused for a new decade, and attempting to engage with issues of development. Its attempts to demonstrate the indivisibility of conservation and development lie in the way they are defined. The key concept here is, of course, sustainability. Indeed, the whole message of the WCS turns on the question of what 'sustainability' and 'sustainable development' mean. The opaque nature and flexibility of these terms have been discussed in Chapter 1. Within the WCS, development and conservation are defined in such a way that their compatibility becomes inevitable. Development is presented as 'the modification of the biosphere and the application of human, financial, and living and non-living resources to satisfy human needs and improve the quality of human life' (IUCN 1980, para. 1.4). Meanwhile, conservation is 'the management of human use of the biosphere so that it may yield the greatest sustainable benefit to present generations while maintaining its potential to meet the needs and aspirations of future generations'.

If it is taken for granted that development ought to be 'sustainable' (meaning that it must be capable of being extended indefinitely for the benefit of future generations), conservation and development are of course 'mutually dependent' and not incompatible (IUCN 1980, para. 1.10). Conservation is essential to every sector (health, energy, industry), for conservation is 'that aspect of management which ensures that utilisation is sustainable' (para. 1.6). Whereas in the past, development practitioners might have seen conservation as irrelevant to the task of development, or even in opposition to it, this was because they failed to understand about 'real' conservation. Properly understood (the WCS argued), 'real' conservation would have helped avoid ecological damage and the failure of development, a failure which demonstrates that much development is not real development at all.

The WCS accepted that environmental modification was a natural and necessary part of development, but argued that not all such modification would achieve the social and economic objectives of development. In suggesting that development planning should not only be socially and economically sound but also fit conservation objectives (Figure 3.1), the WCS established the basic triptych of mainstream sustainable development thinking on the 1990s, of economic, social and environmental sustainability (see Chapter 5). It argued that conflict between conservation and development could be avoided if conservation and development are integrated at every stage of planning (see Figure 3.1). Meanwhile, conservation needed to address the causes and not just the symptoms of environmental change, and avoid coming across as 'anti-development (hence anti-people)' (IUCN 1980, para. 8.6).

This is the core of the *World Conservation Strategy*. It certainly represents a significant repackaging of conservation. It is less clear if it really represents anything more than that, the new principle and purpose that is claimed. The WCS is Janus-headed, addressing the very different worlds of conservation and development at the same time. Within conservation it tries to draw to make a case for conservation in a new way, without losing established priorities. By stressing the possibility that sustainable levels of utilisation of ecosystems can coexist with the preservation of nature, it demonstrates that the ideas presented are familiar and established and not radically new. To environmentalists, the WCS seems to make a logical and effective inroad into the hitherto unfamiliar and inexplicably destructive world of economic development while reaffirming the moral basis for conservation, and while finding ways to justify the protection of land in reserves and parks.

However, at the same time, the WCS actually has to make that advance into the development field effective, and this demands that its arguments are plausible within development. This it tries to do by emphasising that species and ecosystems are resources for human subsistence and development, defending even the establishment of parks and reserves on the grounds of indirect ecological benefits (such as runoff from forested watersheds) or direct economic returns from Third World 'ecotourism' (Young 1986). On this second front the WCS is less successful.

The presentation of the WCS is well rounded and plausible. However, it fails to recognise the essentially political nature of the development process. This matters on two levels. First, conservation – like science – is seen to be above ideology. There is no understanding of the way in which nature and culture interact, such that views of nature are created by society. There is no apparent awareness of arguments about the social production of nature (Smith 1984). Second, the WCS suggests that conservation can in some way bypass structures and inequalities in society. It seems to assume that 'people' can exist in some kind of vacuum, outside the influence of inequality, class or the structures of power. The goal of the WCS is stated as 'the integration of conservation and development to ensure that modifications to the planet do indeed secure the survival and wellbeing of all people' (IUCN 1980, para. 1.12). This is pious, liberal and benign, but it is disastrously naïve.

The WCS avoids the explicit use of the language and ideas of economics and politics. Such ideas are fundamental to the development process, in terms of social and economic structure, the inevitable political implications of economic change, and the constraints of capital, technical knowledge and manpower. This lack of explicit exploration of such ideas in the WCS represents a significant failure of its attempt to place essentially environmentalist ideas in a development matrix. Like so much of the environmentalism of the 1970s, the WCS is therefore blind to political economy. As a result, what it has to say about development is not particularly convincing. The reasons for this, and some of the critiques and alternatives, are examined in more detail later in this book.

The Brundtland Report

The *World Conservation Strategy* was primarily theoretical rather than applied in what it has to say about development. It barely began to address the larger issues of national economic management (questions for example of the relative weight given to different sectors, or the pros and cons of economic globalisation), let alone the burning questions of international political economy. It said nothing beyond bland generalities about the gulf in wealth between North and South, or the dependence of the one upon the other, and, as we shall see later, nothing at all about the various radical theories on the global economy. The global scope of the WCS embraced neither the real world and the practical politics of international development, nor the theoretical ideas being discussed in development studies. Robert Prescott-Allen, who wrote the *World Conservation Strategy*, commented of IUCN in 1998 that 'the problem was that it wanted to sell conservation to the development constituency, but it didn't understand what the development constituency was like. The conservationists didn't see that development was the driving force in human affairs' (quoted in Holdgate 1999, p. 123).

However, as the 1980s wore on, this was to change. Sustainable development was brought into the established political arena of international development through the establishment in December 1983 of the World Commission on Environment and Development at the call of the UN General Assembly. Its report, *Our Common Future* (Brundtland 1987), was presented to the UN General Assembly in 1987.

Our Common Future claimed a very specific heritage. The chair wrote, 'After Brandt's *Programme for Survival* and *Common Crisis*, and after Palme's *Common Security*, would come *Common Future*' (Brundtland 1987, p. x). The environment and poverty were now being treated as a global threat, as war had been. Like its predecessors, it had as its target the promotion of multilateralism and the interdependence of nations: 'the challenge of finding sustainable development paths ought to provide the impetus – indeed the imperative – for a renewed search for multilateral solutions and a restructured international economic system of cooperation' (Brundtland 1987, p. x). The Brundtland Report reflected what Chatterjee and Finger call 'same boat ideology' (1994, p. 80), proposing that global crisis could be staved off by dialogue between enlightened individuals, global environmental awareness and planetary stewardship. In places, Brundtland's vision verged on the sugary, as in the aim of its sustainable development strategy: 'to promote harmony among human beings and between humanity and nature' (Brundtland 1987, p. 65).

The Brundtland Commission was important for several reasons. First, it obviously and rather self-consciously attempted to recapture the 'spirit of Stockholm 1972' which was so celebrated by environmentalists in the early 1970s and whose demise as recession bit was so lamented. The *World Conservation Strategy* tried to do exactly the same thing, of course, but Brundtland achieved this resuscitation far more expertly and effectively, largely because of its origins in the UN General Assembly and not out in the wildwoods of UNEP and IUCN.

Second, *Our Common Future* placed elements of the sustainable development debate within the economic and political context of international development. Its starting point was deliberately broad, and a move to limit its concern simply to the 'environment' was firmly resisted:

> This would have been a grave mistake. The environment does not exist as a sphere separate from human actions, ambitions, and needs, and attempts to defend it in isolation from human concerns have given the very word 'environment' a connotation of naiveté in some political circles.
>
> (Brundtland 1987, p. xi)

Third, Brundtland placed environmental issues firmly on the formal international political agenda. Arguably, of course, it simply reflected the de facto situation created by its predecessors, but nonetheless it achieved something that Stockholm, UNCOD and the WCS failed to do, and got the UN General Assembly to discuss environment and development as one single problem.

The Brundtland Report therefore started from the premise that development and environmental issues could not be separated, recognising that it was futile to try to tackle environmental problems without considering broader issues of 'the factors underlying world poverty and international inequality' (Brundtland 1987, p. 3). This view was at the same time less abstract and less simplistic than that of the WCS. *Our Common Future*'s argument was not presented in the rather general terms of linkages between sustainable use of ecosystems and human wealth and welfare, nor was the spectre of 'ecodisaster' prominent. Instead, the essentially reciprocal links between development and environment were drawn more explicitly. *Our Common Future* recognised that development could 'erode the environmental resources on which they must be based', and hence that environmental degradation could undermine economic development. Furthermore, the links, again reciprocal, between poverty and environment were also recognised, poverty being seen 'as a major cause and effect of global environmental problems' (ibid.).

Our Common Future's familiar definition of sustainable development (see Chapter 1) is based on two concepts. The first is the concept of basic needs and the corollary of the primacy of development action for the poor. The second involves the idea of environmental limits. These limits are not, however, those set by the environment itself, but those set by technology and social organisation. This involves a subtle but extremely important transformation of the ecologically based concept of sustainable development, by leading beyond concepts of physical sustainability to the socio-economic context of development. Physical sustainability could be pursued in 'a rigid social and political setting', but it cannot be secured without policies that actively consider issues such as access to resources and the distribution of costs and benefits. In other words, the sustainable development of *Our Common Future* was defined by the achievement of certain social and economic objectives, and not by some notional measurement of the 'health' of the environment. Whereas the WCS started from the premise of the need to conserve ecosystems and sought to demonstrate why this made good economic sense (and – although the point was underplayed – could promote equity), *Our Common Future* starts with people and goes on to discuss what kind of environmental policies are required to achieve certain socio-economic goals. The answers (perhaps unsurprisingly) are remarkably similar. The premises, however, do differ, and of the two, *Our Common Future* was by far the more effective document in its ability to address and engage government policy-makers.

The elements of the sustainable development ideas in *Our Common Future* are listed in Table 3.3. They represent an interesting blend of environmental and developmental concerns. Among the former is the need to achieve a sustainable level of population. This is a softened version of the neo-Malthusian message which recognises the greater demands on resources made by a First World child, but it still argues that population should 'be stabilised at a level consistent with the productive capacity of the ecosystem' (Brundtland 1987, p. 56). Other rather familiar concerns are conserving (and enhancing) the resource base and

Table 3.3 Critical objectives for environment and development policies that follow from the concept of sustainable development

---

1 Reviving growth.
2 Changing the quality of growth.
3 Meeting essential needs for jobs, food, energy, water and sanitation.
4 Ensuring a sustainable level of population.
5 Conserving and enhancing the resource base.
6 Reorientating technology and managing risk.
7 Merging environment and economics in decision-making.

Source: Brundtland (1987, p. 49)

reorienting technology, particularly with regard to risk. Prominent among the latter is the fundamental concern with meeting basic needs. Prominent too is the need to 'merge' environment and economics in decision-making. Most prominent of all, however, is the focus on growth: economic growth is seen as the only way to tackle poverty, and hence to achieve environment–development objectives. It must, however, be a new form of growth: sustainable, environmentally aware, egalitarian, integrating economic and social development (Brundtland 1987). Above all, the Brundtland Report's vision of sustainable development was predicated on the need to maintain and revitalise the world economy. This means 'more rapid economic growth in both industrial and developing countries, freer market access for the products of developing countries, lower interest rates, greater technology transfer, and significantly larger capital flows, both concessional and commercial' (ibid., p. 89).

However, this prescription was based on an economic and not an environmentalist vision. It used some of the language of 1970s environmentalism, but it did not question growth or technology, and it avoided arguments about eco-disaster. In the classification of Cotgrove (1982) it was firmly cornucopian rather than catastrophist. *Our Common Future* argued that it is poverty that puts pressure on the environment in the Third World, and it is economic growth that will remove that pressure. Furthermore, that growth cannot be conceived of in a geopolitical vacuum; it is only the ending of dependence that will enable these countries to 'outpace' their environmental problems.

The Brundtland Report is therefore built on the need to promote economic growth. But what of the pressures of that growth itself? What about demands for energy and raw materials, or pollution? *Our Common Future* hopes to have its cake and eat it: 'The Commission's overall assessment is that the international economy must speed up world growth while respecting environmental constraints' (Brundtland 1987, p. 89). However, it does not say how this balancing trick is to be achieved.

What does this form of sustainable development require? Quite simply a restructuring of national politics, economics, bureaucracy, social systems, systems of production and technologies, and a new system of international trade and finance (Table 3.4). This is no small agenda. Sustainable development must be global in scope and internationalist in formulation. This demands first that

Table 3.4 Requirements of a strategy for sustainable development

---

1　A political system that secures effective citizen participation in decision-making.
2　An economic system that is able to generate surpluses and technical knowledge on a self-reliant and self-sustained basis.
3　A social system that provides for solutions for the tensions arising from disharmonious development.
4　A production system that respects the obligation to preserve the ecological basis for development.
5　A technological system that can search continuously for new solutions.
6　An international system that fosters sustainable patterns of trade and finance.
7　An administrative system that is flexible and has the capacity for self-correction.

Source: Brundtland (1987, p. 65)

'the sustainability of ecosystems on which the global economy depends must be guaranteed', and second, equitable exchange between nations. It is this latter requirement that lifts the Brundtland Report out of the mould of previous ecodevelopment writing. Environmentalists might have seen it as a logical extension of the *World Conservation Strategy*, and welcomed its apparently new and authoritative tone and the credibility with which it handled development terminology. In fact it was only rather secondarily the fruit of environmentalism. It is better understood as an extension of the thinking of the Brandt Reports *North–South* and *Common Crisis* (Brandt 1980, 1983). Indeed, while the WCS represented the attempt by conservationists to capture the rhetoric of development to repackage old ideas, *Our Common Future* is the result of the reverse process. The existence of a global environmental crisis is cited as evidence for the need for a multilateralist solution:

> At first sight, the introduction of an environmental dimension further complicates the search for cooperation and dialogue. But it also injects an additional element of mutual self-interest, since a failure to address the interaction between resource depletion and rising poverty will accelerate global deterioration.
>
> (Brundtland 1987, p. 90)

In *Our Common Future* the establishment of the sustainable use of resources was shown to demand more than simply dealing with the micro-scale questions of industrial methodology or the issues of refining or reforming project planning procedures. It needed to embrace international trade and international capital flows. Patterns of international trade (for example in hardwood timber products) that impoverish Third World countries also promote unsustainable resource use. Lack of external capital (as aid and rescheduling of debts) limits the improvement of living standards in Third World countries: 'Without reasonable flows, the prospect for any improvements in living standards is bleak. As a result the poor will be forced to over-use the environment to ensure their own survival' (Brundtland 1987, p. 68).

Debt, poverty and population growth restrict the capacity of developing countries to adopt environmentally sound policies. The proposed solutions involve, first, increased capital flow to the Third World, but redirected to promote 'sustainable' projects; second, new deals on commodity trade (particularly attention to hidden pollution costs of industrial processes in developing countries); and third, ending protectionism and reforming transnational investment to ensure 'responsibility' (for example through technology transfer and environmentally sound technologies; Brundtland 1987). This analysis of the potential for productive reform of world trade and finance stems from a very particular vision of the working of the world economy and the pattern of economic and political forces, based on ideas of multilateralism and notions of global cooperation and dialogue.

The Brandt Reports set out much of the ground on which Brundtland built, both in terms of general principles and in the approach to (if not the priority given to) environmental problems. *North–South* discussed measures 'which together would offer new horizons for international relations, the world economy, and for developing countries' (Brandt 1980, p. 64). Those 'new horizons' include the environment, both globally ('the biosphere is our common heritage and must be preserved by cooperation'; ibid., p. 73) and in the countries of the South themselves. The package argued for by the Brandt Commission involved growth (in both North and South), 'massive transfers' of capital, expansion of world trade, the end of protectionism, an orderly monetary system, and a move towards international equality and peace. Like Brundtland, Brandt argued that only the abolition of poverty would bring an end to population growth, and that this was a global problem and not one confined in its impacts to the poor or the Third World alone. Poverty required multilateral action, not just because of the obvious moral imperative to eradicate it, but because of mutual self-interest. Other imperatives 'rooted in the hard self-interest of all countries and people' reinforce the claim of human solidarity (ibid., p. 77).

The Brandt Reports were themselves part of a longer evolution of thinking about economic interdependence (cf. Brookfield 1975). This essentially began with the Bretton Woods system of international financial management (the World Bank family, initially the International Monetary Fund and International Bank for Reconstruction and Development) in 1944. This was based on an essentially Keynesian vision, to create a stable, growing and interdependent world economy, 'an environment for liberal trade and to promote economic cooperation' (World Bank 1981). In the 1950s and 1960s the world economy, and international trade, indeed grew. However, fundamental problems were emerging by the 1960s, and in 1971 fixed exchange rates were abandoned, leading to volatility in currency markets, and eventually the destabilisation of oil prices and the oil crisis of 1973 (Strange 1986).

The oil crisis coincided with the sudden flowering of concern about 'limits to growth'. This too stressed global interdependence, but with an altogether different message about the desirability and possibility of continued growth. The problem of the attitude of such 'global' environmentalism to aid and economic growth in the Third World at the time of the Stockholm Conference in 1972 has been discussed on pp. 55. The reconciliation of these difficulties created formal statements of sustainable development. It is not surprising that those ideas should be based on the resurgence of Keynesian thinking represented by the Brandt Reports, a return to principles of an organised, managed and growing world economy.

The approach taken by Brandt (and Brundtland) to the world economy had many critics (e.g. Seers 1980, Frank 1980, Corbridge 1982). *North–South* argued that the protectionist trade policies of industrialised countries are the root cause of global economic problems and, in particular, persistent slow growth in the South. Tariff barriers and quotas stifled Southern economies and caused stagnation of Northern economies as Southern markets shrank. The solution was to open up the world economy, to pump capital and technical aid into the South to encourage trade, and to accept economic restructuring in the North (Brandt 1980, p. 186).

However, this analysis failed to demonstrate adequately the alleged dependence of the North on Southern markets. Furthermore, it was already clear in the 1980s that on the scale of individual countries, the effect of trade liberalisation is at best difficult to predict and at worst deleterious (e.g. Corbridge 1982). *North–South* also adopted a caricature of development in the South and the constraints upon it. It assumed that actions on the international scale will actually reach and benefit the poor, thus ignoring the problem of political and economic structures within Third World countries (see Frank 1980, Seers 1980). Similarly, *North–South* assumed that political and economic interests in the North are uniform, explicitly articulated and susceptible to rational debate. Northern governments were expected to recognise the mutual benefits from global economic cooperation and change their policies to achieve it. However, their capacity to do this was limited by the international political economy, and the first Brandt initiative foundered in the early 1980s amid growing protectionism. *North–South* adopted an unrealistic picture of the power and logic of capitalism, offering only a 'tepid programme of political action' (Corbridge 1982, p. 263) to achieve ambitious ends. The same epitaph could be justly applied to *Our Common Future*.

Development cannot be achieved by tinkering with world trade, but only by altering the relations of production within Third World countries and globally. Brandt avoided these issues, and as a result his commission's concept of mutuality was too loose to provide an effective basis for international political action. The world proceeded to become ever more interconnected economically and financially, a veritable 'casino of capitalism' with accelerating flows of money and rising uncertainty (Strange 1986). This hectic globalisation of investment, banking and production was very different from Brandt's calm, managed, interdependent world economy (Lash and Urry 1994).

The 'mutuality' of both Brandt and Brundtland was naïve. *Our Common Future* shows greater awareness of the real world of economics and politics than its predecessor, the *World Conservation Strategy*, but it comes no closer to explaining how that system works. Redclift (1987), writing before the publication of *Our Common Future*, suggested that the Brundtland Commission might produce a radical departure from previous writing on sustainable development. This proved not to be the case. Certainly, *Our Common Future* set ideas about sustainable development more credibly within the overall matrix of development thinking, but its untheorised mutualism placed it firmly within the camp of a rather comfortable Keynesian reformism. *Our Common Future* did not change the intellectual landscape of development thinking; it placed sustainable development within it, but in a far from commanding position.

Caring for the Earth

In the year before *Our Common Future* was published, IUCN held an international conference in Ottawa to discuss progress since publication of the *World Conservation Strategy*. This proposed a revision of the WCS, among other things to give it a 'human face', and indeed, through IUCN partners, a 'southern face' (Holdgate 1999, p. 181). Although it was planned that this would be ready by 1988, in fact it was not published until 1991, again by IUCN with WWF and UNEP (*Caring for the Earth: a strategy for sustainable living*, IUCN 1991).

Caring for the Earth was the result of a much more participatory process than its predecessor, and was the fruit of extensive consultation around the IUCN Regions. Martin Holdgate (who had a hand in writing it, with David Munro and Robert Prescott-Allen) described it as 'unashamedly a social and political document' (Holdgate 1999, p. 209). *Caring for the Earth*'s aim was 'to help improve the condition of the world's people' (IUCN 1991, p. 3). It argued that this required two things: first, a commitment to a new 'ethic for sustainable living'; and second, the integration of conservation and development, conservation to keep human actions within the earth's carrying capacity, and development to 'enable people everywhere to enjoy long, healthy and fulfilling lives' (ibid.). It argued that poor care for the earth had raised the risk that the needs of the present generation, and the needs of their descendants, would not be met. That risk could be eliminated 'by ensuring that the benefits of development are distributed equitably, and by learning to care for the Earth and live sustainably' (IUCN 1991, p. 4).

Caring for the Earth picked up themes from the WCS. It presented nine 'principles for sustainable development', and took these as its structure (Table 3.5). It opened with a chapter setting out 'principles to guide the way towards sustainable societies', and these nine principles provided a structure for the rest of the report. They blended the ethical ('respect and care for the community of life'), the humanitarian ('Improve the quality of human life'), the classically environmentalist ('keep within the Earth's carrying capacity' and 'minimise the depletion of non-renewable resources'), the conservationist ('conserve the Earth's vitality and diversity') and the pragmatic (provide a national framework for integrating development and conservation').

Table 3.5 Principles of sustainable development in *Caring for the Earth*

1 Respect and care for the community of life.
2 Improve the quality of human life.
3 Conserve the earth's vitality and diversity.
4 Minimise the depletion of non-renewable resources.
5 Keep within the earth's carrying capacity.
6 Change personal attitudes and practices.
7 Enable communities to care for their own environments.
8 Provide a national framework for integrating development and conservation.
9 Forge a global alliance.

Source: *Caring for the Earth: a strategy for sustainable living* (IUCN 1991)

The central argument of *Caring for the Earth* was much the same as its predecessors', although more carefully and fully expressed:

> we need development that is both people-centred, concentrating on improving the human condition, and conservation-based, maintaining the variety and productivity of nature. We have to stop talking about conservation and development as if they were in opposition, and recognise that they are essential parts of one indispensable process.
>
> (IUCN 1991, p. 8)

The report represented a much more sophisticated presentation of traditional conservationist and environmentalist ideas than its predecessor. Gone is the awkward neo-Malthusianism of the *World Conservation Strategy*, replaced with a homely reminder of the problems of meeting human needs faced with rapid population growth. Human needs have a high profile, as does the ethical basis for conservation ('respect and care for the community of life').

Caring for the Earth shared with the WCS a focus on environmental management, itemising principles for human action in farmland, rangeland, forests, fresh and salt waters, and also in settlements and in the care for the environment required of business, industry and commerce. It shared with the Brundtland Report its emphasis on mutuality and good global management, although it had little to say about the large-scale drivers of the global economy or environmental change. It recognised the importance of tackling poverty, meeting basic needs or food, shelter and health, and addressing issues of quality of life such as illiteracy and unemployment. It recognised the importance of debt (calling for official debt to be written off and commercial debt to be reduced) and the need for reform of South–North financial flows, the possible benefits of trade liberalisation and removal of non-environmental trade barriers, and the need for new and better-targeted aid (Actions 9.5, 9.6, 9.7). General population growth and the notion of global limits to the earth's pollution-absorption capacity were set in the context of gross disparities in levels of resource consumption in North and South. It stated clearly that a 'concerted effort is needed to reduce energy and resource consumption by upper income countries' (IUCN

1991, p. 44). Its vision of a 'sustainable community' was quite radical, almost ecosocialist:

> a sustainable community cares for its own environment and does not damage those of others. It uses resources frugally and sustainably, recycles materials, minimises wastes and disposes of them safely. It conserves life support systems and the diversity of local ecosystems. It meets its own needs so far as it can, but recognises the need to work in partnership with other communities.
>
> (ibid., p. 571)

Caring for the Earth was a cleverly drafted and integrated package, and represented a significant maturing of understanding about development on the part of IUCN. Holdgate (1999) argues that what made it different from other 'green manifestos' was the way it linked principles and suggested action: it listed 132 actions and 113 specific and dated targets, although these still did not engage with the structure of global or local political economy.

The publication of *Caring for the Earth* was a major event: almost 47,000 copies were printed in three languages, and it was launched in sixty-five countries. Its immediate impact was muted by the proximity of the Rio Conference (six months away), by the babble of green rhetoric current in the early 1990s, and by the lack of shock or novelty in its proposals (a direct corollary of their careful practicality). In the longer term, progress in meeting *Caring for the Earth*'s targets proved in many cases disappointing (Holdgate 1996). However, the impact of *Caring for the Earth* was considerable. Critically, its ideas and approach matched thinking in the various meetings of the preparatory committee (PrepComs) leading up to the Rio Conference, and its ideas fed very directly into the statements about sustainable development agreed at the UN Conference on Environment and Development. These are the subject of the next chapter.

Summary

- There is a dominant 'mainstream' to ideas about sustainable development. This 'mainstream sustainable development' (MSD) was formulated and elaborated in a series of documents drafted in the 1980s, the *World Conservation Strategy* (WCS) (IUCN 1980), *Our Common Future* (Brundtland 1987) and *Caring for the Earth* (IUCN 1991). This chapter discusses the creation of these documents and assesses their significance.
- Sustainable development was first explicitly discussed in the context of the UN Conference on the Human Environment at Stockholm in 1972. This was dominated by the environmental concerns of industrialised countries, but in an attempt to meet the fears of Southern countries, the idea was put forward that concern for the environment need not adversely affect development.
- The *World Conservation Strategy* was the culmination of more than two decades of work by conservationists, especially through IUCN, to get

conservation taken seriously in development. It argued for the maintenance of essential ecological processes and life-support systems, the preservation of genetic diversity and the sustainable use of species and ecosystems. It suggested that development and conservation could be made compatible through better and more timely planning.

- The Brundtland Report, *Our Common Future*, attempted to locate the debate about the environment within the economic and political context of international development. It was a successor to the Brandt Commission reports, *North–South* and *Common Crisis* (1980, 1983), and argued that poverty drove environmental degradation and required multilateral (global) action. International self-interest should drive a more equitable world economy, and with appropriate economic growth was necessary to achieve proper environmental management.

- In 1986 IUCN decided to update the WCS, eventually producing *Caring for the Earth* in 1991. This synthesised many of the ideas in the WCS and *Our Common Future*, setting them within the context of a new 'ethic for sustainable living'. It offered an analysis of how change could be made to happen at local, national and global scales, and set out targets. *Caring for the Earth's* impact was absorbed into the preparations for the Rio Conference in 1992.

Further reading

Brundtland, H. (1987) *Our Common Future*, Oxford University Press, Oxford, for the World Commission on Environment and Development.
Holdgate, M. (1996) *From Care to Action: making a sustainable world*, Earthscan, London.
Holdgate, M. (1999) *The Green Web: a union for world conservation*, Earthscan, London.
McCormick, J.S. (1992) *The Global Environmental Movement: reclaiming paradise*, Belhaven, London.
Redclift, M. (1996) *Wasted: counting the costs of global consumption*, Earthscan, London.

Web sources

<*http://www.iucn.org/*> World Conservation Union (IUCN): see for example full-text versions of the Resolutions and Recommendations adopted by IUCN at its General Assemblies from 1948 to 1996.
<*http://www.wwf.org/*> World Wide Fund for Nature: information on the global network of national organisations and programme offices. WWF campaigns are set out in <*http://www.panda.org/*>.
<*http://www.unep.org*> The United Nations Environment Programme (UNEP), fruit of the 1972 Stockholm Conference, with information on programmes, including information on the 'state of the global environment'.
<*http://www.fao.org*> The United Nations Food and Agriculture Organisation (FAO), with details of programmes on agriculture, forestry, fisheries, fisheries, etc.
<*http://www.peopleandplanet.net*> Web home of the magazine *People and Planet*, with a range of items on population, poverty, health, consumption and the environment.

4   Sustainable development: the Rio machine

> The greenhouse effect is the first environmental problem we can't escape by moving to the woods.
>
> (Bill McKibben 1990, p. 188)

Rolling down to Rio

The *World Conservation Strategy* and *Caring for the Earth* fought for attention on the international Stage, but *Our Common Future* had a guaranteed audience. The Brundtland Commission reported directly to the United Nations General Assembly. In December 1989 the UN resolved to convene a conference on environment and development five years after the Brundtland Report, to report progress. This took place at Rio de Janeiro in Brazil in June 1992. Expectations of the United Nations Conference on Environment and Development (UNCED, or more commonly simply 'the Rio Conference') were immense, but the auspices were not good. The secretary-general of UNCED was Maurice Strong, who had filled a similar role at Stockholm two decades earlier. He billed it as a meeting at which decisions would be made 'that will literally decide the fate of the earth' (Pearce 1991, p. 20). However, the Preparatory Commission meetings (styled 'PrepComs') revealed bitter conflicts of interest between industrialised and non-industrialised countries. As at Stockholm, the Rio Conference unleashed a rancorous debate about poverty that was barely contained by the process, and the emollient strategies of international organisations. In 1991, UNCED looked like 'a crunch meeting between management and shop stewards at a company facing bankruptcy' (ibid., p. 21).

UNCED was a massive undertaking, a major outing for the international diplomatic circus. It was attended by 172 states and 116 heads of state or government. There were 8,000 delegates and 9,000 representatives of the press; over 3,000 representatives of non-governmental organisations were accredited (N. Robinson 1993). The cost of the whole process, and the contrast between that cost and the urban poverty of Rio de Janeiro, were much commented upon in the press. Press appetite for such criticism grew as the exaggerated hopes for the conference broke up under the *realpolitik* of international vested interest. It was not helpful that Rio fell in an election year in the USA: the government of George Bush Snr was a reluctant actor in several critical arenas.

The PrepCom met five times, twice in New York, twice in Geneva and once in Nairobi. It was vast, containing all the member states of the UN. After a series of marathon debates, the PrepCom brought an agreed text of twenty-seven principles to the Conference (subsequently adopted as the Rio Principles), but the other documents considered by the conference contained 350 sections of bracketed or disputed language. At the conference itself, no undisputed text was reopened for discussion, but even so, consensus was hard to achieve. The final session of the Main Committee ran from 9.00 p.m. on 10th June to 6.00 a.m. the following morning, but by its end had resolved all but two disputes. Consensus were finally achieved through negotiations at ministerial level on 12th June, and *Agenda 21* and the Statement of Principles on Forest Management were adopted on 14th June (Koh 1993).

The formal procedures of the PrepComs, and the conference itself, were only part of a wider circuit of discussion and negotiation that ran through the five years before the conference. Behind the formal proceedings lay a vast iceberg of international conferences and meetings (such as the Dublin Conference on Water and the Environment in January 1992), national reports on environment and development (172 of which had been received by June 1992), and meetings to coordinate responses by particular groups, such as the World Industry Conference on Environmental Management, and the Business Council for Sustainable Development (Grubb *et al.* 1993).

UNCED was also a major focus of action for non-governmental organisations (NGOs). They were deliberately brought into the UNCED process by Maurice Strong from the first PrepCom in Nairobi in 1991. The Centre for Our Common Future (established in 1988 to carry forwards the work of the Brundtland Commission) also set up an 'International Facilitating Committee' (Chatterjee and Finger 1994). Despite these initiatives, extensive funding of NGOs at Rio, and a specific NGO Conference in Paris in 1991, many NGOs were disappointed by their lack of influence on the UNCED process.

At Rio itself, NGOs were represented at a Global Forum, and had opportunity for networking and debate at an 'Earth Parliament'. However, all this was physically and psychologically distant from the main conference. Some chose to express their views in distinct and forthright ways, for example in Greenpeace's banner above the city of Rio de Janeiro (Plate 4.1). Although some NGOs remained close to government delegations through the conference (to which 1,400 lobbyists were accredited), NGOs were excluded from the official negotiating sessions (Holmberg *et al.* 1993).

Nonetheless, their presence may have influenced the way some issues were approached in *Agenda 21*, for example in the emphasis on 'empowerment'; Chapter 27 of *Agenda 21* explicitly discusses the importance of their role in achieving sustainable development, although it does so through a stream of bland and empty statements. Furthermore, while Rio may have broadened the recognition that grassroots groups, particularly women's groups, were clearly

Plate 4.1 Greenpeace protest at the Rio Conference. A large number of NGOs
attended the United Nations Conference on Environment and Development
at Rio de Janeiro in 1992. They met at a parallel 'Global Forum' that was
physically separated from the conference itself. Many NGOs felt excluded
and that their views were poorly represented. Southern and grassroots
NGOs resented the lobbying power and corporate muscle of big North-
based environmental NGOs. Popular hopes that Rio would usher in a new
environmentally conscious world order were widely disappointed. This sense
of an opportunity missed was well captured by the Greenpeace protest,
hanging a vast banner above the city of Rio. Photo: Greenpeace.

key elements in debates about environment and development (Ekins 1992),
the conference also began to emphasise the distance between the powerful,
wealthy and influential NGOs of industrialised countries and the 'grassroots'
in the sense of groups formed among the poor of the urban and rural Third
World. Chatterjee and Finger (1994) conclude that only the largest and most
globally organised NGOs (almost all of which were North American) had much
influence on the Rio documents. Most NGOs failed to make effective use of
the US-style lobbying process and ended up confused, frustrated and divided.

The documents debated and agreed at UNCED in Rio de Janeiro in 1992
build very directly onto the evolving mainstream of ideas dominating public
debate about environment and development of the 1980s. The same themes
appear in both the Rio Declaration and the much larger text of *Agenda 21*.
However, the creation of these texts was far from straightforward and harmo-
nious. The 'Rio process' was in practice a mutual bludgeoning between teams

of diplomats to produce texts that gave least away to perceived national interests. In particular, the distinction between the views of countries in the industrialised North and the underdeveloped South became steadily more glaring in the run-up to and during the conference. There was difference over the key problems (for the industrialised countries, global atmospheric change and tropical defor-estation; for unindustrialised countries, poverty and the problems that flow from it), and responsibility for finding solutions. As at Stockholm in 1972, there was fear on the part of Third World countries that their attempts to industrialise would be stifled by restrictive international agreements on atmos-pheric emissions. They also feared that their freedom to use natural resources within their boundaries would be constrained by agreement imposed by indus-trialised countries that had themselves become wealthy precisely by squeezing their environments, for example by clearing the vast majority of their forest cover and latterly by allowing industries to develop and operate with limited regard for environmental externalities such as pollution.

The Rio Declaration on Environment and Development

The tensions between Northern and Southern governments are clear in the texts of the UNCED documents themselves. The Rio Declaration was not the strong and sharp 'Earth Charter' originally conceived by the conference chairman, Maurice Strong. Its twenty-seven principles comprise 'a bland decla-ration that provides something for everybody' (Holmberg *et al.* 1992, p. 7). It opens with the statement that 'human beings are at the centre of concerns for sustainable development. They are entitled to a healthy and productive harmony with nature' (Table 4.1). Many of the principles were uncontentious (e.g. 4, the need to integrate conservation and development, or 5 on the erad-ication of poverty). Others were more closely fought over at Rio, particularly those that addressed the central issue of the conference: international action and international responsibility. Thus Principle 2 notes the sovereign right of countries to develop, while Principle 7 establishes the notion of 'common but differentiated responsibilities' for the global environment. Hidden behind a bland comment that 'states shall cooperate in a spirit of global partnership to conserve, protect and restore the health and integrity of the world's ecosystem', responsibility here basically forces the burden of greatest action on developed countries (Holmberg *et al.* 1993). Even the text of the Rio Declaration (compact by the standards of *Agenda 21*) is self-contradictory, and the US delegation released an 'interpretative statement' that effectively dissociated them from a number of the principles agreed. These included the notion of a right to devel-opment in Principle 3 (they argued that 'development is not a right ... on the contrary development is a goal we all hold'; Holmberg *et al.* 1993, p. 30), and also rejecting any interpretation of Principle 7 that suggested any form of international liability.

Table 4.1 The twenty-seven principles of the Rio Declaration on Environment and Development

1   Human beings are at the centre of concerns for sustainable development. They are entitled to a healthy and productive life in harmony with nature.

2   States have, in accordance with the Charter of the United Nations and the principles of international law, the sovereign right to exploit their own resources pursuant to their own environmental and developmental policies, and the responsibility to ensure that activities within their jurisdiction or control do not cause damage to the environment of other States or of areas beyond the limits of national jurisdiction.

3   The right to development must be fulfilled so as to equitably meet developmental and environmental needs of present and future generations.

4   In order to achieve sustainable development, environmental protection shall constitute an integral part of the development process and cannot be considered in isolation from it.

5   All States and all people shall cooperate in the essential task of eradicating poverty as an indispensable requirement for sustainable development, in order to decrease the disparities in standards of living and better meet the needs of the majority of the people in the world.

6   The special situation and needs of developing countries, particularly the least developed, and those most environmentally vulnerable, shall be given special priority. International actions in the field of environment and development should also address the interests and needs of all countries.

7   States shall cooperate in a spirit of global partnership to conserve, protect and restore the health and integrity of the Earth's ecosystem. In view of the different contributions to global environmental degradation, States have common but differentiated responsibilities. The developed countries acknowledge the responsibility that they bear in the international pursuit of sustainable development in view of the pressures their societies place on the global environment and of the technologies and financial resources they command.

8   To achieve sustainable development and a higher quality of life for all people, States should reduce and eliminate unsustainable patterns of production and consumption and promote appropriate demographic policies.

9   States should cooperate to strengthen endogenous capacity-building for sustainable development by improving scientific understanding through exchanges of scientific and technical knowledge, and by enhancing the development, adaptation, diffusion and transfer of technologies, including new and innovative technologies.

10 Environmental issues are best handled with the participation of all concerned citizens, at the relevant level. At the national level, each individual shall have appropriate access to information concerning the environment that is held by public authorities, including information on hazardous materials and activities in their communities, and the opportunity to participate in decision-making processes. States shall facilitate and encourage public awareness and participation by making information widely available. Effective access to judicial and administrative proceedings, including redress and remedy, shall be provided.

11 States shall enact effective environmental legislation. Environmental standards, management objectives and priorities should reflect the environmental and developmental context to which they apply. Standards applied by some countries may be inappropriate and of unwarranted economic and social cost to other countries, in particular developing countries.

12 States should cooperate to promote a supportive and open international economic system that would lead to economic growth and sustainable development in all countries, to better address the problems of environmental degradation. Trade policy measures for environmental purposes should not constitute a means of

arbitrary or unjustifiable discrimination or a disguised restriction on international trade. Unilateral actions to deal with environmental challenges outside the jurisdiction of the importing country should be avoided. Environmental measures addressing transboundary or global environmental problems should, as far as possible, be based on an international consensus.

13 States shall develop national law regarding liability and compensation for the victims of pollution and other environmental damage. States shall also cooperate in an expeditious and more determined manner to develop further international law regarding liability and compensation for adverse effects of environmental damage caused by activities within their jurisdiction or control to areas beyond their jurisdiction.

14 States should effectively cooperate to discourage or prevent the relocation and transfer to other States of any activities and substances that cause severe environmental degradation or are found to be harmful to human health.

15 In order to protect the environment, the precautionary approach shall be widely applied by States according to their capabilities. Where there are threats of serious or irreversible damage, lack of full scientific certainty shall not be used as a reason for postponing cost-effective measures to prevent environmental degradation.

16 National authorities should endeavour to promote the internationalisation of environmental costs and the use of economic instruments, taking into account the approach that the polluter should, in principle, bear the cost of pollution, with due regard to the public interest and without distorting international trade and investment.

17 Environmental impact assessment, as a national instrument, shall be undertaken for proposed activities that are likely to have a significant adverse impact on the environment and are subject to a decision of a competent national authority.

18 States shall immediately notify other States of any natural disasters or other emergencies that are likely to produce sudden harmful effects on the environment of those States. Every effort shall be made by the international community to help States so afflicted.

19 States shall provide prior and timely notification and relevant information to potentially affected States on activities that may have a significant adverse transboundary environmental effect and shall consult with those States at an early stage and in good faith.

20 Women have a vital role in environmental management and development. Their full participation is therefore essential to achieve sustainable development.

21 The creativity, ideals and courage of the youth of the world should be mobilised to forge a global partnership in order to achieve sustainable development and ensure a better future for all.

22 Indigenous people and their communities, and other local communities, have a vital role in environmental management and development because of their knowledge and traditional practices. States should recognise and duly support their identity, culture and interests and enable their effective participation in the achievement of sustainable development.

23 The environment and natural resources of people under oppression, domination and occupation shall be protected.

24 Warfare is inherently destructive of sustainable development. States shall therefore respect international law providing protection for the environment in times of armed conflict and cooperate in its further development, as necessary.

25 Peace, development and environmental protection are independent and indivisible.

26 States shall resolve all their environmental disputes peacefully and by appropriate means in accordance with the Charter of the United Nations.

27 States and people shall cooperate in good faith and in a spirit of partnership in the fulfilment of the principles embodied in this Declaration and in the further development of international law in the field of sustainable development.

Agenda 21

The main output of the Rio Conference, although it is probably the least read, was of course *Agenda 21*. This is a vast document, containing forty separate chapters and amounting to a volume of more than 600 pages. It was drafted and argued over minutely by government officials and lawyers, and is both a fantastic compendium of ideas, issues and principles, and a hard-won agreement that claims to reflect 'a global consensus and political commitment at the highest level on development and environmental cooperation' (Holmberg *et al.* 1993).

The name 'Agenda 21' came from the first PrepCom meeting in Nairobi, when Maurice Strong proposed a document to set out how to make the planet sustainable by the start of the twenty-first century (Chatterjee and Finger 1994). By the time the conference itself began, the document had forty chapters, and was 'quite indigestible and impossible to implement' (ibid., p. 54). *Agenda 21* is a monument to the problems of making the rhetoric of international cooperation about environment and development concrete. It has become an icon of sustainable development, held up to for symbolic veneration for its encapsulation of all possible arguments, a scripture dipped into for proof-texts to legitimate particular points of view but not subjected to detailed analysis.

The chapters of *Agenda 21* are listed in Table 4.2. The scope is enormous, covering issues from water quality and biodiversity to the role of women, children and organised labour in delivering sustainable development. The chapters are divided into four sections (Table 4.2); first, 'Social and Economic Dimensions' (i.e. development, chapters 2–8); second, 'Conservation and Management of Resources for Development' (chapters 9–22); third, 'Strengthening the Role of Major Groups' (chapters 23–32); and fourth, 'Means of Implementation' (chapters 33–40). Chapters vary greatly in length, with those addressing environmental management (Section 2) being the longest, and comprising almost half the total volume.

Commentaries on *Agenda 21* demand a large measure of creativity, for the document itself is so convoluted as to defy straightforward précis. Each chapter seeks to set out the basis for action, the objectives of action, a set of activities and the means to be used to implement them (Grubb *et al.* 1993). In this, each part of *Agenda 21* is in a sense a microcosm of the whole, with a particular emphasis on the means of implementation.

A series of key themes can be identified (United Nations 1993). The first is 'the revitalisation of growth with sustainability'. As at Stockholm, sustainable development at Rio could not be conceived of as questioning the importance of economic growth, for either rich or poor countries. The second theme is 'sustainable living', under which come poverty, health and population growth. The third theme addresses the problems of urbanisation (water supplies, wastes, pollution and health). These concerns had been underplayed in previous 'mainstream' documents, but here they properly come to the fore: the problems of sustainable management of rural resources and the rural poor

Table 4.2 The structure of *Agenda 21*

Chapter 1	Preamble
Section 1	**Social and economic dimensions**
Chapter 2	International cooperation to accelerate sustainable development in developing countries and related domestic policies
Chapter 3	Combating poverty
Chapter 4	Changing consumption patterns
Chapter 5	Demographic dynamics and sustainability
Chapter 6	Protecting and promoting human health
Chapter 7	Promoting sustainable human settlement development
Chapter 8	Integrating environment and development in decision-making
Section 2	**Conservation and management of resources for development**
Chapter 9	Protection of the atmosphere
Chapter 10	Integrated approach to the planning and management of land resource
Chapter 11	Combating deforestation
Chapter 12	Managing fragile ecosystems: combating desertification and drought
Chapter 13	Managing fragile ecosystems: sustainable mountain development
Chapter 14	Promoting sustainable agriculture and rural development
Chapter 15	Conservation of biological diversity
Chapter 16	Environmentally sound management of biotechnology
Chapter 17	Protection of the oceans, all kinds of seas, including enclosed and semi-enclosed seas, and coastal areas and the protection, rational use and development of their living resources
Chapter 18	Protection of the quality and supply of freshwater resources: application of integrated approaches to the development, management and use of water resources
Chapter 19	Environmentally sound management of toxic chemicals, including prevention of illegal international traffic in toxic and dangerous substances
Chapter 20	Environmentally sound management of hazardous wastes, including prevention of illegal international traffic in hazardous wastes
Chapter 21	Environmentally sound management of solid wastes and sewage-related issues
Chapter 22	Safe and environmentally sound management of radioactive wastes
Section 3	**Strengthening the role of major groups**
Chapter 23	Preamble
Chapter 24	Global action for women towards sustainable and equitable development
Chapter 25	Children and youth in sustainable development
Chapter 26	Recognising and strengthening the role of indigenous people and their communities
Chapter 27	Strengthening the role of non-governmental organisations: partners for sustainable development
Chapter 28	Local authorities' initiatives in support of Agenda 21
Chapter 29	Strengthening the role of workers and their trade unions
Chapter 30	Strengthening the role of business and industry
Chapter 31	Scientific and technological community
Chapter 32	Strengthening the role of farmers

Table 4.2 (Continued)

Section 4	Means of implementation
Chapter 33	Financial resources and mechanisms
Chapter 34	Transfer of environmentally sound technology, cooperation and capacity-building
Chapter 35	Science for sustainable development
Chapter 36	Promoting education, public awareness and training
Chapter 37	National mechanisms and international cooperation for capacity-building in developing countries
Chapter 38	International institutional arrangements
Chapter 39	International legal instruments and mechanisms
Chapter 40	Information for decision-making

Source: N. Robinson (1993)

were perhaps more readily recognised, but the deprivation of the urban poor is the more intractable (and more rapidly growing) problem. The fourth theme is 'efficient resource use', under which heading is included everything from combating deforestation and desertification through to conservation of biological diversity. Just as growth is the foundation stone of mainstream sustainable development, efficient resource use is the mechanism for achieving it: Chatterjee and Finger comment, 'in the name of environmental protection . . . *Agenda 21* extends the economic rationality to the most remote corners of the earth' (1992, p. 56). The fifth theme concerns global and regional resources (atmosphere and oceans), the sixth the management of chemicals and wastes. The seventh and final theme is 'people's participation and responsibility' (United Nations 1993).

Agenda 21 bears a strong inheritance from its predecessors. This is evident in various ways. The first is in the centrality it gives to growth. This is the familiar Brundtland agenda re-expressed: in the mainstream interpretation of sustainable development, everything is predicated on economic growth, both globally and nationally.

Second, *Agenda 21* shows a familiar dominance (in volume and position) of straightforward issues of environmental management. In the second section of *Agenda 21* all the familiar environmental issues from the *World Conservation Strategy* appear, developed but unmistakable.

Third, *Agenda 21* is technocentrist. The first six key themes make this quite clear: growth will power and technology will direct the evolution of policy towards more efficient use of the environment and hence towards a more sustainable world economy. The 'essential means' to achieve sustainability also reflect this technocentrism, building on information, science and environmentally sound technology (United Nations 1993).

Fourth, *Agenda 21* has inherited the multilateralism of the Brundtland Report. The dominant mechanism for making any of its provisions happen is the common interest of industrialised and non-industrialised countries, of present generations in both caring about the future. International flows of financial resources and technology will reflect this mutual interest, international agencies will direct and promote these flows and their effectiveness, and international legal instruments will structure and regulate their product.

Fifth, like its predecessors, *Agenda 21* calls for sustainable development through participation. As in *Caring for the Earth*, women, children, young people, indigenous people, trade unionists, businesses, industry, farmers, local authorities and scientists are all summoned to play a role, a rainbow coalition to put flesh on the endless skeleton of the text of *Agenda 21*. Here the text has all the emotive power that motherhood and apple pie statements can render. Chapter 25, 'Children and youth in sustainable development', for example, suggests that 'it is imperative that youth from all parts of the world partici-pate actively in all relevant levels of decision-making processes because it affects their lives today and has implications for their futures' (para. 25.2). Participation is a vital watchword of *Agenda 21*, but like its predecessors, it is much stronger on hopeful sentiments about involvement than political analysis of power.

The Forest Principles

The Rio Conference should have seen a Convention on Forests signed. This did not happen, and instead, a much shorter and lesser document of forest management was agreed. Its title reveals its character, and the problems that beset the convention for which it substituted: a 'Non-legally binding author-itative statement of principles for a global consensus on the management, conservation and sustainable development of all types of forests' (Sullivan 1993). Pressure for specific action on forests came from Northern environmental organ-isations concerned at the rate of clearance of tropical moist forests (rainforests). It followed a series of international initiatives during the 1980s such as the Tropical Forests Action Plan (TFAP, under the UNDP, FAO, World Bank and World Resources Institute) and the International Tropical Timber Organisation (under UNCTAD).

The idea of a global forest convention was made in a review of the TFAP in 1990, and the proposal made at a meeting of the G7 group of industri-alised countries in Houston later that year (Sullivan 1993). However, the North–South divide became very clear, and debate bitter and intransigent. Southern countries (the 'G77', led by Malaysia and India) opposed a global forest convention. (The G77 countries are a group of 128 less developed and less industrialised countries set up as a counter-lobby to the developed 'G7' countries.) They argued that industrialised countries had cleared their own forests during their industrialisation and that non-industrialised countries had a sovereign right to do the same. They could point to both an established history of non-sustainable forestry (notably in the USA) and continuing unsus-tainable practices in the harvesting of old-growth forests, for example in the Pacific Northwest of the USA and in the western and boreal forests of Canada (Maser 1990). Furthermore, if tropical forests served a global benefit (whether through their biodiversity or by locking up CO_2), Southern countries argued that the costs of maintaining them uncleared should be borne globally. If there was to be a global forest convention, it should have a mechanism for compensating Southern countries for revenue forgone in setting aside forest reserves (Holmberg *et al.* 1993).

This debate rapidly overspilled the tight confines of the PrepCom meetings, and by PrepCom 4 it was clear that a legally binding agreement on forests was impossible to achieve at Rio. Energies were focused instead on capturing the high ground and trying to establish some kind of global consensus on forest management. The resulting Forest Principles was a political document and not an operational tool (Holmberg *et al.* 1993). The principles closely reflected chapter 11 of *Agenda 21*, on 'Combating deforestation', and explicitly addressed all forests, i.e. temperate and boreal as well as tropical forests. They avoided specific commitments. They repeated the familiar arguments about the social, environmental and economic importance of forests, and the need for them to be managed sustainably. They mentioned the need for international cooperation and the need for funds from industrialised countries to meet management needs and broadly support free trade in timber and forest products (against calls, for example, for environmentally defined trade bans in the North). They called for scientific assessment and management of environmental impacts of forestry, and they discussed the need for local participation in forest management decisions.

All these principles were widely recognised as desirable, and while they presented a challenge to dominant forest management practices in almost every country (not simply those in the tropics), they also reflected existing ideas within the forestry industry about 'best practice'. Most critically, the Forest Principles emphasised national sovereignty for forests within national borders (N. Robinson 1993, Sullivan 1993). They did not provide a basis for Northern intervention in Southern forest management on environmental grounds. If anything, their evenhandedness reflected the unwillingness of any government (including those in North America and Scandinavia) to constrain logging industries midway through the liquidation of the assets in old-growth forests. The hopes of Northern NGOs that Rio might generate significant constraints on rates of tropical deforestation fell foul of international politics, and the awkward fact of the unsustainability of logging practices in parts of the North (which, of course, Northern NGOs such as Greenpeace and Earth First! also vigorously opposed).

The Biodiversity Convention

The two conventions signed at Rio, the Convention on Biological Diversity and the Framework Convention on Climate Change, were negotiated not through the PrepCom, but through international negotiating committees (Chatterjee and Finger 1994). Both reflect fairly closely the relevant chapters of *Agenda 21*, which considers 'environmentally sound management of biotechnology' and the 'conservation of biological diversity' (chapters 16 and 15), and 'protection of the atmosphere' (chapter 9).

The Convention on Biological Diversity was one of the elements of Rio with the longest pedigree. A draft convention was prepared in the mid-1980s by IUCN, in conjunction with other international organisations (including the WWF, UNEP, the World Resources Institute and the World Bank). This initiative was the fruit of the continuation of the conventional conservation agenda that had inspired the *World Conservation Strategy* a decade before. The idea of a global conservation convention had been mooted at the Second World Congress on National Parks in Bali (organised by IUCN's Commission on National Parks and Protected Areas), and between 1988 and 1992 there was sustained pressure both for the convention itself, and for effective measures to preserve global biodiversity (and hence in large measure Southern biodiversity). All the major international bodies with an interest in environment and development (the World Resources Institute, IUCN, UNEP, WWF, the World Bank, the FAO and UNESCO) contributed to a series of meetings and reports that culminated in the Global Biodiversity Strategy in 1992 (World Resources Institute *et al.* 1992). Completion and adoption of the convention was the priority requirement of this strategy (Holdgate 1999).

Negotiations over a convention were initiated by UNEP in 1990, reflecting essentially Northern concern about rainforest loss. However, at the second Geneva PrepCom meeting, the G77 countries demanded inclusion of the issue of bioprospecting and biotechnology, and the sharing of wealth generated by the exploitation of biodiversity in the South by Northern biotech companies. In this odd hybrid form the convention was agreed at Rio and signed by 156 countries (Chatterjee and Finger 1994). The USA refused to sign at that time, although it did so subsequently.

The aim of the Convention on Biological Diversity is to conserve biological diversity and to promote the sustainable use of species and ecosystems, and the equitable sharing of the economic benefits of genetic resources. It is this last element that sets this convention apart from all previous international conservation agreements. Signatory nations committed themselves to the development of strategies for conserving biological diversity, and for making its use sustainable. Biodiversity conservation can be achieved *in situ* (that is, through conventional methods such as the designation of systems of protected areas) or *ex situ* (for example through captive breeding), but the convention also requires cross-cutting measures (for example affecting forestry or fishing). While all these elements of the convention are qualified by a get-out clause (all is to be done 'as far as possible and appropriate'), this programme is a logical development of the traditional conservationist concern for sustainable ecosystems use, and draws directly on the thinking in the *World Conservation Strategy* and *Caring for the Earth*.

What is novel (and controversial) in the convention is the provisions for the exploitation of genetic resources through biotechnology. In principle, this is no different from the provisions of the earlier mainstream documents that species and ecosystems should provide resources for human benefit, if used in such a way that their availability was sustained. However, by 1992 the rapid development in genetic science had opened up vast new areas of potential exploitation at the sub-specific and molecular level, including the creation of novel organisms (which might perhaps be patented by the corporation that created them) and products (drugs for example) derived from wild species. It was perceived that this technology had the potential to generate vast wealth; however, the biotechnological capacity was almost entirely held by industrialised countries (because of the high costs of research laboratories, research infrastructure and training), and moreover was increasingly held by private corporations within those countries and not by states themselves. Third World countries feared stripping of their genetic resources by bioprospectors, and loss of access to economic benefits derived by First World corporations (Shiva 1997). First World countries (particularly the USA, which dominated in this area of science) feared restriction of economic opportunity if trade in biotechnology were restricted by a benefit-sharing agreement. The convention reflects the balance of these opposite fears, and does contain provision for sharing benefits from commercial exploitation of genetic resources (Article 15). This debate was cross-cut by the desire of those pushing traditional conservation arguments to achieve their conventional goals. Few of those negotiating the treaty could have had direct knowledge of the present (let alone the future) potential of the scientific revolution in biotechnology that took place in the 1990s.

The Convention on Biological Diversity came into force on 29th December 1993, and by 1997 it had been ratified by 162 countries. This rapid entry into force reflected the level of concern about continued biodiversity loss, but also the rapid development of biotechnology in the 1990s, and the potential commercial value of genetic material both in its raw (wild) state and as patentable 'improved' forms. Widespread ratification has not by any means ended controversy, and interest in the Conferences of the Parties has been considerable: more than 130 governments sent over 700 delegates to the first Conference of the Parties in Nassau in the Bahamas in 1994. Debate over the issues embraced by the convention has continued to be fierce, addressing among other things integration with other biodiversity conventions (Ramsar, on wetlands, and CITES, on trade in endangered species for example) and relations with the World Trade Organisation, and its agreement on Trade Related Intellectual Property Rights, TRIPS (Bragdon 1996, Pimbert 1997).

The Framework Convention on Climate Change

The idea of a climate change convention also pre-dates the immediate preparations for Rio. Scientific studies of global warming were stimulated by the International Geophysical Year (1957–8), which also stimulated international biological science and, with the International Biological Programme and related developments, played a role in the stimulation of late twentieth-century environmentalism by fostering awareness of the environment as a global issue (Chapter 2). The notion that countries should take responsibility for transnational pollution was established at the Stockholm Conference in 1972 (see Chapter 3).

Scientific evidence from observations and early computer models was sufficiently strong to lead UNEP, the World Meteorological Organisation (WMO) and the International Council of Scientific Unions (ICSU) to convene a World Climate Conference in Geneva in 1979 (Jäger and O'Riordan 1996). Scientific research continued to develop rapidly through the 1980s. The impact of carbon dioxide concentrations for the 'greenhouse effect' were recognised, as was the potential of other greenhouse gases such as methane, nitrous oxide and CFCs (chlorinated fluorocarbons).

The third of a series of scientific meetings at Villach in Austria in 1985 discussed scenarios for future emissions of all greenhouse gases and began to establish a clear scientific consensus about anthropogenic climate change (Jäger and O'Riordan 1996). This brought the issue clearly onto the political agenda; indeed, it was itself in large measure responsible for the political importance of the environment in that decade. The conference in Toronto in 1988 entitled The Changing Atmosphere produced a statement calling on all developed countries to reduce their CO_2 emissions by 20 per cent from 1987 levels by 2005 (ibid.). The Intergovernmental Panel on Climate Change (IPCC) was established in 1988 (by WMO and UNEP), with three working groups, on scientific evidence, environmental and socio-economic impacts, and response strategies respectively. These reported in 1990.

The Science Assessment Working Group set out to establish a global consensus on the complex science of climate change. In the words of its chair, the IPCC reports 'can be considered as authoritative statements of the contemporary views of the international scientific community' (Houghton 1997, p. 159). Working Group I reported scientific certainty that human action was affecting atmospheric concentrations of greenhouse gases, and estimated that this was responsible for over half the enhanced greenhouse effect, both past and future (Jäger and O'Riordan 1996).

Debate about the Framework Convention on Climate Change reflected the divergent reactions to the IPCC reports of 1990. The IPCC's consensus view of the importance of fossil fuel consumption and carbon dioxide (CO_2) output cut directly at the heart of the interests of the industrialised Northern countries, while also having significant implications for rapidly industrialising countries in the South such as India and China. The International Negotiating Committee on Climate Change began work in 1990, following the Second World Climate Conference in Geneva, with the aim of preparing a convention for signature at Rio in 1992. It rapidly fell foul of fundamental differences between different parties. There was a broad divergence between industrialised and non-industrialised countries, with the North urging the priority of environmental protection and that any measures agreed should be cost-effective, while the South pushed the need for development and industrialisation, and the principle of historical responsibility (Rowbotham 1996). Industrialised countries were unwilling to countenance a significant reduction in CO_2 output. Oil-producing states were also opposed to this, while small island states vulnerable to sea level rise wanted urgent action on precisely this. The EU favoured agreement on targets and a timetable for implementation, the USA was reluctant (the latter even refusing, in the run-up to Rio, to agree to cut back emissions in the year 2000 to 1990 levels). In April 1992, at the last Intergovernmental Negotiating Committee meeting, the compromise was agreed on a non-binding call for an attempt to return to 1990 emissions of CO_2 and other greenhouses gases not controlled by the Montreal Protocol.

The convention was a delicate balance between divergent political and economic interests, and rather full of rather pious intentions (Rowbotham 1996). Like *Agenda 21*, it stressed (in Article 3, 'Principles') the significance of the protection of the climate system for both present and future genera-tions, and stated that there must be equity between industrialised and non-industrialised countries in taking action. This equity must reflect historic responsibility, state of development and capacity to respond (Holmberg *et al.* 1993). The diversity of interest was such that the text was ambiguous, left open to subsequent interpretation at the meetings of the Conference of the Parties (COP). The Framework Convention was weak (arguably 'toothless'; Chatterjee and Finger 1994, p. 45) in that it contained no legally binding commitments for the stabilisation (let alone reduction) of CO_2 emissions. Again, this was left until later. However, the convention's negotiation, in a mere sixteen months, is a remarkable testament to the urgency with which global climate change was viewed in the early 1990s, and is in many ways a major achievement of the Rio process.

The Framework Convention on Climate Change was signed by over 150 states and the European Community (now the European Union) at Rio, and came into force in March 1994 (having received its fiftieth signatory in December the previous year). The First COP took place in Berlin in 1995. The issue of binding targets and timetables for reducing greenhouse gas emis-sions (proposed by the Association of Small Island States) remained controversial, as did the notion of 'joint implementation', under which one country can aid another to implement a project or change a policy that will result in reduced mutual gas emissions (Rowbotham 1996, Bush and Harvey 1997). The IPCC has continued to assess the evidence for and predict the impacts of anthropogenic climate change. It produced the first major new assessment in 1995, drawing on new data and analysis that confirmed the conclusion of the 1990 report: that the balance of the evidence suggested a discernible human influence on climate.

The Third Session of the Conference of the Parties (COP3), in Kyoto in Japan in December 1997, finally established a binding obligation on industri-alised countries ('Annexe 1' countries) to reduce emissions for six greenhouse gases by 5 per cent below 1990 levels by 2008–12. Different countries agreed to different reductions (8 per cent for the European Union and 7 per cent for the USA, for example, while Australia was allowed to *increase* emissions by 8 per cent and Iceland by 10 per cent). Changes in land use and forestry since 1990 that could be held to have a positive impact on net emissions could be counted against national targets. In 1998 a further Action Plan was agreed by the COP at Buenos Aires, focusing on flexible strategies for implementing targets.

Progress towards the emissions targets agreed at Kyoto was set back at COP6 in The Hague in November 2000, where despite two weeks of intensive negotiations, signatories failed to agree on how to operationalise the Kyoto Protocol. There was some progress towards agreement on financial and technical support for Third World countries, but profound disagreement about measures to cut emissions on the part of a handful of wealthy industrialised countries, led by the USA and Australia (*http://cop6.unfccc.int*).

This continuing development in the international regime for the regulation of greenhouse gas emissions reflects the technical and reductionist nature of the global discourse of climate change. It is a classic example of instrumental rationality, applying technical knowledge to an agreed purpose by the most efficient means possible (Cohen *et al.* 1998), an example of the approach of ecological modernisation to be discussed in Chapter 5. The IPCC has become more open than it was to research not based on natural science, but is still dominated on the social science side by the technical procedures of economics (Shackley 1997). There is a divide between scientific and technical approach to understanding and developing policy for dealing with climate change and the much broader set of issues raised about sustainable development at Rio (Cohen *et al.* 1998). The technical demands of the evolving regime on climate change place a considerable and growing burden on the technical and policy capacities of Southern countries, and (like the rest of the Rio process) themselves demand considerable financial support from industrialised countries over and above that required to address wider issues of environment and poverty.

Beyond Rio

UNCED left a legacy of solid achievements – in international law, in new and perhaps refocused international institutions, in the actions of national and local governments, and (above all) in the composition of the political and policy rhetoric that is taken to represent international consensus about environment and development. However, commentators are doubtful concerning the significance of these achievements. Many of the problems on which *Agenda 21* focused have become worse (Brown 1997). Poverty has deepened and widened, and the gap between rich and poor countries has grown. The enhanced resource flows needed to implement *Agenda 21* were not forthcoming, and indeed countries such as the UK have reduced their aid budget. The activities of transnational companies have not been significantly influenced. Moreover, many countries failed to respond to the Framework Convention on Climate Change and curb CO_2 emissions (Parikh *et al.* 1997), and issues of sovereignty over genetic resources remain intractable (Pimbert 1997).

Overall, Brown concludes that Rio did little to promote sustainable development as such; however, she argues that it did at least open the debate about choices in development, about the ways in which the biosphere is restructured in pursuit of profit, of the costs of technology, of the inequalities in wealth, technologies, environmental hazard and life chances (Brown 1997). In many ways, the Rio Conference should not be interpreted as a single event that could be assessed in terms of success or failure. Roddick suggests that it did not set out to create a single regime, but 'an entire framework for the management of environmental problems' (1997, p. 147). This framework included 'soft law': voluntary reporting and action by communities and other non-state actors at local, national and international levels. In practice, however, the established practice of forging formal international agreements has continued to dominate, despite the chronic deficit on enforcement

Probably the chief failure of the Rio Conference was that it did not stimulate the scale of financial support necessary to implement *Agenda 21*. Before the Fourth UNCED PrepCom started in New York, Maurice Strong estimated the cost of implementing the Rio agreements at US$125 billion in new aid every year between 1992 and 2000 (Chatterjee and Finger 1994). At the conference itself, the secretariat estimated that implementation would cost something like $600 billion per year of which $125 billion per year would have to be in the form of gifts and concessional loans. This was more than twice the current total disbursement of official development assistance to Third World countries, and close to the official UN target of 0.7 per cent of GNP from First World countries (Grubb *et al.* 1993). In the event, the money available has been a tiny fraction of that. Of the $125 billion per year needed to take necessary actions, only about $2.5 billion was pledged (Holmberg *et al.* 1993).

The financing mechanism chosen for funding actions arising from Rio was the Global Environment Facility (GEF). The GEF already existed. It was set up in 1990 by the World Bank, UNDP and UNEP. The World Bank administers the GEF and acts as repository of the trust fund. UNEP provides the secretariat for the Scientific and Technical Advisory Panel and supplies environmental expertise on specific projects; UNDP is responsible for technical assistance and project preparation. The idea of channelling funds to implement the Rio agreements through the GEF was strongly opposed by Third World governments (which feared the dismal scientists of the World Bank, and the iron grip of the North on its policies) and by environmentalists (who had maintained a barrage of criticism of the environmental impacts of World Bank lending since the 1980s).

The GEF is widely held to be too small to be effective. It held $1.3 billion in the first phase; by the Fourth GEF Meeting (in December 1992), seventy projects had been approved, amounting to $584 million (Chatterjee and Finger 1994). By September 1998 (seven years after establishment) the GEF had allocated $2 billion (World Bank 2000). While the GEF has brought much-needed funds to some sectors (national parks for example), the overall influence of the GEF's funds, in the context of debt and total resource flows between the First and Third Worlds, has been very limited (Brown *et al.* 1993).

At the same time at it adopted *Agenda 21* (in December 1992), the General Assembly of the UN requested its Economic and Social Council (ECOSOC)

to establish a Commission on Sustainable Development (CSD) (Robinson 1993). This had been suggested in chapter 38 of *Agenda 21*, and was established in 1992. It was seen in many quarters as an opportunity to deal with the 'unfinished business' of Rio, to monitor progress with implementation of *Agenda 21* in member countries, and to oversee (and promote) the other recommendations of the conference. The CSD Bureau has representatives from fifty-three countries, elected from geographical regions and serving for three years each. The CSD has held a series of formal sessions, from 1994 onwards (Bigg 1995), culminating in the Special Session of the UN General Assembly, the so-called 'Earth Summit II' in June 1997, which extended the CSD's work for a further five years (Jordan and Voisey 1998; see discussion below). However, the powers of the CSD are limited; it can report to the General Assembly of the UN only through ECOSOC, and cannot peer-review national *Agenda 21* statements. Overall the CSD has been a disappointment.

One piece of unfinished business at Rio was a Convention on Desertification. This was proposed as a part of the Rio process, but negotiation fell behind in the run-up to the conference. In the event, a formal commitment was made at Rio to negotiate a convention on desertification after the conference was over. This was duly done, and the Convention to Combat Desertification was open for signature by June 1994, coming into force in December 1996. It addresses implementation through four regional annexes that allow for different approaches in different areas (Africa, Asia, Latin America and the Caribbean, and the northern Mediterranean). This convention will be discussed further in Chapter 7 (see also *www.unccd.ch*).

Partly to counter disenchantment at Rio's limited achievements, Maurice Strong argued that UNCED should be seen not as the end of international discussion about the challenge of sustainable development, but as the start of a process of adjustment by national governments. This 'Rio process' was furthered in a succession of follow-up conferences since 1992, including the Conference on Sustainability in Small Island States (1993), the International Conference on Population and Development in Cairo (1994), the World Summit on Social Development in Copenhagen (1995), the World Conference on Women in Beijing (1995), and the World Food Summit in Rome (1996) (Murphy and Bendell 1997, Fresco 1997, Jordan and Voisey 1998). Each of these can be regarded as forming part of a connected process of international engagement with different aspects of the agenda debated at Rio (*www.earthsummit2002.org*).

The General Assembly also resolved in 1992 to hold a special session on sustainable development to review *Agenda 21*. This meeting, often referred to as 'Earth Summit II' (or, less memorably, 'UNGASS'), was held in the summer of 1997 in New York. It attracted even more negative coverage than the Rio Conference itself, from the media, NGOs and from disappointed politicians from the South and from countries in the North such as the UK, which had hoped to use it to launch new initiatives on forests, oceans and freshwater (Jordan and Voisey 1998). The problem was the lack of fulfilment on the 'deal' that emerged at Rio, whereby the South agreed to environmental commitments sought by the North in return for assurances about 'new and additional' financial resources and technical assistance. Aid disbursements in the 1990s mostly fell rather than rising, and with the exception of those from Scandinavian countries were far below the headline figure of 0.7 per cent of donor-country GDP. As noted above, financial flows were far less than the sums calculated as necessary at Rio to implement *Agenda 21*.

Jordan and Voisey (1998) argue that the trust underpinning the Rio 'deal' evaporated in the years following the conference, and although documents were prepared for the 'Earth Summit II' at the preceding CSD meeting, agreement proved possible only on a short six-paragraph 'statement of commitment' and a 'Programme of Action for the Further Implementation of Agenda 21'. Debate was fierce about climate change (the meeting preceded Kyoto), forests (with pressure for the UK, Canada and the EU to establish an international negotiating committee for a forest convention, and Brazil, Malaysia, the USA and many G77 countries opposing the notion on the grounds that forests are a matter for national jurisdiction) and trade (Jordan and Voisey 1998). One interesting development at Earth Summit II was that representatives of civil society participated alongside states for the first time at a UN General Assembly, reflecting the emphasis in *Agenda 21* on the need to strengthen the role of NGOs.

The UN asked its various organisations (including UNEP, UNDP, UNCTAD and UNSO) for specific proposals as to how they would implement the provisions of *Agenda 21* (N. Robinson 1993). An Inter-Agency Committee on Sustainable Development was established to draw together representatives of the nine main UN organisations. In 1993 the UN created the Department for Policy Coordination and Sustainable Development (DPCSD, based in New York) to support the CSD and coordinate UN response to the various conferences that followed Rio (Flanders 1997). However, many observers feel that there has been a lack of strategic leadership within the UN system in the aftermath of Rio, and in particular that UNEP failed to define an effective leadership role. Those commentators, remembering their hopes for a brave new greener world in 1992, see in the many tendrils that have grown from Rio little evidence of change; rather, business as usual. Others, perhaps with more appetite for the brushfires of bureaucratic battles and committee rooms, see evidence of real hope for change. Murphy and Bendell see sustainable development as 'a new organising principle', an 'emerging, positive myth which has the potential to bring together diverse and often competing causes' (1997, p. 35). If so, it may indeed, as they hope, be a catalyst for the formulation and implementation of creative responses to the challenges flung out by Rio.

A meeting to assess progress with the implementation of the Rio Conference ('Rio Plus 10') is proposed for 2002. This meeting, named 'Earth Summit 2002', is charged with achieving what the 1997 meeting (Earth Summit II) failed to do: to 'rekindle the fire' of the Rio accord (*www.earthsummit2002. org*). That will perhaps create the conditions for some substantive move beyond the grindingly slow policy evolution of the 'Rio process' through the 1990s. Since Rio, 'sustainable development' has been freighted with a number of different agendas, and has become one of the most hard-used workhorses of the international system. The documents produced at Rio, and the debates behind them, record the evolving debate between developers and environmentalists, between those who wish to exploit natural resources and those who wish to conserve them. They also highlight the increasingly important debate about the environment between the poor and the rich countries. Rio continued the war of words begun at Stockholm two decades before, and carried forwards through the Brundtland Report and at numerous international meetings. For better or worse, *Agenda 21* and the mountain of other printed words generated by the UNCED hothouse defined the practical contours of sustainable development for the rest of the decade. They developed previous thinking about sustainability, the 'mainstream' of sustainable development that will be the subject of Chapter 5.

Summary

- The United Nations Conference on Environment and Development in 1992 (UNCED, or the Rio Conference) was preceded by protracted international negotiation at meetings of the Preparatory Commission (PrepCom). One hundred and seventy-two governments were represented at Rio, which fell in a US election year. There was a large unofficial NGO contingent.
- A series of documents were agreed at Rio, carrying forward the ideas of the mainstream of sustainable development from the *World Conservation Strategy, Our Common Future* and *Caring for the Earth*.
- The Rio Declaration was a bland and somewhat rambling consensus document, consisting of twenty-seven 'principles' for sustainable development.
- *Agenda 21* was the main output of Rio, a bloated document over 600 pages long, a compendium of environmental thinking and a balancing act between different interests. *Agenda 21* was subject to intense negotiation between national teams of diplomats. It provides the basis for many kinds of action in pursuit of 'sustainability', but it makes none of them mandatory.
- An agreed set of Forest Principles fell far short of proposals for a global forest convention. Agreement was not reached on deforestation between rainforest countries (determined to derive maximum economic benefit from forestry) and industrialised countries driven by preservationist domestic environmental lobbies (but often practising unsustainable forestry at home, and with a long history of forest conversion to other land uses).
- A Convention on Biodiversity was signed at Rio, to conserve biological diversity and promote the sustainable use of species and ecosystems and the equitable sharing of the benefits of genetic resources. It came into force in 1993.

- The Framework Convention on Climate Change agreed at Rio was the fruit of growing recognition of the problem of human-induced climate change through the 1980s, and particularly the scientific consensus achieved in the first report of the IPCC in 1990. The convention came into force in 1994, but it laid no binding commitments to reduce greenhouse gas emissions on individual countries. This continued to be debated and negotiated at meetings of the Conference of the Parties, eventually being agreed at Kyoto in 1997.
- The Rio Conference established mainstream sustainable development thinking, but it failed of itself to achieve binding or timetable commitment to systematic change in national or international policy. Debate about what sustainable development means, and what action should be taken to achieve it, has continued, through the Commission for Sustainable Development, the work of international organisations, and the COPs of the treaties.

Further reading

Chatterjee, P. and Finger, M. (1994) *The Earth Brokers: power, politics and world development*, Routledge, London.

Holdgate, M. (1996) *From Care to Action: making a sustainable world*, Earthscan, London.

Middleton, N. (1999) *The Global Casino: an introduction to environmental issues*, Arnold, London (2nd edition).

Redclift, M. (1996) *Wasted: counting the costs of global consumption*, Earthscan, London.

Robinson, N. (ed.) (1993) *Agenda 21: earth's action plan*, Oceana Publications, New York (IUCN Environmental Policy and Law Paper 27).

UNEP (2000) *Global Environment Outlook 2000*, Earthscan, London, for the United Nations Environment Programme.

United Nations (1993) *The Global Partnership for Environment and Development: a guide to Agenda 21*, post-Rio edition, United Nations, New York.

Web sources

<*http://www.un.org/esa/sustdev/*>   United Nations home page on sustainable development (including information on *Agenda 21* and the Commission on Sustainable Development, and country profiles prepared for Earth Summit +5).

<*http://www.biodiv.org/*>   The website of the Convention on Biological Diversity (the full text is on <*http://www.unep.ch/bio/conv-e.html*>).

<*http://www.igc.apc.org/habitat/agenda21/*>   The full text of *Agenda 21* and other Rio agreements; also links to related conferences and conventions, and to the meetings of the Commission on Sustainable Development.

<*http://www.un.org/esa/sustdev/csd.htm*>   The United Nations Commission on Sustainable Development

<*http://www.unfccc.de/*>   The website of the Framework Convention on Climate Change. The website of the Third Conference of the Parties to the Framework Convention on Climate Change at Kyoto in 1997 (including full text of Kyoto Protocol) is <*http://www.cop3.de/*>. Coverage of the COP6 meeting in the Hague,

November 2000 is provided on *<http://www.climatechange2000.org>* and *<http://cop6. unfccc.int/>*.

<http://www.earthsummit2002.org> The United Nations Environment and Development Forum website; provides background to the Earth Summit 2002 meeting.

<http://www.climatenetwork.org/> The Climate Action Network (CAN), a cooperative network of over 280 non-governmental organisations working on climate change; organised through a series of area networks including networks in Africa, South Asia, South-East Asia and Latin America.

5 Mainstream sustainable development

Can we envisage a more ecologically benign modernity, or is modernity ecologically irredeemable?

(J.S. Dryzek 1995, p. 231)

Inside the mainstream: global reformism

Chapter 4 argued that the Rio documents are part of a continuum of thinking, a mainstream in the diversity of ideas about sustainable development. The definition of that 'mainstream' is obviously in a sense artificial, for there exists a great diversity of thought about environment and development. This diversity (and particularly the radicalism it represents) will be explored in Chapter 6. Of interest in this chapter is the broad consensus of the sustainable development mainstream itself. This is not radical, but rather reformist. It seeks to refocus existing development initiatives and policy action rather than transform their principles or practice. Nicholas Low and Brendan Gleeson comment:

> Joining two positive-sounding words seemed to resolve at a stroke the conflict between an economy based on everlasting growth and a planetary environment of permanent high quality. These goods, it was hoped, could be reconciled if only the economy could be organised around production activities which did not harm the environment.
>
> (Low and Gleeson 1998, p. 12)

Mainstream sustainable development (MSD) shares the dominant industrialist and modernist ideology of the modern world-system, what Aseniero (1985) calls 'developmentalism'. This both underpins ideas about the proper direction of economic change in industrialised countries, and (because 'development' is seen to be a path along which all economies travel, albeit at different speeds) acts as a model for 'developing' countries. The processes of 'modernisation, economic growth and nation-state building' that created the modern world-system have long been held up as the goals for unindustrialised countries (ibid.). Behind the strategies for economic change lies the rich ideological web of the modern world: 'a common corporate industrial culture based on the values of competitive individualism, rationality, growth, efficiency, specialisation, central-isation and big scale' (Friberg and Hettne 1985, p. 231). Mainstream sustainable development does not challenge the dominant capitalist industrialising model, but it does demand debate about methods and priorities. Thus *Our Common Future* focuses on better planning techniques, more careful use of state capital and more careful use of economic appraisal to reduce development that causes ecological disruption. *Agenda 21* addresses the question of retooling the wealth-producing industrial plant of the world economy and changing the priorities of its management team. It does not suggest that its fundamental business, its methods or its products need to be radically re-imagined.

Mainstream sustainable development thinking reflects a very particular ideology that emerged at Rio about global environmental change (sometimes written as GEC; Brown *et al.* 1993). In this view, the chief issue of sustain-able development is seen to be the global environment, and particularly problems of biodiversity depletion and climate change, rather than global poverty or North–South inequality (issues which the Rio Conference failed to discuss effectively; Redclift 1997). The latter are more messy, much more polit-ical, and less amenable to technical solution (Cohen *et al.* 1998). At Rio, sustainable development was interpreted as a concern about possible future economic and social implications of changes in global climate and ecology, reflecting primarily the agenda of industrialised Northern countries. Mainstream sustainable development thus privileges issues of inter-generational equity (i.e. the impacts of environmental change on future generations) over those of intra-generational equity (and the problem of poverty in the South and environmental degradation now). Both these alternatives, of course, represent a shift away from the 1970s, when ideas about sustainable development were dominated by the neo-Malthusian 'limits to growth' debate (e.g. Meadows *et al.* 1972), and were duly reflected in the first 'mainstream' document, the *World Conser-vation Strategy* (see Chapter 3).

The move to centre mainstream sustainable development thinking on the issue of GEC emerging out of Rio positioned it firmly within the bounds of conventional political and economic thinking. Whereas the notion of 'limits to growth' was uncomfortable, challenging conventional capitalist growth, GEC was perceived as a problem 'amenable to mitigation' (Brown *et al.* 1993, p. 573). Furthermore, it could be tackled by the conventional weapons of tech-nological innovation and the market.

Although the issues of climatic change and biodiversity that dominated the Rio Conference, and which now underpin mainstream sustainable development, are vitally important to certain countries (especially those vulnerable to sea level rise), they are not the principal environmental problems faced by most countries of the South (Redclift 1997). However, to industrialised countries, the newly defined problem of GEC did seem to present a serious threat to wealth and welfare, and mainstream sustainable development was developed as a potential solution. This solution was conceived of in such a way that it did not demand radical shifts in corporate or national wealth or power, or shifts in social or industrial organisation. The path of sustainable development could instead be pursued by governments using conventional tools, of environmental and market regulation. Furthermore, it offered a dazzling market opportunity in clean technologies. The OECD report that the global market for environmental goods and services in the mid-1990s was US$200–300 billion (OECD 1996). Mainstream sustainable development was environmentalism reinvented by the Reagan and Thatcher decades: free-market environmentalism (Low and Gleeson 1998).

Mainstream sustainable development became progressively more strongly defined through the 1990s. Following Low and Gleeson (1998), this chapter identifies three important groupings of thought within it: *market environmentalism*, *ecological modernisation* and *environmental populism*. It then explores the potential of the field of environmental economics to contribute to the implementation of a market-led strategy of sustainable development, and discusses some of the policy mechanisms for delivering mainstream sustainable development.

Market environmentalism

The sets of ideas that Low and Gleeson label 'market environmentalism' are utilitarian, individualistic and anthropocentric. These ideas present the market as the most important mechanism for mediating between people, and regulating their interaction with the environment. They involve a political agenda of 'rolling back' the state, deregulating markets, and extending market relations into society and its relations with the environment. This is a familiar recipe to anyone who has lived through the 1980s and 1990s, either in industrialised counties or those enduring the enforced rigours of the economic doctrines of the International Monetary Fund and the World Bank, and their obsession with 'structural adjustment' (e.g. Hanlon 1996). Welfare-utilitarian economists such as Wilfred Beckerman (1974, 1994) argue that the market is the only efficient way to regulate human use of the environment. As Low and Gleeson put this view,

> The 'green economy' will be a capitalist economy. And just as the economy theoretically reaches a level of equilibrium in which social needs are met, so the green economy will theoretically reach a level of 'sustainable development' in which the capacity of the planet to provide raw materials and absorb wastes is not overstretched.
>
> (Low and Gleeson 1998, p. 81)

It is conventionally argued that market prices rise as resources become scarce, and logically people will innovate to find cheaper sources or ways of using resources more efficiently. Beckerman writes, 'insofar as any natural resource does become scarce in a relevant sense its relative price will rise and this will set up a chain of market responses which will tend to discourage its use and the development of substitutes' (1995a, p. 178). Attempts by the state to make rules about resource use are therefore arguably inherently inefficient, and bound to fail both to allow economic welfare to be maximised and to maintain resources at desirable levels. In this analysis, environmental problems are seen to follow from the misallocation of resources. Open-access resources are liable to over-exploitation, while private resources are managed efficiently and conserved. Market-oriented environmental policy in the Third World has therefore focused on the privatisation of resources, for example the privatisation of communally held land and other resources (Bassett 1993).

Market environmentalism argues that the further market exchange penetrates into the environment, the greater the efficiency of environmental management. Policy proposals therefore pursue the commodification of nature, and the setting of prices for environmental 'goods' and 'services'. The optimistic view of conventional economics about future growth is shown in Figure 5.1. This draws a distinction between natural capital (which is created by bio-geophysical processes and not human action) such as clean air, unpolluted water or fertile soil, and human-made capital (created by human skill or agency) such as factories, roads or conventionally measured wealth. This approach will be discussed further below, in the section headed 'Natural capital and sustainability' (pp. 117–23). The economic optimist vision suggests that natural capital is not ultimately necessary to sustain human-made capital, although generated wealth could be used to fund preservation of some elements of natural capital as an amenity (Folke *et al.* 1994).

Market environmentalism is the result of a growing engagement by economists in the sustainable development mainstream. Partly because of the involvement of the World Bank in the development of practical policy responses to

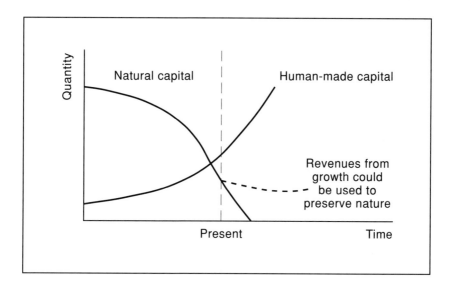

Figure 5.1 The economic optimist vision of the future (after Folke *et al.* 1994).

the Rio Conference, it is the 'dismal science' of economics that has led the way in the greening of governments and development bureaucracies, in some ways a remarkable case of the poacher turned gamekeeper. In particular, there has been reform of the way economies are understood and measured (for example in national accounts), and reform of the economic methods that support development decision-making (for example cost–benefit analysis). The discipline of environmental economics has undergone rapid theoretical development and expansion through the 1980s in universities and government agencies.

Market environmentalism is predicated on continued capitalist growth. Low and Gleeson suggest that it 'defends the status quo the globalising institution of the market, and resists the notion that any fundamental change is needed' (1998, p. 163). It sits therefore in stark opposition to the pessimistic ideas of 'zero growth' or 'limits to growth' that were prominent in the 1970s (Mishan 1969, 1977, Daly 1973, 1977, Cole 1978, see Chapter 2). The environmental pessimist vision of the future is shown diagrammatically in Figure 5.2 (Folke *et al.* 1994). This predicts that over-expansion of the human economy will eventually cause collapse of ecological systems that provide life support and thence in turn the collapse of the economy itself. Some subsequent recovery may be possible (Figure 5.2).

Attacks on the possibility of sustainable economic growth are a familiar element in environmentalist critiques of economics, and radical in the literal sense that they demand fundamental change in economic systems. The 1970s 'limits to growth' environmentalism occasioned considerable hostility from economists (e.g. Beckerman 1974, 1994, Rostow 1978, Kahn 1979, Simon

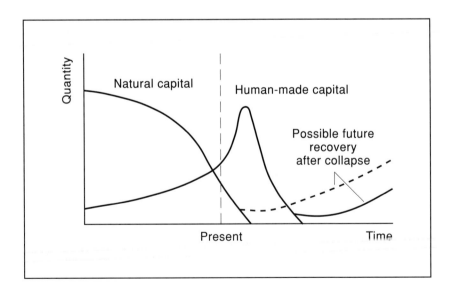

Figure 5.2 The ecological pessimist view of the future (after Folke *et al.* 1994).

1981). Beckerman, for example, stated the orthodox position: 'a failure to maintain economic growth means continued poverty, deprivation, disease, squalor, degradation and slavery to soul-destroying toil for countless millions of the world's population' (1974, p. 9).

In some forms, 'zero-growth' ideas are naïve, and are easily caricatured and dismissed. Other formulations have been better argued, notably perhaps Daly's idea of 'steady-state economics' (Daly 1977). Daly's starting point was his 'impossibility theorem': that a high mass-consumption economy on the US style is impossible (at least for anything other than a short period) in a world of 4 billion people. He pointed out that the rich countries of the world sought to place the burden of scarcity on the poor countries through population control, while the poor wanted it borne by the rich through limiting consumption. Both wanted to pass as much as possible of the burden onto the future. The simple (but as he points out, hopelessly utopian) solution was for the rich to cut consumption, the poor to cut population and raise consumption to the same new reduced 'rich' level, and both to move towards a steady state at a common level of capital stock per person and stabilised or reduced population. This idea fails because of the lack of international goodwill, and internal class conflicts within Third World societies. Thus in the case of Brazil, he said:

> Now that the Brazilians have learned to beat us at our own game of growthmanship, it seems rather ungracious to declare that game obsolete. We can sympathise with Brazilian disbelief and suspicion regarding the motives of the neo-Malthusians. But the dialectic of change has no rule against irony.
>
> (Daly 1977, p. 166)

However, it is market environmentalism and not the radical economic restructuring demanded by zero-growth environmentalism that dominated mainstream sustainable development. In his combative book *Small is Stupid*, for example, Wilfred Beckerman concluded that 'in developing countries there is no conflict between growth and the "quality of life"' (1995b, p. 35). The task for governments and economists is unchanged by the sustainability debate: to seek the economic growth rate that maximises welfare over time – that is, the optimal growth rate. In the *World Development Report* for 1992, the World Bank argues that 'the world is not running out of marketed non-renewable energy and raw materials, but the unmarketed side effects associated with their extraction and consumption have become serious concerns' (World Bank 1992, p. 37). Market-based approaches to environmental problems such as tradable pollution permits have become standard in industrialised countries, and in international environmental regimes (Pearce 1995). They have passionate promoters. Martyn Lewis, for example, writes that 'the economic approach to pollution control deserves the concerted attention of liberal and moderate environmentalists' (1992, p. 180). The development of economic tools with which to understand environmental aspects of development will be discussed later in this chapter.

In practice, development and the environment have a 'two-way relationship' (World Bank 1992, p. 1). Economic growth can have both good and bad effects on the environment. It creates the possibility of 'win–win' opportunities through poverty reduction and improved environmental stewardship in both high- and low-income countries (Figure 5.3). However, growth also brings risks of serious side-effects if markets fail to capture environmental values and deal with externalities, and if the public regime of regulation (local, national and international) is inadequate.

Our Common Future identified sustainable development in terms of economic growth that could yield a broader distribution of economic goods and avoid environmental damage. Its successors in the Rio documents also start from a presumption of the benign possibilities of global capitalism: 'Many of the principles in the Rio agreements are essentially undermined by the structure of the global economy, the drive for economic growth, which is recognised in *Agenda 21* as the prime contributor to unsustainability' (Brown 1997, p. 388). Chatterjee and Finger (1994) argue that Rio provided a unique platform for international business to present its view of the environment development problem, and indeed to present themselves as a part of the solution. The UNCED process favoured the strongest lobbyist, and of course the best-resourced, most skilful lobbyists were global businesses, and moreover they had the clearest message. Maurice Strong specifically involved business in UNCED through the Business Council for Sustainable Development, and a fifth of the costs of the conference secretariat and parts of the programme were funded by corporations (ibid.).

Mainstream sustainable development is predicated on a free market, on the continuation of growth and on the application of technology. Environmental quality is to be achieved through self-regulation, through a revolution in corporate thinking that 'greens' industry from within, combined with a hard-headed (and increasingly global) pursuit of the 'green' consumer dollar. According to market environmentalism, the world can literally grow out of global environment and development problems, and consumption can be the engine through which sustainable environments and livelihoods are to be achieved. Regulation, internationally or by states, is heavy-handed; it is the corporation and the consumer that can deliver on the mainstream agenda of sustainable development.

Mainstream sustainable development is firmly anchored within the existing economic paradigms of the industrialised North. It is reformist, constructed on a platform of continued capitalist growth, 'green growth'. It focuses on the potential for fairly minor reforms of the existing economic system involving further extension of the market to organise social interaction and human engagement with nature. Capitalist markets and regulation by states or by international institutions must be allowed to address the task of cleaning up and 'greening' the planet (Low and Gleeson 1998, p. 81). Martyn Lewis concludes that 'regardless of extremist fantasies, we can expect that once capitalist energies begin to be harnessed to environmental protection, a virtuous spiral will begin to develop' (1992, p. 183). Mainstream sustainable development leaves

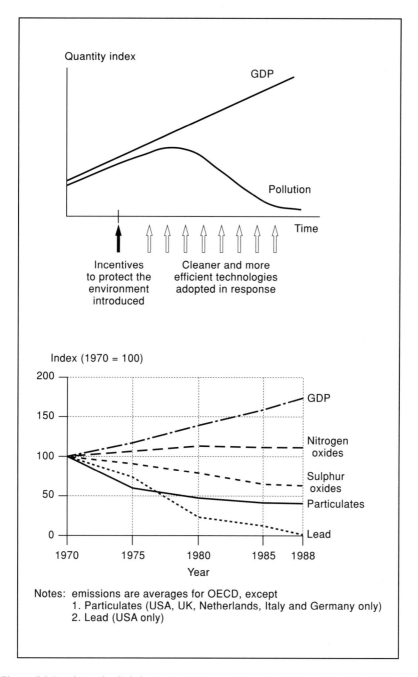

Figure 5.3 Breaking the link between GDP growth and pollution (after World Bank 1992).

little space for alternative economic visions. To Lewis, opposition to a market approach to environment and development suggests that some 'ecoradicals' may have 'more hostility toward capitalism than concern for nature' (ibid., p. 182). The nature and strength of such radicalism are assessed in Chapter 6.

Ecological modernisation

Market environmentalism is the dominant force in mainstream sustainable development. However, it is not blindly cornucopian. Beckerman notes that 'economists have been well aware of the fact that, left to itself, the environment will not be managed in a socially optimal manner. There are too many market imperfections' (1994, p. 205). Furthermore, mainstream sustainable development does not propose an unregulated dependence on the market. The state, and its capacity to regulate the market, is of critical importance.

As will have become clear, market environmentalism does not admit to the existence of a particularly intractable environmental crisis, and beyond proposing to make markets work better, it does not envisage major restructuring of economic or political structures, or the relations between people and non-human nature. It is, quite literally, a 'business as usual' agenda. However, changes have taken place in public environmental policy, both nationally (particularly within industrialised countries) and to a lesser extent internationally, and also in the operational practices of global corporations. It has become commonly argued, especially in Northern Europe, that these comprise a new and systematic phenomenon, the first fruits of a set of technical changes in systems of production and exchange required to avert environmental disaster. It is also suggested that capitalism can and should be steered to make these changes by an enabling state (Low and Gleeson 1998). This view has been labelled 'ecological modernisation' (Spaargaren and Mol 1992, Mol 1996, Christoff 1996), and it comprises the second main strand within mainstream sustainable development.

Low and Gleeson describe ecological modernisation as 'a reformist perspective which, while recognising the ecological dangers posed by unfettered markets, believes in the self-corrective potential of capitalist modernisation' (1998, p. 165). Here is the vision of Brundtland, with economic growth in a capitalist economy working within the constraints of ecological sustainability. Hajer (1995, 1996) describes ecological modernisation as a 'regulation-oriented programme', typified by the German government's response to the impacts of acidification and forest die-back through acidification (*waldsterben*) in the 1980s, or the Japanese response to air pollution in the 1970s (Hajer 1996). He argues that ecological modernisation became the dominant approach from the mid-1980s following Brundtland and Rio, and the acceptance of *Agenda 21*.

Hajer argues that ecological modernisation represents a break with radical environmental critiques of political economy. At its core lies a conviction that the ecological crisis could be overcome by 'technical and procedural innovation' (Hajer 1996, p. 249). In 1981 O'Riordan suggested that environmentalism could be thought of as containing two dimensions, one ecocentrist (romantic, transcendental and concerned for example with the rights of other species) and one technocentrist (rationalist and technocratic, and concerned with the better ordering and regulation of the environment). In these terms, ecological modernisation is technocentrist in its pursuit of rational, technical solutions to environmental problems and more efficient institutions for environmental management and control. It suggests that 'the only way out of the environmental crisis is by going further into industrialisation, toward hyper- or superindustrialisation' (Spaargaren and Mol 1992, p. 336).

The core of technocentrist thinking in sustainable development is a utilitarian view of science, and the application of science to 'solve' human problems. This view of science has been important within mainstream sustainable development since the *World Conservation Strategy*'s agenda, which was not only expressed but also conceived in terms of ecology. Hajer suggests that ecological modernisation needs 'a specific set of social, economic and scientific concepts that make environmental issues calculable and facilitate rational social choice' (1996, p. 252).

Ecological modernisation involves working towards improved and more 'rational' planning, management, regulation and utilisation of human use of the environment. Such approaches would include ideas about the role of the environment in development planning (e.g. the importance of such procedures as environmental impact assessment), and the ways in which economic development can be achieved without undue environmental costs. They draw on many of the principles of the utilitarian 'wise use' philosophies of conservation in the USA in the first decades of the twentieth century (Hays 1959), most notably perhaps in the work of the first US Conservator of Forests, Gifford Pinchot, although many of these ideas are found earlier in the work of George Perkins Marsh. In *Conservation and the Gospel of Efficiency* (1959), Samuel Hays says of the US conservation movement:

> its essence was rational planning to promote efficient development and use of natural resources. . . . Conservationists envisaged, even if they did not realise their aims, a political system guided by the ideal of efficiency and dominated by the technicians who could best determine how to achieve it.
> (Hays 1959, p. 2)

Rationality is a key concept in ecological modernisation. The ordering of non-human nature to suit human ends was a key element in the twentieth-century project of rationalisation (Murphy 1994). The idea of sustainability is a major means by which society has sought to rationalise its interaction with nature: Murphy defines instrumental rationality as 'conscious reasoning in which action is viewed as a means to achieve particular ends and is oriented to anticipated and calculable consequences' (ibid., p. 30). The environmental degradation associated with industrialisation is a 'design fault' (Spaargaren and Mol 1992, p. 329), a failure of rational environmental management: as Murphy points out, 'by not taking ecological factors into account, the development of rationalisation has been inadequate in terms of durability' (Murphy 1994, p. 43). Ecological modernisation might be referred to as 'the intensification of ecological rationalisation under capitalism'.

Spaargaren and Mol argue that ecological modernisation 'indicates the possibility of overcoming the environmental crisis without leaving the path of modernisation' (1992, p. 334). Capitalist economic growth may be reconciled to the requirements of ecological sustainability by a series of strategies. These involve the injection of improved techniques and technologies into production, the refinement and regulation of markets to tune them to ecological constraints, and the 'greening' of corporate ethics and objectives (Low and Gleeson 1998). The new approaches demand a move from reactive 'end-of-pipe' solutions to environmental problems (which tend to be late, unpopular with business and ineffective) to integrated and predictive and holistic frameworks for environmental regulation and management. These frameworks demand new partnerships between state and private enterprise, including market-based incentives and cooperative, voluntary and self-regulatory approaches by industry itself (Christoff 1996). Ecological modernisation also places a premium on new 'partnerships' between state and citizen, and citizen and business: civil society, a green economy and slim efficient governance combining to create a sustainable economy and environment.

The shift in governmental and business practice implied by ecological modernisation of course demands a shift in values, a wider 'greening' of society, and this has undoubtedly taken place in certain industrialised countries, particularly perhaps Germany and the Netherlands. So-called 'green capitalism' and 'green consumerism' became important features of retail marketing in Europe in the 1980s, responding to (and attempting to capitalise on) 'green' political ideas (Capra and Spretnak 1984).

Ecological modernisation is built on the principle that institutions can change, and that actors within them can learn. Hajer attributes the new consensus on ecological modernisation to 'a process of maturation of the environmental movement' (1996, p. 251). In the 1980s and 1990s, environmental groups in countries such as the UK grew and restructured (Rawcliffe 1998). Radical issues became normalised and incorporated into the ideology of dominant political and economic institutions. Some radical pressure groups rethought their strategies, notably perhaps in Greenpeace's focus on 'solutions' campaigning (Rose 1993).

Clearly there is more to ecological modernisation than simply 'greenwash' and politicians' puff. But how much more? Hajer (1996) points to the risks inherent in the rationalisation of ecology through the amendment of existing bureaucratic structures and the invention of new ones. In operationalising sustainable development through ecological modernisation, policy-makers are left with great freedom: 'they can either make a few aesthetic alterations but basically continue with business as usual, or they can use sustainable development as a crowbar to break with previous commitments' (ibid., p. 262). Ecological modernisation has real potential, but offers no magic solution to the self-interest of corporate power and sclerotic government. Christoff (1996) points out that ecological modernisation is 'a strategy of political accommodation of the radical environmentalist critique of the 1970s' as much as a technical response to environmental degradation. It is therefore important to governments and corporations seeking to manage ecological dissent.

There are also important questions about the significance of ecological modernisation outside the comfortable circle of industrialised countries. Without doubt, the growing post-Rio system of international environmental regulation reflects the same trends as those labelled ecological modernisation in Europe. Market reform and efficient enabling governance have become key elements in the post-Rio 'green' development aid agenda. However, it is not clear how useful ecological modernisation is outside the industrial core where it was imagined. Ecological modernisation is predicated on the notion of the nation state, and therefore offers little insight into the significance of globalisation, international trade or the limited power of non-industrialised countries to enforce environmental regulations (Christoff 1996). It is also not clear how far weaknesses in governance, and in the institutions of civil society, in Third World countries may prevent the full flowering of ecological modernisation as envisaged by theorists in Europe (ibid.).

More fundamentally, ecological modernisation is offered as a strategy for making minor adjustments to the conventional growth model. In that sense it is best understood as an extension of (or even a version of) market environmentalism. Developmentalism remains hegemonic as the inevitable and inescapable process of change and improvement. Ecological modernisation is 'the next necessary or even triumphant stage of an evolutionary process of industrial transformation' (Christoff 1996, p. 487). Countries of the global periphery, lacking industrial might and technology, must first transform themselves into facsimiles of the industrialised North. At best, this offers the unhelpful prospect of a perhaps prolonged period of environmental degradation associated with aggressive industrialisation (powered by global capital essentially unregulated by any national jurisdiction) before sufficient affluence is generated to make the 'green' turn to ecological modernisation.

Of course, if critiques of conventional development paradigm are valid, the Third World may never escape its poverty, and may add to it the appalling risks of poorly regulated high-technology production. The issues of sustainability and 'Risk Society' will be discussed later in this book, in Chapter 10. Meanwhile, pursuit of sustainability through ecological modernisation in wealthy industrialised countries in the global core could have the effect of displacing excessive resource depletion, production of wastes and pollution to the countries of the poorer periphery (Low and Gleeson 1998).

Environmental populism

Mainstream sustainable development is predicated on a capitalist economy, is technocentrist in approach and is built on environmental modernisation and

environmental economics in its methodologies. It also involves a very particular political agenda, in environmental populism. It emphasises the capacity for
citizens to take hold of their circumstances and change them for the better.
It proposes strategies for change that emphasise that process of self-generated
change, and that promote capacity for 'participation' by ordinary people (or
'local people') in decision-making. It also emphasises the priority that should
be given to developmental change that 'empowers' people, and promotes
sustainable development explicitly as an approach that achieves this.

Mainstream ideas about sustainable development were strongly influenced
by 'neo-populist' ideas that came to prominence in development thinking in
the 1970s (Byres 1979, Kitching 1982). Definitions of 'ecodevelopment' by
Sachs (1979, 1980) and Glaeser (1984a, b, 1987) had much in common with
the moral arguments in the WCS not only about *what* sustainable development should comprise, but about *how* it should be undertaken. For Sachs,
ecodevelopment was 'an approach to development aimed at harmonising social
and economic objectives with ecologically sound management, in a spirit of
solidarity with future generations' (1979, p. 113). Glaeser and Vyasulu defined
ecodevelopment as 'an alternative vision of a new form of society, squarely
pitted against the dynamic of overdevelopment and underdevelopment' (1984,
p. 25). Ecodevelopment was based on 'a new symbiosis of man and earth' and
'self reliance', and it focuses clearly on the satisfaction of basic needs (Sachs
1979, p. 113).

UNEP emphasised the decentralisation of bureaucracy, a disaggregation of
development focus ('the achievement of sustainable development at local level'),
self-reliance and self-sufficiency, the priority given to meeting basic human
needs, public participation and equitable distribution (*Ambio* 1979). The goal
of ecodevelopment was 'to pursue economic development that relies for the
most part on indigenous human and natural resources and that strives to satisfy
the needs of the population, most of all the basic needs of the poor' (Glaeser
1984a, p. 11). Singh, writing from the perspective of a Malaysian non-governmental conservation organisation, describes it as 'development of the people,
for the people, by the people' (1980, p. 1350).

One reason for the centrality of populist ideas in sustainable development
is their roots in critiques of the inhumane, monolithic and bureaucratic nature
of the development process. The strongest insights of the WCS are borrowed
from critiques of development that emphasise basic human needs, and argue
that those affected by development should participate in decisions that affect
them. These 'populist' critiques stress the significance and importance of indigenous cultures (e.g. McNeely and Pitt 1987), indigenous knowledge (Brokensha
et al. 1980, Chambers 1983, Richards 1985), the need for local participation
in development, and 'development from below' (Stöhr 1981). The WCS
suggested that

> Conservation is entirely compatible with the growing demand for 'people-
> centred' development, that achieves a wider distribution of benefits to

whole populations (better nutrition, health, education, family welfare, fuller employment, greater income security, protection from environmental degradation); that makes fuller use of people's labour, capabilities, motivations and creativity; and that is more sensitive to cultural heritage.

(IUCN 1980, para. 20.6)

These ideas were picked up and developed in the successors to the WCS (notably in *Our Common Future*; Brundtland 1987).

Neo-populism emerged in Russia and Eastern Europe before the First World War. It embraced a range of ideas which argue for 'a pattern of development based on small-scale individual enterprise both in industry and agriculture' (Kitching 1982, p. 19). Neo-populist thinking has been widely applied in the developing world. Bideleux (1987) rehearsed the arguments for 'communal village-based development', considering the work of (among others) Kropotkin, Chayanov and Gandhi, while Kitching discussed Tanzania (and the writings of Julius Nyerere), the concept of 'urban bias' (Lipton 1977), and the 'small is beautiful' thinking of Schumacher (1973). Kitching concludes that populism 'makes good social and moral criticism, and has often produced very effective political sloganeering, but on the whole makes rather flabby economic theory' (1982, p. 140). Despite this flabbiness (or possibly because of it), these ideas have proved attractive to those approaching concepts of sustainable development and ecodevelopment, and a number have been co-opted to strengthen the developmental credentials of thinking about sustainability.

First, there has been a focus on basic needs (O'Riordan 1988): 'Ecodevelopment refers to a process which is geared to the satisfaction of basic and essential human needs, starting with the needs of the poorest and neediest in society' (Glaeser and Vyasulu 1984, p. 25). This is a direct adoption of ideas from the wider debate about basic needs which emerged in the Second Development Decade of the 1970s (e.g. Stewart 1985). Critics of this approach point out its political weakness, resulting from the strength of vested interest and power which oppose wealth redistribution and decentralisation (Lee 1981). Its adoption by ecodevelopment thinkers is evidence of their failure to look beyond morality and the principles of rational planning to address the political economy of the development process.

Second, in mainstream thinking, sustainable development demands participation. Sachs called for 'participatory planning and grass-roots activation' (1979, p. 113). Participatory approaches to development became a prominent element in development thinking in the 1980s, and have since become standard practice (Midgley *et al.* 1986, Cernea 1991, Wright and Nelson 1995). Participatory development demands that 'people who are to be affected by changes which they have decided are desirable cooperate voluntarily in the process of implementing the changes by giving them direction and momentum' (Glaeser and Vyasulu 1984, p. 26). This is in itself a naïve suggestion, but it is typical of thinking about ecodevelopment and sustainable development which advocates non-hierarchical systems of organisation and government. Thus Galtung (1984) differentiated between 'alpha' social systems, which are hierachical, unlimited in size and tending towards uniformity, and beta systems, which are horizontal, limited in scale and inclined towards diversity. Ecodevelopment requires beta structures, but set within a matrix of a benign, flexible, communicating and restraining alpha system. Galtung's ideas owe little to social theory and a great deal to the diversity–stability debate in ecology in the 1960s (Margalef 1968). Nonetheless, they reflect the element of anarchism in environmentalism which provides one of the central themes of 'green' development thinking (e.g. Bookchin 1979, Roszak 1979, Pepper 1993), as well as notions of participation in planning and 'development from below' (Stöhr 1981).

Third, sustainable development picks up the ideas of appropriate and intermediate technology. This is hardly surprising, since *Small is Beautiful* (Schumacher 1973) was one of the definitive books of the 'environmental decade' of the 1970s as well as a successful critique of development. Schumacher wrote a first version of the key chapter of that book, 'Buddhist economics', in 1955 when he was economic adviser to Burma (McRobie 1981). In 1961 he spoke to an international seminar at Poona in India, and the following year wrote a report for the Indian Planning Commission, at the instigation of Nehru, developing the same line of argument. This was coolly received (ibid.), but his ideas on intermediate technology received wider publicity at a seminar in Cambridge in 1964 (Robinson 1971). In May 1965 the Intermediate Technology Development Group (ITDG) was formed, its first major work being a guide to the availability of hand and small-scale tools useful in rural development. In *Small is Beautiful* the environmental movement found both icon and scripture for its crusade to forge a new approach to development.

Participatory approaches assumed a central importance within the sustainable development mainstream in the 1990s. Ghai and Vivian argued that 'people's legitimate interest in the conservation of their resource base must be recognised and supported – not only because this is their basic right, but also because it is a pragmatic course to take in the interests of achieving sustainable development (1992b, p. 17). Incorporation of a participatory approach into mainstream thinking began with *Caring for the Earth* (IUCN 1991). One of the nine principles of sustainable development was to 'enable communities to care for their own environments' (no. 7). As Holdgate (1996) commented, 'the fact is that sustainable development demands partnerships that involve all sectors of the community, from the individual and the small action group through industry to local and central government' (p. 117). In *Caring for the Earth* it was suggested that action in support of 'sustainable communities' should include giving individuals and communities greater control over their own lives (and resources), enabling them to meet their needs in sustainable ways and enabling them to conserve their environment. Achieving this would demand security of resource tenure, exchange of skills and technologies, enhanced participation in conservation and development processes, more effective local government and better financial and technical support (IUCN 1991).

This emphasis on participatory approaches to developmental change (or to conservation), and the related notion of the empowerment of communities *vis-à-vis* the state, was continued in *Agenda 21*. The political principle that those at the receiving end of development initiatives had a right to be involved in the changes to which policy would subject them was combined with growing experience of the adverse environmental and social impacts of projects and programmes, and the recognition that development that did not involve participation could fail (see Chapters 8, 9 and 10). The very obvious failure of 'development from above' led to an often unthinking endorsement of its perceived antithesis, 'development from below'. This neat rhetorical inversion awakes a promise that often proved rather harder to realise in practice. The potential for achieving sustainable development 'from below' will be discussed further in Chapter 12.

This approach was both advocated by and exemplified by Third World NGOs. Locally focused grassroots NGOs established in the 1970s began to open up and organise and even federate internationally in search of influence over the wider institutional context for local change. It was these 'third-generation NGOs' that became engaged by the Rio process, and their agenda of participation and grassroots empowerment became an important element within the wider mainstream (Chatterjee and Finger 1994).

Natural capital and sustainability

As was mentioned on p. 106, a key element in mainstream sustainable development has been the application of environmental economics. Economists began to explore the concept of sustainable development in the 1980s (e.g. Goodland and Ledec 1984, Pearce *et al.* 1989, Turner 1988a). In the UK, for example, the possibility of applying economic calculations to the management of nature and to issues of sustainability spread rapidly in the wake of the work by David Pearce and his colleagues in *Blueprint for a Green Economy* (Pearce *et al.* 1989). The discipline of environmental economics has proved the decisive bridge between environmentalist concern about development and the world of policy and government decisions. It has served to value environmental resources more precisely, and to fit 'the costs and benefits of using such resources into the conventional calculus of economic decision-making' (Munasinghe 1993a, p. 1730). The range of environmental economics is now very wide, and its scope increasingly sophisticated (e.g. Costanza and Daly 1992, Costanza 1991, Barbier *et al.* 1994, Barbier 1998).

Economic approaches to sustainability build on long-established ideas of maximising flows of income while maintaining the stock of assets (or capital) from which they come (Munasinghe 1993a, b). What is new is the distinction drawn between natural and human-made capital (Berkes and Folke 1994). Economists have traditionally defined capital as things people have built that have value, such as roads or factories. This environmental economists define as 'human-made capital'. 'Natural capital' is created by bio-geophysical processes rather than human action, and represents the environment's ability to meet human needs, whether through providing raw materials (fish or timber) or what are called by the rather functionalist term 'services'. Such services would include the role of global bio-geochemical cycles in maintaining ecological conditions suitable for human life, or the more mundane way in which wetlands moderate floods or absorb pollutants (Barbier 1998).

Using these economic concepts, sustainability can be conceived of in various ways. The first, and most obvious, is to stipulate that constant stocks of both human-made and natural capital are maintained over time. Economic development has always implied increasing capital stocks, but the insights of environmental economics suggest the need to account for stocks of both human-made and natural capital separately. Thus development that increases stocks of human-made capital only by the depletion of natural capital of an equivalent value is not sustainable. This view of sustainability is commonly referred to as 'strong' sustainability (Beckerman 1994).

Techniques for the valuation of human-made capital are well established, although in practice even estimates of human-made capital can be problematic. However, one obvious problem with this whole economistic approach to the environment is that while a value can be placed on most forms of human-made capital fairly readily (for example the cost of constructing a road or factory, or the economic benefits that flow from it), it is much harder to place a value on natural capital (Holland and Roxbee Cox 1992). Where a market exists for privately owned facilities or products, valuation is fairly straightforward, and can build on the vast store of expertise in business practice and planning. Public benefits and costs are much harder to gauge; although techniques such as cost–benefit analysis have been extensively developed (Sugden and Williams 1978, Brent 1990), they are far from entirely satisfactory. Natural capital is often not privately owned; indeed, it is commonly shared by smaller or larger groups of people. Under the pressures of market forces and social change associated with development, locally recognised systems for allocating rights to benefits that flow from natural capital can become fiercely contested.

The development of techniques for the valuation of natural capital has proceeded apace (e.g. Turner *et al.* 1992, Tunstall and Coker 1992, Barbier *et al.* 1998), although it continues to present thorny conceptual and practical problems (e.g. Adams 1992). Most environmental goods are not subject to market relations, either because they are held in common (e.g. clean air), because they have only recently become scarce (e.g. clean groundwater, subject to slow and recent pollution penetration), because the structure of existing markets allows certain actors (usually established corporations and economic interests) to treat environmental costs as an externality, or because institutions for organising a market do not exist.

Ecosystem services are not fully captured in commercial markets and are not adequately quantified (or in some instances adequately understood). However, attempts to quantify ecosystem services are growing (Mooney and Ehrlich 1997). Economists recognise direct and indirect use values and non-use values of various kinds (e.g. Figure 5.4), and use a variety of techniques to capture

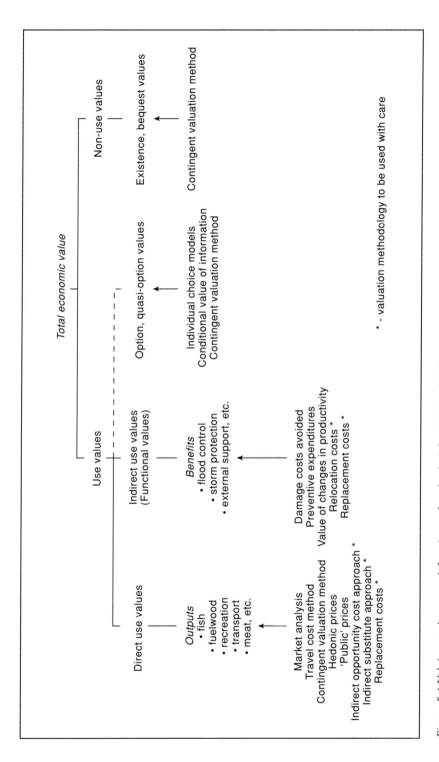

Figure 5.4 Valuing environmental functions of wetlands (after Barbier 1998).

these values (Gouldner and Kennedy 1997). Munasinghe (1993c), for example, used contingent valuation, travel cost and opportunity cost approaches to measure the costs and benefits of a new national park in Madagascar. There have even been attempts to calculate 'the value of everything' (Pimm 1997). Costanza *et al.* (1997) estimate the current economic value of seventeen ecosystem services for sixteen biomes, and on this basis calculate that the value of the entire biosphere is between US$16 and US$54 trillion (10^{12}), with an average of $33 trillion per year. On a yearly basis, the total value from terrestrial ecosystems is $577 per hectare per year, and from terrestrial ecosystems $804 per hectare per year, the latter providing about a third of total global flow of value per year. They regard these as minimum estimates. For comparison, global GNP is about $18 trillion per year. Thus in order for human-made capital to substitute for the natural capital of the biosphere, global GNP would have to grow by at least $33 trillion, but this would of course bring no increase in welfare (Table 5.1).

Many economists (and some environmentalists) regard these calculations as so speculative as to be useless, even dangerous. Be that as it may, they have certainly attracted widespread attention, and made clear that nature has *some* value that can be measured in monetary terms, and that this value is extremely large. For better or worse, the concept of 'natural capital' has changed the way people think about the biosphere, or about 'non-human nature'. Logically, as natural capital becomes scarcer (as ecosystems are transformed and subject

Table 5.1 Estimated annual global value of ecosystem services

	Average total global flow value ($ yr^{-1} × 10^9)
Gas regulation	1,341
Climate regulation	684
Disturbance regulation	1,779
Water regulation	1,115
Water supply	1,692
Erosion control	576
Soil formation	53
Nutrient cycling	17,075
Waste treatment	2,277
Pollination	117
Biological control	417
Habitat/refugia	124
Food production	1,386
Raw materials	721
Genetic resources	79
Recreation	815
Cultural	3,015
Total global flow value ($ yr^{-1} × 10^9)	33,268

Source: Costanza *et al.* (1997), table 2
Note: Area = 51,625 million ha

to greater demands), its dollar value will presumably rise. If human demands cause ecological thresholds to be passed, the value of ecosystem services may jump (conceivably to infinity, if life-support systems collapse). Costanza *et al.* suggest that even their crude estimates present some stark challenges for decision-makers, if patterns of exploitation of ecosystem services are to be brought into balance with their value (for example through carbon or pollution taxes). These calculations provide momentum for those seeking to adjust national accounts to reflect the value of natural capital and ecosystem services.

The simple notion that sustainability demands maintenance of stocks of both human-made and natural capital over time has been fiercely debated. 'Strong sustainability' places a special importance on natural capital and requires that the stocks of both natural and human-made capital are maintained. Economists such as David Pearce have pointed out that a requirement for zero or negative natural capital depreciation places excessive constraints on economic growth (Pearce *et al.* 1989). If this requirement is imposed at the project level it is likely to stultify development, since it effectively makes it impossible to do anything that damages the environment at all. This approach to development, 'strong sustainability', would suit environmentalists opposed to development and economic growth (and these tend to be those living comfortably enough in the North), but is impossibly challenging for governments and business. It is unlikely to appeal to grassroots environmentalists in the South facing the daily human tragedy of poverty. If the requirement for zero natural capital depreciation is set at the *programme* level (i.e. at the level of suites of projects of a particular region, agency or government), there is some flexibility to maximise economic returns from individual projects (ibid.). Lipton (1991) suggests that sustainability should be discussed at the level of the country as a whole, so that damaging and favourable effects of projects can be balanced out.

This notion of 'weak sustainability' therefore involves the principle of trade-offs between losses to natural capital in one project and gains elsewhere and the substitution of either human-made capital or human-induced 'natural capital' for lost natural capital (Barbier *et al.* 1990b). Economists differ as to how far such trade-offs should be allowed to go. Beckerman (1994) argues that '"sustainable development" has been defined in such a way as to be either morally repugnant or logically redundant' (p. 192). He dismisses strong sustainability ('implying that all other components of welfare are to be sacrificed in the interests of preserving the environment in exactly the form it happens to be in today') as 'totally impractical' (1994, p. 203). Jacobs argues that Beckerman's definition of strong sustainability is absurd, suggesting that 'sustainability is the injunction to maintain the *capacities* of the natural environment: its ability to provide humankind with the services of resource provision, waste assimilation, amenity and life support' (1995, p. 62; emphasis in the original).

Beckerman is also scathing about 'weak' sustainability, and the welfare-based definition that allows substitution between natural and human-made capital. He argues that this adds nothing to 'the old-fashioned economist's concept of optimality' (1994, p. 195). He suggests that the only efficient way to proceed is to disregard the distinction between human-made and natural capital alto-gether, and allow gains in the former to replace losses in the latter. Thus the replacement of an unproductive wetland with a productive irrigation scheme should be assessed quite simply in the different economic benefit streams they produce, and development should involve selecting the projects with the greatest economic return because it is this that will maximise human welfare.

Daly (1994) points out that natural and nature-produced services (e.g. atmospheric regulation) and resources (e.g. waste assimilation capacity) are non-replicable. Natural capital is therefore complementary to human-made capital and not a substitute for it. Barbier *et al.* (1990b) recognise this problem of the non-substitutability of human-made for natural capital, and include it within their formulation of cost–benefit analysis as a constraint on the depletion or degradation of the stock of natural capital. El Serafy (1996) goes further in defending weak sustainability, pointing out that it is not a watering-down of strong sustainability but its precursor, as a concern for all capital that sustains future income. Strong sustainability represents a strengthening of this position to hold that *natural* capital specifically be kept intact. The key issue, El Serafy argues, is how much of the income derived from the exploitation of natural capital is *genuine* income, i.e. available for consumption, and how much needs to be reinvested to sustain the same levels of consumption into the future (ibid.). Weak sustainability, like strong sustainability, is arguably therefore a workable theoretical concept.

Among environmental economists who accept that sustainability demands maintenance of stocks of both human-made and natural capital, there is still debate about what kind of trade-offs should be made. One notion is that devel-opments that reduce natural capital should be balanced by others that are specif-ically designed to compensate by creating new natural capital elsewhere. Thus a development programme should contain 'shadow projects' to compensate for degradation other projects cause (Pearce *et al.* 1989). This notion has proved attractive in industrialised economies such as the UK, where developers propos-ing a project that has adverse environmental impacts are now routinely expected to offer as part of the development 'ecosystem restoration' schemes such as new areas of woodland or wetlands to replace those lost (Owens and Cowell 1994).

This notion that trade-offs may be made between one piece of natural capital and another is opposed by conservationists who argue that natural capital is not like human-made capital. It is not made of bricks and mortar. Some forms of natural capital are more like precious pieces of architecture (such as a cathe-dral) than like a dispensable factory that can be taken down and replaced with another, better one as circumstances change. Thus it might be argued that some species (perhaps the mountain gorilla or the blue whale) and some ecosys-tems (perhaps the highly biodiverse *fynbos* vegetation of South Africa or the Great Barrier Reef) are irreplaceable and should not be conceived of as 'capital' to be replaced or exchanged for some other form of natural capital.

There are actually two arguments here. The first is that some (or all) attrib-utes of non-human nature have an 'intrinsic value' that cannot be captured by economic appraisal. This relates to a debate about ethics, and the intrinsic worth of nature, as opposed to the instrumental view of non-human nature as something to be valued only for its utility for humans (Low and Gleeson 1998). Economists try to capture such values using ideas such as 'existence value' (i.e. the non-material value to people of the existence of species; Barbier 1998, and see Figure 5.4). However, the debate as to the validity of intrinsic worth is essentially philosophical and not methodological, and in the field of environmental ethics this approach is not regarded as a solution (Elliot 1997).

The other argument about the non-substitutability of natural capital holds that regardless of the theoretical validity of replacing one piece of natural capital with another, this is in practice not possible. Thus if it were proposed to convert 1,000 ha of rainforest into ranchland, the complexity of the ecological interactions forged through co-evolution would make it impossible in practice to recreate it. Scientific knowledge is too slight for such exchanges to be feasible, and even in the relatively intensively researched and simply structured temperate ecosystems, attempts to move or recreate ecosystems are still in their infancy (Buckley 1989, Cairns 1991).

The concept of 'critical natural capital' is used to refer to those parts of natural capital that cannot be replaced if lost (or at least, not within feasible time frames), and cannot therefore be substituted with human capital or compensated for by positive projects elsewhere (Buckley 1995).

Ecological economics and cultural capital

Two related intellectual movements have allowed conventional economics to make a much more explicit and intelligent engagement with the environment and its management. Both represent a significant departure from standard environmental economics. The first is ecological economics, the second institutional economics. Ecological economics involves a deliberate move away from conventional economics to a transdisciplinary concern for the relationships between ecosystems and economic systems (Costanza 1991, Jansson *et al.* 1994, Berkes and Folke 1998). It claims to take 'a holistic systems approach that goes beyond the normal narrow boundaries of academic disciplines' (Folke *et al.* 1994, p. 3). Ecological economics views socio-economic systems as inextricably linked to environmental systems, and hence seeks to escape the conventional notion that 'resources' can be understood (or exploited) in isolation. It focuses on the linkages between human and ecological systems, and the feedbacks between them. It is also explicitly policy oriented: its aim is the development of practical policies for sustainability (Folke *et al.* 1994, Berkes and Folke 1998).

The ecological economic vision of the future is shown in Figure 5.5 (Folke *et al.* 1994), which can be compared with the alternative visions in Figures 5.1 and 5.2. This draws a distinction between growth and development, and indicates an 'ecological Plimsoll line' reflecting the range of scientific uncertainty about environmental limits and human impacts. This vision suggests that if the physical demands of the human economy are maintained below 'carrying

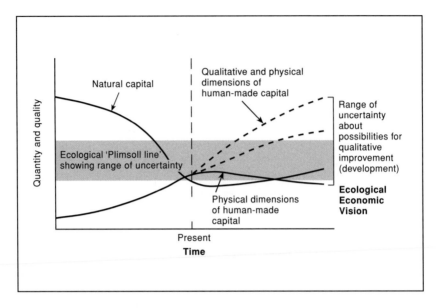

Figure 5.5 The ecological economic vision of the future (after Folke *et al.* 1994).

capacity' (which lies somewhere within the band of uncertainty), then development (measured in terms of human welfare) can continue (Folke *et al.* 1994).

In ecological economics, the zero-growth debate has grown up. Goodland *et al.* (1993) stress the continuing reality of environmental limits. The human economic system has to exist within the biosphere, on which it depends and of which it is part. The biosphere provides the source of all natural resources and is the ultimate sink for all wastes. The throughput of resources between biosphere and economic system is a function of population size and per capita resource consumption, and the capacity of the biosphere to provide resources and sinks demanded of it is finite. Figure 5.6 contrasts the economic subsystem in some notional past 'empty world scenario' with that today. In the 'full world' scenario, the economic system has already started to interfere with the biosphere through excessive demands on sinks, for example through acid rain, greenhouse gas accumulation and ozone depletion (ibid.).

The second slightly unconventional development within the economic field is new institutional economics (NIE), the specific engagement with the economics of institutions (Ostrom 1990, Bromley 1989, Berkes and Folke 1998). Institutional economics has developed to provide theory to describe human interactions in areas such as resource use. Debates about property rights and the management of common property resources have been particularly influential in changing understanding of how people collaborate, or conflict, over the use of resources (Berkes 1989, Berkes and Folke 1994, 1998).

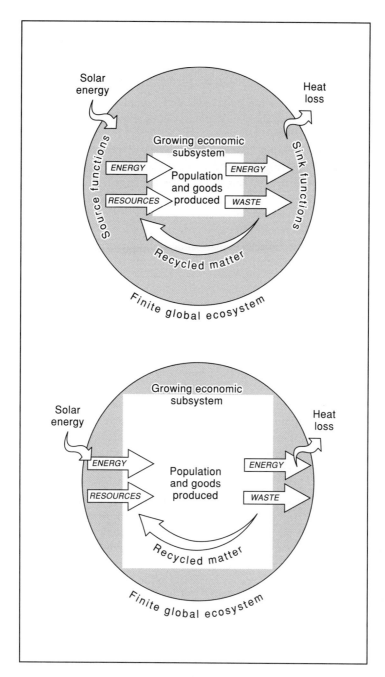

Figure 5.6 The finite biosphere relative to the growing economic subsystem in the 'empty world' (above) and 'full world' (below) scenarios (after Goodland *et al.* 1993).

Ecological economists add to these concepts a third notion, that of 'cultural capital', referring to the factors that enable human societies to adapt to and modify the natural environment (Berkes and Folke 1994). At its simplest, cultural capital can be conceived of as the interface between natural and human-made capital. The idea of cultural capital relates very closely to the ideas of 'institutional capital' (Ostrom 1990) and 'social capital' (Coleman 1990). Cultural capital embraces the conditions and organisation and institutions of collective action, but also the philosophies, values and ethics that underlie them (Berkes and Folke 1994).

The existence of natural capital is the basis and precondition for cultural capital, and is in turn regulated by cultural capital (through the institutions affecting the use of nature). Human-made capital is generated by both natural and cultural capital together. Human-made capital impacts on natural capital (for example when a factory or a city discharges wastes into a river basin). However, it also affects cultural capital, as the creation of technologies (tools, skills and knowledges) affects human understanding of the dynamics of and human demands on nature, and hence the status of natural capital (Berkes and Folke 1994; see Figure 5.7).

Mechanisms for mainstream sustainable development

The question of sustainability can be built into economic appraisal of the costs and benefits of development projects in various ways. Barbier *et al.* (1990b) set out 'sustainability criteria' for both 'strong' and 'weak' sustainability in terms of a requirement to maintain stocks of natural capital at either project or programme scale. Where natural and human-made capital are allowed to substitute for each other, or natural capital is created to compensate for existing natural capital threatened by development, the values of this capital are subject to trade-offs.

Trade-offs are standard fare in development economics, where the relative importance of economic objectives of development is routinely weighed against the importance of social objectives such as poverty alleviation. Conventionally, development strategies involve a mix of economically efficient production (growth to make a larger cake) and targeted social programmes (aimed at more equitable division of the cake). The notion of sustainability adds a third dimension: maintaining the quality or capacity of environmental systems and the resources and benefit streams they generate (Munasinghe 1993b). Sustainable development can then be seen as involving trade-offs between environmental sustainability, economic sustainability and social sustainability (Figure 5.8). A variety of economic tools are conventionally used to aid decisions in development at different scales from the project to the national economy and internationally. These include cost–benefit analysis (CBA), the use of shadow prices and the manipulation of discount rates (Munasinghe 1993b). They demand sectoral or regional studies, national multisectoral economic analyses, and analyses at the international scale of trade and financial flows (Figure 5.9; Munasinghe 1993b).

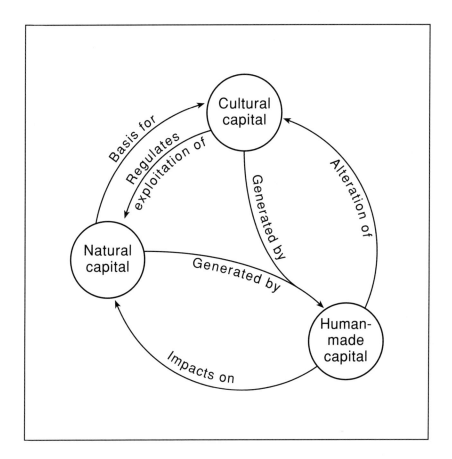

Figure 5.7 Interrelationships between cultural, natural and human-made capital (after Berkes and Folke 1994).

Unfortunately for economists, nature is no respecter of the neat categories of their models and analyses. Although ecological systems can also be thought of as functioning at a hierarchy of scales, the fit with the units of markets, sectors and 'national economies' beloved of economists is poor. Munasinghe argues that environmental economics is capable of providing a bridge between socio-economic and environmental systems, translating the results of environmental analysis into categories that fit the patterns of thinking and analysis used by economists.

Sustainability criteria can obviously be incorporated at the project level through development of techniques such as CBA. It is not so straightforward to incorporate environmental sustainability issues into decisions at national and international scales. The conventional measure of economic growth is gross national product (GNP). This is a measure of the way income flows in an

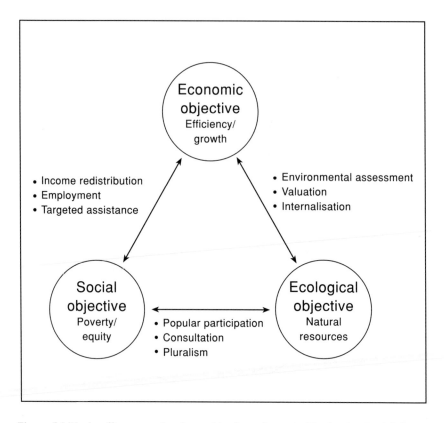

Figure 5.8 Trade-offs among the three objectives of sustainable development (after Munasinghe 1993b, p. 2).

economy, and does not take account of resource depletion or pollution or other environmental costs (Jacobs 1991). Any activity that involves the exchange of money contributes to GNP, so that the production of goods in a polluting factory contributes to GNP, but so do the costs of the resulting clean-up by government: pollution is, by this tunnel-visioned measure, deemed good for the environment. In this instance, the resulting GDP estimate is incorrect because harmful impacts such as pollution are ignored, and beneficial inputs related to environmental needs are undervalued (Munasinghe 1993b). Conventional presentations of national accounts also fail to take account of changes in stocks of natural capital, and the existence of hidden subsidies to certain activities and products, because their impacts on natural capital are not measured (ibid.). The existence of macroeconomic incentives for environmentally destructive behaviour (whether by corporations or small farmers) is of great importance in understanding unsustainable patterns of environmental management, and of course in adapting policy to change them (Barbier 1994).

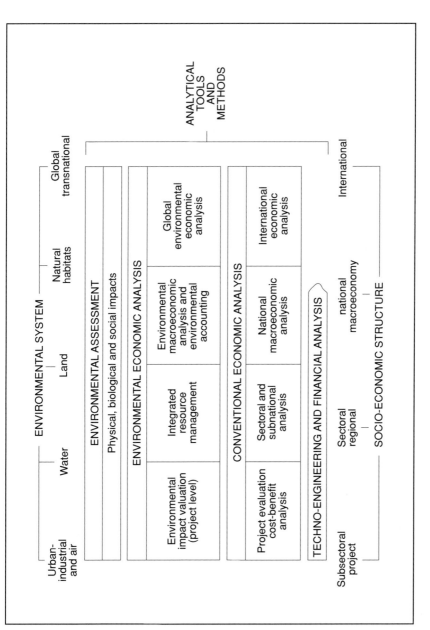

Figure 5.9 The incorporation of environmental concerns into decision-making (after Munasinghe 1993b).

Costanza and Daly emphasise the need to distinguish between growth in the size of an economy and growth in its capacity to deliver solutions to human needs. They write, 'Improvement in human welfare can come about by pushing more matter–energy through the economy or by squeezing more human want-satisfaction out of each unit of matter–energy that passes through' (1992, p. 43). They suggest that increased throughput should be described as growth, and increases in the efficiency with which human needs and wants are met should be described as development. Beyond a certain point, as growth destroys natural capital, because it costs more than it creates, growth has become 'impoverishing not enriching' (ibid., p. 43).

GNP growth measures the size of the economy, but an assessment of the sustainability of that economy depends on what is growing. The scale of the economy is only one factor that determines environmental quality; others include structure (the mix of goods and services produced), the ability to substitute away from resources that are becoming scarce, the ability to use clean technologies and management practices to reduce damage per unit of input or output, and the efficiency of inputs used per unit of output (see Figure 5.10; World Bank 1992).

Jacobs (1991) suggests the need for a concept such as 'environmental impact coefficient' (EIC) of GNP, which will measure the amount of environmental consumption (i.e. damage done) by each unit of national income. All economic activity demands raw materials and energy, the assimilation of wastes and the maintenance of life-support systems such as climatic regulation. As production increases, stress on these three functions will also increase, leading to environmental degradation (Ekins and Jacobs 1995). The 'zero growth' debate was an attack on narrow measures of economic performance such as GNP. The EIC approach, Jacobs suggests, provides a logical escape from the sterility of the automatic association of economic growth and environmental destruction. If there were technological changes in manufacturing and patterns of consumption such that GNP rose and yet the EIC fell, economic growth could be accompanied by *reduced* rates of resource depletion. This would be the platform for a 'green economy' (Jacobs 1991): 'Environmentally sustainable GDP growth thus depends on the achievement of substitution and technical and structural change in order to keep environmental impacts within conditions of environmental sustainability' (Ekins and Jacobs 1995, p. 26).

However, the actions necessary to reduce the EIC are likely to entail costs in terms of new capital and inputs. These actions may result in a 'win–win' situation if they do not entail financial costs (for example if government changes a policy that is economically inefficient as well as environmentally damaging), or if any costs are less than the combination of financial gain and environmental benefits. Ekins and Jacobs conclude:

> It is clearly theoretically possible for GDP growth and environmental sustainability to be compatible. Environmental sustainability is affected by the economy's throughput of energy and materials; on a finite planet there

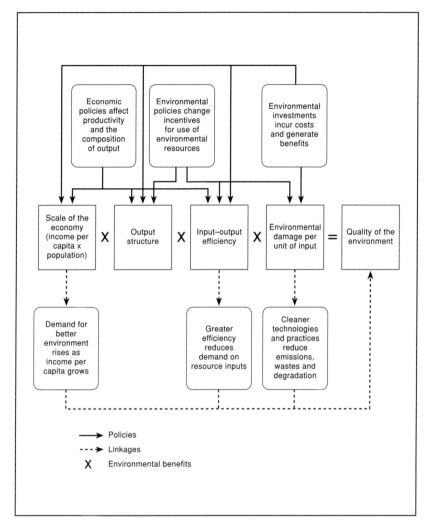

Figure 5.10 Economic factors affecting environmental quality (after World Bank 1992).

is obviously a limit to this throughput. GDP measures value-added. The relationship between value-added and material throughput is variable and can be altered by structural economic change, substitutions between factor inputs, and more efficient use of the same input. These changes are obviously crucially dependent on technological developments.

(1995, p. 40)

Attempts have been made to develop policy measures that might allow the kind of transition that Jacobs outlines. Bartelmus (1994) discusses the integration of the environment into national accounts. One important initiative is the revision of the United Nations System of National Accounts (SNA). The Statistical Division of the United Nations (UNSTAT) developed a System of integrated Environmental and Economic Accounting (SEEA) in 1993 (ibid.). This addresses the depletion of natural resources in production and final demand, and the changes in environmental quality resulting from production and consumption and natural events, and from environmental protection and enhancement (ibid.). It does this by embracing the concept of natural capital, and seeking to measure its stocks and flows. These are measured in 'satellite accounts'. The SEEA therefore seeks to account comprehensively for all impacts of development and link them to the activities and sectors causing them.

The SEEA approach has been used experimentally by the World Bank in studies of Thailand, Papua New Guinea and Mexico. This revealed the potential of the approach, but also the problem of poor data (Bartelmus 1994). In the case of Mexico, two measures of environmentally adjusted net domestic product (EDP) were calculated. The first (EDP1) measured resource depletion (oil, forests, land and living water resources). This amounted to 94 per cent of net domestic product in 1985. The second (EDP2) attempted to measure externalities in the form of the costs of environmental quality degradation, and amounted to 87 per cent of NDP (Table 5.2).

Bartelmus notes that the SEEA is still at an experimental stage. Clearly, too, it is not the only transformation of macroeconomic management required. Environmental issues are also affected by conventional economic mechanisms such as taxation and monetary policy (Munasinghe 1993b). Here too there is a need for environmental issues to be built into economic accounting. The importance of improving national accounting and welfare measures, however, is widely recognised. Daly and Cobb (1990) formulated an 'Index of Sustainable Economic Welfare' (ISEW) that took account of depletion of natural resources, pollution and income distribution. A graph of change in GNP and ISEW in the USA between 1950 and 1985 showed that while GNP rose consistently, the ISEW index did not (Figure 5.11). Similarly, Atkinson *et al.* (1997) showed that growth in smaller resource-rich countries in the 1970s and 1980s was not environmentally sustainable because it was based on the depletion of natural capital.

Table 5.2 Selected indicators comparing conventional and environmentally adjusted national accounts, Mexico, 1985

	Conventional Accounts	EDP 1	EDP 2
NDP	P 42.1 billion	P 39.7 billion	P 36.4 billion
EDP/NDP		94%	87%

Source: Bartelmus (1994), table 2.2, p. 56
Notes:
NDP: net domestic product
EDP: environmentally adjusted net domestic product (EDP 1 measured resource depletion; EDP 2 measured environmental degradation)

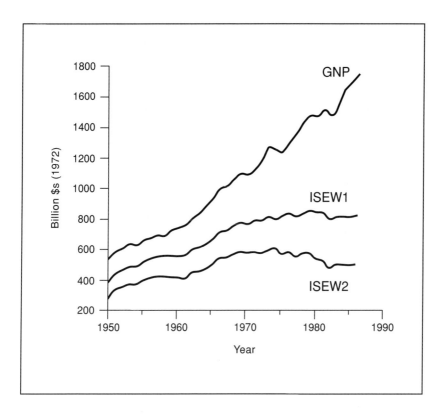

Figure 5.11 GNP and index of sustainable economic welfare (ISEW) in the USA, 1950–85 (after Costanza and Daly 1992). Note: ISEW2 takes account of the depletion of non-renewable resources and long-term environmental damage, ISEW1 does not.

Trade-offs, equity and complexity

A constant theme in accounts of sustainable development is equity: Jacobs, for example, holds that it incorporates 'an inescapable commitment to equity' (1995, p. 60). Thus it must involve not only the creation of wealth and the conservation of resources but their fair distribution between rich and poor. Moreover, it demands not only this principle of *intra*-generational equity but also a commitment to the future and *inter*-generational equity. Mainstream sustainable development has tried to take account of both these dimensions of equity, that between present and future generations and that between rich and poor in the present generation. As we have seen, in the thinking of market environmentalism, these concerns translate into the problems of making trade-offs.

The trade-offs that sustainable development demands can therefore be complex, involving several different scales simultaneously. First, as has been seen, there can be difficult choices between natural and human-made capital or between social, environmental and economic dimensions of development. Second, trade-offs can take place either in space (between losses in natural capital or some other measure here and gains somewhere else) or in time (between losses or gains now, and those coming in the future).

Principles of inter- and intra-generational equity demand that a balance be struck between the needs of present and future generations. This is far from simple. One problem is that of uncertainty about the future (Dovers and Handmer 1992). The adoption of modest economic growth targets in the present generation may make it harder to tackle poverty now as well as having implications for the wealth or well-being of future generations (for example by allowing existing inequalities to endure; Dovers and Handmer 1992, Munasinghe 1993a). Uncertainty about future technologies and future resources makes it difficult to assess to what extent future generations will be able to provide different environmental goods. Uncertainty about the values future generations will hold makes it hard to know how they will value different environmental goods, and how they will view decisions made now on their behalf (Pasek 1992). As Beckerman points out, 'people at different points in time, or in different income levels, or with different cultural and national backgrounds, will differ with respect to what "needs" they regard as important' (1994, p. 194).

There is also considerable uncertainty in predictions of the ways in which ecosystems will behave in the future (because of both scientific uncertainty and anthropogenic environmental change), and hence it is hard to predict the value of flows of benefits from existing and future natural capital. In semi-arid environments such as the Sahel, for example, rainfall variability and drought have severe implications for land use and yet are effectively impossible to forecast (see Chapter 7). Patterns of resource use that are analysed over a short time frame and judged sustainable may well not be sustainable over longer periods (Dixon and Fallon 1989). Dixon and Fallon ask, 'how far into the future do we worry about? Is our concern next week, next year, next century?' (ibid., p. 81).

Because of this uncertainty about the future, a standard formula for sustainability is to demand that capital endowments are kept constant, such that each generation bequeaths a legacy of natural capital no smaller than the one it inherited. The snag with this is that the impacts of development can be delayed far into the future, either because the impact itself is delayed (for example the impacts of poorly stored toxic waste that starts to leak after a few decades), or because the ecological repercussions of a development intervention take time to work their way through the ecosystem (for example the problem of the responses of floodplain ecosystems to dam construction; Thomas and Adams 1997; see also Chapter 8).

Ecosystems function at a range of timescales, and ecological interactions and feedbacks can be complex and often delayed. Environmental response times may be so long as to conceal the link between past and future events, and the true severity of impacts may be revealed only when new economic opportunities come to be developed (e.g. an attempt to irrigate using water from an aquifer contaminated with heavy metals). The importance of timescales in sustainability analysis can also be seen in reverse, in that degraded ecosystems can recover naturally over long time periods, or may be rehabilitated given appropriate management (Thomas and Adams 1999).

The socio-economic impacts of ecosystem change are likely to be even more complex, and may be further delayed. A further complication is that there are likely to be differences between the ways different impacts are viewed. People suffering serious short-term impacts (for example floodplain farmers affected by an upstream dam) may have very short time horizons, while the governments that commissioned the dam may take a much longer view, arguing that eventual benefits to the national economy may outweigh short-term costs (Dixon and Fallon 1989, Thomas and Adams 1997). The definition of social time horizons is inherently political, because the ways people view their resources, and the relative merits of consumption in the present rather than the future, will inevitably vary (Dixon and Fallon 1989). Time horizons set according to political and economic expediency may be too short for the sustainable management of natural systems. As Dovers and Handmer (1993) note, 'a major implication of the moral principle of intergenerational equity is to force institutional systems to think over time-scales that are somewhat closer to those of natural systems' (Dovers and Handmer 1992, p. 219).

Spatial scale is also important in assessing sustainability (Fresco and Kroonenberg 1992), but again the judgement of the sustainability of a development decision is closely dependent on the scale chosen for analysis. Governments routinely trade off sustainability at one location to meet national goals. Thus locally, the benefits of a development project (for example a mine) might be at the cost of reduced sustainability elsewhere (for example pollution downstream; Low and Gleeson 1998). Internationally, industrialised countries seeking to make their policies sustainable may do so at the expense of other places by importing resources or exporting wastes. Both ecological and political boundaries (local, regional, national or international) are relevant to the assessment of sustainability. Debates about the sustainability of particular developments might very easily descend into arguments about boundaries, and different actors (for example governments, NGOs and transnational corporations) may base conflicting assessments of the sustainability of controversial projects on different choices of boundaries for analysis. Sustainability can be determined at a range of scales from local to global (Thomas and Adams 1997).

Natural systems may provide appropriate boundaries for sustainability analysis, but these are not always easy to define. In particular, the ecosystem is effectively an arbitrary analytical category, not a natural entity whose characteristics are endogenously determined. Furthermore, in practice, development planning usually takes place within political jurisdictions, not within natural boundaries. Structures of governance are hierarchical, and human-made boundaries rarely fit the spatial patterns of natural systems (Conway 1985; Munasinghe 1993a). Ecosystems straddle political boundaries, and so do bio-geochemical processes such as trans-boundary acidification, international river basins and oceanic circulation.

Overlapping and conflicting political and environmental management institutions make the practical measurement of sustainability (let alone its promotion through environmental management) highly problematic. So too does the complexity of ecosystem behaviour. Mainstream sustainable development commands an increasingly sophisticated range of methodologies that do much to make the environment a normal and integral issue in development planning. However, such managerialism does not by any means solve all the problems of sustainability. Behind the technical certainties of the mainstream lie other issues. These are addressed by a range of other ideas about sustainability – countercurrents to the mainstream of sustainable development. They are the subject of the next chapter.

Summary

- There is a strong central stream to thought about sustainable development. This 'mainstream sustainable development' runs within dominant capitalist industrialism and developmentalism rather than challenging it. Two key elements within it are market environmentalism and ecological modernisation.
- Market environmentalism suggests that problems of environmental management and degradation should be addressed by extending the institutions of the free market into further dimensions of the environment, setting prices for environmental 'goods' and 'services'. Market environmentalism assumes continued capitalist economic growth and rejects environmentalist ideas of limits to growth.
- Ecological modernisation proposes that capitalist modernisation can be reformed through efficient regulation of markets, governance and technology. It is technocentrist, demanding improved planning, management, regulation and utilisation of human use of nature. Ecological modernisation assumes continued capitalist economic growth, within careful, in many cases technologically regulated, boundaries.
- Mainstream sustainable development is populist, proposing to bring change about through the participation of citizens, use of appropriate technology, and a focus on basic needs.
- A key element in mainstream sustainable development is the development of environmental economics. Critical concepts include the distinction between natural and human-made capital. Natural capital includes stocks from which benefits flow, but which are not the product of technology or human action. 'Strong' sustainability demands that stocks of each are maintained; 'weak' sustainability allows some trade-off between natural and human-made capital.
- Ecological economics is a further development, to take explicit account of relations between economic systems and ecosystems. An important concept here is 'cultural capital', relating to the institutions that regulate human use of the environment, a subject also addressed by 'new institutional economics'.

- The delivery of mainstream sustainable development is being addressed through a number of policy initiatives, including the adjustment of national economic accounts to internalise the environment (for example an index of 'sustainable economic welfare'). In practice, any attempt to factor the environment into economic thinking faces significant problems of trade-offs between different people and interests in space and time, and in predicting future behaviour of environmental systems.
- Behind the technical virtuosity of mainstream sustainable development lies a fairly large degree of uncertainty. Alongside mainstream thinking flow other, more disturbing and radical currents.

Further reading

Barbier, E.B. (1998) *The Economics of Environment and Development: selected essays*, Edward Elgar, Cheltenham.

Barbier, E.B., Burgess, J.C. and Folke, C. (1994) *Paradise Lost? The ecological economics of biodiversity*, Earthscan, London.

Berkes, F. and Folke, C. (eds) (1998) *Linking Social and Ecological Systems: management practices and social mechanisms for building resilience*, Cambridge University Press, Cambridge.

Cernea, M. (ed.) (1991) *Putting People First: sociological variables in rural development*, Oxford University Press, Oxford, for the World Bank.

Costanza, R. (ed.) (1991) *Ecological Economics: the science and management of sustainability*, Columbia University Press, New York.

Costanza, R. and Daly, H.E. (1992) 'Natural capital and sustainable development', *Conservation Biology* 6: 37–46.

Daly, H.E. and Cobb, J.R. Jr (1989) *For the Common Good: redirecting the economy towards community, the environment and a sustainable future*, Beacon Press, Boston, MA.

Ghai, D. and Vivian, J.M. (eds) (1992) *Grassroots Environmental Action: people's participation in sustainable development*, Routledge, London.

Jacobs, M. (1991) *The Green Economy: environment, sustainable development and the politics of the future*, Pluto Press, London.

Jansson, A., Hammer, M., Folke, C. and Costanza, R. (eds) (1994) *Investing in Natural Capital: the ecological economics approach to sustainability*, Island Press, Washington, DC.

Low, N. and Gleeson, B. (1998) *Justice, Nature and Society: an exploration of political ecology*, Routledge, London.

Web sources

<http://www-esd.worldbank.org/>   Environmentally and Socially Sustainable Development network of the World Bank. See especially *<http://www-esd.worldbank.org/eei>* for the World Bank on environmental economics and indicators, including environmental valuation and green accounting, and *<http://www.worldbank.org/data/>* for facts and figures by country and sector on almost everything from environment to governance via poverty and gender.

<http://www.un.org/esa/sustdev/>   United Nations home page on sustainable development.

<http://www.undp.org/>   The website of the United Nations Development Programme information on UNDP's work on crisis prevention and recovery (Emergency Responses Division), and pro-poor policies.

<http://www.fao.org/waicent/faoinfo/sustdev/> Information from the Sustainable Development Department (SD), Food and Agriculture Organisation of the United Nations (FAO).

<http://www.iied.org/>   The website of the International Institute for Environment and Development in London (IIED); see for example the Environmental Economics Programme, *<http://www.iied.org/enveco/index.html>*.

<http://www.oneworld.org/itdg/>   Home of the Intermediate Technology Development Group (ITDG), Schumacher's 'Small is beautiful' for the twenty-first century, still providing 'practical answers to poverty'.

<http://www.ecologicaleconomics.org/> The International Society for Ecological Economics (ISEE), an organisation linking ecological economists and promoting research on and understanding of the relationships among ecological, social and economic systems.

6 Countercurrents in sustainable development

Clearly eco-software will not save the planet if capitalist expansionism remains the name of the game.

(Martin Hajer 1996, p. 255)

Sustainable, ecologically sound capitalist development is a contradiction in terms.

(David Pepper 1993, p. 218)

Beyond the mainstream

Mainstream sustainable development, developed through the 1980s and entrenched at Rio, has begun to acquire the intellectual scaffolding necessary to translate rhetoric into practical policy. The philosophical bases of environmentalist social movements in the so-called new environmentalism of the 1970s (Cotgrove 1982) were complex, eclectic and confused. Sachs (1992b) argues that environmentalism, or the 'ecology movement' as he calls it, combines modernism and anti-modernism, a call for a better science with a critique of the rationality of science. Sustainable development is the uncertain inheritor of this confusion. As we have seen in Chapter 5, the discipline of economics has furnished bridges between normal practice in development planning and concerns for environment. Environmentalism and human rights have been factored into the business spreadsheets of 'Earth plc', enabling trading and planning to continue very much as normal (Pearce 1992).

The enormous success of mainstream sustainable development has therefore been its ability to transcend the uncomfortable claims of environmentalists and critics of development. The Rio Conference epitomised this mainstreaming of critical discourse, its chief triumph being the way in which different interests were intertwined in the negotiated documents, and different governments brought to the table to sign a form of words that implied that they agreed with each other. As was discussed in Chapter 4, some commentators feel this process of accommodation went too far, and see Rio as a sell-out on critical environmental and development issues. Chatterjee and Finger (1994), for example, argue that the non-governmental movement was co-opted to a process that ultimately worked against its interests. Most NGOs were invited in, indeed sucked in by the lure of influence and by generous funding, to lend their support to the 'Rio process' – 'fed into the green machine' as Chatterjee and Finger put it (1994, p. 79). While a small group of mostly US-based NGOs had some influence at Rio, the experience of most NGOs was disorientating and disappointing.

The version of sustainable development expounded at Rio demanded no radical changes in the relations between rich and poor countries, no systematic reorganisation of the control of resources, no reining back of consumption of non-renewable or renewable resources that might harm the delicate constitution of the juggernaut of the world economy. As Brown suggests, 'Rio and the developments since have reaffirmed the South's suspicion that the North is simply not prepared to redefine the international division of labour or its economic, social and political relationship with the rest of the world' (1997, p. 388). More generally, UNCED reaffirmed the unwillingness of the international community to question the nature and direction of development or to consider an alternative to the dominant development paradigm. This is not surprising, for the inertia of the established models of development is very great. Furthermore, as Brown comments, 'any transition from non-sustainable to sustainable development will imply an enormous change in the distribution of winners and losers' (ibid., p. 388). Recasting of the global power-game was not on the agenda at Rio.

More radical voices have, however, been raised about environment and development, and there are countercurrents within the mainstream of sustainable development that do offer a significant critique of the dominant model. Indeed, Martin Lewis outlines five 'schools' of 'extremist' thought (antihumanist anarchism, primitivism, humanist ecoanarchism, Green Marxism and radical ecofeminism) that are 'presently battling it out for the heart of the radical environmental movement' (1992, p. 41). All six, he claims, reject representative democracy, and respond to US government institutions with contempt. Against these variously demonic 'extremisms' he makes the case for mainstream sustainable development, 'a new alliance of moderates from both left and right' (ibid., p. 250), building on 'the clear environmental advantages of the judicious use of market mechanisms' (ibid., p. 182).

Radical Green ideas have not been swept away completely by the rising tide of post-Rio conformism, and still deserve serious attention. Spaargaren and Mol identify a more radical programme beyond the conventional conservative approach of compensation/impact minimisation and that of classic ecological modernisation (clean technologies, valuation of environmental resources and transformation of patterns of production and consumption). Their third programme involves 'a progressive dismantling or deindustrialisation of the economy', and the transformation of industrial structure into small units that link production and consumption more closely (1992, p. 339). This approach, so popular in the 1970s, has substantially lost support. However, it remains a potent element in more radical thinking about sustainable development. It is worth noting also that there is considerable debate, even countercurrents, within the mainstream ideas discussed in Chapter 5. Christoff (1996) identifies several differing and sometimes conflicting versions of ecological modernisation. In a classification paralleling the debate about sustainable development (see Chapter 5), he distinguishes 'weak' ecological modernisation, which is economistic, narrowly technical and focused within national boundaries, from 'strong' ecological modernisation that is ecological, systemic and international (Table 6.1).

Alternative ideas were also presented at Rio in 1992, although they did not find favour with government negotiators. A consortium of NGOs, including Greenpeace International, Friends of the Earth International and the Forum of Brazilian NGOs, offered a '10-point plan to save the Summit' (Chatterjee and Finger 1994; Table 6.2). Few of these issues were tackled effectively at Rio, and few have subsequently found a place in the mainstream of sustainable development.

Table 6.1 Weak and strong ecological modernisation

Weak ecological modernisation	Strong ecological modernisation
Economistic	Ecological
Technological (narrow)	Institutional/systemic (broad)
Instrumental	Communicative
Technocratic/neo-corporatist/closed	Deliberative democratic/open
National	International
Unitary (hegemonic)	Diversifying

Source: Christoff (1996, p. 490)

Table 6.2 'Ten-point plan to save the Summit'

1 Set legally binding targets and timetables for reduction in greenhouse gas emissions, with industrialised countries leading the way.
2 Cut Northern resource consumption and transform technology to create ecological sustainability.
3 Reform the global economy to reverse the South–North flow of resources, improve the South's terms of trade and reduce its debt.
4 End the World Bank's control of the Global Environment Facility (GEF).
5 Regulate transnational corporations and restore the UN Centre on Transnational Corporations.
6 Ban exports of hazardous wastes and dirty industries.
7 Address the real causes of forest destruction, since planting trees as UNCED proposes cannot be a substitute for saving existing natural forests and the cultures that live in them.
8 End nuclear weapons testing.
9 Establish binding safety measures – including a code of conduct – for biotechnology.
10 Reconcile trade with environmental protection, ensuring that free trade is not endorsed as the key to achieving sustainable development.

Source: Chatterjee and Finger (1994, p. 40)

This chapter looks at some of the more radical ideas that run counter to the conformity of mainstream sustainable development. It considers neo-Malthusianism, radical ecologism and deep ecology, Green critiques of development, ecosocialism, ecoanarchism and ecofeminism. This is a diverse and eclectic list. It is a testament to the way in which ideas about environment and development have demanded attention from political thinkers of all persuasions, although it is also a sign of the tendency for environmentalists, ill-informed about political ideas, to pick up fragments of ideologies that catch their eye. Environmentalists setting out political agendas are often magpies with a hoard of shiny ideas they have stolen out of context, and which they barely understand.

Neo-Malthusianism and sustainable development

One source of challenge to the sustainable development mainstream lies in the persistence of the neo-Malthusian notion of limits to growth so dominant in the 1970s. As discussed in Chapters 2 and 5, neo-Malthusian ideas, both about growth and more particularly about population, have been important elements of environmentalist critiques of conventional development and industrialisation strategies, and were central to the 'futures' debate of the 1970s (Cole 1978). This embraced both the directly 'neo-Malthusian' arguments about overpopulation in books such as *Population, Resources and Environment* (Ehrlich and Ehrlich 1970) and *The Population Bomb* (Ehrlich 1972), and the global computer models of Forrester (1971) and Meadows *et al.* (1972).

Of all the ideas current in the early 1970s, only that of population control caught on in terms of policy, largely because it did not threaten the fabric of advanced capitalist countries (Sandbach 1978, p. 29). Concern in the North for population growth in the Third World and the Earth's supposed 'carrying capacity' has been a persistent element in environmentalist critiques of the state of the world (see Chapter 2), and of Northern government discourses about global environmental problems (Kirchner *et al.* 1985). Westing (1981), for example, argued that the maximum sustainable world population was 2 billion, while a major FAO study using computer models to estimate population-supporting capacities predicted severe food shortage problems in a number of countries (Higgins *et al.* 1982).

The neo Malthusian environmentalist vision of the Third World's future was grim. Ehrlich and Ehrlich suggested that because of population growth, many of the underdeveloped countries would 'never, under any conceivable circumstances, be "developed" in the sense in which the United States is today. They could quite accurately be described as the never-to-be-developed countries' (Ehrlich and Ehrlich 1970, p. 2). May spoke of the 'dead-end societies' of the Third World: 'there is no prospect of change in the Third World that would substantially improve the lives of more than a few people' (1981, p. 226). The impact of neo-Malthusian ideas was particularly great in the USA. The 'positive programme' introduced rather apologetically by Ehrlich and Ehrlich (1970) at the end of their book is an American vision, based on a view of the world

Plate 6.1 Children in a village in eastern Bangladesh. Neo-Malthusian ideas about the
        dangers of population growth were an important feature of environmentalism
        and the 'futures' debate of the 1970s. Concern in the North for population
        growth in the Third World and the earth's supposed 'carrying capacity' has
        been a persistent element in environmentalist critiques of the state of the
        world, and of Northern government views of global environmental problems.
        Only in the 1990s did the dominant emphasis in Northern environmentalism
        shift to a concern about resource consumption, and the unequal demands on
        the biosphere made by wealthy countries. In the same decade, research such
        as that in rural Kenya by Mary Tiffen and Mike Mortimore built on the ideas
        of Esther Boserup to challenge the pessimism of neo-Malthusian analysis and
        reveal the possibility of beneficial interactions between population growth,
        environmental improvement and economic output (Tiffen *et al.* 1994; see
        Chapter 7). If they survived, these children will now be grown up, and will
        probably have families of their own.

where US political and economic hegemony is assumed. It discusses what the
US government could and should do nationally and internationally to control
global population. Other organisations, for example the Environmental Fund
(of which Paul Ehrlich and Garrett Hardin, among other things known for his
discussion of 'lifeboat ethics' in 1974, were founder members) and various
groups dedicated to population control such as ZPG and Planned Parenthood,
focused on the population content of USAID giving, with significant impacts
on USAID policy. These were only overturned by attacks in the 1980s by the
anti-abortion lobby in the USA, ironically of course from a position to the far
right of the political spectrum, which was where liberal opponents described
supporters of population control as being.

The political agendas built on ideas of zero population growth often tend towards authoritarianism. Some environmentalists seem to despair of existing political structures for change, and have called for a technocratic global government. Myers and Myers (1982) discussed the notion of technocratic political globalism, with supranational power exercised by some notional 'global community', and Polunin suggested the world needed 'saving from itself – from destruction perpetrated by Mankind, its uniquely intelligent component' (1984, pp. 294–5). The superficially *non*-political call for impartial and expert government is, of course, highly political: scant room for effective democracy, for example, in a world run by an oligopoly of environmental experts and business leaders.

Environmentalists have a tendency to believe that people must be *made* to do the right thing. When Riddell (1981) set out his broadly neo-populist proposals for 'ecodevelopment', he suggested that their implementation would require strong central management and even socially sanctioned coercion. He wanted to outline an alternative to both underdevelopment (the condition of the poor Third World) and 'overdevelopment' (the condition of the industrialised countries of the First World). Riddell more or less wrote off the possibility that the Third World could reach the living standards of the North. He called for reduced levels of material consumption in Northern nations to allow for 'upgrading' in the South. A global strategy of self-reliance in the South and self-constraint in the North, an attempt to 'optimise the balance between population numbers, locally available resources and culturally desired lifestyles' (p. 5), does not immediately suggest itself as a realistic or practicable suggestion, but Riddell, undaunted, goes on to offer a package of actions which could bring it about (see Table 6.3). As they stand, these proposals do not invite serious attempts at implementation. However, the authoritarianism they presume is important. Riddell's ecodevelopment demands a tightly controlled society with 'political cohesion' to 'generate the necessary political will to break with Northern insistence to compete on Northern terms', the 'scotching' of 'entrenched corruption and petty inefficiencies' in bureaucracy, and a 'soundly structured' society within which birth control through 'socio-economic measures' can be implemented (Riddell 1981). Sustainable utilisation of resources and the realisation of basic needs are to be achieved through a strongly authoritarian state.

The politics of neo-Malthusian ideas were fiercely debated in the 1970s and 1980s (e.g. Enzensberger 1974, Harvey 1974, Sandbach 1978, Gregory 1980). Criticism came not only from economists such as Beckerman (1974) and Simon (1981), and scientists (whom Cotgrove (1982) might describe as 'cornucopian') with particular faith in the benign possibilities of technological change (e.g. Maddox 1972), but also from radical thinkers. Marxists made the obvious point that the relation between population and environment is not fixed: the density of people that is economically viable in Manhattan is different from that in the Sahel, and this difference is a function not of climatic conditions but of economic organisation; the population size that can be sustained is therefore determined

Table 6.3 Riddell's principles of ecodevelopment

---

1 Establish an ideological commitment.
2 Sharpen political and administrative integrity.
3 Attain international parity.
4 Alleviate poverty-hunger.
5 Eradicate disease-misery.
6 Reduce arms.
7 Move closer to self-sufficiency.
8 Clean up urban squalor.
9 Balance human numbers with resources.
10 Conserve resources.
11 Protect the environment.

Source: Riddell (1981)

by social relations (Pepper 1993). Marxist analyses of the Sahel in the 1970s stressed the structural causes of famine (Copans 1983, Watts 1983a, b). As Amartya Sen (emphatically not a Marxist) pointed out, people mostly starve in famines not because of a total food shortfall, but because they cannot buy or otherwise get access to food (Sen 1981); as he likes to point out, historically, the best defence against famine has been democracy. Debates about the causes of famine will be discussed further in Chapter 7.

Radical critiques of neo-Malthusian analysis were accompanied by critiques of policy prescription. Harvey argued bluntly that

> the projection of a neo-Malthusian view into the politics of the time appears to invite repression at home and neo-colonialist policies abroad. The neo-Malthusian view often functions to legitimate such policies and, thereby, to preserve the position of a ruling elite.
>
> (1974, p. 276)

Environmentalism offered an essentially determinist analysis based on the principle of unchanging limits on human action, and a pessimistic view of the potential impact of social reform (Pepper 1984, 1993). The 'new barbarism' (the phrase is from Commoner 1972) of neo-Malthusian environmentalism, probably most firmly identified with Hardin's 'lifeboat ethics' (Hardin 1974), therefore has the potential to generate conservative ideology and reactionary and repressive politics (i.e. 'ecofascism', although Pepper (1996, p. 49) warns about applying this emotive label loosely).

In the early 1970s, Hans-Magnus Enzensberger argued that the claimed social neutrality of neo-Malthusian ideas, derived from their ostensible roots in natural science, was a fiction (1974, p. 9). Indeed, he saw ecology as taking shelter in global projection because it was overwhelmed by its inability to theorise sensibly about society, and surrendering 'in the face of the size and complexity of the problems which it has thrown up' (ibid., p. 17). Quite simply,

In the case of man, the mediation between the whole and the part, between subsystem and global system, cannot be explained by the tools of biology. This mediation is social, and its explication requires an elaborated social theory and at least some very basic assumptions about the historical process.

(Enzensberger 1974, p. 17)

The neo-Malthusianism of the environmental debate of the early 1970s was taken up only selectively as the sustainable development mainstream was formed. The Club of Rome picked up the calls for global stability from *The Limits to Growth* but not the notion of zero growth (Golub and Townsend 1977). The problem of population growth was a strong and central message of the *World Conservation Strategy* (IUCN 1980), but not the Brundtland Report. The concern was re-expressed, but in a much more muted form, in *Caring for the Earth* (IUCN 1991). Enthusiasm for the more pessimistic aspects of the neo-Malthusian 'limits' debate rapidly died away (Sandbach 1978). As was discussed in Chapter 5, ideas of limits to growth are not a feature of mainstream sustainable development.

The declining prominence of neo-Malthusian environmentalism reflected a whole range of factors. In part it reflected the embarrassing nature of arguments about Third World population growth on the lips of the rich industrialised Northern conference delegates using the ideas of sustainable development in the search for global consensus. Also important was the realisation that the most immediately pressing limits were in the earth's capacity to absorb wastes (particularly greenhouse-warming, ozone-depleting gases and industrial waste hazards), and that these were more the fruit of excessive consumption in rich countries than of excessive demands by larger numbers of poorer people in the Third World.

There is now a much stronger understanding about the links between population and development (Thomas 1995, Kiessling and Landberg, 1994). Research has shown the inadequacy of earlier Western ideas about population growth. Thus it is now appreciated that the benefits to a rural household of having more hands to work can outweigh the problem of having more mouths to feed – that there can be a strong economic logic in favour of large families; that fertility in most Third World countries is falling, and is related to wealth; and that people make strategic choices about numbers of children, especially where women have the power and education to control their own fertility (Bledsoe 1994, Tiffen *et al.* 1994). Analysts now have both a better understanding of the importance of political factors in food production and distribution, and a more sober appreciation of the prospects (Madely 1995).

The International Conference on Population and Development in Cairo in 1994 embraced a wide diversity of views (from human rights to food), but demonstrated that the crude neo-Malthusianism of the 1970s was a spent force in the development mainstream. Its 'Programme of Action' focused not on the problem of human numbers *per se*, but on reproductive health through better access to family planning and safe motherhood services, and investment in child survival, education and opportunities for women. The logic of this strategy was that social development and small healthy families would together allow a gradual decline in population growth, and this in turn would ease burdens of poverty and pressures on the environment. However, the annual cost of achieving this strategy through to the year 2000 was estimated to be $19 billion, and annual aid investment fell consistently short of this target (Rowley 1999).

Global population growth continues to be rapid, particularly in regions such as sub-Saharan Africa, where fertility remains high. In 1900 there were 1.5 billion people in the world; at the end of the twentieth century there were about 6 billion. Predictions of future growth are problematic, largely because of uncertainty about future trends in fertility. The United Nations Population Division publishes projections every two years, giving five different scenarios (United Nations Population Division 1998). Its middle series projection assumes that the world stabilises on a fertility level of 2.1 children per woman by 2050–5 (Table 6.4). This suggests a world population of 9.4 billion by that time, and a stabilised population of 11 billion. However, small variations in the level of fertility give dramatic differences in stable populations (Figure 6.1); planners face high levels of uncertainty, and cannot be sanguine about either the best direction for

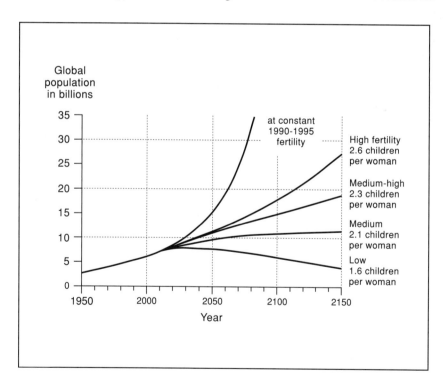

Figure 6.1 Global population projections to 2150 (after UN Population Division 1998).

Table 6.4 Fertility assumptions and projected world populations

Fertility projection	Assumed total fertility rate, 2050–5	Projected population (billions)
Low	1.6	3.6
Medium-low	1.9	6.4
Middle	2.1	10.8
Medium-high	2.3	18.3
High	2.6	27.0

Source: United Nations Population Division (1998)

policy or its effect on overall global population levels (Haub 1999). Leaving neo-Malthusian scares aside, the prospects remain daunting.

Ideas about limits to economic growth are incompatible with business interests in an increasingly interconnected world economy, where market growth is the logic and global capital flows the lifeblood. They are also profoundly unattractive to politicians and policy-makers facing the challenge of providing for growing numbers of people in need of work, shelter and food. At the same time, there is a danger of a false sense of security being created in the comfortable boardrooms and ministries of industrialised countries. Enzensberger was right when he pointed out that all environmental ideas have to undergo 'processing through the sewage system of industrialised publicity' (1974, p. 7). Those who have compiled the various elements in the 'sustainable development mainstream' have too often quite deliberately edited the uncompromising notion of the finite capacity of global ecosystems and the challenge of the growth in consumption out of the message.

Green critiques of developmentalism

Critiques of the standard Western model of industrial development have been an important strand of Western environmentalist thought (Dobson 1990). Martin Wiener describes the shift in attitudes to industrialisation in late Victorian England as the new industrial system began to look 'less and less morally and spiritually supportable' (1981, p. 82). Meredith Veldman (1994) traces a British tradition of romantic protest at industrialisation and large-scale organisation through the twentieth century to the Campaign for Nuclear Disarmament and the early Greens. Veldman argues that the environmentalists of the 1960s and 1970s in Britain moved far beyond their predecessors' concerns for rare species and disappearing habitats to argue for far-reaching social, economic and political change: 'they condemned not only the environmental degradation but also the society that did the degrading' (ibid., p. 210). In other industrialised countries, opposition to industrialism, or at least to its particular local manifestations in polluted or destroyed ecosystems, was a major element in the new environmentalism (Hays 1987).

David Pepper (1996) argues that environmentalism is more than a localised romantic opposition to industrialism, but a rejection of modernism itself – that is, a rejection of the whole project of science, technology and organisation that was ushered in by the Enlightenment of the eighteenth century. He points out that Greens not only have offered a critique of industrialisation's claim to control nature (stressing instead the prevalence of high-technology risks), but also have expressed 'a fashionable mistrust of the grand political theories of the 'modern' period, liberalism and socialism (ibid., p. 5). A case in point: Friberg and Hettne (1985) proposed a 'Green alternative' to both capitalism and Marxism. Against the 'Blue' (market, liberal, capitalist) and 'Red' (state, socialist) strategies they proposed a 'Green counterpoint' that opposed the institutionalisation of the 'modern complex' (i.e. bureaucracy, the industrial, urban system, market and techno-scientific systems, and the 'military-industrial complex'; ibid., p. 207). This Green position is obviously a hybrid, incorporating elements of romanticism, anarchism and utopian socialism, but they argue that its commitment to a just world order means that it cannot be interpreted simply as 'nostalgic conservatism' (p. 207).

It is the opposition to the conventional political strategies of both left and right, and the industrialism and consumerism that support them, that marks out Green political ideas. Paul Ekins opens his critique of conventional development by describing four interlinked global crises: militarisation, poverty, environmental destruction and human repression. He argues that these form 'a single, systemic global *problématique* of great complexity' (1992, p. 13). This *problématique* is maintained by modern technology, world capitalism and state power. These are the fruits of the modern project, and they are sustained by three forces: 'scientism' (exclusive trust in the scientific worldview), 'developmentalism' (belief that economic and human progress depends on an expanding consumer society) and 'statism' (belief that the nation state is the ultimate form of political authority) (ibid., p. 207). Such an analysis is hardly the basis for the reformist policy evolution of mainstream sustainable development.

Radical Green critics of development reject the possibility that capitalism can deliver real development (that is, just, equitable and humane conditions of human life). Addo *et al.* argue that

> the bankruptcy of dominant development models, the deterioration of living conditions virtually everywhere, the sharpening of conflicts within and between nations, and the destruction of the foundations of existence should overwhelm the illusion held for so long of the possibilities of developmental transformation within the capitalist world-system.
>
> (Addo *et al.* 1985, p. 2)

A classic statement of a Green critique of the dominant development paradigm was made by Friberg and Hettne in 1985. The philosophy of the modern world-system is 'developmentalism', whose constitutive processes are modernisation, economic growth and nation state building, its core metaphors being progress, growth and development (Aseniero 1985, p. 51). This paradigm of development was common to both Western capitalist societies and state socialist societies. Now that the 'market' triumphalism of the end of the Cold War has turned sour, it is perhaps more obvious that capitalism and state socialism were two varieties of a common corporate industrial culture based on the values of 'competitive individualism, rationality, growth, efficiency, specialisation, centralisation and big scale' (Friberg and Hettne 1985, p. 231).

The evolutionary assumptions of 'developmentalism' imply that development is directional and cumulative, that it is predetermined and irreversible, and that it is necessarily progressive. As was briefly argued in Chapter 1, these ideas have increasingly come under attack, from social and development theory as much as from environmentalism. Peet and Watts (1996b) describe development as 'modernity on a planetary scale' (p. 19). By the end of the nineteenth century the concept of development had brought about a pervasive ordering of ideas, drawing on 'universal' concepts of science, linearity and modernisation and progress' (Cowen and Shenton 1995). These 'carried the appeal of secular utopias constructed with rationality and enlightenment' (p. 19). After 1945, development also provided the discursive and practical strategies necessary to negotiate the end of European colonialism, the rise of US political and economic hegemony, and the creation of a structured post-colonial world.

Friberg and Hettne rejected the evolutionary notion of progress and the idea that development is a predetermined and irreversible process. People make development, and 'human consciousness and will' are decisive elements within the process (1985, p. 215). The 'so-called developed societies' are neither model nor forerunner for the South (ibid., p. 219). In place of evolutionary developmentalism, Friberg and Hettne set out a direction of development which is radical and humanistic, based on the concept of 'self-realisation', the 'full realisation of the individual in every aspect of its being'; development that is 'endogenous' (1985, p. 216). They defined four Green principles of 'endogenous development' (ibid., p. 220): first, communitarianism (with development rooted in the values and institutions of a culturally defined community); second, self-reliance (at different scales within society, not autarky or national self-reliance); third, social justice; and fourth 'ecological balance' (implying an awareness of local ecosystem potential and local and global limits).

This idea of 'endogenous development' means that development is to be sought in each country's own ecology and culture, not in the supposed 'model' of a developed country. Furthermore, as development is to be through 'voluntary cooperation and autonomous choices by ordinary men and women' (Friberg and Hettne 1985, p. 221), the unit of development is not the state, but people and groups of people defined by culture, indeed 'natural communities' (ibid.). Their notion of social justice goes beyond the established idea of redistribution with growth, and monetary compensation for marginalisation and alienation, to embrace access to wealth, knowledge, decision-making and meaningful work.

The modern world-system, dominated by capitalism and embarked upon first in Western Europe, was imposed upon the periphery through geographical expansion and socio-economic penetration in association with colonialism. It aimed at control, expansion, growth and efficiency, and was legitimised by evolutionist thinking. Friberg and Hettne argued that it also caused the decline and disintegration of previous 'natural communities'. The logic of the modern project is eventually to eradicate all pre-capitalist social formations through continued modernisation in an expanding world economy, whether through the 'Blue road' of capitalism or the 'Red road' of socialist world government. In contrast, Friberg and Hettne proposed a Green strategy of 'demodernisation' (Figure 6.2). This would involve gradual withdrawal from the modern capitalist world economy and the launch of a 'new, non-modern, non-capitalist development project' based on the 'progressive' elements of pre-capitalist social orders, plus their successors, avoiding the exploitative and dehumanising aspects of some small-scale pre-capitalist societies (1985, p. 235).

Friberg and Hettne suggested that the Green project demands circumvention (or in some cases subversion) of the nation state (which they see as 'one of the greatest obstacles to "the Greening of the world"'; Friberg and Hettne 1985, p. 237). The modern project is maintained by 'modern elites' who are in the ambiguous position of increasing involvement in the world economy to increase their resource base, and at the same time effectively fostering the political, military and bureaucratic development of the nation state as a power base. The only force that can counter the power of this modern elite is a non-elite counter-mobilisation by 'counterpoint movements', 'non-party' politics, spontaneous networks and voluntary organisations.

Reactions to capitalist penetration can be violent and revolutionary, but most Third World counterpoint movements have been primarily small-scale movements of protest. Arguably these localised movements can coalesce and challenge the power of the state. There has been increasing discussion of the political significance in development of new social movements, and the effectiveness of the 'weapons of the weak' (Scott 1985, Ekins 1992, Ghai and Vivian 1992a). Ekins (1992), for example, presents a series of accounts of grassroots activism, and suggests that there is a real alternative to conventional modernism and development; he believes that 'another development' (a phrase coined by the Dag Hammarskjöld foundation in 1975) can arise out of networks of resistance against exploitation and struggles for justice.

Friberg and Hettne argued that new social movements are articulating a third (Green) vision of the world (1985, p. 248). First, there are 'traditionalists' who wish to resist capitalist penetration in the form of state-building, commercialisation and industrialisation (including 'non-Western civilisations and religions, old nations and tribes, local communities, kinship groups, peasants and independent producers, informal economies, feminist culture etc.'; ibid., p. 235). Second are 'marginalised people', including unemployed people, those who are mentally ill, handicapped people, and people in dehumanising jobs who have lost 'a meaningful function in the mega-machine' (ibid., p. 264) through pressures for increased productivity, rationalisation and automation. The third group consists of the 'post-materialists' who dominate Western environmentalism (Cotgrove

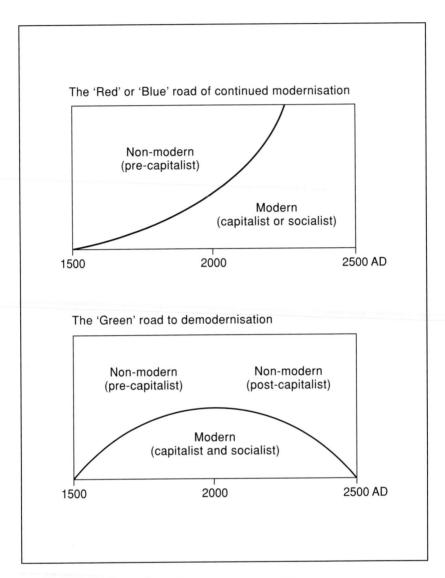

The 'Red' or 'Blue' road of continued modernisation

Non-modern
(pre-capitalist)

Modern
(capitalist or socialist)

1500   2000   2500 AD

The 'Green' road to demodernisation

Non-modern   Non-modern
(pre-capitalist)   (post-capitalist)

Modern
(capitalist and socialist)

1500   2000   2500 AD

Figure 6.2 Friberg and Hettne's 'Green' project (after Friberg and Hettne 1985).

1982), the mostly young and well-educated members of 'the ecological, solidarity, peace, feminist, communal, regional, youth, personal growth and new age movements' (Friberg and Hettne 1985, p. 248). They argue that these 'alternative' movements, from the 1950s anti-bomb protests onwards, have tended to 'converge towards the small-scale logic of functionally integrated communal societies based on direct participation and self-management' (ibid., p. 258), in which capitalist growth, central bureaucracy and 'techno-science' have a small place.

The possibility of pursuing sustainability 'from below' will be discussed further in Chapter 12.

There is no single grand theory of Green development to compare with Marxism or capitalism. However, there is a persistent core of ideas that comprise a critique of conventional modernism and developmentalism. Some of these ideas are echoed within mainstream sustainable development documents (for example neo-populism, or concern for equity and justice); others are beyond the pale. Many of these more radical Green ideas are, however, strongly reflected within socialist thought, in a set of ideas referred to as 'ecosocialism'. These are discussed in the next section.

Ecosocialism and sustainability

Friberg and Hettne tried to demonstrate the distinctiveness of 'Green' and 'Red' development strategies (see Table 6.5). These distinctions are not entirely successful. What validity they have depends on the simplistic and one-sided caricature of 'Red' or socialist or Marxist thinking that they offer. Having pointed out the existence of 'Green' strands within radical thought, for example utopian socialism, they proceed to ignore these in claiming the distinctiveness of 'Green' ideas. Furthermore, Friberg and Hettne more or less dismissed the labour movement as an element in their 'counter-structure', arguing that it has become 'wedded to the large-scale industrial system' and incorporated into the formal-rational complex (Friberg and Hettne 1985, pp. 253, 259). The

Table 6.5 Comparison of Red and Green strategies of development

	Red strategy	*Green strategy*
Oppressive system	The capitalist economy	The technological culture
Enemy	Capitalists	Technocracy
Social vision	Socialist society	Communal society
Method	Macro-revolution, socialism from above by working-class revolution	Micro-revolution, small groups withdraw from system to defend autonomous ways of life
Spatial frame	No socialist islands in a capitalist sea	Local experiments, liberated zones
At stake	Material interests, collective identity, ownership	Existential needs, personal identity, autonomy
The new person	Social transformation before personal	Simultaneous personal and social transformation
Leadership	Intelligentsia	Post-materialistic elite
Social base	Working class	Marginalised people
Institutional base	Big industrial sites	Small neighbourhood communities
Organization	Centralised, formal	Decentralised, informal
Ideology	Abstract, rational	Concrete, intuitive, open

Source: Friberg and Hettne (1985)

labour movement is therefore judged to have lost its momentum as a creative force and itself reproduces the features of the modern system. This dismissal of socialism is unhelpful, for ecosocialism is, in fact, both diverse and more interesting in its engagement with Green critiques of developmentalism.

There are strong links between 'Green' and 'Red' thinking. Rudolf Bahro, for example, has occupied a distinctive position on this boundary, both personally and intellectually. He left East Germany after the publication of *The Alternative in Eastern Europe* (Bahro 1978). In West Germany he became involved in the Green Party, because he believed that the ecological crisis would bring about the end of capitalism (Bahro 1984). In *Socialism and Survival* (Bahro 1982) he offered a double critique, of capitalism and 'actually existing socialism'. To Bahro, socialism had failed to address the ecological crisis, which brought questions onto the agenda which were 'already there before the first class society took shape' (Bahro 1984, p. 148). He attacked industrialisation in both capitalist and socialist systems, arguing that 'the increase in material consumption and production, with the inbuilt waste, pollution and depletion of resources ... is enough to destroy us in a few generations' (ibid., p. 179). Bahro criticised the 'huge structures' of industrialisation and bureaucracy, and questioned not only the kind of products made (i.e. the organisation of industrial output), but the 'reproductive process itself' (i.e. the nature of industrial society; ibid., p. 147). Bahro concluded that 'the time has come when the utopian communist and socialist visions are no longer utopian. We have reached the limit. Nature will not accept any more, and it's striking back' (ibid., p. 184).

Ecosocialists do not agree that socialism lacks the power to explain the cause of development and environmental crises, and argue that it alone can deliver solutions. David Pepper argues that historically, the labour movement *was* essentially an environmental movement, concerned with the living and working conditions, health and life-chances of the poor. He comments, 'Marxism reminds us that for most people, nineteenth century environmental problems were *clearly* socially inflicted' (1993, p. 63; emphasis in the original). The environmentalism of the 1970s onwards in industrialised countries such as the UK emphasised more strongly the more bourgeois concerns of species extinction and loss of landscape heritage, and exported these concerns to the Third World through international ideologies of conservation. However, the core concern with the quality of the environment in which people live is common both to the environmental struggles of the Third World (Gadgil and Guha 1995, Guha and Martinez-Alier 1997) and to the arguments of analysts of industrialisation concerned about the 'Risk Society' (Beck 1992, 1995; see also Chapter 10).

The relationship between environmentalism and socialism is not simple. Pepper (1993) points out the rather confused blend of Marxist and anarchist ideas in Green thought. Amin argued that environmentalism is not a single social movement, 'organised and homogeneous' (1985, p. 279), but simply part of a huge diversity of 'alternative' movements, including for example feminism. Some of these are perfectly compatible with capitalism, while others subscribe to the logic of socialism. Pepper (1993) is unimpressed with ecosocialists' enthusiasm for 'new social movements' rather than labour as the basis for a revolution in both consciousness and material social relations. He argues that these movements (the Green movement prominent among them) are 'idealistic and superstructural' (ibid., p. 136), meaning that they do not address the material basis for exploitation of people or nature; he concludes that 'they have more to do with ahistorical postmodernism than with Marxism's historical materialism' (ibid., p. 136). In particular, he argues that in the Third World, environmental struggles are still about the basic requirements for an environmentally secure life. This is a distant world from the obsession with aesthetic aspects of the environment, the politics of 'NIMBYism' and the rhetoric of 'sustainable development planning' in industrialised countries.

Socialist thought offers a significant radical critique of development, moreover one that is in many ways at least as Green as it is conventionally Red. Pepper argues that ecocentric thought is inherently anti-capitalist (1993, p. 70). In particular, there is common ground between some environmentalist visions of the future and those of decentralism, communalism and utopian socialism within the socialist tradition (for example the ideas of William Morris, with his vision of production for use and not exchange value, and production to meet human need). These grade into (but should be kept distinct from) social anarchism or anarchocommunism (Pepper 1984, 1993). Environmentalism (or 'ecologism' as Pepper calls it) is idealist rather than materialist in its approach, viewing humankind as part of a global ecosystem and subject to 'natural' laws (Pepper 1993). Ecocentrism starts with a view of nature, and attempts to develop a human response to it. Socialists start with social concerns, particularly wealth distribution, social justice and quality of life, and see the environment as an issue that vitally affects those concerns, and is in turn affected by social action (ibid., p. 46).

Environmentalist critiques of Marxism tend, as we have seen, to argue that Marx, and early Marxists, assumed that resources were inexhaustible; they have taken the state capitalism of the Soviet Union in particular as proof that Marxism has been woefully blind to the environment (Pepper 1993). However, as Redclift (1987) points out, some Marxist writers have addressed environmental and resource depletion issues. Notable among them was Hans Marius Enzensberger, who argued in 1974 that there were real scientific problems lying behind the bourgeois packaging of the environmental movement, and reiterated the 'commonplace of Marxism' which environmentalists had highlighted, the 'catastrophic consequences' of the capitalist mode of production (1974, p. 10). The editorial in that 1974 volume of *New Left Review* asserts that 'to identify and combat these has become a central scientific and political task of the socialist movement everywhere'.

It is probably true that Marxist thinkers have historically underplayed the importance of the environment and any 'environmental crisis'. Enzensberger argued that the left in Europe remained sceptical and aloof from environmentalist groups, simply incorporating selected elements of the environmental debate in their repertoire of anti-capitalist agitation (1974, p. 9). On the other hand, it can be argued that the rise of Green thinking reflected the failure of the strategies of the European left over several decades (Amin 1985). The record of Marxism on the environment is not as silent as it might have seemed in the 1970s, for diligent excavation has brought to light considerable sensitivities to the environmental dimensions of the impact of capitalism among Marxists, particularly Friedrich Engels (Parsons 1979), and as the environmental revolution unfurled, and ideas of mainstream sustainable development began to be laid out, the theoretical framework for a Marxist theory of nature was duly discussed, for example by Burgess (1978) and Smith (1984). The relevance of Marxism to environmentalism has subsequently been more fully explored, for example by Redclift (1984, 1987) and Pepper (1984, 1993). Michael Redclift called for 'a fundamental revision of Marxist political economy, to reflect the urgency of the South's environmental crisis' (1984, p. 18). Although the American and British literatures were rather parochial, he believed that the environmentalist consciousness developing in Europe offered 'a more radical and enduring critique of underdevelopment than any other currently on offer' (ibid., p. 55). Arguably, in fact, Marx and Engels were forerunners of human political and social ecology, favouring 'active and planned intervention in nature but not its triumphant and ultimately irrational destruction' (Pepper 1993, p. 62). Marx was no ecocentric, but his view of the instrumental values of nature embraced aesthetic, scientific and moral values as well as straightforwardly economic or material values (ibid., p. 64).

Marxist analysis of capitalism underlies most Green critiques of economic development. Capitalism emerged in Europe out of feudalism as a means to allow new wealth (from slavery, agricultural production, mining and simple manufacture) to be invested. Richard Peet comments that

> capitalism was made possible by the raiding of stored wealth, the reorientation of trade routes, the imposition of unequal exchange, the forceful movement of millions of people in world space, and the conversion of the people and territories of whole continents into colonies where all aspects of existence were subject to the purposes of the Europeans.
>
> (1991, p. 145)

Merchant capitalism gave way to industrial capitalism, and increasingly this has evolved through Fordism (the division of labour into specialised tasks and their integration and routinisation on a production line), and into various more flexible forms of capitalism (Harvey 1990, Peet 1991, Pepper 1993).

Capital is the result of the surplus derived from the employment of labour (that is, the difference between the value of what labour produces and the price that has to be paid to workers to persuade them to work). This 'surplus value' (profit) accrues to whatever individual or group owns the production process and the 'means of production', and 'the motive force for capitalism is the accumulation of wealth derived from profits' (Johnston 1989, p. 52). For capitalism to work, the desire to accumulate profits has to be made to seem 'natural'. Ideas about how society should be organised (forms of 'social consciousness') are closely related to (and influenced by) the way production is organised. Capitalist relations of production therefore relate to a particular way of understanding how the world works (a particular 'form of consciousness'; Pepper 1996, p. 68). Institutions (both formal institutions such as laws and informal institutions such as ideas and values) emerge that support the capitalist system, and also support the class interests that chiefly benefit from that system – that is, the owners (and increasingly the elite managers) of capital and the means of production (ibid., p. 69).

An important feature of Marx's account of the transition to capitalism is that it links the removal of people from the land (an economic alienation) with their separation from nature and consciousness of human dependence on, and impacts on, the environment (Pepper 1993, p. 72). The romanticism about nature and 'the countryside' which in due course provided a powerful root of environmentalist thinking in industrialised countries (e.g. Veldman 1994, Adams 1996) thus itself stemmed from the impact of capitalism. Under capitalism, people sell their labour power, and relate to nature similarly as an object, to be bought and sold. Increasingly nature is also a product, physically refashioned by state or business corporation, paid for at point of consumption, or is packaged as an image or a product in cyberspace (A. Wilson 1992, Adams 1996). Capitalism therefore commodifies both labour (and hence relations between people) and nature (and hence relations between people and non-human nature). There is more Marx in environmentalist thought than might at first appear.

Marxism suggests that in various ways, the capitalist system is unsustainable: it 'contains within itself the seeds of its own destruction' (Johnston 1989, p. 58). One problem is the power of large corporations to create monopolies (which negate the claimed 'efficiency' of the market). The growing internationalisation of capital and the vast size of the largest global corporations have severely restricted the capacity of national governments to restrain the profit-seeking behaviour of capital. A second problem is that increased productivity can lead to production outstripping the capacity to consume, and overproduction and reduced profitability. David Harvey argues that creative destruction is embedded within the circulation of capital itself: 'innovation exacerbates instability and insecurity, and in the end, becomes the primary force pushing capitalism into periodic paroxysms of crisis' (1990, p. 106). This means that as modern industry goes through periods of boom and stagnation, labour is subjected to instability and uncertainty as a 'normal' part of life (Harvey 1990) – the inevitable result of what Margaret Thatcher described as 'the laws of economic gravity' (Pepper 1996, p. 89). In response, enterprise managers seek to reduce costs, through cheaper raw materials, cheaper labour and better machines, stimulation of demand by advertising and finding new markets and products.

The search for cheaper materials, cheaper labour and new markets is of course the engine that drives the development/modernisation process in the Third World. The post-Second World War 'crisis' of Fordism in Europe and North

America led to both the transformation of industrial production to more flexible systems of labour organisation in the First World (deskilling, extension of automation, longer and flexible working hours, de-unionisation, reduction of job security) and the extension of Fordist production systems to the Third World, creating 'peripheral Fordism' (Harvey 1990, p. 186). Relocation of production to the Third World was a strategy aimed at reducing production costs and maximising profit. Important elements in these costs were the costs of employment (wages and related living costs, and measures to protect employees from sickness and to provide job security) and the costs of taking account of 'externalities' of production such as pollution. Relocation to countries with low wage rates, weak labour and environmental laws, and weak enforcement of those laws made perfect sense in terms of global business strategies of maximising returns on investment. David Harvey argues that while this capitalist penetration of the periphery promised emancipation from want and full integration into Fordism, it delivered instead destruction of local cultures, oppression, and various forms of capitalist domination in return for rather meagre gains in mass living standards and services (except of course for the small and soon super-affluent indigenous elites, who collaborated with and profited from the penetration of capital; Harvey 1990, p. 139).

Many analysts, including some Marxists, have lost confidence in the capacity of capitalism to 'develop' the Third World to the level of the industrialised West, thus preparing the way for an advanced socialist society (Peet 1991). In his book in 1972, Sutcliffe argued for industrialisation in large-scale units under autarchic conditions. By 1984 he had changed his view. He now suggested that capitalism produced inappropriate products (cars, weapons, obsolescent goods) and it used the wrong methods (centralised, deskilled, totalitarian and alienating; Sutcliffe 1984). Capitalist industrialisation could not therefore be the material basis for what we might now call a sustainable world. Sutcliffe added that it had also created its own mirror image in the 'centralised, statist and bureaucratic' view of state socialism. In his 1984 paper Sutcliffe urged a recapture of utopian traditions by socialist thinking on industrialisation and development. He admitted certain strengths in the critiques of industrialism by populists and intermediate technologists, and called for a search 'for a more humane alternative to economic development than the rocky path represented by actually existing industrialisation' (1984, p. 133). Rudolf Bahro also regarded established development strategies for the Third World (through increasing trade and industrialisation) as 'a tunnel without exit' (1984, p. 211), arguing that industrialism in the Third World would mean 'poverty for whole generations and hunger for millions' (ibid., p. 184). The poverty of the proletariat created by capitalist industrialisation in eighteenth- and nineteenth-century Europe was 'made bearable' by the prospect of escaping it through exploitation of the periphery. However, 'for the present periphery there is no further periphery to be exploited, no way of attaining the good life of London, Paris or Washington' (ibid., p. 211), making the prospect of proletarianisation in the contemporary Third World 'a horrific vision' (ibid., p. 184).

Pepper argues that capitalism is '*inherently* "environmentally unfriendly"' (1993, p. 91; emphasis in original): it 'continuously gnaws away at the resource base that sustains it' (ibid., p. 92). It externalises its costs, leaving them to be met by the state or in the bodies of the poor in terms of sickness or reduced life expectancy, or by future generations. Capitalism as a system (like individual enterprises) can reward those in a structural position to profit from it, but only at the expense of others elsewhere. Low and Gleeson discuss the problem of 'environmental racism', and the racially discriminatory distribution of polluting industries and hazardous waste within the USA. The same kinds of inequalities in the distribution of environmental risk and other externalities of production exist at a global scale. They comment, 'A critical aspect of this system of flows, involving the circulation of money, products and risks, is the fact that it permits developed countries to externalise risks by moving hazardous forms of production beyond their borders' (1998, p. 123).

The 'environmental crisis' is not uniform, for it affects some people much more than others. Low and Gleeson comment that 'the impacts of environmental degradation are always socially and spatially differentiated. They may end up affecting the global environment, but first they damage small parts of it' (1998, p. 19). As capital is increasingly managed globally, investing and disinvesting to maximise profit, its beneficiaries are the employed of the First World, particularly the self-regenerating class of managers; the losers are the unemployed workers of 'rustbelt' regions in the North and the vast numbers of the Third World poor. As was repeatedly pointed out at Rio, 'sustainable development' for the First World (and for wealthy areas within those countries) is all too easily built on selective unsustainable extraction of resources, unequal trade of commodities, and inhumane and polluting manufacture of products in the South.

Socialism clearly provides a powerful critique of the environmental and developmental impacts of capitalism, notwithstanding the collapse of the coercive state socialism of the USSR and Eastern Europe in the 1990s. Developing without strong institutions of civil society, the profoundly corrupt version of the 'free market' that has developed in these 'countries in transition' gives much food for thought. Ecosocialism remains an important influence on debates about sustainable development.

Ecoanarchism

Environmentalist critiques of development also draw extensively if not always explicitly on anarchist thinking, for example in the work of people such as Kropotkin and his ideas of social anarchism or anarchocommunism (Kropotkin 1972, 1974, Breitbart 1981, Galois 1976), or in the work of writers such as Murray Bookchin (1979) and Theodore Roszak (1979). In particular, Murray Bookchin's 'social ecology' was influential in the development of Green thought, for example in *Post-Scarcity Anarchism* (1971) and *The Ecology of Freedom* (1982). Bookchin argues that the domination of non-human nature by humans arises directly from the existence of hierarchy in society, and the domination of humans by humans (Eckersley 1992). There is a particular focus on the coercive power of the state: without the state and other structures of exploitation associated with hierarchy among people and structure of domination, environmental problems would not arise (Pepper 1993). Carter's attempt to devise a 'Green political theory' drawing on anarchist (and other) thought maps tight links between centralised state, inegalitarian economic relations, 'hard' technologies and militarism (Figure 6.3). As he comments, this dynamic is 'environmentally hazardous in the extreme' (1993, p. 45). An alternative dynamic can be imagined, drawing on ideas of decentralisation, participatory democracy, self-sufficiency, egalitarianism, alternative technology, pacifism and internationalism (Figure 6.4). Such a shift would be resisted, the suggested route towards it being through non-compliance with the state.

A central feature of anarchism is the importance of viewing development from the perspective of the individual; indeed Pepper (1993) suggests that

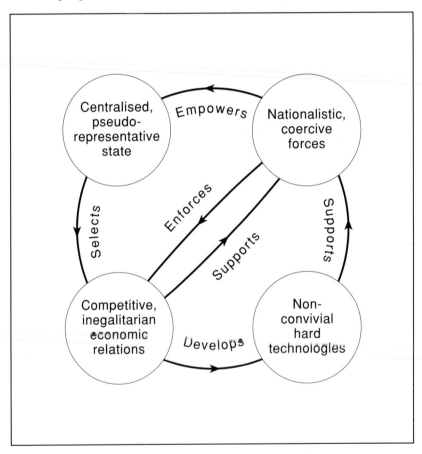

Figure 6.3 An environmentally hazardous dynamic (after Carter 1993, figure 3.2).

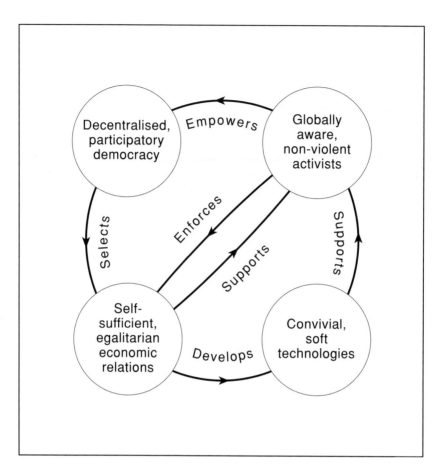

Figure 6.4 An environmentally benign dynamic (after Carter 1993).

anarchism can be viewed as a form of extreme liberalism. Kropotkin believed that 'true individualism can only be cultivated by the conscious and reflective interaction of people with a social environment which supports their personal freedom and growth' (Breitbart 1981, p. 136). Pepper (1984) points out the difference between this starting point in the individual and that of environmentalist thinking such as *Blueprint for Survival* (Goldsmith *et al.* 1972), which suggests an ecological imperative for action to achieve utopia, the human dimensions of which are secondary. Bookchin notes the environmentalist urge to protect nature from destructive societies, but argues that the social root of the destruction of nature is

> our particular civilisation, with its hierarchical social relations, which pit men against women, privileged whites against people of colour, elites

against masses, employers against workers, the First World against the Third World, and, ultimately, a cancer-like, "grow or die" industrial capitalist economic system against the natural world and other life forms.

(Bookchin in Bookchin and Foreman 1991, p. 31)

In *Person/Planet*, Roszak argues that as long as Western society remains locked into 'the orthodox urban-industrial vision of human purpose', there is no hope that poverty, and the injustice it brings with it, can be 'more than temporarily and partially mitigated for a fortunate nation here, a privileged class there' (1979, p. 317). Bookchin argues that 'in the final analysis, it is impossible to achieve a harmonisation of people and nature without creating a human community that lives in a lasting balance with its natural environment' (1979, p. 23).

Bookchin based his arguments quite extensively on ecological ideas, seeing ecology as one science which might avoid assimilation by the established social order, and he mourned the absorption of environmentalists into governmental institutions, warning that 'ecological dislocation proceeds unabated, and cannot be resolved within the existing social framework' (1979, p. 22). Ecology is integrative and reconstructive, and this 'leads directly into anarchist areas of social thought' (p. 23).

Clearly, 'nature' is being treated here as a source of values, and political ideas are being drawn from particular interpretations of the organisation or nature (Dobson 1990; Table 6.6). To ecoanarchists, collaboration (Kropotkin's 'mutual aid') is 'natural' (Pepper 1993). In this, ecoanarchism shares with other forms of ecologistic thinking the fallacy of treating human understandings of non-human nature at a particular point in time (including the notion of separation of 'human' self and the non-human 'other') as true, and using that understanding as the basis for political argument. If understanding of nature (whether through science or other means) is seen to be contingent on particular historical moments and social processes, the 'naturalness' of that arrangement is at once seen to be also a social construct.

Pepper (1993) argues that ecoanarchists tend to be vague about why the problems they identify in society (and its treatment of nature) emerged. For example, if states are unnatural, what forces brought them into being; if people 'naturally' cooperate, why is altruism so rare? He believes that Green political writers often fudge the boundary between anarchism and socialism. Eckersley

Table 6.6 Political ideas drawn from nature

Perceived attribute of nature	*Political idea or principle*
Diversity	Toleration, stability and democracy
Interdependence	Equality
Longevity	Tradition
Nature as 'female'	A particular conception of feminism

Source: Dobson (1990, p. 24)

(1992) distinguishes between two strands in ecoanarchism, social ecology and 'ecocommunalism'. Pepper (1993) suggests that fuzzy Green thinkers fasten loosely onto the latter, proposing a form of anarchocommunalism that claims to attempt to bypass or subvert the state through communal living and lifestyle change. Such ideas have played a major role in the mainstream Western environmental movement (captured perhaps in the now tired slogan 'think globally, act locally' and in the dubious logics of 'green consumerism'), although as environmental organisations in the North have adopted increasingly complex corporate structures in the 1990s it is more than ever clear than anarchism is uncommon and perhaps unsustainable in the structure of environmental organisations, whether environmental pressure groups or Green political parties. Dave Foreman comments, 'I guess if you organise yourself like a corporation, you start to think like a corporation' (in Bookchin and Foreman 1991, p. 38). Of course, most Northern countries can show peaceful (and sometimes long-lived) 'alternative' communities, but Pepper reminds us to be suspicious of utopias, anarchist or otherwise, and the belief that changing minds and vocabularies can change material relations (noting that 'the world cannot be restructured by moral example'; 1993, p. 150). He also notes the danger of anti-urban (and anti-working class) romanticism.

Anarchists also commonly advocate political tactics that include strikes, boycotts and demonstrations (predominantly non-violent, although commitment to pacifism and non-violence varies, and is hotly debated). It is easy to see the influence of anarchist ideas in the work of First World environmental organisations such as Greenpeace and Earth First!, and the self-avowedly 'eco-anarchist' opposition to roads and other projects in countries such as the UK (Rawcliffe 1998). It is also possible to trace related traditions elsewhere, for example in the techniques of peaceful direct action developed by Mahatma Gandhi in India, and in more recent struggles such as the Chipko movement and Narmada Bachao Andolan in India (Guha and Martinez-Alier 1997). These will be discussed further in Chapter 12.

However, Pepper admits that anarchism, anarchocommunism and decentralist socialism lie close alongside each other, and they have been drawn on extensively by those developing radical Green critiques of society. All demand drastic changes to 'business as usual', and all provide a very different picture of sustainability from mainstream sustainable development:

> The anarchist concepts of a balanced community, a face-to-face democracy, a humanistic technology and a decentralised society – these rich libertarian concepts – are not only desirable, they are also necessary. They belong not only to the great visions of a human future, they now constitute the preconditions for human survival.
>
> (Bookchin 1979, p. 27)

To Bookchin, the 'ecological crisis' was just part of a larger malaise: 'Humanity has produced imbalances not only in nature but, more fundamentally, in relations amongst people and in the very structure of society.' Therefore, 'what we are seeing today is a crisis in social ecology', in which Western society is 'being organised round immense urban belts, a highly industrialised agriculture and, capping both, a swollen bureaucratised, anonymous state apparatus' (1979, p. 25). Bookchin's 'ecological anarchism' is therefore essentially anti-industrial, anti-bureaucratic and anti-statist. In some ways it shares ground with Schumacher's *Small is Beautiful* (1973), but for Bookchin that approach demands adaptation to the norms of society, not the 'revolutionary opposition' he believes is required.

Deep ecology and sustainability

Much of the environmentalist critique of development has drawn on 'ecocentric' or 'biocentric' ideas which contrast with the technocentrism of mainstream sustainable development (O'Riordan 1981). The core of biocentrism is the ascribing of intrinsic values or moral status to non-human nature. Dobson (1990) describes the political ideology of 'ecologism' as seeking 'to persuade us that the "natural" world has intrinsic value: that we should care for it not simply because this may be of benefit to us' (p. 49). A conventional distinction is drawn between 'Deep Ecology', which is biocentric in this way, and 'Shallow Ecology', which is anthropocentric, and concerned about the values of nature for the human species. A parallel distinction has commonly been drawn between 'dark' or 'light' Green thinking (e.g. Wissenburg 1993). It should be noted that the word 'ecology' here does not represent in any direct way the practice of the experimental natural science of ecology (e.g. McIntosh 1985), but rather the looser 1970s American meaning of an 'ecology movement' (e.g. Chisholm 1972). Clearly, mainstream sustainable development as described in Chapter 5 draws chiefly on the 'shallow' end of the ecology movement.

This distinction, and the phrase 'deep ecology', come from the writings of the Norwegian philosopher Arne Naess. Arne Naess began to consider ecological philosophy in the late 1960s, drawing on philosophy of Spinoza and following his own previous work on Gandhian thought. He first wrote about deep ecology in 1973 in a paper explicitly contrasting 'shallow ecology' and 'deep long-range ecology' (Naess 1973). He suggested that the shallow ecology movement had as its central objective the health and affluence of people in developed countries, and fought against pollution and resource depletion. Deep ecology differed in two ways. First, it rejected this anthropocentrism and the separation of 'human' and 'environment' (and indeed the separation of 'thing' and 'its milieu') in favour of a 'total field model', of organisms as knots in a field of intrinsic relations. Second, deep ecology was based on the principle of 'biospherical egalitarianism', and a recognition of the equal rights of organisms to live and blossom. Naess argued that the anthropocentric restriction of this right to humans was detrimental to life quality for humans themselves, and that the attempt to ignore the interdependence between humans and other organisms and to establish a master–slave role had contributed to the alienation of humans from themselves (Reed and Rothenberg 1993).

Naess has sought to set out a philosophical system that relates self to nature, which he calls an 'ecosophy', a personal philosophy or a code of values and a view of the world that guides personal decisions about relations with the natural world (Reed and Rothenberg 1993). He called his own version 'ecosophy T', and it was offered not as a finished system of thought, but as a means for other people to develop their own personal ecosophies. His ideas continued to evolve, and in 1984 he produced the 'deep ecology platform' (see Table 6.7), to establish a common ecophilosophical ground for deep ecology (Sessions 1995).

This platform was not an attempt to define a 'deep ecology' dogma. George Sessions, and other commentators, have pointed out their various proposed modifications or elaborations. This is important, for the very openness of Naess's account of deep ecology has allowed an enormous diversity of thinking to develop and claim this label. By the mid-1980s, discussion of 'deep ecology' had begun to develop through a varied mix of writings by people such as Warwick Fox, George Sessions and Bill Devall (Fox 1984, 1990, Devall and Sessions 1985). This writing tended to divide into a development of ecosophical thinking, notably in the 'transpersonal ecology' of Warwick Fox (1990), and 'a range of normative and sometimes radical visions of the human relationship with nature' (Reed and Rothenberg 1993, p. 2), for example in the work of Devall and Sessions (1985).

Deep ecology emphasises the transcendental attributes of nature. Graber suggested that the 'wilderness ethic' is strongly religious in character (1976,

Table 6.7 The deep ecology platform of Arne Naess

- The well-being and flourishing of human and non-human life on earth have value in themselves (synonyms: inherent worth; intrinsic value; inherent value). These values are independent of the usefulness of the non-human world for human purposes.
- Richness and diversity of life forms contribute to the realisation of these values and are also values in themselves.
- Humans have no right to reduce this richness and diversity except to satisfy vital needs.
- The flourishing of human life and cultures is compatible with a substantial decrease of the human population. The flourishing of non-human life demands such a decrease.
- Present human interference with the non-human world is excessive, and the situation is rapidly worsening.
- Policies must therefore be changed. The changes in policies affect basic economic, technical and ideological structures. The resulting state of affairs will be deeply different from the present.
- The ideological change is mainly that of appreciating life quality (dwelling in situations of inherent worth) rather than adhering to an increasingly higher standard of living. There will be a profound awareness of the difference between big and great.
- Those who subscribe to the foregoing points have an obligation directly or indirectly to participate in the attempt to implement the necessary changes.

Source: Reed and Rothenberg (1993)

p. 111). 'Wilderness purists' draw on the works of Thoreau, Muir and Aldo Leopold for inspiration and group definition. Deep ecologists too reference themselves by the writings of such people and their sense of moral order in nature, and of the continuity between humans and other organisms (and indeed inanimate nature). Deep ecology calls for a new relation with nature that challenges both established anthropocentric utilitarian ideas (i.e. conventional 'development') and the managerialist reformism of mainstream sustainable development.

In his book *Simple in Means, Rich in Ends*, Bill Devall attempted to outline the basis of a *practice* of deep ecology, taking the phrase to refer to 'finding our bearings, to the process of grounding ourselves through fuller experience of our connection to earth' (1988, p. 11). One element in such thinking is the notion of bioregionalism, a deliberate focusing on the 'homeland of ecological self' (ibid., p. 58). Devall and Sessions argued that 'many individuals and societies throughout history have developed an intuitive mystical sense of interpenetration with the landscape and an abiding and all-pervading "sense of place" (1985, p. 241). Katz and Kirby (1991) speak of 'constructs of the Native American lifeworld' – a system in which 'there exist no dualities between humans and nature, or necessarily between animate and inanimate' (p. 262). This shift in consciousness reflects the wider angst at modernity and globalisation that has fed Northern environmentalism, and reflects the relatively narrow social and political base of that movement in terms of education, wealth and employment (see Cotgrove 1982). It has significance for debates about sustainable development because of the continuing influence of Northern environmentalism on ideas about nature (and its 'development') in the Third World, and because of the increasing global exchange of such ideas (Guha and Martinez-Alier 1997).

Another element of deep ecology thinking is the notion of 'ecodefence' ('monkeywrenching' or 'ecotage'; Devall 1988, named from Edward Abbey's novel *The Monkey Wrench Gang*, 1975). This is epitomised in the industrialised world by the ecoradicalism of Earth First!, and the development of direct-action protests against environmentally destructive development, for example clear-felling of old-growth forests in western Canada and the north-west USA, or against roads and other developments in the UK (Bookchin and Foreman 1991, Devall 1988, Rawcliffe 1998). An important (although not universal) element in such deep ecology-inspired protest against ecosystem transformation is non-violence: Arne Naess himself took part in peaceful non-violent protests against the construction of a dam at the Mardøla Falls in Norway in the 1970s (Reed and Rothenberg 1993).

The resonance between such protests and those of Third World organisations such as Narmada Bachao Andolan in India (Guha 1989, Guha and Martinez-Alier 1997) is obvious, although the depth of their similarity is less clear. Certainly the radical 'eco-warriors' of the USA in particular have alarmed many observers. Murray Bookchin argued that radical ecologism (for example that of Earth First!) was potentially anti-social and anti-human (Bookchin and

Foreman 1991), and clearly that movement lacked a consistent and clear social analysis of ecological crisis, and a consistent commitment to humane social ethics. Strategies such as 'tree-spiking' (driving nails or metal rods into timber trees as a protest against clear-fell logging) threatened the lives and safety of timber workers, and took such actions well beyond conventional 'monkey-wrenching' (cf. Abbey 1975). The eventual renouncement by Earth First! of tree-spiking in 1990 and the search for collaboration with timber workers against the unsustainable pursuit of profit by logging companies took them back towards pacifism and non-violence, although the move was itself met with violence by the US state (Rowell 1996).

Pepper (1993) notes the dangers of reactionary political ideas as part of Bioregionalism. The extreme right-wing politics associated with 'survivalist' groups in the American West, and the violence of the US 'Wise Use' movement against radical environmentalists mark a significant departure from the more demure symbolic protests that have characterised environmental movement in the past (Rowell 1996). Critics of deep ecology have been quick to criticise the misanthopy and glorification of violence that have been part of the history of (and remains an element within) this strand of radical Green thought (Lewis 1992).

The biocentrism of deep ecology potentially undermines the moral basis of most development action. It is easy to identify a possible alliance between the neo-Malthusian science-based critique of population growth (with its apparently sound concepts such as 'carrying capacity') that sees famines as somehow 'natural' and biocentric ideologies that identify people as organisms with no special rights, and that see intervention to sustain human lives at the expense of other organisms and inanimate objects as unacceptable. This kind of 'Green' thinking is perhaps what Amin has in mind when he speaks of Green ideas as 'a form of religious fundamentalism' (1985, p. 281). Such ideas can generate the deeply conservative ideology and reactionary and repressive politics of 'ecofascism' (Pepper 1984). Lewis excoriates 'harsh deep ecology' as 'primitivism', accusing it of advocating 'the active destruction of civilisation' (1992, p. 28).

Ecofeminism and sustainability

The biocentrism of deep ecology is both echoed and challenged by ecofeminism. Awareness of the importance of gender in relations between people and non-human nature has grown apace through the 1980s and 1990s (Nesmith and Radcliffe 1993, Jackson 1994). In particular, 'ecofeminism', or 'ecological feminism' (Warren 1994), has become an important challenge both to mainstream sustainable development thinking, and at the same time to other radical streams of thought such as deep ecology. However, the world of ecofeminism is (once again) rather complex. Rocheleau *et al.* (1996a) identify a series of schools of thought within which environment and gender engage, including feminist environmentalism, feminist poststructuralism and socialist feminism. Environmentalists have begun to engage with liberal feminist agendas, to consider women as actors in environmental management, and (in terms of policy) as active agents in environmental projects. This approach mirrors wider changes in development thinking, and recognition of the significance of gender as a factor in the management of the environment (argued for example by Townsend 1995). However, in practice a simplistic and functionalist focus on gender can lead to a policy straitjacket that simply sees women as cost-effective 'target groups' in development (Elmhirst 1998), and the erroneous notion that policy can be devised that is synergistic in addressing problems of population, environment and development (Jackson 1993).

Feminist poststructuralism (Rocheleau *et al.*) builds on poststructuralist ideas about development (e.g. the ideas of Arturo Escobar, 1995) and on feminist critiques of science, for example by Donna Haraway (e.g. 1991). Haraway writes about research on primate behaviour, and shows that, over time, competing scientific 'stories' in the academic literature (based on 'scientific' data) mirror the social and political world in which the scientists moved: ideas about what women were like, and how they ought to behave, were inextricably linked to the conclusions scientists drew about non-human apes. Haraway's analysis suggests that science cannot produce meanings that are free of their context. Nature cannot exist without social meaning (Fitsimmons 1989).

Socialist feminists have addressed the importance of gender in an understanding of political economy, and hence have focused on gender divisions of production and reproduction (Rocheleau *et al.* 1996a), and issues of access to and rights over land and other environmental resources (e.g. Carney 1993). Thus Bandarage (1984) fiercely criticised liberal feminism and the 'Women in Development' (WID) school, which, while it correctly identifies the systematic impoverishment and disempowerment of women, fails to escape the Western modernisation model of development: WID is a movement *for* women, not *of* women. By contrast, she argued that a 'Marxist–feminist synthesis' could 'situate sexual oppression historically as it interacts with class oppression and imperialism' (1984, pp. 504–5).

Ecofeminists argue that the coercive relations between humans and non-human nature are the result of an essentially gendered process of exploitation. Ecofeminist analysis argues that the same structured oppression of women by men (patriarchy) is reflected in patterns of imperialism and capitalist accumulation; there is a dual subjugation of both women and nature (Mies 1986, Shiva 1988). Furthermore, this oppression extends to science and other forms of hegemonic 'Western' knowledge. In her influential book *Staying Alive*, Vandana Shiva portrayed science as a 'masculine and patriarchal project' that 'necessarily entailed the subjugation of both nature and women' (1988, p. 15). She argues that a gender-based ideology of patriarchy underlies ecological destruction, and the definition of nature as passive and requiring taming and 'development':

> From the perspective of Third World women, productivity is a measure of producing life and sustenance; that this kind of productivity has been rendered invisible does not reduce its centrality to survival – it merely

reflects the domination of modern patriarchal economic categories which see only profits, not life.

(ibid., p. 5)

Thus Shiva argued that in the forests of India, reductionist Western science, driven by capitalist profit maximisation, had marginalised women and degraded their environment. The destruction of the forest and the displacement of women were both structurally linked to the reductionist (and capitalist) paradigm of science. Attempts to deal with deforestation that stem from the same patriarchal science and capitalism exacerbate both the crisis of human survival and that of environmental degradation. Salvation, the recovery of 'the feminist principle in nature', and of the view of the earth as sustainer and provider, is to be found in the capacity of women, specifically Third World women, to challenge established male-dominated modes of thinking about knowledge, wealth and value. Shiva argues that 'the intellectual heritage for survival lies with those who are experts in survival' (1988, p. 224). Only they have the knowledge and experience 'to extricate us from the ecological cul-de-sac that the western masculine mind has manoeuvred us into' (ibid.). She called for a 'non gender-based ideology of liberation' (ibid., p. xvii).

However, at the core of Shiva's analysis is an essentialist proposition: that men and women are essentially different, and that women are closer to nature than men. Thus 'the reductionist mind superimposes the roles and forms of power of Western male-oriented concepts on women, all non-Western peoples and even on nature, rendering all three "deficient", and in need of "development"' (p. 5). What Nesmith and Radcliffe call 'spiritual environmental feminists' have affirmed the female–nature connection. However, many other feminists have challenged that essentialism, while agreeing with the history of patriarchy and the twin domination of nature and women (Nesmith and Radcliffe 1993).

Jackson (1994) argues that ecofeminism, and ecocentric environmentalism as a whole, is essentialistic in its understanding of both women and environments. The problem is that women are conceived of as a category with universal characteristics that transcend place, time and material circumstances. Ecofeminists attempt to revalue the feminine, and emphasise 'feminine' values such as caring and nurturing as universal attributes of women, and the possible basis for a strategy for reworking human relations with non-human nature. However, if 'feminine' has been defined in distinction from 'masculine', can a feminine essence be defined? (ibid.). Ecofeminist discourse is 'innocent of gender analysis in which masculinity and femininity are relational, socially constructed, culturally specific negotiated categories' (ibid., p. 125). The acceptance of a fundamental similarity between women and nature invites the continued oppression of both.

Val Plumwood (1993) critiques the notion of 'special' connection between women and non-human nature, and argues for an anti-dualistic 'critical ecological feminism' that challenges, among others, the dual categories of human/

nature and man/woman. She suggests that the fierce debates within 'Green theory', particularly the two-sided argument (largely between male protagonists) between social ecology and deep ecology, have obscured the potential for explanations of human domination of nature that are compatible with older critiques of hierarchy and human social domination (ibid.). She describes a 'web of oppression' whose source is the ideology of the control of reason over nature: oppressed groups have been counted as part of the 'chaotic and deficient' realm of nature by the mastering and ordering 'reason' of Western culture (ibid., p. 74).

Feminist views on the environment and development are diverse and lively, evolving very rapidly. Almost all, however, share one characteristic, which is the fierce challenge they offer to developmentalism and the status quo of the world systems, and hence to the conformism of mainstream sustainable development.

Sustainability and radical environmentalism

Radical streams in writing on the environment are rich and strong. They are of great importance to understanding sustainability. A coherent understanding of how society and nature relate must go beyond the simple oppositionism of conventional Western environmentalism, and the limited reformism of mainstream sustainable development thinking. That thinking provides no answers to the proposed 'environmental crisis' beyond reform of the procedures and organisation of development planning. Its view of society and environment is restricted, untheorised and naïve.

In order to move beyond this, it is necessary to take account of political economy, and move outside environmental disciplines, and outside environmentalism, to approach the problem from political economy and not environmental science. There is great power in an approach to the understanding of environmental aspects of development which uses the insights of both natural and social science, as Blaikie (1985) and Blaikie and Brookfield (1987) do, in the process building bridges between environmental issues and radical social studies. Their work has led to the burgeoning field of *political ecology* (Bryant and Bailey 1997, Bryant 1998, Rocheleau *et al.* 1996b, Peet and Watts 1996b, Haila and Levins, 1992), which will be discussed further in Chapter 9.

All relations between environment and people are political, just as all development is ideological. In a preface to the 1979 reprint of his paper 'Ecology and revolutionary thought' (originally written in 1965), Murray Bookchin argued that 'the domination of human by human lies at the root of our contemporary Promethean notion of the domination of nature by man' (p. 21). The way people relate to non-human nature around them (their environment) – as well as the way they understand it – is created by culture and bounded by social relations, by structures of power and domination. Development itself is a product of power relations, of the power of states, using capital, technology and knowledge, and the market to alter the culture and society of particular groups of people. States co-opt cultures while the world-system engages indigenous economies. We call the result of this process development.

Development is about control, both of nature and of people. It is at this level that mainstream sustainable development thinking is seen to be so profoundly unradical. For although it is a major strength of a great many of the environmental critiques of development (including many of those that are naïve and limited in scope) that they adopt an essentially moral critique of the development process, they offer in its stead simply another version of the same paradigm. Mainstream sustainable development depends on arguments about capitalist mutualism at the scale of relations between nation states. Mainstream sustainable development is also deeply dependent on a vision of managerialism drawn from ecology and, in turn, from systems theory (McIntosh 1985). This offers an ideology of control that has attained a wide currency within environmentalism, and it has an important position in the thinking of environmentalists about development. The fact that the ideological component of systems theory is often cryptic does nothing to lessen its significance (Gregory 1980).

The view of sustainable development as essentially a managerial process, where reform of procedures will ensure some 'optimal' outcome, fails to address the central issue of the ideology of the 'modern project' (to use Friberg and Hettne's 1985 phrase). Michael Jacobs reminds us not to forget that sustainability is necessarily about ideology:

> It needs to be remembered after all that sustainable development and sustainability were not originally intended as 'economic' terms. They were, and remain, essentially political objectives, more like 'social justice' and 'democracy' than 'economic growth'. And as such their purpose, or 'use', is mainly to express key ideas about how society – including the economy – should be governed.
>
> (Jacobs 1995, p. 65)

Development creates losers as well as winners. It is something that is done to people, usually by governments, often backed by ideologies and resources from elsewhere and driven by the market. To the recipient, the target, this 'development' holds many terrors, many costs as well as benefits. States make decisions on behalf of people regarding proper balances of costs and benefits, and about where those costs should be borne and those benefits enjoyed. Sometimes (as environmentalists devastatingly point out) the two do not balance. Even more often, the costs and benefits are unequally shared.

In a sense, then, development is a double-edged phenomenon, a tiger that governments ride, and to which they attach (sometimes by force; Crummey 1986) the lives of individuals and groups of people. This is a perspective that surfaces in some environmentalist thinking on development (such as voiced concern for indigenous people), but is rarely included in theoretical work.

In development the domination of nature is part of wider political and economic processes. It is impossible to understand the relation between development and nature without considering and comprehending political economy. However, once we escape from the straitjacket of evolutionary thinking about 'development', which argues that it necessarily involves a progression towards 'better' conditions, it is possible to start to consider anew the persistent questions raised by environmentalist critiques of development practice. Specifically, these are: How do poverty and environmental quality relate to each other? What impact do development actions have on the environment? Why is the reform of development practice so difficult to achieve? These will be considered in the remaining chapters of this book.

Chapter 7 will focus on the central issues in debates about sustainable development, the question of environmental degradation, particularly dryland desertification. What do we know about threats to drylands? How are ideas about degradation affected by ideology? Is it possible for farmers or pastoralists to manage their environments sustainably, and if so, what impacts do policies to 'combat desertification' have on them? Chapter 8 will then go on to discuss the environmental impacts of development, looking at water resources, and particularly at dams and irrigation schemes. Do such projects on balance work, and if not, why not? What implications do their failures have for our understanding of sustainable development? These two chapters will conclude that the interactions of environment and society need to be understood in political terms; this theme will be picked up in Chapter 9 in a discussion of the political ecology of sustainability, and in Chapter 10 in a discussion of Risk Society and environmental class. The final chapters (11 and 12) will look at policy for promoting sustainability, discussing the mainstream sustainable development approach to environmental risk, and the feasibility of alternative strategies to seek sustainability through development from below.

Summary

- A number of radical ideas about environment and society form countercurrents to the ideas of mainstream sustainable development discussed at Rio in 1992.
- Neo-Malthusian ideas about the dangers of global population growth were an important element in 1970s environmentalism. Views on overpopulation have evolved considerably, but neo-Malthusian ideas remain a significant theme in sustainable development thinking.
- Radical Green critics of developmentalism reject the possibility that capitalism and industrialism can deliver justice, equity and humane conditions of human life. They suggest that the modern world-system is flawed, generating interlinked crises (militarisation, poverty, human repression and environmental destruction): mainstream sustainable development cannot bring about the changes required.
- There is an important stream of radical thought in socialism, although historically much socialist writing (particularly in Marxism) underplayed the importance of the environment. Ecosocialism includes a range of polit-

ical ideas, including decentralisation, communalism and utopian socialism, and the contribution of socialist ideas to Green thinking is greater than is commonly recognised. Socialism provides a powerful critique of the environmental and developmental impacts of capitalism.

- Anarchist thinking also provides a basis for radical Green ideas about development, particularly in Murray Bookchin's 'social ecology'. Ideas with anarchist roots such as decentralisation, participatory democracy, self-sufficiency, egalitarianism and alternative technologies are important elements in environmentalism, and of direct relevance to debates about sustainable development.
- Biocentrism (the recognition of intrinsic values or moral status in non-human nature) is an important theme in environmentalism. Deep ecologists reject anthropocentrism, and their ideas offer a profound challenge to many of the assumptions on which mainstream sustainable development is based, although they also raise critical political questions.
- Ecofeminism offers a diverse range of ideas that challenge mainstream sustainable development, for example in emphasising the patriarchal roots of developmentalism and capitalism. There are significant differences between separate strands of feminist thought, for example the nature of any special connection between femininity and nature.
- Radical ideas about environment and development have by no means been suppressed by the routinisation of environmental concern in mainstream sustainable development. They emphasise in particular the political dimensions to decisions about society and nature, and challenge the notion that sustainable development can be achieved through technical and managerial responses alone.

Further reading

Bookchin, M. (1982) *The Ecology of Freedom: the emergence and dissolution of hierarchy*, Cheshire, Palo Alto, CA.

Dobson, A. (1990) *Green Political Thought*, HarperCollins, London.

Eckersley, R. (1992) *Environmentalism and Political Theory: toward an ecocentric approach*, UCL Books, London.

Ekins, P. (1992) *A New World Order: grassroots movements and global change*, Routledge, London.

Guha, R. and Martinez-Alier, J. (1997) *Varieties of Environmentalism: essays North and South*, Earthscan, London.

Lewis, M.W. (1992) *Green Delusions: an environmentalist critique of radical environmentalism*, Duke University Press, Durham, NC, and London.

Morse, S. and Stocking, M. (eds) *People and Environment*, UCL Press, London

Pepper, D. (1993) *Eco-socialism: from deep ecology to social justice*, Routledge, London.

Pepper, D. (1996) *Modern Environmentalism: an introduction*, Routledge, London.

Redclift, M.R. and Benton, T. (eds) (1994) *Social Theory and the Global Environment*, Routledge, London.

Reed, P. and Rothenberg, D. (eds) (1993) *Wisdom in the Open Air: the Norwegian roots of deep ecology*, University of Minnesota Press, Minneapolis.

Sessions, G. (ed.) (1995) *Deep Ecology for the 21st Century: readings on the philosophy and practice of the new environmentalism*, Shambhala, Boston.

Web sources

<http://userpage.fu-berlin.de/~jdingler/ecofem.html>　This webpage of Johannes Dingler (Frei Universität Berlin) gives a vast list of publications on ecofeminism. There is a similar list for deep ecology on *<http://userpage.fu-berlin.de/~jdingler/deepeco.html>*.

<http://www.heureka.clara.net/gaia/deep-eco.htm>　An enthusiast's account of deep ecology, by Keith Parkins.

<http://www.ecotrust.org/>　Bioregionalism in action: Ecotrust is a non-governmental organisation seeking to promote a 'conservation economy' along the western North American coast, from San Francisco to Anchorage. Read about the 'bioregional pattern language' of the coastal region on *<http://www.conservationeconomy.net/>*.

<http://homepages.together.net/~jbiehl/>　This site, of the Social Ecology Project of Burlington, Vermont, presents the work of a group of writers and activists (including Murray Bookchin) working on social ecology. Also of interest is the website of the Institute for Social Ecology: *<http://www.social-ecology.org/>*.

7 Environment, degradation and sustainability

Desertification is caused by the excessive pressures of overuse on productive ecosystems that are inherently fragile.

(Mustafa Tolba 1986, p. 260)

Governments and their advisers have very stereotyped views about what development in dry areas should consist of.

(Jeremy Swift 1982, p. 168)

Environmental degradation

When the forester E.P. Stebbing visited West Africa in 1934, a representative of the Emir of Katsina was among those who showed him evidence of recently deteriorated conditions in the far north of Nigeria. He suggested that West Africa was undergoing progressive desiccation and that the Sahara was moving southwards: the 'silent invasion of the great desert' (Stebbing 1935, p. 518). He saw the savanna as a form of open deciduous forest, progressively degraded by burning and shifting cultivation, grazing and browsing, and then destructive pollarding by graziers: 'Under this system of treatment the final extinction of the savanna forest takes place, when the weakened roots and vanishing rainfall result in the death of the trees' (ibid., p. 513).

What Stebbing described in that now famous (or infamous) passage is the phenomenon that has become known as desertification, the degradation of productive land in dry regions. Concern about desertification has evolved a great deal since Stebbing's work in the 1930s, but it remains vigorous. Fifty years later, Williams and Balling could write in the *Desertification Control Bulletin* that desertification was 'a direct threat to 250 million people around the world and an indirect threat to a further 750 million people' (1995, p. 8). The United Nations Environment Programme (UNEP) estimated that 70 per cent of all agriculturally used drylands were to some extent degraded, some 3.5 billion hectares in over 100 countries (ibid.).

Desertification is the most obvious example of a much wider range of concerns about the misuse of the environment by people (and more specifically by poor people in harsh parts of the rural Third World), and has formed an important element in environmentalist thinking about 'ecological crisis' for more than a century (Anderson 1984, Grove 1987).

However, it is not just environmentalists who have preached about the evils of desertification: for two decades it has been an obsession of international organisations, national governments and a wide variety of drylands researchers. In the semi-arid Third World, the desert stalks green fields, pastoralists' livestock nibble inexorably away at the basis of their own subsistence, and burgeoning human and animal numbers threaten the long-term sustainability of soil productivity and rural economies. Elsewhere, small farmers in forest clearings, or clinging to steeply terraced mountain land, teeter on a slender cusp between survival and both economic and ecological disaster.

Desertification shares with one or two other issues (notably the destruction of tropical rainforest) the distinction of being widely assumed to prove the unsustainable nature of human relations with the environment on a global scale. These are what were described in *Our Common Future* as 'environmental trends that threaten to radically alter the planet, that threaten the lives of many species upon it, including the human species' (Brundtland 1987, p. 2).

Environmental degradation by definition reflects a failure of sustainable management, and there is a vast literature that argues (sometimes with evidence and sometimes without that burden) that many environments are threatened by piecemeal degradation. In places this may be the case, but it is increasingly being realised that it is not everywhere so. Very often, research reviews and oft-repeated generalisations have been based on false understandings of environmental dynamics, poor data and sloppy thinking. The spectre of environmental degradation is so compelling that too many observers have been swept along by it, and have not demanded to see the evidence.

The question of environmental degradation is central to the discourse of mainstream sustainable development. In the industrialised world, concern about environmental degradation has centred on the impacts of urbanisation and industrialisation, for example in the problems of pollution, toxic and other waste, and loss of wildlife and biodiverse habitats. These issues are increasingly important in the Third World also, particularly in those countries that are industrialising, and will be discussed in Chapter 10.

However, the literature of mainstream sustainable development has given the central place in debates about environmental degradation in the Third World to the degradation of rural environments through the actions of the poor themselves (a constant refrain). Third World environments are repeatedly portrayed as suffering degradation at the hand of the teeming multitudes of the Third World poor, being eroded and destroyed by the multiple problems of overgrazing, soil erosion and desertification. Furthermore, unlike environmentalist complaints about pollution or resource depletion by industrial society, concern about human-induced environmental degradation and the supposedly rapacious appetites of the poor has been substantially absorbed into mainstream environmental discourse and development policy.

This chapter considers the issue of environmental degradation, and explores why scientists and policy-makers hold the views they do about it. It focuses in particular on the problem that Stebbing highlighted in the 1930s, desertification, although it also ranges more widely to discuss studies of persistent expert misunderstanding of the environment elsewhere, in forest and mountain environments.

The desertification industry

The links between environmental degradation, poverty and development are seen most clearly in the case of arid land desertification. Desertification became one of the recurrent themes of environmentalist writing on the Third World in the 1980s (e.g. Grainger 1982, Brown and Wolf 1984, Timberlake 1985). The close links between drought, land management, environmental conditions and poverty have been the subject of endless speculation and considerable field research for decades, and particularly since the Sahel drought of 1972–4. A typical set of relationships identified by researchers is shown in Figure 7.1, in this case referring to rain-fed agriculture in the Gambia (Baker 1995). The key question, however, when one is faced with a nexus of processes and phenomena of this kind, is to identify correctly cause and effect, and to separate hypothesis from empirically observed conclusion. Dominant ideas about desertification offer so all-inclusive a model of human action and environmental degradation that their hypotheses are all too easily mistaken for fact.

Desertification has long been an integral theme within mainstream sustainable development. The *World Conservation Strategy* saw desertification as 'a response to the inherent vulnerability of the land and the pressure of human activities' (IUCN 1980, para. 16.9). It described the way in which 'overstocking' has severely degraded grazing lands in Africa's Sahelian and Sudanian zones: farmers 'moving onto land marginal for agriculture' displace pastoralists onto land marginal for livestock rearing, while in the Himalayas and Andes 'too many improperly tended animals remove both trees and grass cover . . . and erosion accelerates' (para. 4.13). The Brundtland Report took as read the alarmism of UNEP estimates of the problem of desertification, stating that 'each year another 6 million hectares of productive dryland turns into worthless desert' (Brundtland 1987, p. 2). *Agenda 21* devoted a chapter to the problem of 'combating desertification and drought', as one of two under the rather emotive title 'Managing fragile ecosystems' (the other was on 'sustainable mountain development').

International awareness of desertification burgeoned following the 1972–4 Sahelian drought. Following debate on the floor of the United Nations, the United Nations Conference on Desertification (UNCOD) was held in Nairobi in 1977, organised by UNEP. In that year UNEP was made responsible by the UN General Assembly for coordinating implementation of a Plan of Action to Combat Desertification (PACD). UNEP established a desertification branch, an Interagency Working Group and a Consultative Group on Desertification

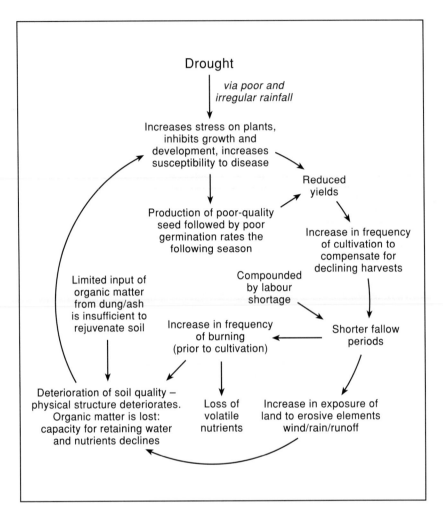

Figure 7.1 Drought and dryland agriculture and environmental degradation in the
Gambia (after Baker 1995).

Control, and began publication of the *Desertification Control Bulletin*. There
have been a series of editions of a *World Atlas of Desertification* (Middleton
and Thomas 1997). This institutional interest, and the persistence of Sahelian
and African drought through the 1970s and 1980s, kept desertification on the
international agenda (see for example Mabbutt 1984, Tolba 1986, Rapp 1987,
Agnew 1995, Warren 1993, 1996). In fact, the issue was so prominent that
Thomas and Middleton described desertification as 'a concept out of hand'
and 'an institutional myth' (1994, pp. 63 and 50). As a result of the Rio
Conference, desertification now has an inviolable existence, at least in inter-

Plate 7.1 An active sand dune, Kaska Village, northern Nigeria. The commonest
popular description of desertification is 'the advance of the desert'. Sand
dunes do move, but this is a poor basis for understanding arid ecosystem
degradation. This dune, slowly overwhelming a small village near the
northern border of Nigeria with Niger in West Africa, has become something of a PR celebrity, much frequented by people with visitors whom
they wish to convince about the perils of desertification. In fact the dune
forms part of the Manga Grasslands, a relict of the former sub-Saharan
dunefields active at the height of the last glaciation over 12,000 years ago.
Mike Mortimore has studied them intensively, and points out that the
grassland landscape today is much as described by the Anglo-French
Boundary Commission in 1937, when it had much the same borders, and
was also treeless and with active dunes (Mortimore 1989).

national law. A resolution was passed to draw up a convention to combat
desertification (entitled, rather laboriously, 'the UN Convention to Combat
Desertification in Those Countries Experiencing Severe Drought and/or
Desertification, especially in Africa'). This was adopted in 1994 and entered
into force in December 1996 (International Institute for Sustainable
Development 1997).

Concern about desertification in mainstream sustainable development literature has been the result of (and to an extent the stimulus for) the stark
statistics on the severity and global extent of desertification that have been
prevalent in the two decades since the Sahel drought of 1972–4. Most of these
have appeared through the work of UNEP. In 1983 UNEP estimated that
about 35 per cent of the terrestrial globe was vulnerable to erosion (some 4.5

billion hectares); this land supported about a fifth of the world's population, some 850 million people. Of this area, about 80 per cent of rangelands (3.1 billion hectares) were affected by 'overuse' (Tolba 1986). Of the 4.5 billion hectares at risk of desertification globally, 30 per cent was severely or very severely desertified, and the rest was already moderately desertified. In the 1980s the extent and severity of desertification were seen to be increasing everywhere in the Third World (Mabbutt 1984).

The most often quoted (and misquoted) statistics on desertification relate to the rate at which deserts, and particularly the Sahara, are 'spreading'. The classic work here was a study by Lamprey in 1975 on vegetation change in the southern Sudan (Lamprey 1988). A comparison of the boundary between desert and sub-desert grassland and scrub between a 1958 vegetation map and a 1975 satellite image suggested that the boundary had shifted 90–100 km in the intervening 17 years (5.5 km a year). This statistic was taken up, repeated and reinvented by politicians and many others in the ensuing decades, and the process that it implies (of desert expansion) still dominates debate about desertification, despite a careful follow-up study by Hellden (1988) that questioned Lamprey's findings. The 1958 vegetation boundary was not surveyed but interpolated from a rainfall isohyet (itself likely to be notional in such a remote area, and of course subject to considerable inter-annual variation); comparison of 1975 and 1979 satellite imagery showed no change in vegetation boundaries. Other Swedish studies, including field surveys, showed short-term impacts of drought between 1965 and 1974, but no systematic decline in crop production, no shift in dunefield positions, and no major changes in vegetation cover (Thomas and Middleton 1994). Elsewhere, careful empirical research reveals similar patterns, Sullivan (1999), for example, demonstrates that despite entrenched official fears of resource overuse in the savanna woodlands of northwest Namibia, vegetation surveys show no degradation, except very locally within settlements.

These careful studies are something of an exception. The factual base for supporting or challenging the dominant dire prognosis about desertification is still surprisingly slim. One reason for the lack of data has been a profound uncertainty about exactly what desertification is. Given that the term has been widely known and used by environmentalists, politicians, policy-makers and journalists alike for some twenty-five years, it might be surprising that there should be confusion over what exactly it means. However, as so often, the word has become so widely used as to attract a great many linked, but divergent, meanings. The French ecologist Aubréville coined the word 'desertification' in 1949 to refer to an extreme form of 'savannisation', the conversion of tropical and subtropical forests into savannas. Desertification involved severe soil erosion, changes in soil properties and the invasion of dryland plant species. The geographer A.T. Grove subsequently also took a broad view, defining it as 'the spread of desert conditions for whatever reasons, desert being land with sparse vegetation and very low productivity associated with aridity, the degradation being persistent or in extreme cases irreversible' (1977, p. 299). However, definition has been confused by the coining of rival and overlapping terms such as aridisation, aridification or xerotisation (Verstraete 1986).

Clearly, desertification embraces the phenomena of drought, soil erosion, the destruction of vegetation cover, and the degradation of human living conditions, but it cannot be reduced to these things alone (Verstraete 1986). The UNCOD Plan of Action on desertification defines it as 'the diminution or destruction of the land', which 'can lead ultimately to desert-like conditions' (Mabbutt 1984, p. 103). Warren and Maizels suggested in one of the UNCOD preparatory papers that it was 'sustained decline in the yield of useful crops from a dry area accompanying certain kinds of environmental change, both natural and induced' (1977, p. 173). UNCOD focused specifically on degradation within drylands, and incorporated new phenomena such as the rise of salinity on irrigation schemes (ibid.). The definition of desertification adopted at UNCED, and in the UN Convention to Combat Desertification (in 1994), was a bland political catch-all. It defined desertification as 'land degradation in arid, semi-arid and dry subhumid areas resulting from various factors including climatic variations and human activities' (N. Robinson 1993).

Without a clear definition, it is perhaps not surprising that past scientific measurements of 'desertification' have left something to be desired. Attempts to assess the extent of desertification use indicators such as the growth and encroachment of moving sand, the deterioration of rangelands, the degradation of rainfed croplands, waterlogging and salinisation of irrigated areas, deforestation (usually specifically in arid environments), and declining ground or surface water supplies (Mabbutt 1984). Verstraete (1986) suggests that albedo, vegetation cover, soil depth, livestock composition, human health and average revenues might all be taken as indicators of desertification. Notwithstanding the relative ease with which the requirements of desertification monitoring can be set out, quantitative field studies are unusual. Perceptions of the severity of desertification are to a large extent self-confirming, but have a scanty scientific base. Relatively detailed studies have been done in certain areas, for example attempts to estimate present rates of water and wind erosion and land salinisation in the Nile Basin (Kishk 1986). However, systematic data on the extent of the desertification problem based on actual observations of real places over time are few. Despite the use of satellite remote sensing over several decades (e.g. Tucker *et al.* 1985, 1986, Justice *et al.* 1985), data are still lacking on the extent of land degradation on a global scale. Agencies such as UNEP lack the money or the remit to undertake scientific research on an international scale, and the urgency of policy concern about 'desertification' is based on loosely conceived concepts that provide no clear and consistent theoretical basis on which scientists might be attracted to build their research programmes and careers.

Estimates of the global extent of desertification have continued to be problematic. UNEP undertook a 'global assessment of progress' on desertification control (using a complex questionnaire to governments) in 1984, and repeated it in 1992 using a GIS-based global database of human-induced soil degradation

(GLASOD). This used a consistent and replicable methodology and was an improvement on previous analyses (Thomas 1993). On this basis, UNEP published in 1992 a *World Atlas of Desertification* containing maps of susceptibility to and incidence of desertification (Figure 7.2). UNEP estimated that between 1,016 and 1,036 million hectares of land were experiencing soil degradation (less than one-third of the area estimated at UNCOD in 1977, or in the 1984 survey), a total of perhaps 2,556 million hectares if vegetation change is included (Thomas and Middleton 1994, Middleton and Thomas 1997).

It is clearly facile to suggest that desertification is a distinctive problem in its own right (Warren 1996). The environmental problems of drylands (epitomised by those of the African Sahel) can instead be conceived of as the result of an interlocking set of processes, consisting of drought, desiccation and dryland degradation (Warren and Khogali 1992). 'Drought' here means a dry period that lies within the range of ecosystem response: ecosystems (and the economic systems that depend on them) are affected, but survive, and can return to their former state when the dry spell is over. 'Desiccation', on the other hand, is a longer dry period, long and intense enough to bring about permanent (destructive) change in ecological and/or human communities. The logic here is that while one or two years of drought will be accompanied by loss of annual vegetation cover, this will return quickly from buried seed and deep-rooted trees, and with it will return the economic productivity of the rangeland (or fields). More prolonged drought will kill seed-bearing plants, and have a deeper destructive signature on the landscape. The third element of the triptych of dryland troubles is 'dryland degradation', defined as 'a process that reduces land productivity to the extent that natural recovery can only happen over many decades or where artificially accelerated recovery is beyond the capital and technical resources of existing communities' (Warren 1996, p. 346).

Of course, this neat separation is something of an intellectual sleight of hand, for while to an analyst the difference between drought and desiccation may be clear, to a pastoralist or Sahelian farmer looking at ruined crops, hungry cattle and intruding neighbours, the problem has a holistic force not necessarily clear to the outsider (Mortimore 1989, 1998); as Warren (1996) points out, these three problems never come singly.

Desertification and climate change

The word 'desertification' is now in near-universal use. With time, both popular and academic accounts of desertification have become progressively more sophisticated with regard to causation. Climatic variation and the occurrence and persistence of low-rainfall periods in the Sahel, the nature of social and economic activity and the interaction of people with the environment are all important.

In the past three decades, during which Sahelian desertification has become a global issue, understanding of the nature of climatic variability in the Sahel has grown a great deal. However, at the time of the Sahel drought of 1972–4, understanding of climatic variability was severely limited, particularly in national

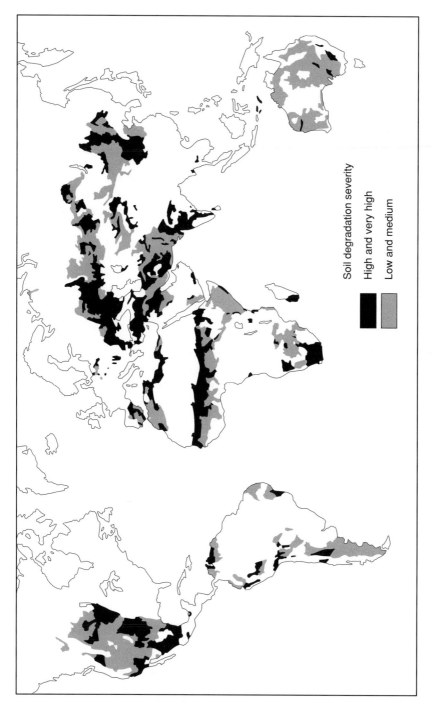

Figure 7.2 Desertification in the world's susceptible drylands, UNEP World Atlas of Desertification (after Middleton and Thomas 1997).

governments and international organisations. The existence of a series of dry years in tropical Africa beginning at the end of the nineteenth century was identified by the 1950s, but the existence of climatic variations on such a timescale was little understood by climatologists, at least until the 1970s (Lamb 1979). It is now clear that the 1950s and 1960s were a rather wet period in the Sudan and Sahel zones of Africa (Hulme 1996; Figure 7.3). Years of lower than average rainfall began in 1968, and rainfall was particularly low between 1972 and 1974 (Kowal and Adeoye 1973). However, drought did not end in 1974. Hare comments, 'the sense that the Sahelian drought ended in 1974, and with it the need for political concern, was illusory' (1984, p. 15). The recurrence of drought in the Sahel and famine in the Horn of Africa in 1984 reawakened international concern, as did the drought of 1992 in Southern Africa, during which 20 million people needed food relief (Hulme 1995).

Inevitably, greater understanding of past climatic variations led to attempts to elucidate causation (Lockwood 1986, Flohn and Nicholson 1980). Of greater interest to those concerned about desertification, however, has been research on the possible role of people in causing or exacerbating climatic change through bio-geophysical feedback (Nicholson 1988). Atmospheric variability is driven by both internal and external 'forcing' factors. External factors include earth surface conditions such as soil moisture, vegetation cover, sea surface temperature and the amount of snow and ice. Surface forcing is of particular importance in tropical regions (ibid.).

Interest in the influence of the land surface on climate has focused on questions of earth surface temperature, albedo, vegetation cover and atmospheric dust. All these have been influenced to some extent by human impact on a global scale. Although the global impacts of processes such as deforestation on albedo are relatively trivial (Henderson-Sellers and Gornitz 1984), regional effects may have been larger, for example in West Africa (Gornitz and NASA 1985). Other studies within the Sahel have identified seasonal and inter-annual variations in albedo on a small scale (Courel *et al.* 1984), while satellite imagery from the Advanced Very High Resolution Radiometer on the NOAA polar orbiting platforms demonstrated an apparent decrease in vegetation cover between 1981 and 1984 (Tucker *et al.* 1985, 1986). There have also been studies of the occurrence of atmospheric dust in the Sahara in relation to drought (Prospero and Nees 1977, Middleton 1985). One of the various hypotheses relating dust to rainfall suppression links it to the creation of warm inversion layers at altitude that suppress convective rising of humid surface air. Photography from the US Space Shuttle missions in the 1980s provided some support for this hypothesis.

Links between albedo changes and climate were drawn by Otterman (1974), who reported a temperature difference of 5°C between the heavily grazed Sinai desert and the less grazed Negev. He suggested that the baring of soils through overgrazing might be a possible mechanism of desertification, in which higher albedo would cause lower surface temperatures and reduced atmospheric convection. These observations were subsequently backed up by modelling experiments

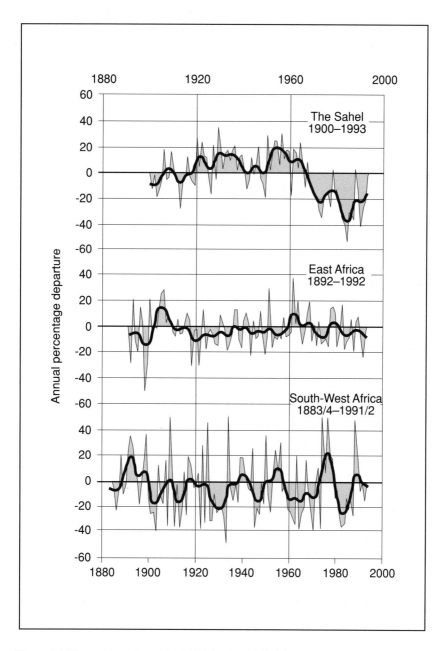

Figure 7.3 Time series of annual rainfall in the Sahel (after Hulme 1996).

(e.g. Charney 1975, Charney *et al.* 1975) that produced interesting results, and some controversy (Idso 1977 and the debate in *Science* volumes 189 and 191). The notion that, in Charney's memorable phrase, 'the desert feeds on itself' was widely taken as fact. It is better viewed as a hypothesis because the evidence for it remains inconclusive. The effect of vegetation on albedo cannot easily be separated from its other effects, for example soil moisture (Nicholson 1988). A study in the Sonoran Desert (Jackson and Idso 1975) found bare land to be warmer than a vegetated surface, in contrast to Otterman's results (1974). However, vegetation types in the two areas are not directly comparable (Nicholson 1988), and the controversy serves to demonstrate the difficulty of producing clear and simple conclusions from a small number of empirical (and often remotely sensed) observations and powerful computer models.

Clearly, surface conditions are an important input into climate dynamics in the tropics, and human action does influence those conditions. Locally at least, that influence can be significant. In addition to albedo and atmospheric dust, there has been increasing interest in the contribution of atmospheric aerosols from biomass burning to global climate change, and particularly emissions of soot, carbon, nitrogen and ozone (Williams and Balling 1995). Smoke emissions of tropical burning have been estimated to be 25–80 billion kilograms a year, of which the contribution of dryland (savanna) burning may be 10 per cent (ibid.).

However, the linkage between human agency and climatic variation remains extremely complex. Research since the 1970s has made it quite clear that an understanding of the causes of climatic variation in the tropics must embrace the global atmospheric system. The best evidence of this is the demonstration of links between Sahelian rainfall and global sea temperature anomalies (Folland *et al.* 1986), with the implications that this has for the linkage of tropical droughts and atmospheric carbon dioxide, or the 'greenhouse effect'. Negative rainfall anomalies in the Sahel are closely associated with warm sea surface temperatures in the southern oceans and Indian Ocean, and cool temperatures in northern oceans (Hulme 1996). There is also some association between regional rainfall anomalies in Southern Africa and indices of El Niño/Southern Oscillation (ENSO) phenomena (ibid.); maize yields in Zimbabwe can be correlated with the previous year's Southern Oscillation Index (Hulme 1995).

On a longer timescale, there is also an increasingly good, if still incomplete, picture of climate change over recent centuries (Schove 1977, Grove 1981, Nicholson 1980, 1996). Nicholson (1978) shows from historical records that the Saharan margins were wetter throughout the sixteenth to eighteenth centuries, but that a series of major droughts occurred within this period, between 1681 and 1687, between 1738 and 1756, and between 1828 and 1839. In addition there were other, more localised, droughts. By the end of the eighteenth century, general conditions were becoming more arid, and dry conditions persisted through the 1828–39 drought until wetter conditions began in the third quarter of the nineteenth century. Precipitation was continuously above the present level between 1875 and 1895 on both margins of the Sahara, in the central Sahara and in much of East and North Africa. This period, of course, was that in which colonial expansion began in much of Africa. The Europeans arrived at the end of a period exceptionally blessed with rainfall. Historical research in Southern Africa has established regional climate chronologies that confirm that within recent centuries there have been periods of wetter and drier conditions than the present ones across Africa, although the timing of these periods is regionally specific. Maps of rainfall anomalies in twentieth-century Africa reflect those of the nineteenth century and earlier; regional anomaly patterns are complex, and changes between wet and dry conditions can occur rapidly (Nicholson 1996).

The variability in Sahelian rainfall in recent centuries must be seen in the context of much larger climatic variations over timescales of thousands of years (Hulme 1996, Nicholson 1996, Roberts 1998). Attempts to understand past and present climate change now depend to a considerable extent on complex numerical models (Henderson-Sellers 1994). Concern about the contemporary issue of global warming has focused attention on the radiative forcing of climate change in past periods, for example the warming that took place following the last glacial maximum (18,000–6000 BP; Street-Perrott and Roberts 1994).

The tropics were arid during the last glacial maximum (Williams 1985), and dunefields formerly extended far into the Sahel and elsewhere in Africa (Grove 1958, Grove and Warren 1968, Thomas 1984). Former lake levels (see for example Washbourn 1967, Street and Grove 1976) demonstrate the variability over millennial timescales of prolonged periods of extremely low rainfall in the Sahel and other parts of semi-arid Africa. Palaeoenvironmental studies in South America and Asia support the view of variability in past climates in the tropics, and widespread (but not universal) aridity at the peak of the last glaciation (Street-Perrott and Roberts 1994, Roberts 1998).

Drought must therefore be understood as a persistent feature of tropical Africa. Its occurrence is complex in both space and time. Rainfall variability is not uniform through the wet season in the Sahel (Dennett *et al.* 1985, Hulme 1996). Declines in August rainfall have a critical impact on agricultural crop growth, and thus total or partial crop failure can occur without necessarily being revealed in aggregate statistics. Hulme (1987) analysed the date of onset and termination of the wet season in the central Sudan. In areas with 200–300 mm of rainfall, wet-season failures (seasons too short for crop growth) occurred for the first time last century in the 1970s, and shorter wet seasons and mid- and late-season breaks in the wet season occurred more frequently in drought periods. The spatial complexity of environmental conditions is also important (Raynaut *et al.* 1998). Agnew (1990) demonstrated the spatial complexity of patterns of drought (both declines in rainfall amount, or 'meteorological drought', and shortages of water availability for crops, or 'agricultural drought') in the Sahel in the 1970s. Wetter southern areas did not experience continuous drought.

Climatic models are starting to become sophisticated enough to be able to predict future changes in climate with reasonable accuracy (Hulme 1996), although they are not yet sufficiently detailed to predict small-scale temporal variations in rainfall (especially the start and end of rainy seasons). Attempts to use models to predict the impact of global climate change on global food supplies suggest that scenarios with high levels of atmospheric carbon dioxide would have only a small net negative impact on global food supplies once the benefits of enhanced CO_2 on crop growth and the effects of farm-level adaptations and technological change are taken into account (Rosenweig and Parry 1994). However, this global figure hides a disparity: whereas production in developed countries would increase under this scenario (because they could take advantage of increased CO_2 levels), that in developing countries would decline. Contemporary global climate change holds little comfort for the rural poor in the Third World.

The impacts of future climate change as a result of global warming are likely to be particularly severe in areas where rain-fed production is marginal. Changes in the global radiation budget caused by greenhouse gas emissions (much of them of course from First World industrial urban regions) are quite likely to have a significant impact on the timing, consistency and duration of rainy seasons in semi-arid parts of the Third World, and particularly in sub-Saharan Africa (Glantz 1992); the possibility of increased climatic variability needs to be seen in the context of past climatic variations. Regional climate impact scenarios suggest that impacts on agricultural and pastoral economies, and on livelihoods, are likely to be considerable. They will also be extremely complex, with changes in ecosystems and wild species, surface water availability (amount and variability), and disease, in addition to the obvious effects on food production (Hulme 1995). The socio-economic consequences of climate change for a region such as Southern Africa are likely to be considerable (Hulme 1995).

Desertification and the scientific imagination

Concern about environmental degradation in arid and semi-arid areas is not new. Over time, both the focus of attention and the kinds of explanations canvassed have shifted. In the case of the Sahel, for example, arguments about the southwards invasion of desert conditions into new regions have given way to ideas of degradation of semi-arid areas from within. Regional discourse about the Sahel has been relentlessly globalised, initially through the donor response to the Sahel drought and latterly through UNEP, the Rio process and the desertification convention. The remarkable features of desertification consciousness, however, lie not in the way it has changed, but its persistence, continuity and geographical spread. In areas such as West Africa it is possible to trace a continuous thread of environmental concern in colonial and post-colonial states, and latterly in the international arena, from early in the twentieth century to the present day. The nature of this West African experience reveals a great deal about the ideology of environmentalism, and hence of thinking and policy aimed at non-degrading or sustainable development.

The first focus of concern in West Africa concerned the creation of desert conditions by the physical advance of the desert into more moist areas, a process of desiccation. The fact of desiccation in Africa over geological time was well established (e.g. Hobley 1914), but that was not the point at issue here. Rather, it was the fear of increasing aridity over the short term, and specifically the encroachment southwards of the Sahara (e.g. Bovill 1921). Droughts in the early decades of colonial rule (for example in Hausaland in 1913; Grove 1973) appear to have had significant effects on the thinking of colonial administrators. For example, low rainfall and low river floods in Sokoto Province in northern Nigeria in 1917 and 1918 led to a remarkable level of colonial government concern at 'desiccation' and consequent interest in small-scale flood irrigation in the valley of the River Sokoto between 1919 and 1921, although these experiments ultimately failed (Adams 1987).

The apocalyptic views of E.P. Stebbing on the spread of the Sahara desert southwards were mentioned at the start of this chapter. His hypothesis was substantively refuted by Jones (1938) on the strength of the fieldwork of the Anglo-French Forestry Commission to northern Nigeria and Niger in 1936 and 1937. Jones argued that there was no evidence of southward encroachment of sand, retrogression of vegetation, permanent reduction of rainfall, the recent shrinkage of streams or lakes, or the lowering of the water table. He concluded, 'there is no reason to fear that the desiccation through climatic causes will impair the habitability of the West African colonies for many generations to come' (ibid., p. 421). Subsequently, Stebbing (1938) went some way towards this position, stressing instead the spoliation and erosion of land due to population increase and to agricultural and pastoral practices. The geographer Dudley Stamp concluded that there was 'no need to fear desiccation through climatic causes' (1940, p. 300). However, if climatic change was not the problem, Stamp agreed with Stebbing that 'the spread of man-made desert from within is quite another matter' (ibid.).

This shift in concern from the expansion of deserts on a broad front to a concern about land management and processes of deterioration *in situ* involving soil erosion was obviously linked to alarm at the experience of the American Dust Bowl. Soil erosion and dust storms in the American Midwest, and particularly the experience of dust over the eastern seaboard, generated huge public concern in the USA, with concomitant theoretical and institutional responses regarding the position of ecology and the establishment of the United States Bureau of Reclamation (Worster 1985). The report of the Great Plains Committee in 1936 ushered in 'a radically new environmental outlook' (ibid., p. 232), one based on the ecological thinking of men such as Clements, whose notion of vegetation succession to climatic climax, Worster argues, came to dominate the US national imagination as well as subsequent ecological thinking. Perception of this environmental crisis in the USA passed via newspapers and other routes (including the Imperial School of Tropical Agriculture in Trinidad; Anderson 1984) to inform and feed concern about soil erosion in arid and semi-arid environments elsewhere. These took firm root in Africa (see for example Beinart 1984, Anderson 1984, Harroy 1949, Aubréville 1949), although the scope of concern was essentially worldwide (Jacks and Whyte

1938, Furon 1947). Stamp suggested, 'there now seems little doubt that the problem before West Africa is not the special one of Saharan encroachment but the universal one of man-induced soil erosion' (1940, p. 300). Similarly, Vogt concluded, 'Whether or not Africa is actually suffering a climatic change, man is most effectively helping to desiccate the continent' (1949, p. 248).

Clearly to some extent at least, desertification is in the eye of the beholder; historically people have seen what they wanted to see. Preconceptions about ecological processes, an exaggerated ability to infer process from form, and therefore make assumptions about what is happening in a landscape on the basis of what can be seen at any given moment, and above all the ideological baggage of the observer all contribute to the perception of desertification. The importance of ideology in affecting environmental perception is well demonstrated by Anderson (1984) in the context of colonial East Africa. He describes four factors that encouraged a move by the colonial states to undertake direct intervention in African agriculture over the 1930s: the Depression, the American Dust Bowl, population growth and drought.

The Depression threatened both settler and African farmers. African farmers responded to the slump in agricultural prices by expanding the area under production, of cotton and coffee in Uganda and maize in Kenya and Tanganyika. White farmers facing bankruptcy responded defensively, seeking to entrench their position, in the process claiming (notably to the Kenya Land Commission) that African farming practices were damaging to the environment and unproductive. The impacts of the Depression and the political response to it were compounded by images of the American Dust Bowl, which 'caused Agricultural Officers all over British Africa to examine their own localities for signs of this menace' (Anderson 1984, p. 327). The prevention of erosion, particularly as this was perceived to occur primarily on African land, was a major stimulus to intervention in African agriculture. Meanwhile, by the 1930s the growth of the African population had become a matter of concern, partly because of the problems of overcrowding and landlessness in the Kikuyu reserves (Throup 1987). European settlers farming in the White Highlands perceived soil erosion both on the reserves and by squatters on their own farms. Finally, the years from the mid-1920s to the mid-1930s were dry in East Africa, and drought created periodic local food shortages and sometimes extensive deaths of cattle. Famine relief by the state was expensive, and drought heightened the perception of environmental degradation caused by African husbandry.

It is not perhaps surprising to find that by 1938 soil conservation had become a major concern of government in the East African colonies, the cutting edge of a policy of state intervention in African agriculture. The details of the implementation of policy are diverse, but what is most important here is the complexity of the context in which they emerged. Political and economic concerns of white settlers were combined with the occurrence of natural phenomena (drought), wider economic concerns (the Depression and the growth of African agriculture) and the existence of a powerful and emotive image of environmental decline in semi-arid regions under the pressure of agricultural misuse. The result was a persuasive and self-confirming perception of environmental degradation, or what would now be called desertification.

If we look back at the records, there are few data with which to assess scientifically the severity of any such problems. Such data were not collected. They were not necessary; the ideology of degradation was sufficient to generate policy. Where such data do exist, and researchers are innovative enough to find them, they can tell a very different story. Areas such as Machakos in Kenya were portrayed in the 1930s as degraded wastelands, where human survival was at risk from soil erosion. When historical data from this period were used to examine changes in land use over time, they revealed a remarkable phenomenon, one of progressive improvements in soil conservation (Tiffen *et al.* 1994). This work is discussed further below.

Development planners have a regrettable habit of depending on self-referencing stories, convincing each other that their understanding of problems is correct and their choice of solutions appropriate. Emory Roe (1991) calls such stories 'narratives', and points out how persistent they are: they cannot be overturned by simply showing that they are untrue in a particular instance, but only by providing a better and more convincing story. Leach and Mearns (1996) suggest that such narratives dominate thinking about the environment in Africa, nowhere more so than in discussions of environmental degradation (e.g. in Ethiopia; Hoben 1995) and desertification (Swift 1996).

Swift argues that the desertification narrative served the interests of three specific constituencies: national governments in Africa, international aid bureaucracies (especially UN agencies) and scientists. In the 1970s, recently independent African governments were restructuring their bureaucracies and strengthening central control over natural resources. Drought, and the assumptions about human-induced environmental degradation linked to them, legitimated such claims and made centralised top-down environmental planning seem a logical strategy. Pastoralism, in particular, could be portrayed as doomed and self-destructive, and its replacement by sedentary agriculture made to seem necessary and beneficial. Aid donors, meanwhile, found in desertification a problem that seemed to transcend politics and legitimated 'large, technology-driven international programmes' (Swift 1996, p. 88). To scientists developing new fields such as remote sensing, desertification offered fertile terrain for expansion: satellite imagery seemed to offer answers without the need for lengthy and tedious fieldwork, and desertification became a source of funding and legitimacy for new cadres of technicians and researchers.

Despite the apparent power of ecological science to sidestep partial knowledge and offer a definitive statement about environmental change, and despite their faith in their science, ecologists of recent decades have suffered from problems similar to those afflicting their colonial predecessors. Data are without doubt too few to refute or sustain the hypothesis that serious desertification is taking place on any scale, but that does not justify the dominant position of the ideology of human-induced environmental decline.

The neo-Malthusian narrative

Underlying the desertification narrative, with all its weight of scientific and pseudo-scientific argument, lies the familiar logic of neo-Malthusianism. A large fraction of the power of the narrative derives from the wider support for neo-Malthusian thinking within environmentalism, and within debates about people and environment in certain parts of the globe, particularly Africa. As long ago as 1949 Vogt wrote:

> The European in Africa has temporarily removed the Malthusian checks. He has put down tribal wars, destroyed predators, moved enough food about the continent to check famine – but he has not substituted constructive measures to balance his destruction of the old order.
>
> (1949, p. 260)

Vogt's central concern is carrying capacity, and he argues that 'man has moved into an untenable position by protracted and wholesale violation of certain natural laws' (ibid., p. 264). This idea has persisted in thinking about desertification. For example, forty years later, Curry-Lindahl argued that

> the fact that each area has a carrying capacity beyond which it cannot be utilised without causing damage, deterioration, and decreased productivity, is an ecological rule that is almost always forgotten by those who plan and carry out land-use development schemes.
>
> (1986, p. 125)

Neo-Malthusian thinking about overpopulation, overgrazing and land degradation was present in muted form in the *World Conservation Strategy*. Here desertification was seen to be caused primarily by human transgression of ecologically defined limits. Although there are problems of 'unwise development projects' and the 'inadequate management' of irrigated areas, the key problem is seen to be 'pressure of human numbers and numbers of livestock' (IUCN 1980, para. 16.9).

The neo-Malthusian view is readily found in starker form in writings on desertification, particularly in studies of Africa. Thus Curry-Lindahl suggested that

> for decades past, ecologists have been warning us about the serious environmental consequences and ecological undermining of present land-use practices [in semi-arid lands]: oversized stocks of cattle, goats and sheep, as well as unwise agricultural methods, coupled with extensive collecting of firewood and excessive burning, had long been paving the way for desertification and 'death' of productive lands.
>
> (1986, p. 106)

The Sahelian drought of the 1970s focused attention on the problem, but governments and aid agencies drew the wrong conclusions: 'they called the effects of the drought a "natural disaster" and blamed the climate! The real root of the problem, that Man himself was responsible, was not recognised or admitted: it was too unpopular a message' (ibid., 107).

Yet neo-Malthusian blinkers do not help the environmentalist make sense of the Sahel. Sinclair and Fryxell (1985) identified two contrasting views of the crisis in the Sahel, and particularly that of Ethiopia in 1984. They argued that the dominant view, the 'drought hypothesis', which suggests that problems are due to the failure of the rains exacerbated by warfare, was misconceived. Instead they suggested a 'settlement–overgrazing hypothesis'. They argued that until about the middle of the twentieth century the 'normal' land use pattern in the Sahel was based on migratory grazing using seasonally available resources. They suggested that this system had been operating in a 'balanced and reasonably stable' way for many centuries, possibly since domestic cattle first appeared in the Sahel five thousand years ago. It broke down 'through well-intended but short-sighted and misinformed intervention through aid projects'. Problems began after the Second World War exacerbated by population growth, overgrazing, and agricultural practices aimed at short-term profit not sustained yield (ibid., p. 992): overgrazing brought about the 'regression of plants' around boreholes and wells, and as these 'piospheres' (Warren and Maizels 1977) expanded and joined up, extensive areas of the Sahel became desertified. Arguments about the possible feedback effects of bare desertified soil on climate then suggest that the Sahelian ecosystem 'is being pushed into a new stable state of self-perpetuating drought' (Sinclair and Fryxell 1985, p. 992). It is a stark story of humans degrading the land, although in their analysis the real blame is laid on the aid agencies, which fund projects that break down the older and sustainable migratory pattern. Their conclusion is that short-term food aid by itself will 'only make the situation worse', since 'simply feeding the people and leaving them on the degraded land will maintain and exacerbate the imbalance and not allow the land to recover' (ibid.).

Overpopulation or intensification?

Empirical research in Africa in the 1990s has started to call into question neo-Malthusian assumptions about the inevitability of environmental degradation as population density rises. Neo-Malthusian thinking forms the basis of a powerful policy narrative that is increasingly under fire (Roe 1991, 1995, Leach and Mearns 1996). In Ethiopia, for example, a neo-Malthusian environmental narrative exaggerates the rate of degradation and misrepresents the role of human agency in causing it (Hoben 1995). Both Western donors and the post-revolutionary Ethiopian government needed a clear explanation of famine and its fundamental causes that could be expressed in narrowly technical terms (free of political ideology) and would provide a rationale for a large food-aid programme. A neo-Malthusian explanation (too many people causing a degraded environment) fitted a selective reading of the facts.

Rural population densities in Africa are low compared to those in equivalent drylands in Asia, and historically the lack of labour for agriculture has been a critical factor in the evolution of farming systems and environmental management (Iliffe 1995). Comparative study of agricultural farming systems in a range of countries shows increases in agricultural output per head, quite contrary to the customary wisdom of agrarian crisis and falling food production per capita (Wiggins 1995). As more careful studies have been undertaken, it has become clear that under some circumstances population growth in sub-Saharan Africa is leading to sustainable intensification of agriculture, not degradation. In northern Nigeria high population densities have been maintained for centuries in the close-settled zone around Kano City. This agricultural landscape is referred to in the literature as 'farmed parkland', with closely packed fields set with economic trees. By 1913 no more than one-third of the land was fallow, and by 1991 87 per cent of it was cultivated, and rural population densities were 348 people per square kilometre (Mortimore 1993). The farming system is complex, with several crops (particularly millet, sorghum, cowpeas and groundnuts) of a wide range of local varieties grown together in different inter-cropping and relay cropping mixtures (Adams and Mortimore 1997, Mortimore and Adams 1999; see Plate 7.2). The key to the sustainability of cultivation without prolonged fallow periods, however, lies in the maintenance of soil fertility through the close management of nutrient cycles, use of legume crops and the integration of agriculture and livestock-keeping, particularly in the use of crop residues as fodder for small stock (sheep and goats; Harris 1998). Some soil nutrients also arrive in the form of dust deposits.

It could be argued that the Kano close-settled zone is a remarkable and untypical place, and that circumstances there are unlikely to be repeated else-where. However, studies in drier Sahelian farming systems further north-east in Nigeria suggest that similar patterns of intensification may be developing as population densities rise (Mortimore and Adams 1999, Harris 1999). For rural households the allocation of household labour to different tasks in cultivation, livestock-keeping, off-farm activity and household work is a critical factor in their ability to achieve sustainable livelihoods (Mortimore and Adams 1999).

The possibility of a positive relationship between rural population growth and environmental sustainability has started to become conventional wisdom as the implications of a major historical study of Machakos District in Kenya have been disseminated within the policy community. This book, provocatively titled *More People, Less Erosion* (Tiffen et al. 1994), presents a detailed and empirically backed argument that there has been a beneficial interaction between population growth, environmental improvement and economic output per capita: they reiterate the optimism of Esther Boserup's (1965) arguments and challenge the pessimism of neo-Malthusians.

What gives this story its impact is that in the 1930s, Machakos was held within colonial Kenya, and more widely, to epitomise the threat of soil erosion under relentless pressure of population. It was the very public target of late-colonial projects to tackle the menace of land degradation, an 'experiment in

Plate 7.2 Sustained intensive agriculture in the Kano close-settled zone, Nigeria. A path between fields intercropped with late millet and sorghum. Land around Kano City has been farmed continuously for several centuries. Cropping systems are diverse, and population densities are high. Soil fertility is maintained through careful nutrient management (particularly the integration of livestock-keeping and crop production, and use of compound sweepings as manure.

colonialism' (Huxley 1960). Not only was disaster averted, but Tiffen and Mortimore argue that population growth allowed an astonishing level of invest-ment in land (particularly terracing) and the wholesale transformation of agriculture to highly intensive production systems (Tiffen and Mortimore 1994, Tiffen *et al.* 1995, Mortimore and Tiffen 1995).

Machakos lies south-east of Nairobi, and stretches from relatively high and well-watered land (2,000 m above sea level, 1,200 mm rainfall) to lower dry rangelands (600 m above sea level, 700 mm rainfall). Like the rest of Kenya, it has been subject to a high rate of population growth, ranging up to 3.7 per cent per year in the 1970s. The population of the district was 240,000 in 1930 and 1.4 million in 1990. The value of agricultural output rose three times per capita and eleven times per unit of area between 1930 and 1990 as farmers invested off-farm incomes in land, intensified production (Plate 7.3), turned to cash crops such as coffee, harnessed labour to terrace hillsides, and made use of the denser networks of contacts to learn new ideas and sell their produce. Tiffen *et al.* conclude that

The Machakos experience between 1930 and 1980 lends no support to the view that rapid population growth leads inexorably to environmental degradation. It is impossible to show that a reduced rate of population growth might have had a more beneficial effect on the environment. On the contrary, it might have made less labour available for conservation technologies, resulted in less market demand and incentives for development, and reduced the speed at which new land was demarcated, cleared and conserved.

(1984, p. 284)

This is a remarkable and sustained argument, and it has had a significant impact on the way development policy-makers think about African agriculture. There are, inevitably, other arguments that need to be heard. For example, Murton (1999) demonstrates that not all Machakos households can cope with and effectively harness extra labour: those with buoyant off-farm income (particularly in nearby Nairobi) can accumulate land and innovate as farmers; those dependent on agricultural labour opportunities struggle. Terraces on the hillside may represent a control of erosion, and perhaps environmental sustainability, but they do not necessarily translate into sustainable livelihoods for all, especially in the longer run. Jones (1996) draws similar conclusions

Plate 7.3 Stall-fed cow; intensified agriculture, Machakos District, Kenya. Machakos, south-east of the city of Nairobi in Kenya, was known in the 1930s as the classic threatened environment, where overpopulation was causing drastic soil erosion. Sixty years later, slopes are terraced and cloaked with trees and cash crops such as coffee. Cows like this one are stall-fed, and provide manure for intensively managed fields.

about the ability of richer and poorer households to sustain soil fertility in the Uluguru Mountains in Tanzania.

There is a need, also, to be careful about running away with policy conclusions. Rocheleau *et al.* (1995) look at Machakos in a longer time frame, and quite rightly draw attention to the way in which for over a century it has been the subject of repeated study and intervention by outsider 'experts' (and clearly even this most recent enthusiastic endorsement of indigenous skill and enterprise falls into that category). They emphasise the dangers of the policy search for rapid diagnosis and the 'quick fix', pointing out how this tends to produce problems further down the line. They emphasise the importance of taking account of the multiple stories that local people have about places and people.

There is no doubt that the pessimistic neo-Malthusian narrative about dryland farming has been convincingly and widely challenged. Comparable findings to those in Nigeria and Kenya have been reported from other areas. Lindblade *et al.* (1998), for example, examined land use change in Kabale District in Uganda between 1945 and 1996, finding that despite population growth (and population densities of 265 per square kilometre) and a history of colonial concern about overpopulation, a higher proportion of land was now being fallowed, and evidence of land degradation was limited. Farmland was carefully terraced (Plate 7.4). Steep slopes had been turned over to woodlots, and valley-bottom wetlands drained for grazing, while soil fertility was being maintained by using animal manure, household compost and mulching.

Overgrazing and new range ecology

Many analyses of ecological change were (and still are) based on problematic assumptions about livestock management or indigenous agriculture. The conventional view of rangeland management and mismanagement has been built around ideas of range conditions, class and carrying capacity. The logic was that the environment is capable of supporting a certain fixed number of livestock (or biomass), and that for any given ecosystem this could be calculated primarily as a function of rainfall. There is a general relationship between rainfall and the productivity of herbivores (Coe *et al.* 1976), and similar plots of livestock biomass against annual rainfall can be drawn (e.g. Jewell 1980, Bourn 1978; see Figure 7.4). If these regressions are taken to represent 'carrying capacity', it can be argued that at stocking levels lower than the solid curve, pasture resources are being underused, and that at stocking levels above the solid curve, resources are being overused, such that there is likely to be adverse ecological change (for example extinction of palatable species and eventually loss of vegetation cover) and eventually the death of excess stock.

The concepts of overgrazing and carrying capacity have underpinned conventional pastoral policy in many parts of the Third World, and particularly in Africa. Policies aimed at confining, controlling and often settling nomadic pastoralists in sub-Saharan Africa have reflected what Horowitz and Little (1987)

Plate 7.4 Agricultural terracing in Kabale District, south-west Uganda. Land in
     Kabale District is in short supply and rarely flat. Indigenous soil conserva-
     tion practices created unusually receptive conditions for a colonial terracing
     programme, and land in the district is now everywhere carefully and
     thoroughly terraced.

call an 'intellectual tradition of anti-nomadism'. Governments have tended to
distrust people who are mobile and difficult to locate, tax, educate and provide
with services. Conventional rangeland science has added to this a particular
distrust of their apparently feckless management of seemingly fragile rangelands. Recommended management strategies have been stereotyped (Swift
1982). They have involved, first, the control of livestock numbers to match
range conditions, through destocking and the promotion of commercial offtake,
to improve stock health and weight; second, fencing and paddocking to allow
grazing pressure on particular pieces of land to be closely controlled, and provision of watering points to allow optimal livestock dispersal; third, manipulation
of range ecology through controlled burning, bush clearance and pasture
reseeding; and fourth, disease control and stock breeding. None of these strategies fits with nomadic or semi-nomadic subsistence livestock production, so

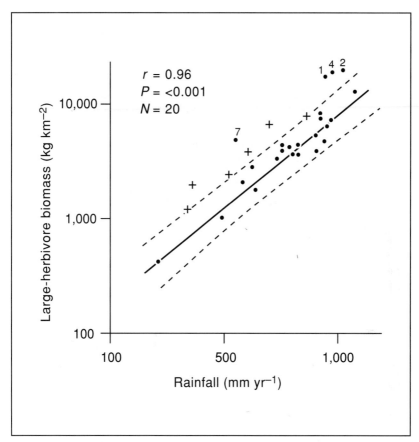

Figure 7.4 Graph of livestock/herbivore biomass against rainfall (after Coe *et al.* 1976). Pastoral areas (+) are omitted from the regression, as were areas with atypical soils or drainage (numbered).

government pastoral policy has tended to emphasise sedentarisation, formal (i.e. freehold or leasehold) land tenure and capitalist production.

There are wide gaps between pastoral policy prescriptions and the ways pastoralists actually manage their herds and rangelands. By implication, conventional pastoral policy has tended to emphasise the production of animal products from the slaughter of stock (i.e. meat and hides) rather than the products from live animals (milk or blood) that typify many indigenous pastoral systems. Similarly, pastoral development planning tends to focus on cattle, whereas indigenous production systems typically involve a mix of species. Indigenous pastoral ecosystems seem well adapted to exploit the spatial and temporal variability in production, adapting herd composition and using movement to maximise survival chances. The Turkana in Kenya, for example, have mixed herds, with camels, which use browse resources that are available even in the dry season, and cattle, which are more productive in the wet season but have to move out of the plains into the hills in the dry season (Coughenour *et al.* 1985, Coppock *et al.* 1986). Such systems offer a relatively low output compared to modern capitalist systems such as ranching. However, they are remarkably robust in terms of providing a predictable, if limited, livelihood.

Researchers have increasingly expressed reservations about the universal applicability of the concept of overgrazing, and the unreflective links drawn between it and desertification (Sandford 1983, Horowitz and Little 1987, Mace 1991). Hogg (1987a), for example, argued that 'overstocking' and 'overgrazing' are rarely defined, and that judgements about carrying capacity are subjective, although that subjectivity is rarely admitted. Homewood and Rodgers (1984, 1987) questioned the whole concept of carrying capacity.

The key to the growing disquiet with concepts of carrying capacity, overgrazing and desertification is the failure of commentators, and major international organisations (the FAO, UNESCO, UNEP, the World Bank), to define these terms and processes clearly or to produce adequate scientific data on their occurrence in space and time. The data necessary to assess 'long-term degradation' of vegetation or desertification in most cases simply do not exist. Furthermore, there are real problems of obtaining adequate quantitative data either on the responses of vegetation to different stocking levels or on livestock numbers. Figures of 'carrying capacity' can take no account of seasonal variations in fodder availability or annual variations. They concentrate on absolute numbers of livestock and not densities, and rarely consider spatial mobility. They are therefore of little value in understanding either rangeland ecology or pastoral practice. Although they have become both entrenched and self-reinforcing, they are an unsatisfactory and sometimes dangerous basis for management.

The productivity of semi-arid rangelands varies both seasonally and between years. The primary cause of this variation is rainfall. The high spatial and temporal variability of precipitation, particularly in sub-Saharan Africa, has increasingly been recognised. Other factors affecting productivity also vary over time and space, notably soil nitrogen and phosphate (Breman and de Wit 1983) and the impact of fire (Norton-Griffiths 1979). Ecological studies tend to be of short duration (often confined to the fieldwork period associated with a Ph.D. thesis) and localised. Yet semi-arid grazing ecosystems undergo considerable and important fluctuations from year to year and place to place. Analyses of real or predicted degradation tend to be built on estimates of regional stocking rates. Not only are such estimates notoriously unreliable (because livestock are hard and expensive to count, particularly if their owners do not want you to), they also fail to take account of spatial and temporal variations. More seriously still, the assumption of mechanistic relations between stock numbers and biological productivity completely fails to take account of the social processes of herd management, such as, for example, stimulated the boom in cattle numbers that took place in the 1960s in the north-west part of the Niger Inland Delta in Mali (Turner 1993).

Plate 7.5 Groundwater-fed *Acacia* trees along streams, Turkana, northern Kenya. Rangeland resources are not uniformly distributed in either space or time. This photograph shows mature *Acacia tortilis* trees along seasonal drainage channels. These trees can obtain water from the shallow aquifer in the sands below the stream bed long after the rains have ended. These trees produce seed pods that are an important and nutritious food resource for Turkana livestock. Riparian *Acacia* woodland is an important rangeland resource, and access to it is carefully controlled. The survival of the woodlands of the main river of north-west Kenya, the Turkwel, is threatened by construction of the Turkwel Gorge Dam (Adams 1992).

Biologically based estimates of carrying capacity tend to be arbitrary, based on limited data and deriving from 'rule of thumb' and the experience of range ecologists rather than empirical scientific fieldwork (Mace 1991). Arguably, the concept of carrying capacity is not appropriate in areas with great annual variation in primary productivity, and most estimates fail to take account of the variability and resilience of savanna ecosystems (Homewood and Rodgers 1987). Furthermore, it is fairly meaningless to seek to identify a single 'carrying capacity' for an ecosystem. Carrying capacity should be defined in terms of the density of animals and plants that allows the manager to get what they want out of the system; in other words, that it is possible to speak of a carrying capacity only in the context of a particular management goal. What suits a nomadic pastoralist may not suit a rancher; many African systems have a subsistence stocking rate higher than commercial ranchers would adopt, giving low rates of production per animal but high output per unit area (Homewood and

Rodgers 1987). Actual stocking levels can and do exceed 'carrying capacity' for decades at a time (Behnke and Scoones 1991).

Overgrazing is a classic 'institutional fact'; as Ruth Mace comments, 'sometimes we are so sure of something we don't need to see the evidence' (1991, p. 280). Despite the volume of literature on overgrazing and carrying capacity, researchers now conclude that there is no one simple ecological succession towards an overgrazed state, but complex patterns of ecological change in response to exogenous conditions (especially rainfall) and stock numbers and management. Such ecological changes can take many forms, not all of them serious, and they can proceed by diverse routes, some of which can be reversed more easily than others, and some of which are more sensitive to particular management than others. There are no 'naturally' stable points in semi-arid ecosystems that can usefully be taken to define an 'equilibrial' state.

Through the 1980s and 1990s, conventional thinking about carrying capacity and overgrazing began to be challenged by so-called new range ecology (Behnke *et al.* 1993). In drier rangelands, with greater rainfall variability, ecosystems exhibit non-equilibrial behaviour. Ecosystem state and productivity are largely driven by rainfall, and pastoral strategies are designed to track environmental variation (taking advantage of wet years and coping with dry ones), rather than being conservative (seeking a steady-state equilibrial output). This awareness of the non-equilibrial nature of savanna ecosystem dynamics reflects a wider understanding of the importance of non-linear processes in ecology as a whole (e.g. Botkin 1990, Pahl-Wostl 1995).

Whatever the implications for ecological thought, it is certainly the case that non-equilibrial behaviour is characteristic in the dry tropics. Once this is appreciated, much of what appeared to be perversity or conservatism on the part of pastoralists is revealed to be highly adaptive (Behnke *et al.* 1993, Prior 1993, Scoones 1994).

Figure 7.5 shows the changing relationship between the amount of plant material available to livestock and livestock numbers in a succession of years. The point of this diagram is that there is no single 'carrying capacity' of the ecosystem, represented by an equilibrium number of livestock, but a changing balance of livestock and range resources, with drought years first reducing the condition of stock and then (through disease, death and destitution-forced sales) reducing stock numbers. Good rain years then allow pastures to recover, allowing a lagged recovery of herd numbers as pastoralists track environmental conditions. Not only do herd managers need extensive knowledge of environmental conditions and opportunities in different areas open to them, and resilient multi-species herds, to survive under such conditions, but they also need institutions for the exchange and recovery of stock through kinship networks. Development strategies must support indigenous capacity to track rainfall and maintain social and economic networks, rather than demand a shift to a static, equilibrial capitalist form of production.

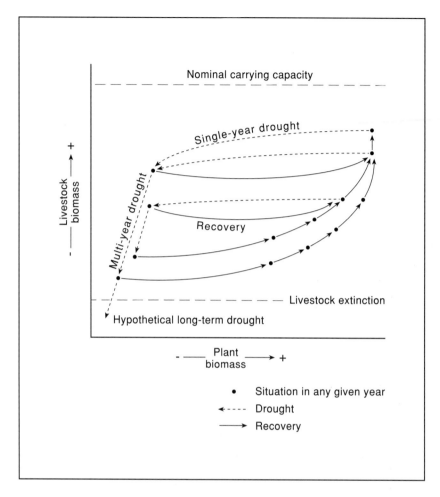

Figure 7.5 Plant–livestock interactions under drought perturbation (after Behnke and Scoones 1991).

Landscapes misread

Mainstream sustainable development grew from scientific assessments of human impacts on the environment (Chapter 3). Through the 1980s the scientific basis of environmentalist critiques of development was rarely challenged. In the 1990s that changed, as the above discussion of the re-evaluation of the concept of overgrazing has made clear. This reassessment of the record of science and scientists to make sense of complex landscapes is by no means confined to drylands, but is part of a much wider re-evaluation of accounts of environmental change (Leach and Mearns 1996). Much of this reassessment involves a reconsideration of the histories of particular places and landscapes, by doing new, innovative and better research. However, it is also influenced by debates about the intrinsic limits to what science can reveal, and to broader ideas drawn from postmodernism. It is impossible to consider these latter developments here, but two examples will serve to demonstrate how new work is challenging conventional thinking about degradation.

The first example is the work of James Fairhead and Melissa Leach on the dynamics of forests over time in Guinea, West Africa, *Misreading the African Landscape* (1995a, 1995b, 1996). The focus of this study is Kissidougou Prefecture, which ecologically lies on the boundary between moist forest and savanna. The landscape is one of dense forest patches and corridors, around villages and along streams, set in a matrix of grassland. The rainfall is variable, but about 1,900–2,000 mm per year. Throughout the twentieth century a succession of outside experts and administrators have regarded the Kissidougou landscape and drawn a single conclusion: that people were clearing the forest at an alarming rate, and the mosaic landscape was subject to rapid degradation, particularly due to fire. Thus a French colonial botanist, Auguste Chevalier, reached this conclusion in 1909, as did Kissidougou's senior administrator and a European Commission watershed rehabilitation project in the 1990s. The forest patches around villages and along streams were presumed to be the fragments of a once-continuous forest cover, requiring urgent and draconian measures for their conservation (for example prohibition of tree-cutting and fires, and attempts to persuade local people to plant trees in open patches).

In fact, Fairhead and Leach and their team demonstrate that exactly the reverse is the case. With the perfect timing of experienced raconteurs (and with convincing data) they demonstrate from historical written accounts, maps and sequential air photographs that if anything forest cover has increased in the past century; that the landscape that greeted the first French intruders was substantially the same as it is today; that agriculture improves and does not degrade land; that high population densities allow fire management and control rather than reducing it. The savanna was not 'derived', as so much research in Guinea and elsewhere in Africa had assumed for so long, and the forest islands were not 'relics' of former forest cover. Kissi informants made clear that they did not see the area as a forest landscape that was progressively losing its trees, but as a savanna landscape filling with forest: policy-makers were 'reading forest history backwards'. This reading of landscape led to and legitimised the intrusion of the state into the lives of Kissi villagers, taking away control of resources, imposing taxes and fines, and diverting funds better spent in other ways. Inspired by this discourse of degradation, outsiders to Kissidougou have

> accused people of wanton destruction, criminalised many of their everyday activities, denied the technical validity of their ecological knowledge and research into developing it, denied value and credibility to their cultural forms, expressions and basis of morality, and at times denied even people's consciousness and intelligence.
>
> (Fairhead and Leach 1996, p. 295)

A second example that demonstrates the unreliable nature of accepted understandings of environmental degradation is the debate over 'environmental supercrisis' in the Himalayas. In *The Himalayan Dilemma*, Jack Ives and Bruno Messerli (1989) questioned the alarmist reports of massive environmental degradation in the Himalayan region. They argue that problems are overdramatised and distorted, in large measure because of the inherent uncertainty that surrounds the debate.

The 'Himalayan environmental degradation theory' which they address suggests a vicious circle of degradation driven ultimately by population growth, which is 2–3 per cent per year in Nepal. It is suggested that this population growth, plus immigration from India into the Terai lowlands, has increased demand for fuelwood, timber, fodder and land. This has brought about a reduction in forest cover on the hills (half Nepal's forest reserves being lost between 1950 and 1980), and the extension of agricultural terraces onto steeper and more marginal slopes, unprotected by trees. This has caused more soil erosion, loss of land to landslides, and disruption of hillslope hydrology. In turn this has increased runoff in the monsoon and caused in turn flooding and sedimentation downstream in the Ganges valley (burying fertile land, filling reservoirs, and reducing dry-season flows because less water is retained higher up river basins). As land is eroded, agriculture reaches yet further onto more marginal land, and as trees are eliminated, people turn to animal dung for fuel, and soil fertility declines. This tightens a spiral of environmental overuse and degradation.

In a paper discussing 'uncertainty on a Himalayan scale', Thompson and Warburton (1988) argue that there is uncertainty about what is really happening in the Himalayas not only because of the biological and physical properties of the ecosystems and the difficulty of monitoring them, but because uncertainty is being generated *by* and *for* organisations involved in the debate. They comment, 'the survival of an institution rests ultimately upon the credibility it can muster for its idea of how the world is; for its definition of the problem; for its claim that its version of the real is self-evident' (ibid., p. 21). There is no doubt that many, perhaps most, of the problems comprising the 'Himalayan environmental degradation theory' are real. However, Ives and Messerli (1989) argue that the way in which they have been wrapped up together into a 'supercrisis' is unhelpful: it homogenises spatially specific issues, and it paralyses policy. Furthermore, as in Kissidougou, an inability to understand the processes going on means that any attempt to develop policy is liable to have unforeseen and perhaps deleterious effects on the poor. Thompson and Warburton suggest that the real problem is that there is not simply *a* problem, but 'a plurality of contradictory and contending problems' (1988, p. 46): diversity of view and contradiction are a resource, not a threat. As Ives and Messerli comment, even the best-planned projects face a sea of uncertainty, endless problems of implementation and many surprises (ibid., p. 255); recognising uncertainty may be the only alternative to enslavement to flawed narratives of degradation, and the opening to plural (and modest) solutions.

Sustainability and desertification policy

Dominant narratives of desertification and overgrazing, linking land management and climatic change through bio-geophysical feedback loops, have provided the foundation stone of dryland policy in areas such as the Sahel, aimed at the re-establishment of particular sets of environmental conditions. In 1984 Hare concluded that desertification 'harshens the microclimate', and that measures to control desertification not only 'restore plant cover and soil conditions', but 'repair the microclimate' (p. 19). To achieve this, he suggested land use control, meaning fencing and the control of the movement of animals. Sinclair and Fryxell (1985) also called for more control, although they offered a somewhat different prescription for the Sahel, arguing for the re-creation of free migratory movements and the control of where and how people could live. They suggested that the regeneration of degraded environments could take place only if the land were 'rested' until vegetation could recover, and this required population resettlement and control, destocking and the establishment of a new migration or rotational grazing system (Table 7.1).

Above all, Sinclair and Fryxell proposed that 'uncoordinated and ecologically damaging aid-giving' by international donors be ended, and that developmental interventions be drastically altered. These policies are naïvely conceived and draconian. They suggest the establishment of 'a new and suitable rural economy', without apparently appreciating the difficulties of even conceiving of such a thing, let alone the politics and economics of achieving it. They fail to appreciate the practical difficulties or the associated human costs involved in induced population resettlement in the Third World (see Chapter 8). The political, economic and human costs of extensive involuntary government resettlement in the Sahel would be vast. Sinclair and Fryxell suffer the blindness common to many commentators to the ecological and economic logics of African grazing systems, and assume that destocking is ecologically necessary, and economically and practically feasible. They propose turning the Sahel into a massive controlled grazing scheme, a breathtakingly misconceived proposal.

Repeatedly, state intervention in rangelands, often triggered or legitimised by concerns about desertification, has involved attempts to make nomads settle,

Table 7.1 Policy recommendations for the Sahel

- Move people from the degraded land to new areas.
- Educate and help people in those new areas to establish a new and suitable rural economy.
- Institute education and family planning to control population growth.
- Severely restrict and cull cattle herds.
- Monitor vegetation succession on degraded land.
- When recovered establish a modified migration or rotational grazing system.
- Only construct wells if they do not harm the migration system.
- Encourage African governments to institute these measures.

Source: Sinclair and Fryxell (1985)

often as part of a wider attempt to transform the economy of semi-arid areas the livelihoods of their people. Aid donors responded to the African droughts of the 1970s with commitments of funds to support and transform the region. In 1973 seven Sahelian nations established the Comité Inter-États de Lutte contre la Sécheresse Sahélienne (CILSS), and the Secretary-General of the United Nations established the UN Sudano-Sahelian Office (from 1976 under UNEP). The US Congress supported the notion of a long-term development programme in the Sahel in December 1973 under the Foreign Assistant Act (Derman 1984). Political changes in Africa (independence of the Portuguese colonies and revolution in Ethiopia) may have accelerated the tempo of aid-giving. In 1976 an organisation, the Club du Sahel, was formed to coordinate aid-giving, jointly sponsored by CILSS and certain aid donors; Henry Kissinger made a public commitment of $7.5 billion to 'roll back the desert' (Franke and Chasin 1980, p. 137).

The thrust of the new aid was to solve the problems of the Sahel, not simply to succour the destitute. A strategy for drought control and development in the Sahel produced by the OECD (Organisation for Economic Cooperation and Development) was adopted by CILSS and the Club du Sahel in 1977 (Derman 1984). Its aim was, first, to alleviate drought and re-establish food production and 'ecological recovery' (Franke and Chasin 1980, p. 148), and then to enable the Sahelian countries to achieve food self-sufficiency by raising agricultural production. The programme would require massive increases in the production of wheat (seventyfold), rice (fivefold), cattle (twofold) and fish (twofold) on 1970 figures. In large part this was to be achieved by improved rain-fed farming (doubling production by the end of the century, working on drought-tolerant crops and off-farm employment), but a massive fivefold increase in the area irrigated (to 1.2 million hectares) was also planned. Other initiatives were to include transport infrastructure, well construction, animal health, work on crop storage and reforestation.

Franke and Chasin (1980) criticised this plan on a number of grounds, including its excessive focus on 'cash' rather than 'staple' crops (and a failure to appreciate the fluidity of such categories), and the excessive speed with which development is proposed, particularly in the field of irrigation. The plan accepted too readily simplistic assumptions about the nature and causes of over-grazing ('traditional' livestock-raising practices), and failed to recognise the potential impacts of initiatives such as well-digging. Derman (1984) concluded that despite the New Directions mandate, new USAID lending in the Sahel simply turned the 'familiar arsenal of modernisation' on the problem of small farmers. Policy sought simply to get the commercial structure right, with no effective appreciation of the links between ecology and the way production is organised. Franke and Chasin (1980) concluded that the vaunted 'new' approach to development in the Sahel would simply reproduce existing vulnerability to drought. Development, in this instance at least, would bring no solution to environmental degradation.

Many other approaches to 'desertification control' have proved equally costly. For example, in Mauritania the UN Sudan-Sahelian Office has explored physical and biological methods of controlling sand movement and encroachment on fields and facilities (Grojean 1991). Work began in 1983, identifying biological protection techniques (live fences of introduced species such as *Prosopis juliflora*, or local species such as *Leptadenis pyrotechnica*), and 'raising awareness among the insufficiently motivated, newly sedentary population' (ibid., p. 6). On average, mechanical stabilisation cost $1,000 per hectare and biological stabilisation $400 per hectare. This aim of this 'curative sand encroachment control project' was technical, not developmental; the technology could be made to work, but not at a feasible economic cost.

There is a great deal of evidence of the failure of external plans to transform dryland production systems, and the lives of farmers and pastoralists. One example is the history of attempts to settle pastoralists in northern Kenya destituted by drought in the 1970s and 1980s, often on irrigation schemes. These were expensive, unpopular and unsuccessful (Hogg 1983, 1987a, b). Since the 1980s substantial numbers of Turkana pastoralists have been settled on irrigation schemes along the seasonal Kerio and Turkwel rivers, and on the shores of Lake Turkana as fishermen (Hogg 1987b). A Ministry of Agriculture appraisal in 1984 calculated that the total development cost of the Turkana cluster of small-scale irrigation schemes was fifteen times the cost of setting up each family with a herd of replacement livestock, and the equivalent to famine relief for two hundred years (Adams 1992). Hogg argues that irrigation development in northern Kenya was a deliberate strategy to combat what the government perceived to be the destructive effects of pastoralism. It made no sense in terms of the normal socio-economic objectives of development; indeed, irrigation contributed to the further marginalisation of already poor pastoralists and significantly increased pressure on key grazing lands through clearance of riverine forest vegetation near settlements (a process Hogg sees as 'policy-induced desertification'; Hogg 1987a). The root causes of desertification in Turkana are, then, first poverty, second induced settlement and third modernisation. Development itself is implicated in the causation of environmental degradation.

Recognition of the non-equilibrial nature of savanna ecology presents a considerable challenge to policy-makers. It has cast increasing doubt on the validity of traditional management of rangelands aimed at maintaining a range in a specified condition. Alternative strategies have been proposed that emphasise the opportunism of pastoral management (Sandford 1983, Behnke *et al.* 1993, Scoones 1994). New ideas recognise that opportunistic strategies are long established, that husbandry systems may well not need drastic reform (let alone abandonment), and that development strategies can be gradual and fully participatory, leading to piecemeal and not wholesale change (Scoones 1994). The insight that ecologists can offer is confirmation of the herders' need to balance fodder supply and stock numbers. Strategies to help herders track environmental change include a focus on enhancing feed supply (maintaining exchanges with farming communities, supplying feed), supporting mobility (supporting tenure of key dry-season grazing sites and access to trekking routes) and promoting human rights. Animal health is important to stock survival in drought, and mobile vaccination facilities can be important, while there is still a role for the stock-breeding beloved of government livestock researchers, but the focus needs to be on the capacity of animals to survive disease, drought and poor dry-season grazing, in preference to milk or meat yield under favourable conditions.

It is also now widely recognised that pastoralists need help to endure crises such as drought. Innovative policies include provision for purchasing stock at reasonable prices in droughts (when supply of poor animals rises and prices crash) and for helping pastoral families restock, or communal grain banks for pastoralists (thus enabling them to weather spiralling grain prices during droughts). Most important of all is the provision of security to rights in key areas of rangeland, particularly wetlands patches that support communities in surrounding drylands, and particularly in crisis years (Scoones 1991).

Finally, there is a need for more support for herders to move into and out of stock-keeping; not through mass resettlement and retraining campaigns (of the kind that have, for example, sought to convert Somali herders to coastal fishermen), but by supporting a diversity of livelihood options between which people can choose. Diversity and flexibility are cornerstones of survival in both pastoral and agricultural production in drylands, and policy-makers must recognise and foster these, rather than seeking to sweep them away in the pursuit of higher productivity and a cash income (Mortimore and Adams 1999).

The Convention to Combat Desertification

The most important result of the United Nations Conference on Desertification (UNCOD) in Nairobi in 1977 was the adoption of the Plan of Action to Combat Desertification (PACD). This was designed to 'prevent and to arrest the advance of desertification, and, where possible, to reclaim desertified land for productive use' (Walls 1984). The PACD envisaged both transnational projects (for example a 'transnational green belt in North Africa') and action by national governments (Table 7.2). It was not a success. A major reason for this was that the investment in the form of aid from donor nations did not materialise. UNEP estimated in 1980 that about US$90 billion would be required to finance a programme to meet the demands of the core of the PACD over twenty years: $4.5 billion per year. This would have rehabilitated all desertified irrigated land, 70 per cent of rain-fed cropland and 50 per cent of rangelands. The UN General Assembly did establish a 'special account' for anti-desertification project finance, despite opposition from certain donor nations. However, the UN 'special account' attracted only $48,500 in its first six years. UNEP itself spent $20 million on desertification between 1974 and 1983, but overall only $7 billion was spent between 1978 and 1983 ($1.17 million per year, or 0.2 per cent of that needed). Furthermore, of that $7 billion, only about $400,000 was actually spent directly on projects aimed at 'desertification control', the rest going on infrastructural projects such as roads,

Table 7.2 National government actions under the 1977 Plan of Action to Combat Desertification

---

- Establish or designate a government authority to combat desertification.
- Assess desertification problems at national, provincial or sub-provincial levels.
- Establish national priorities for actions against desertification.
- Prepare a national plan of action against desertification.
- Select priorities for national action.
- Prepare and submit requests for international support for specific activities.
- Implement the national plans of action.

---

water supplies, buildings, research and training (Walls 1984, Dregne 1984). While such infrastructure might be necessary, it had little to do with desertification control as such, and appeared to result in existing projects in semi-arid areas being relabelled as 'desertification' projects, either by aid agencies eager to show they were doing something or by governments hoping to make projects more attractive by jumping on the latest bandwagon.

National planning to combat desertification also faltered. By 1984 only two countries (Sudan and Afghanistan) had drawn up anti-desertification plans, although there were nine others in draft (Walls 1984). Few countries had designated agencies for desertification control. Assessments of desertification in 1983 suggested it to be increasing, not diminishing (Mabbutt 1984, Berry 1984, Tolba 1986). At the UNEP Governing Council in 1980 the Executive Director highlighted lack of cooperation within the UN system, inadequate external funding of projects, and the low priority assigned to desertification control by afflicted countries. Beneath these lay basic problems of lack of money and lack of political will (Walls 1984). A review in 1991 of UNEP's Desertification Control Programme Activity Centre revealed a continuing suite of problems, including lack of focus and overambitious goals.

Desertification underwent an institutional renaissance in the run-up to the UN Conference on Environment and Development in Rio de Janeiro in 1992. In a manner reminiscent of the dissatisfaction of non-industrialised countries at Stockholm twenty years before, Southern countries resented the sidelining of the environmental problems relevant to them, and desertification came to embody their dissatisfaction. Southern Africa was also in the grip of severe drought. As a result, the issue was discussed at length in the final PrepCom meeting before the Rio Conference, and a chapter on desertification was included in *Agenda 21*. A formal commitment was made at Rio to negotiate and agree a Convention on Desertification by 1994, although this did not go through without opposition (Carr and Mpande 1996). Arguments included the question of whether desertification was actually a global issue, and the question of whether Southern demands for a desertification convention would be traded off against the desire from the USA and the EU for a forest convention.

Following Rio, an Intergovernmental Negotiating Committee was established rapidly, meeting in Geneva in 1993. It worked through a series of issues, including scientific uncertainty about the definition of desertification, and the extent to which it was a global problem (Carr and Mpande 1996). After five sessions, a text of the Convention with four regional annexes (on Africa, Asia, Latin America and the northern Mediterranean respectively) was complete for signature by the deadline in June 1994 (although the Intergovernmental Negotiating Committee continued to meet to clarify the meaning and implication of certain articles). The final convention is an interesting reflection of both the politics of the Rio process (discussed in Chapter 4) and several decades of confused thinking about environmental degradation and development.

The convention came into force in December 1996, the first Conference of the Parties taking place in Rome in 1997. A permanent secretariat was established in Bonn, Germany, and (unusually) the conference included a plenary meeting for dialogue with NGOs (Toulmin 1993, European Commission 1997, International Institute for Sustainable Development 1997). By 1997 the convention had been ratified by 113 countries (twice as many as the other two Rio conventions, on Biological Diversity and Climate Change), although several key countries had yet to ratify it, notably Japan, Russia and the USA.

The Convention to Combat Desertification was proposed by Southern countries primarily as a way to focus financial resources on real problems of some of the world's poorest people, but in the end it contained only the weakest of commitments by donor countries to provide extra funds (Carr and Mpande 1996). As with the PACD, lack of funds is a key constraint on implementation. A 'global mechanism' (administered by IFAD, the International Fund for Agricultural Development) was agreed to mobilise and channel funds (a mechanism similar to that of the Framework Convention on Climate Change), but the flow has been slight. It is not clear whether, in the long run, Africa will be able to persuade other regions to let it be a 'special case' deserving privileged attention, and whether the broad focus of the convention (embracing environmental management, poverty, democratisation and governance) will prove workable, or will actually have any impact on the lives of the poor in arid areas (International Institute for Sustainable Development 1997).

Meanwhile, some aid donors have geared up to support the convention, notably perhaps the European Union, which is recognised as having desertification within its own region, in the Mediterranean. Between 1990 and 1995–6 some 524 million ECU was dedicated to desertification projects in developing countries through the European Development Fund, cooperation agreements with Asian and Latin American countries, and thematic budget lines. These supported 237 projects relating to desertification (European Commission 1997). The EC has had a particular commitment to Africa since the 1970s. It launched the 'EC action plan for Africa concerning the protection of natural resources for combating desertification' in 1986, and spent 280 million ECU between 1990 and 1996 in twenty-six countries of sub-Saharan Africa, 51 per cent of which went to West Africa. From 1990 the EC agreed to spend 23.2 million ECU over eight years on a project to rehabilitate common lands in the Aravali Hills, Haryana, India. The project sought to restore vegetation over 33,000 ha of village-owned hill land, conserving soil fertility, reducing soil erosion, 're-establishing natural hydrological balance', and enabling villagers to meet needs for fuelwood, fodder and timber (ibid., p. 27). The work of the project ranged from tree- and grass-planting, contour-trenching and wall-building through to work on land management institutions (to encourage effective community control of these lands and the involvement of women in land management decisions) and the introduction of new technologies such as fuel-efficient stoves, grass-harvesting and silage-making (European Commission 1997).

Since the 1970s, anti-desertification projects have become increasingly multi-disciplinary and diverse, reflecting the growing perception that the problem of 'desertification' is not simple, and certainly not conducive to narrow technical solutions. Poverty, economy and social organisation are an integral part of the challenge facing development planners, and those they seek to help. Concern about desertification, which may have started by addressing 'the inexorable advance of the desert', has ended up addressing questions of poverty and sustainable livelihoods at household scale. It is quite proper that issues of environmental degradation should be central to thinking and policy in sustainable development, for the development process is both a response to, and too often the cause of, harsh and degraded environments endured by the poor.

This chapter has argued that experts and planners have a very mixed track record in their attempts to define and identify environmental degradation, and often a frankly poor record in trying to overcome it. The poor experience degradation not as an aggregate phenomenon of ecological change, but directly, in the form of challenges to welfare and livelihood sustainability (Carney 1998b). Chapter 8 will extend this discussion by looking at the problem of development planning itself as a source of environmental impacts, and the irony that the attempt to exert control over nature in the name of human benefit can so easily generate instead human and environmental costs.

Summary

- Environmental degradation is a central issue in sustainable development. While reference to degradation (and particularly 'desertification') in dryland areas abound in the literature and in policy, considerable caution is warranted in thinking about them. Definitions are confused, and strong scientific evidence on long-term environmental change is often lacking. Discussions of desertification in particular are often more dependent on commonly accepted wisdom among so-called experts than on hard field evidence.
- Climate change is a long-established feature of regions such as dryland Africa, and rainfall varies in space and time in complex ways. Although scientific understanding is growing, evidence does not encourage simplistic conclusions about causes of drought or the links between climate and land use change.
- Fear of desertification and soil erosion has been a repeated theme in African development thinking, but neo-Malthusian analyses of population growth

and environmental degradation have been challenged by studies (e.g. in Machakos, Kenya) of agricultural intensification and sustainability.

- Studies of dryland pastoralists also now challenge conventional wisdom about the inevitability of environmental degradation. New range ecology involves a recognition of the way indigenous pastoral systems are adapted to variations in rainfall and grazing productivity in space and time.

- Studies of other environments, such as the forest–savanna transition zone of West Africa or the Himalayan hills, also offer challenges to narratives of ecological crisis.

- Policies devised to 'combat desertification' are unlikely to be successful if they are based on an erroneous understanding of the relations between people and environment (which they often have been). The Convention to Combat Desertification was negotiated following the Rio Conference, and came into force in 1996. It embodies much of the confused thinking about dryland degradation that has dominated international debate since the 1970s.

Further reading

Behnke, R.H. Jr, Scoones, I. and Kerven, C. (1993) *Range Ecology at Disequilibrium: new models of natural variability and pastoral adaptation in African savannas*, Overseas Development Institute, London.

Botkin, D.B. (1990) *Discordant Harmonies: a new ecology for the twenty-first century*, Oxford University Press, New York.

Fairhead, J. and Leach, M. (1996) *Misreading the African Landscape: society and ecology in a forest savanna land*, Cambridge University Press, Cambridge.

Ives, J. and Messerli, B. (1989) *The Himalayan Dilemma: reconciling development and conservation*, Routledge, London.

Leach, M. and Mearns, R. (eds) (1996) *The Lie of the Land: challenging received wisdom on the African environment*, James Currey/International African Institute, London.

Mortimore, M. (1989) *Adapting to Drought: farmers, famines and desertification in West Africa*, Cambridge University Press, Cambridge.

Mortimore, M. (1998) *Roots in the African Dust: sustaining the drylands*, Cambridge University Press, Cambridge.

Mortimore, M. and Adams, W.M. (1999) *Working the Sahel: environment and society in northern Nigeria*, Routledge, London.

Pahl-Wostl, C. (1995) *The Dynamic Nature of Ecosystems: chaos and order intertwined*, Wiley, Chichester.

Roberts, N. (1998) *The Holocene: an environmental history*, Blackwell, Oxford (2nd edition).

Scoones, I. (1994) *Living with Uncertainty: new directions in pastoral development in Africa*, IT Publications, London.

Thomas, D.S.G. and Middleton, T. (1994) *Desertification: exploding the myth*, Wiley, Chichester.

Tiffen, M., Mortimore, M.J. and Gichugi, F. (1994) *More People, Less Erosion: environmental recovery in Kenya*, Wiley, Chichester.

Web sources

<http://www.fao.org/desertification/>   The FAO desertification website: look no further for desertification data.

<http://www.undp.org/seed/unso/>   UNSO, the UNDP Office to Combat Desertification and Drought.

<http://www.unccd.int/main.php>   The United Nations Secretariat of the Convention to Combat Desertification, including information on the meetings of the Conference of the Parties.

<http://www.iied.org/drylands/index.html>   IIED Drylands Programme home page: sensible research and networking on drought and land degradation.

<http://www.isnar.org/ilri/>   The International Livestock Research Institute (ILRI) in Nairobi is the international CGIAR (Consultative Group on International Agricultural Research) centre for livestock research, and the website gives information on its ideas about issues such as livestock nutrition, health and livestock policy.

8 The environmental costs of development

I am always bothered by the Western arrogance, by its assurance that it knows all the answers and can quite readily fix everything so that the tropical peoples can live happily ever after, if only they will listen.

(Marston Bates, 1953, p. 239)

The costs of environmental control

Environmental variability and environmental change present significant challenges to sustainability, particularly the sustainability of rural livelihoods. In the case of arid ecosystem degradation, human impacts on natural patterns of climatic variability are implicated at two scales, that of global climatic change and (slightly less certainly) that of local resource depletion and degradation. Both these forms of human environmental impact are more or less accidental, or at least incidental to human endeavour of wealth creation, economic growth or economic survival. However, much human modification of the environment is quite deliberate, and this, too, can bring significant challenges to sustainability.

This chapter discusses the ways in which formal development schemes can themselves be unsustainable. It does this by taking one particular area of development, the use of water resources, and analysing the environmental and socio-economic impacts of dams and irrigation schemes. Such projects represent a particular approach to development, demanding intensive technical planning and massive environmental and socio-economic transformation. Dams in particular affect large and dynamic natural systems (rivers and floodplains) on which large numbers of people depend, and where human uses of the environment are intricately linked to ecosystems and the ways in which they change. This dependence of people's livelihoods on patterns of river flooding has not been widely understood in the past, and many dams have had serious environmental impacts.

Environmental modification is an inherent part of the development process. Cowen and Shenton distinguish between immanent development (the changes to economy and society – and one might add environment – that take place) and intentional development (the 'active practice of the state'; 1996, p. 61). Both dimensions of development involve environmental transformation, but it is the second kind that has produced the most dramatic environmental changes, and for which deleterious environmental and socio-economic impacts might be the most surprising, and are certainly the most shocking and regrettable. This is the subject of this chapter.

Intentional development was formed as an ideology and a practice of global international relations and a strategy of national governance in the ex-colonial world in the decades that followed the end of the Second World War (Chapter 2). It was founded on an instrumental-rationalist approach to planning, to the notion that social, economic and environmental resources should be assessed, harnessed and brought to bear on the systematic improvement of human welfare. Rationalisation has four dimensions (Murphy 1994): first, the development of science and technology, the 'calculated, systematic expansion of the means to understand and manipulate nature' (ibid., p. 28) and the related belief in the possibility (and desirability) of the mastery of nature through increased scientific and technical knowledge; second, the expanding capitalist economy (and the rationality of the market); third, formal hierarchical organisation (to translate social action into rationally organised action); and fourth, the elaboration of the legal system to manage social conflict and promote the predictability and calculability of the consequences of social action (Murphy 1994).

The place of rationalisation in recent moves towards ecological modernisation was mentioned in Chapter 5. In development, its role is fundamental. Rationalisation underpinned the formalised process of planning social, economic and environmental change, and the ideology of developmentalism that drove it: the view that environments and societies must be transformed in an all-out drive to modernise and achieve economic transformation. The spirit of the age was technocratic, optimistic, modernist and Promethean. This was the era of ecological managerialism, from which mainstream sustainable development emerged (Chapter 2). Across a wide range of disciplines (agriculture, health, veterinary science, fisheries, forestry and education, among others), colonial regimes in Africa and elsewhere adopted new ideas, employed staff with new skills, and invented new institutional forms for planning and delivering change.

The technocratic strategies and institutions of war were adopted and retooled to deliver development. Swords were beaten into ploughshares, sometimes almost literally: the Groundnut Scheme in Tanganyika tried to convert Sherman tanks for stump clearance (unsuccessfully as it turned out – a reflection of the disastrous performance of the scheme as a whole; Wood 1950). The industrial successes of Fordist factory organisation in the 1930s, developed so successfully in the Second World War in the production of machines of war, were now applied conceptually to the delivery of development. Development projects adopted large-scale and mechanised farming, an attempt to mass-produce food (Jones 1938).

Alas, the urge to rush development led in many instances to failure. Baldwin, reviewing the Niger Agricultural Project in the Nigerian Middle Belt (another unsuccessful groundnut production scheme) commented that 'the removal of limitations on money led almost inevitably to removal of limitations on the size of project. Hence there grew up a fallacious notion that the bigger the scheme the better the results likely to be obtained' (Baldwin 1957, p. 2).

Planning water resource development

Nowhere has what Herbert Frankel called 'the twin dangers in all development of grandiosity and arrogance' (in Baldwin 1957) been better developed than in the control of river flows behind dams (McCully 1996). The juxtaposition of large rivers, unurbanised and unindustrialised economies and large numbers of relatively impoverished subsistence populations has repeatedly led engineers and planners to dream of harnessing the waters and energy of rivers for irrigation or hydroelectric power generation. River water unharnessed for power or other purposes is all too easily seen to be running to waste (Adams 1992, Rosenberg *et al.* 1995, Usher 1997a, World Commission on Dams 2000).

Grand river development schemes date back to the ancient Mesopotamian kingdoms, but in the modern era they have become one of the hallmarks of colonial expertise. British Imperial India saw major experiments in irrigation development from the early nineteenth century, when existing irrigation systems, for example in the kingdoms of Delhi and Tajore, and Madras, were found profitable, and were refurbished and replicated (Singh 1997), while the importance of the annual flood of the Nile to the prosperity of Egypt was a major factor in the imperial ambitions of the major European powers in the second half of the nineteenth century.

Canals and barrages were built in the Nile Delta early in the nineteenth century to allow perennial irrigation, and two major barrages were built north of Cairo between 1843 and 1861 (Waterbury 1979, Collins 1990). These works were improved and extended in the last two decades of the nineteenth century, and the first Aswan Dam was built on the Nile in 1902, and heightened in 1912 to double its storage. These works were simply the springboard for more extensive studies of the upper Nile in the Sudan, and in due course for more dramatic project proposals. Sir William Garstin, Under-Secretary of State for Public Works in the Anglo-Egyptian condominium of the Sudan, published a series of technical studies of the Nile in 1904 that discussed ideas for dams and water storage on the Blue Nile, on the River Atbara and in the East African Great Lakes. Once launched into government consciousness, these ideas were studied and debated repeatedly in successive attempts to bind the flows of both Blue and White Niles more efficiently to the task of irrigation. Egypt and Sudan had concluded the Nile Waters Agreement, which allocated 94 per cent of the available flows (48 billion out of an estimated 51 billion cubic metres) to Egypt. Dams were built at Sennar on the Blue Nile in 1925, and upstream at Jebel Aulia in 1937. The Aswan Dam itself was heightened again in 1934.

More dramatic plans were developed. In 1931 the Egyptian Ministry of Works began to publish a series of volumes, *The Nile Basin*, containing ideas and proposals for the development of the Nile. Volume 7 (1947) proposed the Equatorial Nile Project, involving storage in the Great Lakes and a vast canal to carry water past the extensive swamps and wetlands of the Sudd (where evapotranspiration significantly reduced river flow) to deliver the 'saved' water downstream to the northern Sudan and Egypt. The Jonglei Investigation Team undertook extensive hydrological and environmental research to assess the feasibility of the project (and its environmental impacts) between 1946 and 1954. However, largely for political reasons (not least the Suez crisis and a new relationship with the Soviet Union), Egyptian attention switched to the Aswan High Dam, built within Egypt's own borders with the services of Soviet engineers. This was begun in 1960 and finished in 1971 (Greener 1962, Waterbury 1979, Collins 1990). Over the same period the Sudan built the Roseires Dam on the Blue Nile (completed in 1966) and the Khashm el Girba Dam on the Atbara (completed in 1965), the latter solely to provide a refuge for evacuees from Nubia, whose land beside the Nile was lost below the waters of the upper end of Lake Nasser. Meanwhile the idea of the Jonglei Canal lurked still in the official mind in Egypt, and it surfaced again in the 1970s (Waterbury 1979, Howell *et al.* 1988). Construction of the canal eventually began in 1976, although the project eventually became an issue in the civil war, and all construction work was brought to a halt in 1983 (Collins 1990).

The Nile was by no means a lone example of the growing ambitions and competence of river basin planners and engineers. The US experience with development planning as part of Roosevelt's 'New Deal' in the 1930s, and particularly the work of the Tennessee Valley Authority (TVA) and state intervention in planning and creating development infrastructure that extended far beyond water resources, was influential across the globe. Planning in the Mekong Basin began in 1957 with the formation of the Committee for the Coordination of the Investigations of the Lower Mekong River (the Mekong Committee) by the UN Economic Commission for Asia and the Far East. A cascade of seven dams from northern Laos down to Cambodia's Tonle Sap were envisaged (Usher and Ryder 1997). Planning continued in the Mekong Basin to some extent throughout the Vietnam War. A basin plan was completed in 1972, soon triggering concern among environmentalists (Bardach 1973).

The 'TVA model' for integrated river basin planning (or rather, a selective interpretation of what the TVA represented in terms of planning) was adopted by a United Nations Panel of Experts in 1958 and disseminated widely. In Africa, river basin authorities were created in many countries in the 1960s, for example in Ghana (the Volta River Authority) and in Nigeria (the Niger Delta Development Board in 1960 and the Niger Dams Authority in 1961, prior to the commencement of construction of the Kainji Dam on the River Niger for hydroelectric power in 1964). The river basin model continued to expand, for example in Nigeria (where there were eighteen multifunctional River Basin Development Authorities (RBDAs) by 1984) and in Kenya (where the Tana and Athi River Basin Development Authority was created in 1974 and the Kerio Valley Development Authority in 1979; Adams 1992). International river basins within Africa were also made subject to planning agencies, for example the Central African Power Corporation, concerned with development on the Zambezi, particularly the dams at Kariba (between Zimbabwe and Zambia) and Cahora Bassa (Mozambique), and the Organisation pour la Mise en Valeur du Fleuve Sénégal (OMVS) established with international aid donor support by Mali, Mauritania and Senegal in 1972.

In addition to planning, the 1960s also saw the first rash of large dam projects in the Third World. By 1986 China alone was reported by the International Commission on Large Dams to have 18,820 large dams (over 15 m high), part of a global total of 40,000 large dams (McCully 1996). In India, 1,554 large dams had been built by 1979, for both hydroelectric power and irrigation, absorbing almost 10 per cent of total public-sector investment (Singh 1997).

In Africa the 1960s was the decade of rapid decolonisation, and as part of that process a series of major dams were built: the Aswan High Dam on the Nile (built with Soviet expertise, as mentioned), the Akosombo on the Volta in Ghana (the USA's riposte, a concrete demonstration of the powers of democracy; Hart 1980, Mabogunje 1973), the Kariba Dam on the Zambezi, and the Kainji on the Niger (Adams 1985b). The reservoirs created by these dams became the subject of considerable international research attention (Lowe-McConnell 1966, Obeng 1969, Rubin and Warren 1968, Ackerman *et al.* 1973). Furthermore, their ecological impacts became one of the foci of environmental concern at that time about development in the Third World, for example in the volume *The Careless Technology* (Farvar and Milton 1973). Tropical dams have never quite lost the notoriety acquired at that time (e.g. Goldsmith and Hildyard 1984, Pearce 1992, McCully 1996, Usher 1997a).

Concern about the environmental effects of dams grew rapidly and with an astonishing unison, as critics identified and explained failings in the way dam projects were conceived and designed. Lagler (1969) described the problems of economic loss and human suffering which could arise where planning failed to look far enough ahead. Sulton described problems of 'myopia in the planning process, misinterpretation of ecological signs, poor timing and indifference to human suffering' (1970, p. 128) which observers suggested accounted for many of the unfortunate yet often avoidable consequences of the construction of dams. The problem was one of planning failure or, to use Peter Hall's memorable phrase, 'planning disaster' (Hall 1980). It was seen to be caused by the failure to integrate existing knowledge of ecosystems holistically into the planning process (Lagler 1971). Such ecosystem thinking was still relatively new in the 1960s (McIntosh 1985). Although foreign to engineers, hydrologists and economists, it proved very suited to the elucidation of the various complex relationships involved in river and floodplain environments, as the substantial body of knowledge on the ecological effects of river control that accumulated showed (e.g. Oglesby *et al.* 1972, Baxter 1977, Ward and Stanford 1979, Petts 1984).

Aquatic and riparian (floodplain) ecosystems are enormously complex. Freshwater habitats associated with river systems include both static water bodies (such as floodplain pools and meander cut-offs) and flowing water environments.

Floodplain environments are ecotonal, ranging from dryland environments to low-lying wetland areas (distant from the river itself) (Malanson 1993). Aquatic and floodplain ecosystems are each subject to the dynamic flow patterns of the river, in terms both of the annual discharge regime, and the size and longevity of shorter-term flood events, and the groundwater regime is subject to the distribution of groundwater in space and time that these river flows support. Changes in river flow regimes can therefore obviously have very large potential effects on river and floodplain environments (Malanson 1993, Hughes 1997). These can usefully be separated into impacts on the physical environment and impacts on the ecology of river and floodplain environments, and these will be discussed in turn. However, it should be noted that physical and ecological systems are closely linked. It is the physical processes driven by the flows of the river that create and destroy the physical contexts within which ecological change takes place, and to some extent ecological processes (for example the establishment of high floodplain forest) influence future evolution of landforms within the floodplain and the evolution of channel patterns.

In most tropical floodplains, both rivers and floodplains provide economic resources of vital importance to both local and regional (and sometimes national) economies. Changes to river flow patterns can therefore also have significant socio-economic impacts on floodplain people and economies. Incautious river basin development can present a serious threat to sustainability.

The physical impacts of river control

The impacts of dam construction on downstream hydrology and erosion have been intensively investigated (e.g. Petts 1984). Most research in these technical fields has taken place in the temperate rivers of industrialised countries. However, there are studies of other areas, for example by Chien (1985) on the effects of dam construction on the regime of the Han River in China, and Pickup (1980), who modelled hydrology and sediment yield of the proposed Wabo scheme on the Purari River in Papua New Guinea. Sagua (1978) identified a 60 per cent reduction in downstream flows between 1970 and 1976 following construction of the Kainji Dam on the River Niger, although the Sahel drought undoubtedly had an effect during this period.

The nature of hydrological effects varies with the purpose of the dam and the seasonal regime of the river. However, moderation and delay of the incoming flood peak frequently takes place because of the flood-routing effect of the storage impoundment. Such effects can be particularly significant where river regime is flashy and such peaks are common, for example in rivers in the semi-arid tropics. Flood control dams exacerbate peak flow moderation effects, particularly in such seasonally torrential rivers. Dams for all-year irrigation moderate variations in flow regime on a longer timescale, storing water at seasons of high flow for use at times of low flow. However, unless they also have a flood control function, discharge beyond storage capacity is usually spilled, allowing some flood flows to pass downstream, albeit in a routed and hence attenuated form. Hydroelectric dams are designed to create a constant flow through turbines, and therefore tend to have a similar effect on discharge patterns. However, if the intention is to provide power at peak periods, variations in discharge of considerable magnitude can occur over short timescales, creating artificial freshets or floods downstream. These can have disastrous consequences, particularly (as in Mozambique in 2000) if management of upstream dams exacerbates natural rainfall and flood patterns (Pearce 2000).

Links between river regime and geomorphological processes are complex. Sediment is often deposited in the relatively still waters of impoundments, leading to considerable problems of loss of storage capacity. Allen (1973) presented a post-construction analysis of the Anchicaya hydroelectric project in Colombia, highlighting sedimentation problems among many others. Reservoir siltation is particularly a problem in river basins with high rates of subaerial erosion and sediment transport, such as China (Yuqian and Qishun 1981). The Tarbela Dam in Pakistan, for example (3 km long and 143 m high, the twelfth largest hydroelectric scheme in the world), had lost 12 per cent of its live storage by 1992, after eighteen years of operation (McCully 1996). Siltation has seriously reduced the storage capacity of the Khashm el Girba Reservoir on the Atbara River in Sudan, built to supply water for irrigation by Nubian evacuees from the Aswan Dam (Khogali 1982). Silt accumulated at a rate of 50–55 million cubic metres a year, and by 1977 the original storage capacity had dropped by 59 per cent, with predictions that by 2025 there would be little water for irrigation (ibid.). A possible solution is the construction of upstream dams simply to act as silt traps, but such remedial development would obviously drastically alter the original economic and social assessments of the costs and benefits of the Khashm el Girba and possibly even the Aswan dams.

Sediment deposition within reservoirs can lead to clear water releases below the dam and consequent erosion immediately downstream (Rasid 1979). There can be changes in the balance of coarse and fine sediment in the river bed (with impacts on instream ecology, particularly affecting fish-breeding) and changes in the rate of channel erosion and lateral movement on the floodplain (with impacts on infrastructure such as bridges, on riverside settlements and agricultural activity). There are engineering solutions to some of these impacts (for example further dams or barrages, or artificial armouring of banks or river bed), but they are expensive, and effectively confined to rich industrialised countries.

The Tarbela Dam in Pakistan began to erode its spillway soon after full operation began in 1976, requiring a massive and costly project to stabilise the plunge pool: the dam had cost $1.5 billion by 1986, almost twice the 1968 estimate (McCully 1996). Further downstream the geomorphological impacts of dams become more complex as patterns of erosion and sedimentation adjust in complex ways to changed discharge and sediment load. Olofin (1984) showed that the channel of the Kano River in Nigeria stabilised following construction of the Tiga Dam in 1973, becoming a vegetated perennial channel. In this instance the socio-economic impacts were on balance favourable, because the changes made possible year-round cultivation in and adjacent to the river bed. By contrast, Attwell (1970) and Guy (1981) described increased river bank erosion on the Zambezi downstream of the Kariba Dam, in the Mana Pools Game Reserve, due to unseasonal high flows, high flow levels, rapid fluctuations in flow levels, and the fact that water released by the dam is low in sediment. The dam hastened normal river system dynamics and affected the regeneration of riparian *Acacia albida* woodland.

The physical impacts of dam construction can extend downstream into delta and estuary environments. The significance of such effects has been argued in a number of cases, notably that of the Nile Delta following the closure of the Aswan High Dam. The original dam was built in 1902 and raised in 1912 and 1933. Engineering studies began for the new dam in 1953–4, and it was completed in 1969 at a cost of $625 million (Farid 1975). Planned storage included silt-trapping capacity sufficient for an estimated five hundred years. Questioning of the wisdom of the new development at Aswan began early (see, for example, van der Schalie 1960). Against gains which included hydroelectric power, increased water availability (which allowed the expansion of the irrigated area in Egypt and extensive conversion to double-cropping) and flood protection, it was argued that there were problems of reduced soil fertility in the Nile Valley because of the lack of sediment in floodwaters and consequent erosion of the delta, and reduced flows which led to saline penetration of coastal aquifers (Farid 1975, Biswas 1980, Shalash 1983). Various studies suggest that erosion in the Nile Delta caused ultimately by the restriction of sediment supply by the High Dam is a serious problem, particularly for fishing villages and highly productive coastal lagoon fisheries (Kassas 1973, Sharaf el Din 1977).

The impacts of dams can also stretch into the shallow marine environment. Patterns of sediment movement seem to have changed off the Nile Delta (Murray *et al.* 1981). Bathymetric surveys offshore identified a new sand ridge system acting as a sink for eroded material. George (1973) identified links between marine fisheries in the eastern Mediterranean and the construction of the Aswan Dam. The closure of the Cahora Bassa Dam on the Zambezi in Mozambique changed the seasonal flow regime of the river, and the associated supply of nutrients to the shallow coastal waters. This has had a significant negative impact on the recruitment of shrimps on the Sofala Bank and the lucrative inshore shrimp fishery (Gammelsrød 1996).

Physical impacts downstream of dams are common and clearly highly complex. They can extend for many hundreds of kilometres downstream, and well beyond the confines of the river channel. It is only too easy for such impacts to be missed or underestimated when planners attempt to identify environmental costs. The lumpy nature of expenditure on major projects means that secondary and tertiary environmental impacts often go unrecorded (except by those left coping with environmental change alongside rivers downstream of dams), and few studies have monitored downstream degradation following dam construction in the Third World in a systematic way. This lack of research does not reflect the importance of the problem. The significance of physical impacts is increased because they can have complex knock-on effects on aquatic and flood-plain ecosystems. Thus for the Tucurui Dam on the Tocantins River in Brazil, physical and hydrological impacts have in turn had impacts on riverine and floodplain ecosystems, and on the sustainability of riparian cultivation on the *varzea* land owing to reduced inputs of silt (Barrow 1987). These will be discussed in the next section.

The ecological impacts of river control

The most direct ecological effect of dam construction is obviously the loss of upstream terrestrial environments. Singh *et al.* (1984) describe the significance of the loss of 530 ha of species-rich rainforest in the Silent Valley Forest reserve on the Nilgiri Plateau, India, and the attendant disturbance of surviving forest during project construction. Beyond this crude and direct impact, however, there are a series of other, more complex and sometimes subtle effects of dam construction on the ecosystems of the controlled river.

A considerable body of knowledge has been built up over the years on the ecology of running waters (e.g. Hynes 1970, Whitton 1975, Davies and Walker 1984). Transformation or modification of discharge patterns and stream environments has a range of significant effects on those ecosystems. The most obvious of these concern the ecological succession of the newly created impoundment as the organisms of flowing water (lotic ecosystems) are replaced by those of still (lentic) ecosystems, and planktonic and littoral species arrive (Baxter 1977). The limnology of reservoirs is distinctive, and the evolution of tropical impoundments in particular has been of considerable interest to ecologists. In 1969 White argued that research effort had been on too small a scale and had started too late to provide 'either useful predictions or firm conclusions, about the course of biological events consequent upon impounding water in the Tropics' (White 1969, p. 37). To an extent this complaint is still valid, despite over three decades of research, because of the complexity of succession in tropical ecosystems.

The development of the reservoir ecosystem depends partly on the chemistry, turbidity and temperature of inflowing waters, and partly on the nature of the land flooded. Rotting of flooded vegetation can lead to releases of methane and carbon dioxide, important greenhouse gases (Rudd *et al.* 1993). At Kainji on the Niger, 37,000 ha of vegetation was cleared and burned, and a diverse lake bottom fauna developed. The Volta Lake behind the Akosombo Dam, on the other hand, filled much more slowly (over a seven-year period), and there was no clearance of vegetation. As a result, there was initially considerable deoxygenation at depth as vegetation decayed. The lake became poor in nutrients, with a low crop of phytoplankton (Baxter 1977). In neither of these cases was there a severe infestation of algae or macrophytes of the kind that has occurred elsewhere, for example in Cabora Bassa or Kariba on the Zambezi (Balon and Coche 1974, Davies *et al.* 1972; Plate 8.1). Floating plants such as the Nile cabbage (*Pistia stratiotes*) or the water hyacinth (*Eichornia crassipes*) can create problems for hydroelectric turbines, and increase evapotranspiration losses from the reservoir surface. On the other hand, they can, to an extent, provide substrates where fish can find food sources. But enhanced mercury levels have been recorded in reservoir fish, for example in Thailand (Rosenberg *et al.* 1995).

The establishment of reservoir fish populations (and reservoir fisheries) also depends on the way in which the reservoir ecosystem develops, and there are no simple rules (Lowe-McConnell 1985). There is often an initial peak in fish population as nutrients from flooded areas feed into the ecosystem, followed by a slump to a lower level (Jackson 1966, Petr 1975). Calculations of potential catches from new reservoirs based on data from small (and often much shallower) natural lakes are rarely of any value. Furthermore, reservoir fish stocks will be at least in part pelagic, and may require fishing techniques very different from those used by previous fishing inhabitants of a river floodplain. Predicting possible future benefits from a reservoir fishery is far from simple.

The impacts of dam construction on running-water ecosystems are relatively well understood in temperate rivers, particularly where there are important sport fisheries. The impacts of dams on salmonid fish, for example, were studied

Plate 8.1 Floating mats of water hyacinth (*Eichornia crassipes*) on Lake Kariba, Zimbabwe, 1990s. The Kariba Dam on the Zambezi, between Zimbabwe and Zambia, is an important source of hydroelectric power, but caused forced resettlement of large numbers of Tonga people. Water hyacinth restricts navigation on the lake (which has an important fishery), and can clog hydroelectric turbines, as well as increase evapotranspiration losses from the lake surface.

in both Britain and North America in the 1960s and 1970s. The result of this work is a relatively detailed understanding of the migratory patterns and in-stream habitat requirements of migratory game fish, and the development of responses in dam design (e.g. fish ladders), dam operation (the release of arti-ficial floods) and downstream river management (e.g. in-stream flow diversion structures) to minimise adverse impacts of control on fish stocks. Knowledge of tropical rivers is less complete (but see Payne 1986), although there are numerous data on freshwater fisheries, particularly in Africa (Hickling 1961, Lowe-McConnell 1975, Welcomme 1979), and on the limnology of certain rivers such as the Nile (Rzóska 1976), the Niger (Lowe-McConnell 1985) and others in tropical Africa (Davies 1979).

Passage through a reservoir has a number of effects on water quality. Water released from low outlets in a dam may be deoxygenated, and sometimes rich in hydrogen sulphide or mercury, and it may also be cold. There may be influ-ences on invertebrate drift, and river bed degradation downstream can lead to the loss of important in-stream spawning grounds. The simple barrier effect of a dam can severely curtail movement in active aquatic species, and can be a serious threat to endangered species such as the Indus river dolphin (Reeves and Chaudhry 1998). In tropical floodplain rivers, however, it is the impact of dams on natural flood regimes that is the most significant. Many fish exhibit fairly short 'lateral' migrations or longitudinal migrations of greater length in response to seasonal fluctuations in the river (Lowe-McConnell 1975, Welcomme 1979). As water spills out onto the floodplain it becomes enriched with organic matter from decaying vegetation, animal manure and other mate-rials. This creates a flush of algal, bacterial and zooplanktonic growth, which forms a food source for invertebrates. Aquatic and emergent vegetation growth is also rapid and extensive. As a result there is abundant food for fish which follow the floodwaters out of the river channel, and in particular both food and shelter for young fish. Growth rates are very rapid in these flood condi-tions. As the flood subsides, fish move back to the river channel, and in many cases eventually to the small and deoxygenated pools of largely dry river beds. This season sees high mortality, both of fish stranded in evaporating floodplain pools and of those taken by birds, predatory fish and people (Lowe-McConnell 1985).

Reduction in flooding due to drought or dam construction can significantly cut recruitment, fish population numbers and the economic return to the fishing people, as for example in the Yaérés, the extensive floodplain of the Logone River systems above Lake Chad in Cameroon (Benech 1992). Several fish species failed to spawn in the Phongolo floodplain in South Africa following river control and a reduction in the annual flood that isolated floodplain pools from the main channel (Jubb 1972). On the Niger in West Africa, the Kainji Dam acts as a complete barrier to fish movement, and although catches at the foot of the dam are high, studies further downstream reported significant reduc-tions in fish catches between 1967 and 1969 (Lelek and El-Zarka 1973, Adeniyi 1973, Lowe-McConnell 1985), and associated reductions in fishing activity.

Similar reductions in fish catches and fishing effort below dams were recorded elsewhere in Nigeria, for example on the Sokoto (Adams 1985a). A more complex response was shown by the *Egeria* clam fishery of the lower Volta. Breeding in the clams is triggered by a rise in salinity following reduced river flow. During construction of the dam at Akosombo, the critical salinity conditions moved 30–50 km inland, and once it was operating, the constant flows pushed the fishery down to within 10 km of the river mouth (Lawson 1963, Hilton and Kuwo-Tsri 1970, Chisholm and Grove 1985). The further impacts of reduced flows in the Volta in the mid-1980s on the clam fishery are not known, nor is the importance (for better or worse) of the dams at Akosombo and Kpong.

Floodplain vegetation communities, both those immediately bounding river channels and those of larger and more extensive river-fed wetlands, are influenced by flooding patterns in much the same way as aquatic ecosystems. Among the most extensive studies of such impacts are those of the University of Zambia under the Kafue Basin Research Project, begun in 1967 (Williams and Howard 1977, Howard and Williams 1982, Handlos and Williams 1984). The development of a hydroelectric dam in the Kafue Gorge was mooted in the 1950s at the time of the Central African Federation, and would in many ways have been more logical than the Kariba project that was eventually adopted (Williams 1977). In 1967, eleven months after the unilateral declaration of independence by Rhodesia (now Zimbabwe), Zambia announced that development would go ahead, and work began on the Kafue Gorge Dam. It was completed in 1972, and the second phase of the project, further generation capacity in the Kafue Gorge Dam and a second dam at Itezhitezhi above the Kafue Flats, was completed in 1982. There has been considerable concern about the environmental impacts of the developments at Kafue, particularly concerning a wetland antelope, the Kafue lechwe (Rees 1978). However, the Itezhitezhi Dam was built to allow releases of water for downstream wetland maintenance, and an annual release of an artificial flood of 300 m^3 s^{-1} over four weeks each March was agreed (Scudder and Acreman 1996). Although releases have not mimicked natural flows very closely, and concern about environmental impacts continues (Sheppe 1985), environmental impacts have not been as serious as some had feared.

If the impacts of dams on downstream environments are predicted, planning can take them into account. One example of this is the Southern Integrated Water Development Project in Botswana. This involved deepening the Boro River at the lower end of the Okavango Delta and the construction of three reservoirs for flood recession agriculture and urban water supply. Local outcry at these proposals led the government to cancel the construction contract in 1991 and invite an independent review. Subsequent development by the Botswana Department of Water Affairs has been much less drastic, involving conjunctive use of surface and groundwater (Manley and Wright 1996).

Some floodplain ecosystems respond quite rapidly to changes in river flows. In other cases, impacts are delayed, and these are particularly problematic. In semi-arid Africa the river floodplains frequently support woodland vegetation in areas of dry savanna bush (Hughes 1987). These riparian or riverine woodlands are supported by high groundwater tables fed by river flows, and the outer edge of the forest is determined in part at least by the depth to water table. However, forest regeneration depends on periodic high flood flows. The maintenance of the forest therefore depends in a complex way on the annual and inter-annual pattern of river discharge. Alterations in that discharge pattern, for example from dam construction, can have significant impacts on the viability of the forest ecosystem. These impacts may only become clear decades after dam construction (ibid.).

The implications of such river control are understood for few rivers or wetland areas, but have been studied in detail on the Tana River in Kenya (Hughes 1984). The Tana River flows for about 650 km from the slopes of Mount Kenya to enter the Indian Ocean north of Malindi. A series of dams has been built in the headwaters, notably at Kindaruma in 1968, Kamburu in 1975, Gitaru in 1978 and Masinga in 1982. Below these the river flows across a plain of low elevation with Sahelian bush savanna vegetation. Rainfall is low, rising from 300 mm to 1,000 mm at the coast. Groundwater recharge by the river, particularly at high flow stages, supports a narrow belt of forest 1–2 km wide. This now exists in a series of discontinuous blocks, partly because of clearance by Pokomo and Malekote people who live along the river. Some of these blocks support two endemic primates, the Tana River Red Colobus and the Tana Mangabey, which have received protection in the Tana River Primate reserve.

The floodplain forest is diverse in structure and floristics, reflecting the complex geomorphology of the floodplain itself (Hughes 1984, 1987, 1990). Evergreen species occur on heavier soils, trees such as figs are important on sandy levees, there is an endemic poplar (*Populus ilicifolia*) which forms a succession on point bars, and oxbows exhibit a complex succession of low thorny scrub vegetation. Drier parts of the forest are dominated by species of *Acacia*. The dynamics of succession in the forest are complex, but the importance of river flooding patterns is clear. The Tana floods twice a year, with low-flow periods between February and March and between September and October. There are periodic high flows, particularly in May, which inundate extensive areas of the floodplain. There is considerable inter-annual variation in flows, and the record shows high flood years every few decades (for example 1961 saw three times the average annual maximum flow). Studies of tree girths and growth rates suggest that past regeneration has been associated with extreme flows of this kind. Upstream dams reduce both the height and the frequency of high flows in the lower Tana, and are likely to bring forest regeneration to an end (Hughes 1984). Other pressures of development local to the forests, notably the cutting of construction timber and fuelwood for the irrigation scheme at Bura, also appear to be having serious impacts on the forests of the lower Tana (Hughes 1984, 1987).

Tropical floodplain ecosystems are closely adapted to existing patterns of discharge, and hence are vulnerable to environmental change caused by dam

Plate 8.2 Floodplain forest, Tana River, Kenya. The Tana River flows from the slopes of Mount Kenya through dry bush to the sea. The river's water supports a narrow belt of riverine high forest, parts of which include these impressive *Sterculia appendiculata* trees. The forest is important both economically to Pokomo and Malekote people and ecologically for endemic primate species, but it depends on high river flood flows for its regeneration, and these have effectively been prevented by the construction of dams in the river's headwaters, calling into question the ability of the forest to replace itself. Photo: F. Hughes.

construction. However, they are also both ecologically productive and economically important environments. Impacts on natural processes, particularly the ecology of fish production, can therefore have significant socio-economic implications. These will be discussed in the next section.

Dams, people and floodplains

While it is useful to analyse the impacts of projects such as dams on physical and biological systems, it is their significance for economy and society that gives them particular importance in debates about sustainability (World Commission on

Dams 2000). In many parts of the Third World, people depend in a very direct way on the productivity of natural or semi-natural ecosystems for their livelihoods. The intensity of human use, and hence the potential severity of socio-economic impact, is particularly great in extensive floodplain wetlands, for example those of arid or semi-arid Africa which are used for agriculture, hunting, fishing, grazing and gathering. Floodplain wetlands are among the most ecologically productive of global ecosystems and are acutely vulnerable to the impacts of narrowly conceived and poorly planned development project (Hollis 1990).

The economic importance of tropical rivers and their associated wetlands can be very great. The Niger Inland Delta in Mali, for example, supports over half a million people, and in the dry season provides grazing for about 1–1.5 million cattle, 2 million sheep and goats and 0.7 million camels; there are some 80,000 fishermen (Moorhead 1988). These economic functions may overlap in time and space, or may be used by different communities in different ways through a year. Hunting, gathering and grazing and fishing activities are closely linked to the seasonal cycle of river discharge. Thus, for example, the seasonal grazing resources of the Niger Inland Delta are based on the perennial aquatic grass *bourgou* (*Echinochloa stagnina*), which can yield up to 25 tonnes per hectare of forage, and is accessible to livestock once seasonal floodwaters have retreated (Skinner 1992).

The economic values of rivers and wetlands are dependent on the interconnection of geomorphological, hydrological and ecological processes. Thus floodplain agriculture in the West African Sahel may take the form of farming on the rising flood (planting before the flood arrives), or on the falling flood, using residual soil moisture left by retreating floods. Farmers usually have extensive knowledge of crop ecological requirements and flooding patterns, and a detailed appreciation of the variation in land types in the floodplain. Such agriculture is ancient (West African rice (*Oryza glaberrima*) was domesticated three thousand years ago in the Niger Inland Delta), and is still widespread (Adams 1992). Indigenous use of water resources also extends to irrigation, including in West Africa and the Nile Valley the use of simple wells dug into floodplain sediments, sometimes with a shadoof. Human-powered water-lifting for irrigation is being rapidly replaced by small motor-powered pumps, for example in northern Nigeria, where compact, portable and relatively cheap petrol pumps began to be introduced in the early 1980s (Kimmage 1991, Adams 1992).

Wetlands are also important in other ways, for example in sustaining regional groundwater levels and as an ecological and economic resource for very extensive surrounding drylands, particularly in times of drought (Scoones 1981). The wetlands of the Sahel, for example, such as the Inland Delta of the River Niger in Mali, and Lake Chad, not only provide dry-season grazing for huge numbers of livestock, but are important staging-posts for Palaearctic birds on migration to and from wintering grounds in tropical Africa. The 'functions' of wetlands, as the IUCN Wetlands Conservation Strategy terms them, also embrace flood control, food production and wildlife conservation (Barbier *et al.* 1998; see Figure 5.4, p. 119).

Major water resource projects can have serious impacts on floodplain people and their economies. One classic example is the Jonglei Canal, proposed to reduce the volume of water lost from the White Nile in the swamps of the Sudd (mentioned on p. 218). It was recognised in the 1940s and 1950s by the Jonglei Investigation Team that the Equatorial Nile Project as then conceived had significant adverse environmental impacts (Howell 1953), and with the building of the High Dam at Aswan was not pursued. Division of the Nile's flow between Egypt and the Sudan under the second Nile Waters Agreement of 1959 (following a study published in 1958, using computer analysis for the first time) granted Egypt rights to 55.5 billion cubic metres (an increase of 7.5 billion) and Sudan to 18.5 billion cubic metres (a rise of 14.5 billion cubic metres over its previous paltry 3 billion). The Jonglei Canal seemed an obvious way to obtain more water for downstream irrigation, and it was proposed again in 1974.

The canal, whose construction actually began in the 1970s, was smaller than the original scheme. Nonetheless, it was to be 54 m wide and 360 km long, carrying 20 million cubic metres of water per day. Extensive adverse environmental impacts were feared (Tahir 1980, el Moghraby 1982), but new surveys suggested that owing to a fortuitous (and not wholly understood) increase in the discharge of the White Nile in the 1960s, impacts on the Sudd would be small. It was predicted that if high flows in the White Nile persisted, the area of seasonally flooded *toich* grassland would be reduced by 25 per cent, with impacts on fish populations and serious problems of access across the canal for Dinka pastoralists (Howell 1983, Howell *et al.* 1988, el Moghraby and el Sammani 1985). However, the plans of the Sudanese state for the 'development' of the southern Sudan extended well beyond the Jonglei Canal itself, to include irrigation, ranching, improved road communications with cities in the north of the Sudan. All these threatened their own impacts on the economic, social and cultural bases of Dinka society (Lako 1985). In the event, debates about the impacts of development were made redundant by the onset of bitter civil war in the southern Sudan in the 1980s and 1990s. All development investment ended, while the war inflicted its own grim destructive impacts on the environment, economy and society.

The impacts of dams on floodplain resource users can be direct and immediate. In the Sokoto Valley of northern Nigeria, as in the floodplains of many West African rivers, the farming of seasonally flooded land has long been integrated with dryland cultivation in the local economy. There is a single rainy season, from about the end of May to the end of September, and the River Sokoto has a strongly seasonal regime with high peak flows in July, August and September. Rice and flood-resistant sorghum are planted in up to 90 per cent of the floodplain in the rains, with rain-fed millet and sorghum followed by relay crops of cotton, cowpeas or groundnuts on upland areas. Different local varieties of rice and sorghum are grown, adapted to particular conditions of flooding, soil waterlogging and desiccation across the floodplain. Both rain-fed and floodplain crops are harvested from October onwards, and the floodplain then comes into its own because a second crop (for example peppers of various kinds, onions and sweet potatoes) can be grown using residual soil moisture and, sometimes, shallow groundwater irrigation.

In 1978 the Bakolori Dam (Plate 8.3) was completed on the Sokoto River, and in subsequent years it brought about a significant reduction in peak flows in the Sokoto (Adams 1985a). This in turn halved the average depth, duration and extent of flooding in survey villages downstream in the 120 km stretch before the next major confluence, that with the River Rima. The area cropped fell from 82 per cent to 53 per cent of plots in one village, and there was a particularly marked fall in the area under rice (from about 60 per cent to 14 per cent of fields) and in the amount of dry-season farming. The proportion of households undertaking dry-season cultivation fell from 100 per cent to 27 per cent in one village (ibid.). It was estimated that of a total of 19,000 ha of floodplain land, the dam caused the loss of 7,000 ha of rice and 5,000 ha of dry-season crops.

To an extent these losses were compensated for by increases in the area under millet and sorghum, but the new uncertainty about flooding patterns and the intolerance of millet to waterlogging following rain on heavy soils meant that farmers were not able to adapt wholly to the new conditions. Furthermore, although irrigation was a known and tested technology in the Sokoto Valley, the costs of well-digging and the labour demands of water-lifting were both

Plate 8.3 The Bakolori Dam, Nigeria. The Bakolori Dam was built in the late 1970s on the Sokoto River in northern Nigeria, to store water to supply an irrigation scheme on floodplain and terrace land downstream. The reservoir caused the resettlement of 12,000 people, and disrupted floodplain agriculture and fishing for several hundred kilometres downstream.

increased because reduced floods were accompanied by increased depth to water table. As a result, irrigation became a more specialised technique, accessible only to larger producers. The magnitude of lost production in the Sokoto floodplain needs to be seen against predictions that the flood control effects of the dam would allow increased production of rice from downstream areas. The value of lost downstream production can be estimated, and shown to have a significant effect on the benefit/cost ratio of the Bakolori Project as a whole (Adams 1985a).

The impacts of dams on wetlands have been similar elsewhere in West Africa. In the north-east of Nigeria, the Hadejia and Jama'are rivers join, forming the Komadugu Yobe, and drain towards Lake Chad. Around their junction is an extensive floodplain complex of seasonal wetlands and pools, fed by the seasonal flood flows. This area is of enormous economic and ecological importance, supporting a large human community engaged in extensive rice farming, grazing and fishing. The combination of upstream dams (to supply rather unsuccessful large-scale irrigation schemes) and low rainfall through the 1970s and 1980s has caused significantly reduced flooding in the wetland. This in turn has had

Plate 8.4 The Hadejia-Jama'are wetlands, Nigeria. These wetlands in north-east Nigeria consist of an extensive complex of seasonal wetlands and pools, formed where the rivers flow through an ancient dunefield. The area has many villages engaged in rice farming, grazing and fishing. The wetlands have progressively dried out owing to a combination of low-rainfall years and upstream dams, with serious environmental and economic impacts. These have been offset in some instances by the successful adoption of small petrol irrigation pumps and shallow tubewells, but these cannot be used in all areas.

considerable impact on flood recession farming, fishing and grazing, and has had measurable economic costs (Hollis *et al.* 1994, Polet and Thompson 1996, Barbier *et al.* 1998).

The Nigerian experience is by no means unique. Similar impacts have been recorded in every continent, wherever dams have affected floodplain river people (McCully 1996, Singh 1997, Usher 1997a). Environmental impacts not only impinge on the livelihoods of the poor, but can be highly relevant to the cost calculations of developers as well. For this reason, environmental impacts have quite rightly become a major concern of project planners. The rise of such concerns in the 1970s has been mentioned in Chapter 2 as one of the factors that preceded the formulation of mainstream sustainable development. The success of attempts to avoid or mitigate such impacts is discussed in Chapter 11.

The socio-economic impacts of dam construction may be drastic, but it is important to note that floodplain people are not always passive victims of those impacts. They tend to be ingenious and industrious in their attempts to adapt to their new circumstances. Environmental impacts themselves evolve over time, as physical and ecological systems adapt to changing flood frequency and duration. People also respond to changed environmental conditions by changing their patterns of resource use to take account of reduced flooding. If external conditions are favourable, they may be able to adapt successfully. Thus, for example, farmers in both the Sokoto and Hadejia-Jama'are floodplains were able to join in a boom in small-scale irrigation in the 1980s and 1990s based on small petrol pumps and shallow tubewells that were originally introduced by World Bank agricultural projects (Kimmage 1991). These technologies allowed some floodplain farmers (those in favourable locations and with cash resources) to take up commercial dry-season irrigation. Relatively cheap and abundant petrol supplies, the existing local experience with small two-stroke motorcycle engines, and the investment in tarred roads to allow access to urban markets were all factors in the success of the irrigation. Using pumps and tubewells, some floodplain people were able to improve their economic position in the face of environmental degradation (Kimmage 1991, Thomas and Adams 1999). However, not everyone could invest in this way, and some floodplain people lost out seriously. The boom itself also had costs, for example in shutting Fulani pastoralists off from their former dry-season grazing land. The critical point is that the impacts of the upstream dams need to be understood in the context of complex chains of impacts and responses that can interact with various other kinds of positive or negative change.

Dams and resettlement

The environmental and social costs of dams are not confined to downstream environments. Development programmes and projects create both winners and losers (Scudder 1991a, Hardjono 1983). In the case of dams, those forced to resettle as a result of reservoir creation bear the largest burden of costs, and may not share the benefits of development (see, for example, Brokensha and

Scudder 1968, Barrow 1981). The problem of resettlement was recognised early in the period of enthusiasm for large dams in the 1960s (Colson 1971, Scudder 1975). Despite some extensive planning exercises, forced resettlement of reservoir evacuees (or 'oustees') has often been far from satisfactory. As Gosling comments in the context of the Mekong Basin, 'reality is harsher than dreams, and the opportunities for evacuees are more limited in reality than on paper' (1979, p. 119).

Very substantial numbers of people have been moved to make way for reservoirs. McCully (1996) suggests a figure of 2.2 million people displaced by just 243 completed dams in a range of countries. However, this figure excludes China and India. Chinese government figures suggest that 1 to 2 million people have been displaced by dams, and alternative (but still conservative) estimates suggest that a figure of 30 million may be more accurate (ibid.). China's controversial Three Gorges Dam will displace a cool 1.3 million people. In India perhaps 4 million people have been displaced by reservoirs and irrigation schemes (ibid.). While many Indian schemes have few oustees, several (e.g. Polaran, Kangsabati, Kumari and Bansagar) have over 100,000, while the Sardar Sarovar Dam and Narmada Sagar Dam on the Narmada River will flood 265,000 and 170,000 respectively (Singh 1997). A disproportionate number of oustees in India have been people from scheduled tribes, or landless people (43 per cent of the oustees from the Narmada Sagar Dam for example are landless; ibid.). The human cost of dam construction varies greatly between countries, but globally it is significant, and locally it can be devastating (World Commission on Dams 2000).

Scudder (1991a) identifies a four-stage model of resettlement projects. Stage 1 is the stage of planning, infrastructural development and settler recruitment. Stage 2 is transition, a period of one to five years during which people actually move and seek to re-establish livelihoods in a new location, making use of whatever investment has been made for them (e.g. health facilities, roads, housing or employment). In stage 3, settlers ideally start to become more risk-taking, making investment strategies to increase productivity through diversification of family labour (investing in education, livestock, off-farm income). In stage 4, resettlement project activities are handed over to local organisations, and a generation of settlers takes over.

It is in the relocation stage, however, that evacuees face the greatest costs, particularly where it is rushed, as in the case of refugee relocation and the resettlement following the construction of dams (Scudder and Colson 1982). Compulsory resettlement is traumatic, causing 'multidimensional stress' (Scudder 1975, p. 455). This stress arises from the way in which people are uprooted from homes and occupations and brought to question their own values and behaviour, and the power of their leaders (Lumsden 1975). Land flooded by reservoirs can be of direct economic importance for production, but it can also have cultural or religious significance, as Colchester (1985) described in the case of the Gond tribal people losing land beneath hydro-electric reservoirs in Maharashtra and Madhya Pradesh in India. Evacuees stranded in isolated settlements on the edge of new reservoirs, or decanted into the urban squalor of dam construction towns, can face large cultural and socio-economic costs, and severe challenges in establishing new livelihoods. People who own land that is flooded, but who are not resettled, or whose lives are disrupted by the dam construction process, can also lose out severely as a result of new projects.

As Sutton (1977) points out, there is little difference between forced resettlement for 'developmental' purposes and anti-guerrilla resettlement such as the *regroupement* policy undertaken by the French army in Algeria between 1954 and 1961. Both are externally imposed and in the end both often involve coercion. State violence against evacuees, up to and including murder, is widely recognised, for example in the case of Guatemala's Chixoy Dam in the 1970s (McCully 1996) and the Sardar Sarovar Project on the Narmada River in India in the 1990s (Baviskar 1995). Such violence usually represents an attempt to quench protest by citizens and civil disobedience and non-compliance with orders by evacuees. Protest at resettlement at Kariba in 1958 led to deaths (in Northern Rhodesia, present-day Zambia; Howarth 1961), as did protest about compensation at the Bakolori Project in Nigeria in the early 1980s (Adams 1988a). These are by no means isolated incidents. Opposition to forced removal for dam construction has grown in a number of countries, particularly through the high-profile protests such as Narmada Bachao Andolan (the Movement to Save the Narmada) in western India (McCully 1996, Baviskar 1995). Such opposition will be discussed further in Chapter 12. In the 1980s the World Bank recognised 'the hardship and human suffering caused by involuntary resettlement', and made a policy commitment to avoid or minimise it, and explore alternative solutions (Cernea 1988, p. 4).

The impacts of resettlement linger long after the relocation phase. The Gwembe Tonga relocated from the Zambian portion of Lake Kariba in the late 1950s suffered enormous initial dislocation, but the development of a gillnet fishery on Lake Kariba, the eradication of tsetse fly (and sleeping sickness in humans and cattle) and new roads into and out of the area meant that living standards rose between 1956 and 1974 (Scudder and Habarad 1991). However, they fell (with those in the rest of Zambia) with the collapse of copper prices in 1974, and have remained depressed thereafter as infrastructure degraded and economic opportunities within and outside the area remained limited.

The key problem in reservoir resettlement is the unequal way in which project costs and benefits are allocated. The Bakun Hydroelectric Project in the rain-forests of Sarawak involves a dam on the Balui River impounding 4 billion cubic metres of water in a reservoir covering 695 square kilometres, and demands resettlement of as many as 4,300 Kenyah, Kayan and Kajang people (Mohun and Sattaur 1987). The Bakun Dam is one of a series of four dams planned in Sarawak to generate power for Peninsular Malaysia, supplied through an undersea cable. As so often, the benefits of development (hydroelectric power) are being enjoyed by one group of people, the costs (resettlement) by another. This is highly inequitable. In theory, development planning should be done in such a way that evacuees do not bear disproportionate amounts of the costs of dam projects (Gosling 1979). In practice, they often do bear these costs, and it is this failure of planning that has been the focus of attention by environmentalist critics of dam construction (e.g. Goldsmith and Hildyard 1984, Roggeri 1985, McCully 1996).

Very often, resettlement projects begin with high hopes of re-establishing evacuees in conditions no worse than those they have left. Thus the Resettlement Working Party convened by the Volta River Authority aimed to use resettlement to 'enhance the social, cultural and physical conditions of the people' (Chambers 1970). Such high hopes are rarely realised. Lightfoot argues that the literature shows that 'most reservoir resettlements have been badly planned and inadequately financed, and that most evacuees have become at least temporarily and in many cases permanently worse off as a result, both economically and socially' (Lightfoot 1978, p. 63). To an extent the litany of failure may be a function of a lack of published research, and the syndrome that social scientists get more kudos from researching failure than success (Chambers 1983). Nonetheless, there are well-documented case histories, for example from Ghana (Chambers 1970) and Nigeria (Mabogunje 1973).

The reasons for the poor record of reservoir resettlement projects are manifold. The most fundamental is the disciplinary bias within dam-building organisations. Technical disciplines such as engineering, geology and hydrology dominate the dam project planning field, and the appraisal process concentrates on technical problems relevant to these disciplines. The characteristics of the population of the resettlement area, their economy and society, are neither recognised nor understood. Field investigations concentrate on the dam site and not the inundation area. Frequently the only research in the reservoir area itself concerns bedrock geology and a perfunctory check on topography to confirm the integrity of the reservoir's proposed storage level. The engineering companies called in at each successive stage of project appraisal lack the skills necessary to comprehend resettlement planning problems.

A second cause of the failure of resettlement projects is the fact that project appraisal rarely allows sufficient time for effective planning to be done. Socioeconomic planning is both more difficult and more time-consuming than many technical experts assume. It is therefore rarely made part of the technical planning process. It is usually introduced too late to be effective. In the case of the Akosombo Dam in Ghana, for example, although the Preparatory Commission had made recommendations about surveys for resettlement, uncertainties over the future of the project meant that nothing was done until the construction contract for the dam was awarded in 1961. The first resettlement staff were appointed in that year, but effective resettlement work did not begin until nine months after the start of dam construction. The stress of resettlement is exacerbated when resettlement is rushed. The human costs of a crash resettlement programme are great, particularly on the elderly and infirm. Furthermore, speed brings errors, and problems of food shortage and poor water supplies are often increased (Scudder 1975).

A third problem with resettlement planning is its inherent complexity. To many observers, the hard part about building a dam would probably seem to be the geotechnical or engineering challenge, or the sheer logistics of building a massive artefact in an area remote from supplies of petrol, engineers and cement. In fact, resettlement planning has often proved the Achilles heel of reservoir projects, seemingly straightforward but in practice fiendishly complicated. Resettlement planning typically involves a series of tasks, including a population survey and an inventory of property and land within the reservoir area, and surveys to locate possible new settlement sites either near to or further away from the home area. Only at this stage can the scale of the resettlement problem be assessed. Knowledge of the costs of resettlement is of course a vital element in the assessment of the practicability and acceptability of the project as a whole. Without it, all the technical planning and design may prove useless. However, data on resettlement needs are rarely available in a suitable form in sufficient time to influence decision-making about dam construction. Resettlement is regarded as a secondary problem, to be addressed once the technical feasibility of the project is known. The sunk costs of technical engineering appraisal are such that it can be hard to stop a project once the full human (and environmental) costs are finally factored in. There are clearly quite specific skills associated with population resettlement planning that have been known for many years (e.g. Butcher 1967). Goodland (1978) outlined the measures necessary in the fields of social and cultural ecology with respect to the Tucuruí Dam in the Tocantins River basin in Brazil. He urged that mitigation of the project's impact on indigenous Amerindian people should be allocated time and resources commensurate with its importance, and that the associated costs should be considered an integral part of the costs of the whole project. Guidelines have been available for some years, not only outlining the essential elements of a resettlement project, but discussing their integration into the operational procedures of the project development cycle (Cernea 1988).

A fourth problem is that of cost. Resettlement is expensive, even if done badly, and it is usually under-resourced. This is partly because costs can only be calculated once preliminary surveys have been carried out. But these surveys begin only after the decision to go ahead with the project has been made. Resettlement costs are therefore seen as somehow an added extra, additional to the costs of construction. It is as if they were in some way optional. An interesting example of this is the development of the Volta Scheme in Ghana. The Akosombo Dam was begun in 1961, but was preceded by an extensive study published in 1956 by a government Preparatory Commission. This recommended that extensive further studies be carried out into resettlement problems, but suggested that self-help resettlement with cash compensation might be the best solution. Partly because of its concern with resettlement and the resulting costs, the work of the Preparatory Commission made the economic prospects of the dam seem sufficiently unattractive that work did not proceed. Within a few years, however, in the aftermath of the agreement between the USSR and Egypt over the Aswan High Dam, and in an independent Ghana, the scheme took on a new aspect. Revisions of the Preparatory Commission's figures were made which excluded a number of costs, and work began (Hart 1980, Lanning and Mueller 1979).

Resettlement costs are often underestimated, owing to failure to specify the basis for resettlement, lack of data on the affected population, and inadequate budgetary provision, with the lion's share of this being eaten up by the survey and planning elements of the resettlement process. This was the experience in the case of resettlement of 26,000 people in another Nigerian scheme, the Dadin Kowa Dam. Planning for resettlement only began once construction had started in 1980. It soon became clear that the cost of compensation at the rates set by the federal government would be about 60 million naira (then about US$50 million), more than an order of magnitude greater than the total resettlement budget. On top of this, further expenditure would be needed on infrastructural development (for example roads and water supplies in new villages). Even then, all that would have been achieved would be relocation, since such a resettlement was far from recreating a viable new economy for evacuees. Obviously resettlement was a very significant element in total project costs; if tackled seriously it would have altered the cost/benefit calculations of the dam irrevocably. In the event, attempts to seek additional funds from the federal government were overtaken by the fall in world oil prices and severe financial stringency at federal level. Resettlement planning effectively became a paper exercise (Adams 1985b).

A typical example of poorly organised and implemented resettlement planning is provided by the Bakolori Dam, completed on the River Sokoto in Nigeria in 1978, flooding the homes of 12,000 people. In this case there had been no attempt to involve evacuees in planning, and no specific provision at design stage to set out a specification for resettlement. Survey in the reservoir area was left to the river basin authority, which lacked people with technical skills. Most of it was in fact done by teams of students, and surveys were only narrowly completed in advance of the waters of the filling reservoir (Adams 1988a). Complaints about the inaccuracies and haste of the survey, the inadequacy of the compensation and the poor resettlement site (on a barren hilltop remote from the river valley) escalated, and, together with the similar grievances of those in the irrigation area served by the dam, led to blockades of the project area by protesters, and violent repressive action by federal police (ibid.).

The lack of consultation at Bakolori (and the consequences of its ill-planned, hasty and coercive implementation) may be contrasted with the method adopted in the Volta Resettlement Project in Ghana ten years before (Chambers 1970). The scale here was much larger, involving some 80,000 people in 739 villages, about 1 per cent of the population of Ghana. Nonetheless, the selection of new village sites attempted to take account of the views of evacuees. As part of an extensive social survey, enumerators recorded the positive and negative preferences of each village about relocation. Seventy-two sites were identified in this way; of these, twenty-seven were rejected by the Volta Resettlement Authority on technical grounds, although some were accepted after further discussion. Fifty-two village sites were eventually agreed (Chambers 1970). By the standards of many large dam projects, the Volta resettlement was exemplary; it is unfortunate that its lessons have so repeatedly been ignored in subsequent schemes elsewhere. However, even here the project fell short of its own targets, for example in its attempt to clear land for mechanised farming. It was estimated that 42,000 ha would be needed to support evacuees, but land clearance was slow, and by 1967 only 2,500 ha was being cultivated, over half of it manually. In 1968 only 52 per cent of the adult males resettled could farm at all, and food relief had been necessary for three years (Hart 1980).

Perhaps the best measure of the 'success' of these resettlement projects is the loyalty of those resettled. The Kainji Dam in Nigeria was completed in 1968, creating a lake of 1,200 square kilometres, flooding 203 villages containing 44,000 people, the towns of Bussa and Yelwa, and 15,000 ha of farmland (Adeniyi 1973). Resettlement planning began before independence in Nigeria, and surveys began in 1962 with a view to providing cash compensation to evacuees. By the end of 1963, 2,338 people in eighteen villages had been moved. However, resettlement was slow and inefficient. In November 1964 the policy on housing was reversed, and the authority began to build houses for evacuees using a design by British architects using sandcrete block walls and ferro-cement roofs (Atkinson 1973). Cash was paid for farmland and trees of economic importance, and agricultural advice was offered to evacuees. The main immediate focus of complaints by Kainji evacuees was poor village location and the design of the new houses, particularly the layout of rooms, and the houses' thermal properties and leaking roofs. By 1989 some houses had been abandoned, but most resettlement villages were still inhabited (Roder 1994).

Like the Volta project, Kainji might claim a modest success for its new settlement planning, but it is eloquent testimony that of the 67,500 evacuees on the Volta project only 25,900 were still present in resettlement villages in 1968. Other evacuees had moved elsewhere, while outsiders, including fishermen displaced by declining catches downstream, had arrived. Other resettlement schemes exhibit similarly high turnover rates. Evacuees from the Aswan Dam were resettled on the New Halfa Agricultural Scheme (originally called the Khashm el Girba Scheme in Sudan). Of 300,000 people in the area in 1977, about 24,000 had left by 1980. Both the productivity and cultivated area under irrigation began to fall, owing to a combination of technical and management problems (Khogali 1982).

To achieve parity between economic conditions before and after resettlement requires fair systems of compensation. Not only is it technically complex and expensive to organise effective surveys of land and household effects, but there is a need for probity, transparency and even-handedness in the bases for compensation and the payment process. These are often lacking. Singh (1997) comments that in many Indian projects, compensation is arbitrary, and depends on lawyers and middlemen to whom only richer oustees have access, and the payment of bribes. Compensation is therefore often least available to those most in need. It is also usually available only to those who are registered land title-holders, excluding many tribal households, and those dependent on informal use of lands such as forests (Singh 1997). The attempt to fit the informal systems of resource access and the complex dynamics of human need into the bureaucratic rationality of a resettlement planning process is rarely wholly successful.

Both the Kainji and the Volta resettlements began with self-managed resettlement based on cash compensation and abandoned it. In the Kainji case, for example, self-managed settlement was too slow, and villages were sometimes located in sites unsuitable for water supply or other infrastructure. Perhaps more fundamentally, evacuees were using compensation money for purposes other than house-building. Presumably this reflected either the existence of economic opportunities other than re-establishment of a farming household and land clearance, or (more probably) hidden costs facing evacuees. However, it was interpreted as a shortcoming of the laissez-faire approach, and the centralised planning input was strengthened.

In most cases, resettlement planning is based on this kind of top-down centre-outwards approach to planning. Planners assume that their expertise allows them to 'understand and manage the interests of the farmers better than the farmers do for themselves' (Lightfoot 1979, p. 30). They tend to favour direct control of the resettlement process, despite the fact that among evacuees from the Nam Pong Project (5,012 households resettled in 1964), those who had been resettled in planned schemes were worse off than those who had resettled themselves (Lightfoot 1978, 1979, 1981). Resettlement planning without public participation forms part of a development process that is imposed from outside, and meets an agenda which may be little influenced by local experience of past or present. There is increasing recognition of the need to revise and improve upon dam planning procedures (World Commission on Dams 2000).

The environmental impacts of irrigation

Dams are perhaps the most dramatic examples of human capacity to transform nature in the name of development, but they are but one of a very large number of such initiatives. Some of these will be described in Chapter 9, but one deserves mention here because of its organisational links with river control. Irrigation has been the basis for real plans for future development in many parts of the Third World. As a technology, irrigation encapsulates the developers' determination to control the variability of existing environments and transform their productivity (Adams 1992). Particularly in semi-arid areas prone to drought, the lure of irrigation is very strong: the provision of just the right amount of water at the right time to the growing crop; the encouragement of skilled and market-orientated groups of farmers; the provision of the infrastructure to train farmers, to introduce new technologies (improved seeds, fertilisers, pesticides, machinery) and to remove crops to markets; the efficient location of people in settlements where health, education and clean water can be conveniently supplied; all the costs of investment paid for by double- and triple-cropped irrigated fields, and each investment gaining from and contributing to the next.

Large-scale irrigation projects also have a problematic record in terms of both social and environmental impacts (Worthington 1977, Adams and Hughes 1990). Irrigation has also earned an unfortunate reputation for economic inefficiency. This is particularly true of those areas where large investments were made in new smallholder schemes in the 1970s. The most prominent failure of irrigation to perform adequately in economic terms has been in sub-Saharan Africa, for example in Nigeria and Kenya (W.M. Adams 1991, 1992, Moris 1987, Moris and Thom 1990). In Africa, capital costs are high for a number of reasons, including difficult terrain, poor soils and erratic hydrology, poor communications and remote locations, and lack of skilled staff (particularly engineers). By 1982 the Bura irrigation scheme in Kenya was two years behind schedule and costs were 187 per cent of those predicted in 1977 (Adams and Hughes 1990). Once built, irrigation schemes faced acute problems of poor transport infrastructure, limited markets for produce, overvalued currencies, lack of skilled people and poor public-sector management (World Bank 1990).

Despite the bold claims made for it and the wholesale transformation of arid environments and dryland production systems that it offered, irrigation has not delivered development on the scale planned, and has certainly not done so sustainably. The inability of irrigation to live up to its promise is by no means confined to Africa. Even once built, many, perhaps most, irrigation schemes, particularly large-scale canal systems built and run by the state, suffer from poor management (Chambers 1988a). Some of the African schemes constructed in the 1970s (the Bakolori Project in Nigeria, or Bura in Kenya for example) failed to become effectively established (Adams and Hughes 1990, Adams 1991), but even once built and working, irrigation schemes have proved expensive to run and severely limited in their capacity to deliver sustained increases in livelihoods for farmers.

One reason for this is the persistent notion that irrigation is a technical discipline. It is true that there are many complex engineering and hydrological challenges involved in harnessing a water source, bringing water to the field, applying it efficiently to crops and draining any surplus out again. However, these tasks merely set the context for the engagement between people, water and crops from which the productivity of the scheme will flow. The critical constraints are very often sociological or economic rather than engineering ones. The disappointing performance of many irrigation projects may be traced to 'their conception as exercises in applied hydraulics on a large scale, rather than as a facility for providing a reliable water input to the farmer' (Rydzewski 1990).

In places where formal large-scale irrigation is long established, as in dryland India, the most fundamental management problem is a failure to guarantee regular and adequate water supplies to all parts of a scheme (Chambers 1988a). Unpredictable water supplies reduce yields, which reduces project benefits. Poor water management thus has a significant economic impact, both on production at scheme level, and hence returns to project investment, and also on the ability of individual farmers to survive within the scheme. The causes of poor water supplies lie both outside and inside the scheme. Seasonal rainfall (characteristic of the semi-arid tropics) demands water storage (dams), with all the costs and problems such as evaporation this involves. Inter-annual variability in rainfall makes predicting sustainable water yields from a dam or river source difficult. A striking example is the South Chad Irrigation Project in north-eastern Nigeria, which was crippled by the shrinkage of Lake Chad due to low rainfall in the early 1980s. The area irrigated has been but a small fraction of that predicted, and the economic benefits that were supposed to meet construction costs have never materialised (Kolawole 1987). Where dam sites are few, or in flat terrain, gravity irrigation requires long supply canals that are expensive to build and maintain, and where transmission losses (to seepage or evaporation) can be high. Schemes supplied by pump demand the importation of costly equipment and spares, and high standards of skilled maintenance (all of which have to be paid for out of the production of the scheme, much of it as foreign exchange).

Within a project, water distribution depends on effective maintenance of canals and control structures (weirs or gates). If water bailiffs (or farmers) open gates when they are not supposed to, or allow sediment and weeds to grow in canals, or cut holes in canal banks to water outside their turn, water does not move around the scheme as intended. This 'inefficiency' in water use is exacerbated if there is significant water loss from canals (it can be stopped by canal lining, but that is expensive). Farmers respond to unpredictability in water supply by ignoring recommended water use practices, overwatering when they can, stealing water and bribing technical staff (Wade 1982, Chambers 1988a). Such actions by farmers near the source of water ('top-enders') mainly serve to make the supply of water to those further down the system ('tail-enders') even less predictable (Chambers 1988a). If farmers cannot get water when they need it, their incentive to take it outside the rules is increased.

Individual farmers may be able to obtain water they need, if they are rich enough or powerful enough to bribe the water manager or to escape censure for breaking the rules, or if they live next to the main canal. However, the more they maximise their personal interests at the expense of others, the more the overall efficiency of water use goes down, and the overall yields and economic returns of the project fall. Inefficiency in water use can be related to overuse as well as underuse of water. Overwatering can produce waterlogged soils, particularly where drainage systems are inadequate or badly maintained, and these in turn can lead to problems of salinity or alkalinity (Kovda 1977). These are major problems in areas with high rates of evapotranspiration and poor drainage (for example Pakistan, Iraq, Egypt), and may affect up to half the world's irrigated area. Once soil has become saline or alkaline, further crop growth is effectively impossible without expensive rehabilitation.

Operation and maintenance ('O and M') is recognised as the Achilles heel of many irrigation schemes. Problems include siltation of canals and structures, particularly where water is being drawn from a river draining a semi-arid catchment with high rates of soil erosion, blockage of canals and structures by floating and rooted plants, leakage of canal lining leading to high transmission loss of water, and the failure of engineering structures. Irrigation schemes require continuous maintenance to deal with such difficulties as they arise. Management costs on the large-scale irrigation schemes tend to be high, in some cases exceeding the gross value of production.

Farmers face paying for the high capital costs and running costs of irrigation schemes. However, threatened by low yields, uncertain water and input supply and costly services, farmer incomes tend to be low and uncertain. Farmers face a production cost squeeze due to poor yields and fixed production costs (for example paying for tractor hire or fertiliser). Those who cannot make ends meet may leave the scheme, or shift over to crops with which they are familiar and which will survive drought, such as sorghum. These may meet subsistence

Plate 8.5 Waterlogging on an irrigated field corner, Bakolori Project, Nigeria. Many established irrigation schemes suffer from poor management, or what are called 'operation and maintenance' ('O and M') problems. Physical problems include the blockage of canals and structures by accumulated sediment or vegetation, water losses through leakage from canals and evaporation from open water bodies, and poor drainage. This can lead to some parts of an irrigation scheme receiving too little water, while others receive so much that crops suffer from waterlogging and die. Poor water allocation can be made worse by farmers taking water out of turn, with those at the tail end of the canal system suffering disproportionately when water distribution is unfair. Waterlogging and associated salinisation due to poor drainage is a major global problem on irrigated land. This photograph shows maize suffering in a poorly drained field corner.

requirements, but are unlikely to meet the predicted economic returns of the project. They also tend not to fit into irrigated cropping timetables, so that, for example, a farmer growing sorghum in the wet season in a northern Nigerian irrigation scheme will not be ready to grow a dry-season crop such as wheat at the time necessary to avoid heat damage. As a result, planting is late, there is a severe bottleneck in demand for ploughing and land preparation services, and dry-season yields are reduced. Thus the poor performance of the scheme is exacerbated, and the economic noose around farmers, particularly poorer farmers, tightens (Wallace 1980, 1981, Palmer-Jones 1984, Barnett 1978, Andrae and Beckman 1985).

Debt is therefore a major factor in the socio-economic impact of irrigated cropping. It can take two forms: first, debt to the project (as farmers are expected to pay for inputs despite low yields), and second, debts to others (for example where farmers have to borrow grain where grain harvest or cash income are too small). Tail-enders' farms are often smaller and the farmers themselves poorer. In general, the ability of farmers to command sufficient water is closely related to wealth. Poor farmers are much more completely at the mercy of the system, and they suffer disproportionately when the water distribution system is unfair or inefficient.

The ability of farmers to make a living from an irrigation scheme, and their ability to give that scheme a positive economic rate of return, demands good health. Unfortunately, irrigation also brings certain very specific problems of disease, particularly malaria and bilharzia or schistosomiasis (Amin 1977, Farid 1977). Standing water provides excellent breeding conditions for the mosquitoes that carry malaria, and shallow, slow-moving irrigation canals, particularly where they are badly maintained and overgrown with plants, provide conditions for the snail that is the intermediate host of the parasite that causes schistosomiasis (bilharzia). It is estimated that 60 per cent of adults and 80 per cent of children on the Gezira Project in the Sudan have bilharzia. Poor health and nutrition affect the ability of farmers to work their land and meet the targets of irrigation planners. Poverty among farmers is closely linked to other problems, particularly family health and nutrition. At Bura a 1985 survey showed that 52 per cent of children on the project suffered from malnutrition (Vainio-Matilla 1987). Rift Valley fever and malaria increased following construction of the Manantali Dam and the expansion of irrigation in the lower Senegal River valley (Verhoef 1996).

There are also vitally important gender dimensions to irrigation development. Irrigation planners often assume that farmers are male, but this is often not the case: in Africa, for example, up to 80 per cent of farm work is done by women, and women farmers are important even in Muslim areas such as the Sahel. Gender cross-cuts irrigation planning in many ways. Women characteristically bear a heavy domestic labour burden (including for example water, fuelwood, childcare), and any assumption about labour availability based on observations about male labour is unlikely to be very relevant. Irrigation development can also alter gender divisions of labour, and create strong tensions between men and women for labour power, land and the products of agricultural work. Studies of smallholder rice production on the Jahaly-Pacharr Scheme, begun in 1984 in the Gambia, reveal the nature of such pressures. In the floodplain of the Gambia River, swamp rice has long been grown by Mandinka women, while men work dryland fields away from the river. Most attempts to develop rice irrigation schemes along the Gambia River had focused on giving household heads (almost all men) green revolution rice technology. These had not been a success. Men were not experienced in rice-farming, and anyway gave priority to wet-season crops on surrounding upland areas. Women were not allocated plots in the irrigation projects, and continued to farm in the surrounding swamps. As a result, the formal rice schemes were under-utilised in the dry season and underperforming. The Jahaly-Pacharr scheme explicitly targeted women and attempted to harness their labour, and their indigenous skills in rice production (Carney and Watts 1990, Carney 1992). Plots were allocated on a twenty-year lease, but despite plans to grant plots to women, 90 per cent of plots ended up owned by men, even when they had once belonged to women. The project was large, and took over most of the available swampland, leaving nowhere for women to maintain independent production outside the scheme. The coming of irrigation shifted tenure of rice land from women to men, yet demanded that women continue to do the work of cultivation. Women therefore lost rights to their individual plots (and lost control over their produce) and were locked into intensified work routines controlled by male household heads (Carney and Watts 1990, Carney 1992, Watts 1993).

The management failings of large-scale irrigation schemes are now widely recognised, and may in retrospect seem obvious. They were not taken account of in planning in the 1970s and 1980s (and in general are still underestimated now) because of the unreasonable optimism of economic forecasts. There are various reasons for this. First, yields under irrigation tend to be overestimated. Figures for likely yields under irrigation from experimental farms or 'pilot' projects where crops are grown under careful management of an agronomic experiment, or by selected farmers well supplied with inputs such as fertiliser and pesticide (and water) by a numerous, competent and highly motivated management team, are used to predict yields on a working irrigation scheme where none of these conditions is met. Real-world yields are much less than those predicted. On the Bakolori Project in Nigeria, yields ranged from 17 per cent of those predicted for cotton to 56 per cent for rice and 69 per cent for groundnuts (Adams 1992). If soil surveys are incomplete before project development (as was the case, for example, with the Bura Scheme in Kenya; Adams and Hughes 1990), the limitations of poor soils (saline, sodic, waterlogged or infertile) go unrecognised: pilot farms and experimental farms are rarely located on such soils, and data on crop performance can therefore be highly misleading.

Second, yields of existing crops which the scheme will replace (grown under rain-fed or flood recession conditions) tend to be underestimated. Few economists are in a position to predict yields under indigenous cropping, partly because of a lack of either experimental work by agriculturalists on indigenous crops, or extensive surveys of yields farmers actually obtain. It is often assumed that yields under indigenous cropping are low, although the net economic return on such crops can be high. Overestimates of yields with irrigation and underestimation of yields without irrigation can bias project appraisal in favour of formal irrigation over rain-fed or indigenous irrigated cropping because it makes the economics of a proposed irrigation scheme look more favourable than they should. A related problem is the underestimation of other economic uses extinguished by irrigation. Where wetland environments are developed, these impacts can be serious. In particular, pastoralists may lose dry-season grazing resources, or may find their access to dry-season water supplies or their migration routes blocked.

Third, if economic assessments of irrigation schemes include assumptions about investments parallel to the project that might be necessary to realise the full value of production (e.g. construction of processing plants or roads to reach remote markets, as in the case of the Bakolori Project, Nigeria), wild estimates of anticipated production and revenue may ensue (Adams 1992). In the real world of cost overruns, delays, bureaucratic ineptitude and corruption, once quick-look surveys of soils and water resources are replaced with detailed experience, the neat picture created by the economists' analysis is often drastically changed.

Fourth, economic appraisals of irrigation projects tend to focus on the performance of the scheme as a whole, whereas if the economics are to make sense, there has to be a reliable and good return for each farm household individually. This requires understanding of the economic importance of non-farm activities essential to household economic survival (e.g. labour migration, craft work or movement with livestock) that compete for labour time with low returns from irrigated cropping. The new activity of irrigation must replace the income available from these established and familiar activities, and do so with equal or reduced levels of variability and risk.

While donors and national governments continue to see an important place for irrigation in development strategies, particularly for countries in the semi-arid tropics, irrigation policy has evolved considerably. The failures of conventional large-scale smallholder irrigation schemes, and the associated recognition of the inefficiency of government irrigation bureaucracies, have been recognised. Donors and irrigation planners have turned to small-scale projects and the possibility of developing (or 'rehabilitating') indigenous irrigation (e.g. Kimmage and Adams 1990, Vincent 1992, 1995, Turner 1994, Watson et al. 1997). Indigenous irrigation falls within a wider paradigm of 'locally managed irrigation' or 'farmer-managed irrigation'. There has also been considerable investment in the 'turnover' of government irrigation schemes to farmers, usually in the form of water users' associations (e.g. Geijer et al. 1996, Kloezen and Slabbers 1992). These approaches may have much more success than their predecessors.

Locally, irrigation may perform well, and provide both sustainable livelihoods and valuable foreign exchange. Globally, it may be a necessary element in a package of intensive agriculture necessary to sustain people and economies. It is, however, far from a magical technology, and in many parts of the Third World it provides an environmentally damaging subsistence income at a net cost to national economies. Irrigation and dam projects have held a Promethean promise as dramatic strategies for disciplining nature and intensifying its exploitation to produce economic benefits. In many cases these benefits prove persistently elusive.

Development projects are the product of a planning process called into being as a result of a search for betterment of the human condition. On the face of it, development projects should not cause significant persistent adverse environmental and socio-economic costs, and certainly not costs that are not properly compensated from the benefits elsewhere. It is an unavoidable fact, however, that they often do. National or regional interests may conflict with the interests of those people immediately and adversely affected by the immediate context of development (Cernea 1988, 1991). It is commonplace that impacts on these people are not predicted, not recognised or not compensated. It is still standard practice that those bearing environmental and social costs are not consulted about whether large-scale developments should go ahead, even if increasingly they are consulted about options consequently available to them. Development is still imposed from above, the balance of costs and benefits weighed only by ranks of technical experts, dispassionately viewed by planners and decision-makers who are themselves insulated from both the consequences of their actions and the consequences of any failure to plan well.

In the final analysis, development is a political and not a planning issue. Questions of equity and justice are fundamental to sustainability. Whatever the planning process, environmental impacts are experienced by particular people in particular places. The nature of those impacts, and any steps taken to compensate or reduce them, reflects patterns of power, wealth and influence as well as the geography of the environment. There is a political ecology to issues of sustainability; this is the subject of the next chapter.

Summary

- Development itself can be a significant source of *un*sustainability. Projects such as dams and irrigation schemes can generate a chain of environmental and socio-economic impacts that are serious and complex. These result in particular from the urge to regulate the environment, and rationalise and modernise society and the human use of nature.
- Dams and reservoirs alter the pattern of river flows, typically lowering and extending flood peaks; they can affect sediment loads, alter patterns of erosion downstream, and cause channel movement and damage to infrastructure.
- Downstream physical changes in dammed rivers can cause changes in aquatic ecosystems (for example to fish and hence fishing people) and in downstream floodplain wetlands (for example floodplain forests and farmlands). Socio-economic impacts on people fishing, farming and grazing can be significant.

- Upstream of dams, serious cultural, social and economic costs are imposed on those forced to evacuate from reservoir areas. Despite decades of experience of reservoir resettlement planning, these costs remain serious, and are often underestimated and inadequately compensated.
- Irrigation development is also capable of delivering fewer benefits and more costs than planners anticipate. The problems of new large-scale irrigation schemes in areas such as sub-Saharan Africa have been particularly acute, but even in areas of established irrigation such as South Asia, problems of disease, inefficient and inequitable water distribution, poor yields, farmer debt and poor economic performance can be significant.

Further reading

Adams, W.M. (1992) *Wasting the Rain: rivers, people and planning in Africa*, Earthscan, London.

Andrae, G. and Beckman, B. (1985) *The Wheat Trap: bread and underdevelopment in Nigeria*, Zed Press, London.

Chambers, R. (1988) *Managing Canal Irrigation: practical analysis from South Asia*, Cambridge University Press, Cambridge.

Collins, R.O. (1990) *The Waters of the Nile: hydropolitics and the Jonglei Canal, 1900–1988*, Clarendon Press, Oxford.

McCully, P. (1996) *Silenced Rivers: the ecology and politics of large dams*, Zed Press, London.

Moris, J.R. and Thom, D.J. (1990) *Irrigation Development in Africa: lessons from experience*, Westview Press, Boulder, CO.

Murphy, R. (1994) *Rationality and Nature: a sociological inquiry into a changing relationship*, Westview Press, Boulder, CO.

Pearce, F. (1992) *The Dammed: rivers, dams and the coming world water crisis*, The Bodley Head, London.

Usher, A.D. (ed.) (1997) *Dams as Aid: a political anatomy of Nordic development thinking*, Routledge, London.

World Commission on Dams (2000) *Dams and Development: a new framework for decision-making*, Earthscan, London.

Web sources

<*http://www.dams.org*>   The World Commission on Dams. This site contains the texts of the reports that were prepared for the commission, and information about and reactions to the final report in November 2000.

<*http://www.irn.org/*>   The International Rivers Network, leading international non-governmental organisation campaigning on dams.

<*http://www.narmada.org/*>   Friends of the River Narmada. A support organisation for Barchao Narmada Andolan (Save the Narmada Movement), campaigning against the Indian government's Narmada Valley development project.

<*http://www.icid.org*>   The International Commission on Irrigation and Drainage, established in 1950 as an international technical organisation devoted to improving water and land management and the productivity of irrigated and drained land through

irrigation, drainage and flood management. A place to look for hard engineering approaches to water management problems and conservative responses to criticisms of dams.

<*http://www.icold-cigb.org*> The International Commission on Large Dams (ICOLD), the major global advocate of large dams, with interesting insights into its efforts to support 'environmentally and socially responsible development and management of the world's water resources'.

<*http://www.cgiar.org/iwmi/*> The International Water Management Institute (formerly International Irrigation Management Institute) is a CGIAR international research organisation on water and irrigation. Its work has developed from having a specific focus on irrigation (including farmer-managed irrigation) to a broader concern for water management. The website reports research reports and other information, tools and software models.

<*http://www.worldwatercouncil.org*> The World Water Council is an international water policy think-tank, founded in 1996, with members from public institutions, UN organisations, private firms and non-governmental organisations, and it tries to take an independent view of water issues.

9 The political ecology of sustainability

All critical examinations of the relation to nature are simultaneously critical examinations of society.

(David Harvey 1996, p. 174)

Poverty, environment and degradation

The poor are not only frequently blamed for their role in causing environmental degradation (see Chapter 7), but are often chief among its victims, and frequently the losers in the game of balancing costs and benefits of major development projects, environmental refugees in the face of anthropogenic environmental transformations (Chapter 8). There are links between poverty and environmental degradation, which provide a clear demonstration of the centrality of social, political and economic issues in questions of environment and development.

Many of the poorest people in the Third World both live in and suffer from impoverished and degraded environments. They may create or exacerbate environmental degradation because their poverty forces them to do so. Piers Blaikie comments, 'small producers cause soil erosion because they are poor, and in turn soil erosion exacerbates that condition. A set of socio-economic relations called underdevelopment is at the centre of this poverty' (1985, p. 138). Thus agropastoralists enduring drought and poor soil fertility in the Sahel, or women forced to travel miles to collect fuelwood, or fishing people suffering the effects of pollution from mine tailings in tropical rivers are all experiencing environmental degradation that relates directly to their ability to achieve sustainable livelihoods and an acceptable quality of life.

Environmental degradation is integral to the hazards of life experienced by the poor. Very often those affected have neither the freedom to stop causing degradation, nor the opportunity to move elsewhere. Poverty and environmental degradation often form a trap from which there is little chance of escape. Farmers in the highlands of Ethiopia or Nepal farm steep and eroded hillsides, and agropastoralists in the Sahel suffer the consequences of shrinking common pastures and declining productivity not through perversity but through necessity (Blaikie 1988, Ives and Messerli 1989, Mortimore 1998): people mostly live and work degraded landscapes because they have nowhere else to go.

The new field of political ecology has begun to challenge established approaches to understanding the links between human action and environmental change. In particular, it has emphasised the centrality of political economy to the explanation of why particular people experience the environment in particular ways, and to challenge understandings that draw narrowly on science and claim to be impartial and apolitical. This work has diverse roots (Bryant 1998), but came to prominence in Piers Blaikie's classic book *The Political Ecology of Soil Erosion in Developing Countries* (1985). This has been followed by a considerable literature exploring the political ecology of environment–society relations (e.g. Peet and Watts 1996b, Rocheleau *et al.* 1996b, Bryant and Bailey 1997, Dietz 1996, Stott and Sullivan 2000). This chapter will describe these developments, and the insights into the nature of environmental degradation and sustainability that they bring.

The starting point for analysis must be the centrality of reciprocal links between poverty and environmental degradation, which bring about what Blaikie described as the 'desperate ecocide' of the poor, small producers who 'cause soil erosion because they are poor and desperate, and for whom soil erosion in its turn exacerbates their condition' (1985, p. 138). Obviously the environment is not neutral in its effects on the poor: environmental quality is mediated by society, and society is not undifferentiated. Access to and the distribution of environmental 'goods' (be they cultivable land, fuelwood or clean air) are uneven. Piers Blaikie argued for a political-economic understanding of the phenomena of environmental degradation:

> It is when the physical phenomenon of soil erosion affects people so that they have to respond and adapt their mode of life that it becomes a social phenomenon. When this response affects others and brings about a clash of interests . . . it becomes a political problem as well.
>
> (Blaikie 1985, p. 89)

Blaikie's central argument is that it is vital to see the links between environment, economy and society. However, each of these things is itself highly complex. It is 'development' which very often provides that dynamic. In a crude sense, development both affects and is affected by environmental quality: 'land degradation can undermine and frustrate economic development, while low levels of economic development can in turn have a strong causal impact on the incidence of land degradation' (Blaikie and Brookfield 1987, p. 13). More importantly, because the development process involves the transformation of social and economic relations, it relates to the ways in which individuals and groups within a society experience their environment, and the ways in which they use it.

Political ecology

The field of political ecology is broad and intellectually eclectic (indeed, rather sprawling). At its core is the observation of the centrality of politics in attempts to explain the interactions between people and the environment, or the 'dynamics and properties of a "politicised environment"' (Bryant 1998, p. 82). Specifically, political ecology emphasises the importance of asymmetries of power, the unequal relations between different actors, in explaining the inter-action of society and environment (Bryant and Bailey 1997).

Blaikie (1995) suggests that since the 1970s, political ecology, like the wheel, has been repeatedly reinvented. Bryant and Bailey (1997) identify two phases in its development. The first, from the mid-1970s to the mid-1980s, built on neo-Marxism, and emphasised structural explanations of human–environment rela-tions. Development in this period was slow and piecemeal, partly because of the lack of interest by Marxist scholars in the environment, and partly because of radical aversion to the neo-Malthusian explanations of environmental 'problems' associated with ecological approaches to understanding environmental change (see Chapter 6). Nevertheless, in the 1980s, critiques of neo-Malthusianism emphasising the political economy of the environment (for example work on famine in the Sahel by Watts, Copans and others, see later in this chapter) made a major contribution to the development of political ecology. The second period in the development of political ecology, from the later 1980s, has been more complex, with a greater focus on the role of grassroots actors and social move-ments (Bryant and Bailey 1997); it has seen in particular a greater awareness of discursive dimensions of environment–society interactions (Peet and Watts 1996a).

The critical innovation of political ecology has been the search for explana-tions that take account of both the natural and social sciences (Blaikie and Brookfield 1987). The appeal of such a synthesis is obvious, although the intel-lectual coherence is more questionable (Peet and Watts 1996a). Bryant (1998) suggests two key themes running through the political ecology literature. The first is the way in which unequal power relations relate to conflicts over access to and use of resources; the second concerns the ways in which power rela-tions are reflected in conflicting discourses and knowledge-claims about the environment and development. Peet and Watts point to new work in four areas. The first seeks to make explicit the causal connections between the logics and dynamics of capitalist growth and specific environmental outcomes. The second specifically addresses the politics of social action about the environ-ment. The third explores social movements and civil society. The fourth addresses discursive dimensions of the way the environment is defined and its dynamics described (Peet and Watts 1996a).

Conventional analyses of environmental 'problems' tend to draw heavily on scientific explanations, with social and political dimensions ignored or down-played. And yet, as we have already seen in Chapter 6, questions of epistemology (of theories about knowledge) make the assumption of the possibility of impartial or neutral definitions of environmental problems or analyses of envi-ronmental change problematic. The definition of environmental problems is also socially mediated. As Jarosz (1996) points out, environmental 'problems' tend to be defined in ways that make them amenable to technical solutions;

and technical interventions do not appear from nowhere, but are themselves highly political (Blaikie 1985, 1995).

Before 'environmental issues' can be articulated, the environment is experienced by people and made the subject of conscious thought (whether by scientists or pastoralists). Moreover, the ways in which different people will experience the environment, derive their understandings and develop discourses about it will vary, and differences of view (whether between people with different bundles of rights, or between those with different claims to environmental knowledge such as scientists and lay people) will interact in a political process. The status of actors will be reflected in the power of their arguments: 'hard' natural scientists may disparage the insights of 'soft' anthropologists; bureaucrats or urban businesspeople may seek to ignore the views of Indian *adivasis* or African pastoralists; men may downplay the environmental knowledges and understandings of women. In political ecology the exercise of power (over nature, over other people) therefore has to be understood at the discursive as well as the material level (Bryant and Bailey 1997).

The social construction of knowledges about the environment can itself be the means by which power is exercised over both nature and society. In the book *Liberation Ecologies*, Peet and Watts (1996b) explore the engagement between political ecology and poststructuralism, emphasising the importance of the politics of meaning and the construction of knowledge. They argue that political ecology is moving (and should move) in response to shifts in social science theory towards poststructuralism. Rather than 'environmental problems' generating 'single understandings to which people respond, there are multiple realizations about all levels of environmental problems' (Peet and Watts 1996a, p. 37), and these stimulate creative reactions that may (or may not) emerge as social movements. While the struggles that emerge may be material struggles about survival or livelihoods, they are also struggles about the ways in which people speak about and think about and organise understandings of human and non-human nature. They propose a new term to embrace this new openness to debates about imagination and discourse, 'liberation ecology'. Debates about the social construction of nature – and about the material reconstitution and 'invention' of life-forms possible through genetic engineering – suggests that political ecology should be open to new understandings of culture and nature, and embrace 'poststructural Political Ecology' (Escobar 1996).

Political ecology also seeks to integrate explanations across spatial (and temporal) scales (Blaikie and Brookfield 1987). Thus local 'environmental problems' (soil erosion is now the classic example) need to be seen as the product not only of local processes (farming practices) but also of political economy at local, national and international scales. Thus soil erosion might be said to happen because it rains hard just after fields have been dug, but is also influenced by a series of other factors – among others, for example, the fact that labour shortages prevent timely land preparation, that men have migrated out in search of wage employment owing to a depressed agricultural economy, that national fiscal policy subsidises imported agricultural products, that of patterns of national indebtedness affect local areas. An explanation that leaves out the wider political economy leaves out a vital part of the story. In 1981 Blaikie argued the importance of linking the circumstances of decision-makers to micro-level, national and worldwide political economy. In 1995 he developed this into the concept of a 'chain of explanation' (Figure 9.1). Thus in considering soil erosion, Piers Blaikie suggested that physical changes in soils and vegetation were linked to economic symptoms at particular places at particular times, and in turn to land use practices in that place, to the resources, skills, assets, time horizons and technologies of land users, the nature of agrarian society and finally to the international political economy (Figure 9.2).

The multidisciplinary, multilevel scope of political ecology has commended itself to an increasing number of researchers wanting to explain environmental degradation or environmental change, and understand their significance for different groups within society. This chapter explores some of this work, looking in particular at problems of deforestation, conservation and famine.

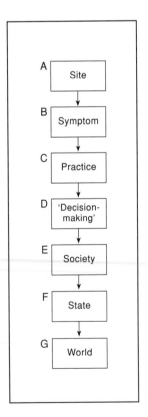

Figure 9.1 Political ecology's chain of explanation (after Blaikie 1995).

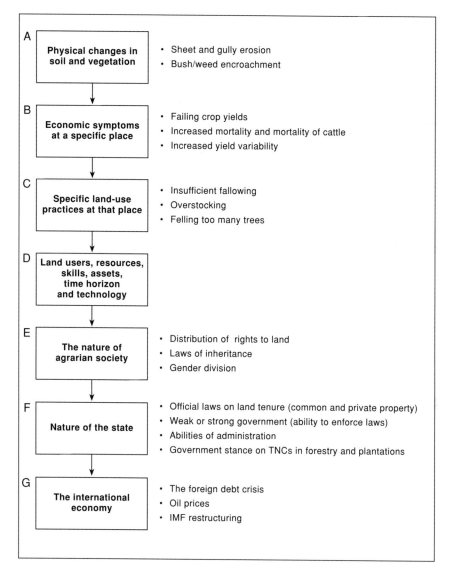

A
Physical changes in soil and vegetation
- Sheet and gully erosion
- Bush/weed encroachment

B
Economic symptoms at a specific place
- Failing crop yields
- Increased mortality and mortality of cattle
- Increased yield variability

C
Specific land-use practices at that place
- Insufficient fallowing
- Overstocking
- Felling too many trees

D
Land users, resources, skills, assets, time horizon and technology

E
The nature of agrarian society
- Distribution of rights to land
- Laws of inheritance
- Gender division

F
Nature of the state
- Official laws on land tenure (common and private property)
- Weak or strong government (ability to enforce laws)
- Abilities of administration
- Government stance on TNCs in forestry and plantations

G
The international economy
- The foreign debt crisis
- Oil prices
- IMF restructuring

Figure 9.2 The chain of explanation of land degradation (after Blaikie 1995).

Understanding tropical deforestation

Tropical moist forests (or more colloquially, tropical rainforests) are the subject of a stark and insistent picture of destruction ably and emotively painted by environmentalists and many scientists (e.g. Myers 1984, Caufield 1985, Groombridge 1992, E.O. Wilson 1992). In 1973 Denevan suggested that development was bringing about 'the imminent demise of the Amazonian rainforest' as a result of a 'pell-mell destructive rush to the heart of Amazonia' (1973, p. 137). His analysis was criticised, although he argued in 1980 that further data confirmed its conclusions (Denevan 1980). By that time, the loss of rainforest was becoming a significant element in First World environmentalism. In *The Times* in 1980, an advertisement for the World Wildlife Fund showed cleared forest in Sumatra with a six-page special report on the *World Conservation Strategy* that stated (rather wildly) that 'the earth's lungs are being destroyed at the rate fifty acres [20 ha] a minute'. Statistics on the rate of loss of forest land were stated and extrapolated to emphasise the scale and speed of crisis: the Friends of the Earth Tropical Rainforest Campaign which began in 1985 argued that 7.5 million hectares of undisturbed tropical moist forest was destroyed or degraded annually, and calculated therefore that 14 ha was being cleared every minute, and that by 1990 the rate of extinction of species would have risen globally to one per hour (Secrett 1985).

Although there can be no doubt of the fact of rapid forest cover change in many humid and sub-humid tropical areas, particularly from the 1980s onwards, there remains much debate about its rate and extent. Reliable data on rates of forest loss are surprisingly limited, and difficult to interpret. This is partly because of difficulties of definition, both of the forest types and what is meant by 'deforestation'. There is by no means universal agreement on the extent of tropical moist forests. In the 1980s the FAO distinguished between closed forest (of which there is some 1.20 billion hectares globally) and open forest (0.73 billion hectares; Lanly 1982). However, the FAO also distinguished fallows of both closed forest and open forest (0.24 billion hectares and 0.17 billion hectares respectively), and scrubland (0.62 billion hectares). The composition of forests varies considerably across the globe: Latin America held over half the world's closed forests, while Africa had about two-thirds of all open forests, fallow open forests and scrub. The FAO's *Forest Resources Assessment 1990* suggested that the total extent of tropical forest was 1,756 million hectares, of which 718 million hectares was tropical rainforest, 587 million hectares moist deciduous forest, 204 million hectares montane forest and 246 million hectares dry forest (Grainger 1996). However, these classifications were not wholly sound, and forest area estimates (from ninety tropical countries) included data from modelled adjustments to outdated national surveys. The categories are not fully compatible with previous surveys (ibid.).

Confusion over the extent of different kinds of forests is greatly compounded by differing estimates of the extent of deforestation (Allen and Barnes 1985). One cause of the discrepancies that exist is lack of a clear and universal definition of deforestation. Variations in terminology lead to what appear to be substantial differences in estimates of rates of forest disturbance, although it is argued that these can be reconciled (e.g. Melillo *et al.* 1985). Certainly, where large tracts of land are cleared permanently of forest and turned over to pasture, definition is easy. However, it is far more difficult, for example, to decide when to count internal changes caused by shifting cultivation practices as forest clearance.

Data on deforestation have been improved, but also made more complex by the use of satellite imagery (Green 1983, Myers 1980, Turner and Meyer 1994). However, problems such as cloud cover and the problem of detecting vegetation change from forest to other land covers (involving subtle changes in spectral signature) in small areas of irregular shape (typically the result of forest clearance by small farmers) are significant. Singh (1986) found data from the Landsat Multi-Spectral Scanner (MSS) inadequate to assess the extent of small-scale clearance of forest in Manipur State in India. Tucker *et al.* (1984) used data from the Advanced Very High Resolution Radiometer (which has a coarse spatial resolution but is available on a daily basis) with some success to study strip clearance of forest in Rondônia in Brazil for roadside settlement, and Fearnside (1986) analysed Landsat data to derive estimates of the extent of forest clearance in the Brazilian Amazon, showing rapid acceleration in rate of clearance between 1975 and 1980 in certain key areas, notably strategic highways, for example that through Mato Grosso and Rondônia.

Comparison of FAO data on forest cover change in seventy-six countries between 1980 and 1990 suggests increasing rates of loss (0.9 per cent, or 169,000 square kilometres per year for the period 1980–90); however, Groombridge (1992) and Williams (1994) conclude that the true extent of remaining moist forest cover, and rates of loss, remains uncertain at the global scale. Nationally or locally, patterns of forest cover reduction are unequivocal, and in many instances well documented. Thus in Costa Rica, 67 per cent of the country was under primary forest in 1940, and only 17 per cent by 1983; in Madagascar, 7.6 million hectares of eastern rainforest remained in 1950 (from a former total of 11.2 million hectares), but only 3.8 million hectares remained in 1985 (Groombridge 1992). The story elsewhere is broadly similar. For example, in Sumatra the area of unlogged forest declined dramatically in the last two-thirds of the twentieth century. Large areas were cleared for industrial plantations, subsistence agriculture, for settling transmigrants from elsewhere in Indonesia, and logging (ibid.; Figure 9.3).

There are marked regional variations in the factors causing deforestation (Myers 1980). Logging for timber production (largely for European and South-East Asian markets) is the dominant factor in deforestation in South-East Asia, particularly Malaysia, Indonesia, Papua New Guinea and the Philippines. The international timber trade is much less important in Latin America, although there is a considerable internal trade in timber in Brazil. Logging has overtaken pasture creation as the leading cause of deforestation in the Brazilian Amazon (Parayil and Tong 1998). In Africa, internal markets for timber are also large, and rates of logging (usually clear-felling) are locally high. Elsewhere, settlement schemes are important in Amazonia, Indonesia and Malaysia, ranching and pasture development in Central and Latin America, fuelwood and charcoal collection in Africa and India (Williams 1989, 1994).

Debate about the sustainability of tropical deforestation reflects three concerns. The first is the loss of species (particularly in recognised centres of high species diversity; Groombridge 1992), which has been particularly significant

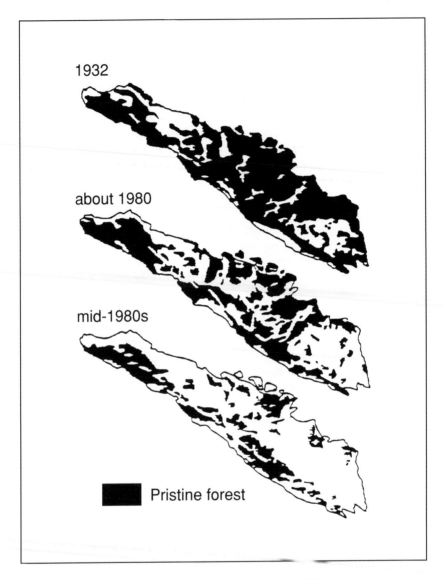

1932

about 1980

mid-1980s

■ Pristine forest

Figure 9.3 Deforestation in Sumatra (after Groombridge 1992).

since the Rio Conference and the signing of the Convention on Biological Diversity. It was with the rainforest that Edward Wilson began his celebration of biological diversity and his warning of a massive anthropogenic extinction spasm in *The Diversity of Life* (E.O. Wilson 1992); rainforest still has a central place in the iconography of global environmentalism. The second concern is the role of forest loss in the context of the global carbon balance, and hence anthropogenic climate change (Sutlive *et al.* 1980, 1981, Sioli 1985, Gash *et al.* 1996). Debate over the Climate Change Convention at the Rio Conference and afterwards, and over carbon dioxide emission targets, revealed the complexity of the scientific evidence, and sharp differences in views in North and South over the fate of the forests. The third dimension of concern about forest loss is in many ways the most significant, and it relates to the political ecology of deforestation, and particularly the long-term capacity of forested land to yield resources for human use, particularly for people depending on it for their livelihood.

Conventional approaches to the explanation of deforestation see it as a direct result of pressure on resources, with population growth the primary driving force (e.g. Myers 1980). This analysis summons well-worn neo-Malthusian arguments in its support, and may be locally appropriate. However, such an analysis tends to suggest purely technical responses to forest loss, focusing on people's capacity to use the forest for production (by establishing forest reserves where 'rational' and 'scientific' forest management can be practised under controlled conditions), and tackling perceived problems of over-reproduction by tackling human population growth through birth control. Data for twenty-eight African, Asian and Latin American countries showed that rates of deforestation between 1968 and 1978 were correlated with population growth, fuelwood production and wood export, and agricultural expansion (Allen and Barnes 1985).

However, although there is increasing sophistication in socio-economic models of the causes of deforestation (Jepma 1995), it is still not easy to explain forest loss satisfactorily in a statistical sense in response to factors such as population growth or debt (e.g. Palo 1994, Kahn and MacDonald 1994, Gullison and Losos 1993). Cross-national data on forest loss for the period 1975–90 suggest that rural population growth and poverty (lack of other economic options) were key factors where forest extent was already limited, while where forests were extensive, loss was due to the actions of entrepreneurs, small farmers and companies working in concert (Rudel and Roper 1997). Mather (1992) identified a 'forest transition', linking deforestation and industrialisation and the growth of a national economy. He pointed out that forest area fell in industrialised countries in the past (as it is doing in non-industrialised countries now), but that it has subsequently grown again (e.g. in North-West Europe). Grainger (1995) argues that a lag is to be expected in the transition from net forest loss to replenishment. Nonetheless, it is something of an act of faith (not least in historical precedent and models) to believe that tropical deforestation will be reversed. Aggregate analyses of this kind certainly do not explain why or how particular pieces of forest are cleared, and who gains or loses by the process.

An economic analysis of forest conversion suggests that it happens because the flow of benefits from the forest is not matched by benefits from other uses (Pearce and Brown 1994, Barbier *et al.* 1995). However, there is a clear problem that those who gain the benefits are often different (and often more powerful) people than those who lose benefits from uncleared forest. This is clearly the case where logging companies and the politicians they pay for felling licences clear forest land occupied by shifting cultivators or hunters. It is also true where small farmers wish to move to the forest frontier as settlers, and are able to do so and clear the forest without the consent of existing forest people.

The most common problem is that forest land is an open-access resource – that is, either one to which access is wholly unregulated, or one where regulations exist but are not implemented (or where the modern state disregards indigenous peoples' systems of tenure and rules of resource management). In many areas of the Third World the state has taken powers to regulate the use of forest land, timber and other forest products, but lacks the authority and power to implement them. Forests are thus wide open to any entrepreneur able to negotiate unofficial access to forest land independently of an official legal regime (for example by bribery), and convert a slow, sustainable trickle of economic benefits into a one-off windfall of timber or cleared land. Much commercial logging in rainforests takes place in contravention of formal plans and regulations, again often because logging contractors are able to arrange de facto access even to reserved forests.

A more interesting question is why forests are treated as open-access resources. Pearce and Brown (1994) identify three factors, all reflecting economic failure. The first is the conventional problem of local market failure: those who convert forest land take the benefits from doing so (for example timber revenues) but do not have to pay the costs (loss of subsistence livelihoods, soil erosion, loss of biodiversity, loss of future revenues from timber if forests do not regenerate). The second factor is a failure of government intervention, whereby governments either fail to address market failure (and make those who gain from forest conversion pay the costs), or offer perverse incentives (for example failing to tax or regulate logging companies, or allowing inefficient logging practices that encourage wastage), so that the economic balance is pushed towards forest conversion. These problems are exacerbated by poor governance, with inefficiency and corruption tending to favour the short-term profits of large corporations at the expense of forests (and very often the people who live in them). The third dimension of economic failure is 'global appropriation failure', or the fact that forests support all sort of attributes and economic 'functions' that are poorly represented by the market. Tropical forests have a role as a carbon store and hence as a factor in the rate of global warming, and there are global values too in their biodiversity (whether this is seen to have intrinsic value or use value as a reservoir of pharmaceutical products or a destination for future ecotourists). There is no direct market for many of these values (although a growing system of tradeable carbon permits is starting to produce one, leading among other things to industrial corporations funding afforestation in the tropics), and as a result the values of standing forests are under-represented when choices are made about clearance.

West Africa provides an example of significance of poor policy for unsustainable forest management, as Pearce and Brown (1994) describe. The agricultural policy environment in West Africa (characterised by overvalued exchange rates and pricing that effectively taxes agricultural exports, subsidises agricultural imports and discourages market production) has prevented farmers from investing in their farms, and particularly from adopting new technologies and inputs (Cleaver 1994). Poor rural services and infrastructure (no roads to get to market, no clinics to tackle sickness, no access to inputs and new agricultural knowledge) have further hampered innovation. As a result, farmers have not intensified, but have extended production – among other things expanding onto new forest land, and clearing trees to plant crops.

The political ecology of deforestation

Political ecology is central to the issue of deforestation. Jarosz comments that explanations of deforestation are 'socially and politically constructed to the advantage of powerful people' (1996, p. 148). Analyses must take this into account. As mentioned in Chapter 7, Fairhead and Leach (1996) demonstrate that French colonial foresters and their successors in independent Guinea have systematically misunderstood the role that people play in relation to the forest in West Africa. The conventional view is that the patches of forest in the savanna landscape (along streams and around villages) represent relics of a once-continuous forest cover, cleared by past human action and threatened by contemporary forest products and wood use and burning. The foresters saw their job as protecting the forest against the Kissi people, which they did with some fierceness. However, research showed that people have historically fostered forest development, particularly in patches around villages, for a variety of entirely rational reasons (for example protection from grass fires). Far from deforesting the landscape, Kissi farmers have forested it: government forestry scientists have read forest history backwards (ibid.). Indeed, Fairhead and Leach (1998) go on to suggest that very similar rather disastrous mistakes have occurred in many other West African countries.

Deforestation is not, therefore, as simple to explain as might first appear (Parayil and Tong 1998). The search for the factors responsible must embrace political and discursive dimensions of the problem. The two are closely linked. Jarosz (1996) describes the lasting power of colonial discourse about 'irrational' peasant farmers in eastern Madagascar, whose shifting cultivation (*tavy*) was banned by the colonial state. In contrast, the state established a 'rational' approach to forest management, creating forest reserves. These were opened up to exploitation in 1921. Suppression of *tavy* removed indigenous institutions that regulated how and where the forest could be cleared. Forest cover fell dramatically in response to the spread of cash cropping (for coffee), demands for timber (for example for the railways) and an effectively uncontrolled mixture of cultivation, grazing, burning, and extraction forest products. Population, initially, remained static: neo-Malthusian explanations of deforestation will not wash.

Deforestation is the result of structures and decisions by actors at a range of levels. Wood (1990) suggests that the politics of deforestation can be

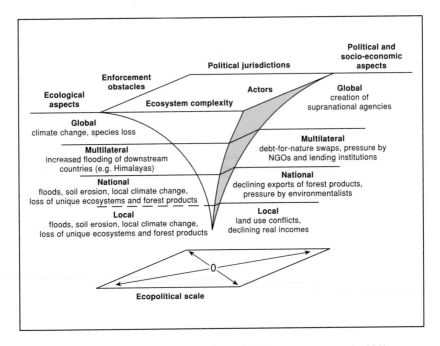

Figure 9.4 Ecopolitical hierarchy of tropical deforestation (after Wood 1990).

imagined as an upside-down pyramid of increasing scale (local, national, multi-lateral and global) and increasing ecological and political complexity (Figure 9.4). At the local scale, from the neighbourhood to first-order administrative units (provinces for example), environmental crises are source specific and directly concern local economies, for example the links between deforestation and fuelwood shortage, or the downstream effects of soil erosion. However, while local, these impacts relate to complex and disintegrated government bureaucracies, and are often beyond the influence of local people themselves. Any benefits from unsustainable resource use are lost to remote cities, or inter-nationally, or the pockets of business or an often corrupt bureaucratic elite.

At the national level, environmental issues may reflect different goals, for example loss of ecosystems and desire for cash revenue from standing forests. National politics will reflect strategic differences between government ministries, and between national and regional or local government. Even where laws are passed to promote sustainable forest management, capacity for implementation may be very limited. Internationally, agreements on forest management can be controversial, and the strategic interests of countries may differ (for example between forest countries such as Pacific states and countries hosting major logging interests such as Japan). Globally, issues such as climate change and species loss become dominant, and as the Rio Conference demonstrated, political debate can focus on differences between poor rainforest countries and the environmental sensibilities of delegations of rich countries.

The interplay between these levels can be complex. Internationally, the governments of industrialised economies can seek to apply pressure on national policies through 'environmental conditionality' and targeting in aid-giving. National governments can also seek to change policy regimes, as Brazil, for example, did in 1989, with a new programme to remove tax incentives for ranching, to ban timber exports and to establish new reserves (Wood 1990). These interactions will in due course have some kind of impact on the ground, moderated by complex political and economic feedbacks at intervening levels.

Forest conversion cannot therefore be explained simply in neo-Malthusian terms (as a product of rural population growth), or in terms of economic demand for export crops, or the inappropriateness of agricultural technology (Hecht 1984, 1985). The Brazilian Amazon, for example, is an important case in point. Although the social, economic and environmental dimensions of settlement in the Amazon are diverse and complex (Furley 1994), the penetration of settlers and ranchers into the forest must be seen in the context of the political economy of Brazil (Hecht and Cockburn 1990, Parayil and Tong 1998). Deforestation has been driven primarily by subsidy provided by the Brazilian state (e.g. Goodland 1980, Moran 1983, Caufield 1985, Hecht and Cockburn 1989).

Brazil's region of Amazonas covers 5 million square kilometres (58 per cent of the country), but only 3.5 million square kilometres is (or recently has been) actually forested, and only 70 per cent of that is tropical moist forest (Barraclough and Ghimire 1995). Extensive forest clearance began to occur in the early 1970s, and has run at about 20,000–30,000 square kilometres per year. Following the military coup in 1964, the state sought legitimacy through economic growth. However, the astonishing growth in the Brazilian economy in the 1960s and 1970s (9 per cent annually from 1965 to 1980) was accompanied by increasingly industrialised methods in the rural sector, and a crisis of access to land by the poor. The opening up of Amazonia (through the construction of roads to central Brazil and the coastal cities and other infrastructure) obviated the need for land reform by providing land for unemployed agricultural workers. Operation Amazonia was launched in 1965. The population of Amazonia was about 1.5 million in 1940, but this had grown to 8.6 million by 1989, mostly through immigration from the south and north-east, the settlement of small farmers being actively promoted by the colonisation agency INCRA (Ghimire and Pimbert 1996).

Laws passed in 1966 provided tax reductions for companies (including foreign corporations) investing in Amazonia, and additional credit was available both from within Brazil and internationally. These tax incentives and cheap credit encouraged the annexation of land by ranchers and speculators (Hecht and Cockburn 1990, Ghimire and Pimbert 1995). Land allocation was chaotic and disputes became common, their outcomes determined by 'political connections, economic power, bribes and sheer physical force' (Ghimire and Pimbert 1995,

p. 56). New technologies of seeded pastures opened up profitable possibilities for industrial agribusiness to move into beef ranching, thus providing an important (and largely new) export product, new forms of production, and an outlet for investment. The incentives for the development of Amazonia were supported by international concerns and interests, for example the views of the World Bank and the FAO.

In 1970 the New Integration Programme brought a new focus on small-scale settlement, but the policy was again reversed in 1975 and ranching predominated once more. Cattle-ranching achieved notoriety through critical accounts of the 'hamburger connection' (Myers 1981), and the linkage between Amazonian forest clearance and American fast food outlets. American beef prices rose in the 1970s, and by 1978 Central American beef was less than half the wholesale price of beef raised in the USA. American beef imports rose rapidly, the lean grass-fed beef of rainforest pastures being suited to the fast food industry, which comprises 25 per cent of US beef consumption. Between 1961 and 1978, beef exports rose by five times in Costa Rica and by fifteen times in Guatemala (Myers 1981). In Brazil the dry *terra firme* became the subject of intensive development both by transnational companies such as Volkswagen, King Ranch and Armour-Swift, and by large investors from Brazil (Fearnside 1980). Deforestation was driven by both domestic structural factors ('possibly the most extensive, destructive and chaotic private land enclosure in history'; Dove and Noguiera 1994, p. 492) and macroeconomic disequilibrium (inflation and foreign debt).

Environmental myths, or narratives (Leach and Mearns 1996), also legitimate forest clearance and settlement. In Brazil there was a pervasive belief among both scientists and policy-makers through the 1980s that cattle-ranching was a valid and effective long-term land use, and indeed that the creation of pasture improves soils (Fearnside 1980). This was based on the argument that clearance of forest releases nutrients held in rainforest vegetation, and there is a rapid increase in nutrient availability. In fact, critics of cattle-ranching in rainforest-derived pastures argue that this view ignores serious longer-term ecological impacts that make cattle-ranching far from sustainable (Fearnside 1980, Hecht 1980, 1984, 1985). Forest clearance does generally bring about short-term falls in soil acidity, and a rise in the amount of calcium, magnesium, potassium and phosphorus, although there is a great deal of variability, and larger samples tend to show that increases are less dramatic than is often claimed. However, the pulse of nutrients is short-lived. In particular, phosphorus levels decline after about five years to something very close to the levels in soils under undisturbed forest, and over longer-term five- to fifteen-year periods nutrient status and pasture productivity decline, sometimes drastically. Soil erosion can also become a serious problem: cattle-ranching is not usually a sustainable land use in rainforest areas (Fearnside 1980). Ranch economics dictate the sale of pasture after about five years; indeed, Hecht (1985) reported that 85 per cent of ranches in Paragominas had failed by 1978. In ecological terms this seems to make little sense, but of course such short-term exploitation of land can be good economics. Entrepreneurs are attracted to rainforest ranching not by the long-term productivity of the land but by short-term returns on investment (Hecht 1984, 1985). As Hecht comments, 'if the productivity of the land itself is of low importance, cautious land management becomes irrelevant and environmental degradation is the inevitable result' (1984, p. 393). Lack of secure land tenure meant that landowners, tenants and squatters had no incentive to do anything but mine soil fertility, forest resources and minerals as rapidly as possible, a process exacerbated by the way in which land clearance was taken as evidence of land improvement and effective occupancy (Ghimire and Pimbert 1995).

The social processes of resource definition, extraction, control and distribution are central to the understanding of deforestation (Jarosz 1996). The attitudes to indigenous resource users on the part of foresters in colonial Madagascar that Jarosz describes, and the resort to a rational forest exploitation policy of reservation and commercial exploitation, were repeated throughout the colonial world, for example in India (Gadgil and Guha 1995, Jewitt 1995). In Burma a discourse of 'forestry as progress' supported a view of the forest that focused on teak and the revenue that could flow from it, effectively making alternative uses invisible (and illegal). In India a Forest Department was established in 1864, and the Forest Act of 1878 allowed for the closing or 'reservation' of forests to allow 'scientific forestry' to concentrate on efficient timber production. Existing use rights for timber and non-timber forest products were extinguished. In the Uttarakhand (the Himalayan hills of Kumaon and Garhwal), demand for railway sleepers led to the closure of vast tracts of forests to subsistence use, and the banning of practices such as burning. The resulting hardship began a long history of protest against state forest policy (Guha 1989). The idea that forests should be places for timber production alone, and dismissive or hostile attitudes to the wider range of uses typical of forest dwellers, persist today in many parts of the Third World.

The carrot of the market and the stick of colonial taxation stimulated forest clearance for cash crops in many places (for example South-East Asia in the nineteenth century; Parnwell and Bryant 1996), and much unsustainable timber extraction takes place under legal licensing agreements. The political ecology of deforestation is complex, and a single issue such as rural population growth is but a tiny element in a much larger equation.

Forest clearance and forest people

Globally, it is farmers who are the most significant clearers of tropical moist forests, and are the dominant influence in Africa and Latin America. In the mid-1970s forest farmers numbered perhaps 140 million, and occupied over 20 per cent of the area of tropical moist forests, 20 million hectares (Myers 1980). Thus in East Kalimantan (Indonesia), shifting cultivation had created 2.4 million hectares of secondary forest and 0.4 million hectares of grassland dominated by one species, *Imperata cylindrica* (Kartawinata *et al.* 1980). This amounted to 16 per cent of the total forest area.

The impacts of forest clearance for agriculture are complex (Nye and Greenland 1960). Ecological succession following clearance as part of slash-and-burn agriculture leads to a flush of nutrients after clearance, followed by a fairly rapid decline in soil nutrient status and yields (Kartawinata *et al.* 1980, Herrera *et al.* 1981). Forest farmers therefore characteristically move on once the flush of nutrients begins to be depleted. This can be sustainable, at least at low levels of intensity, particularly if cropping systems mimic the physical structure of rainforest.

However, where access to land is restricted, so that population pressure rises, or farmers are trapped by institutional constraints such as indebtedness to a settlement scheme authority, inappropriate agricultural techniques lead rapidly to exhaustion of soil nutrients, low economic returns and impoverishment of farming households (Trenbath 1984). The economic and ecological collapse of such systems is obviously quite possible, particularly where agriculture is

Plate 9.1 Forest clearance for agriculture on the edge of the Bwindi Impenetrable
        National Park, south-west Uganda. Forest clearance for small-scale agricul-
        ture can be rapid and permanent, although without appropriate farming
        systems cultivation may not prove sustainable. Bwindi Impenetrable
        National Park was declared in 1991 to protect a range of species, including
        most notably the mountain gorilla. The agricultural frontier has advanced in
        Kigezi throughout the twentieth century; rural population densities in the
        region are still high and poverty a major problem. The interests of
        conservation and farmers are in many ways diametrically opposed, although
        the generation of foreign exchange income from gorilla-watching tourism
        provides some prospect of common ground for those who see a sustainable
        future for 'community conservation' strategies (see Chapter 12).

based on continuous cropping of nutrient-demanding annual crops such as maize. Thangam (1982) discusses the feasibility of such cropping in place of shifting cultivation among tribal people in forests in north-east India. There is increasing interest in the development of agroforestry cash-cropping regimes for forest settlers that allow sustainable continuous cropping (Nicholaides *et al.* 1984, Jepma 1995).

Settlement schemes in rainforest areas have proved difficult to plan, and are of limited success even in narrowly economic terms (Schmick and Wood 1984). Hiraoka and Yamamoto (1980), for example, argued that new settlements in north-east Ecuador were ill-suited to the environment. Many projects, such as the enormous Transmigrasi Project of Indonesia, to resettle people from Java (Hardjono 1977, Otten 1986), appear to derive their justification primarily from powerful political support (Budiarjo 1986).

State support for forest settlement, whether directed or spontaneous, results not only from political and economic interests, but also from powerful ideologies. Allen (1983) argued that the supposed 'virgin lands' of the Amazon acquired an image that owed little or nothing to the substantive success (or more usually lack of it) shown by any actual development projects. They are sustained by fantasies woven about them and sustained by government rhetoric. Skillings (1984) also cited the psychological effects of the sheer extent of Amazonia, which formed 57 per cent of Brazil, held only 4 per cent of the population and generated 2 per cent of GDP. He suggested that this acted as a challenge to political leaders, one backed up by perceived strategic implications.

Small forest farmers are commonly regarded as villains in accounts of forest clearance. Government forestry departments, in particular, tend to regard rural people as the self-deluded destroyers of a valuable resource of timber, cutting minor forest products (canes, rattans, etc.), hunting and gathering, and above all clearing forest for farmland, only to abandon it and seek to move on. The image of the forest farmer is either of the rapacious chancer, knowingly clearing land and managing it unsustainably, or of the ignorant victim, condemned by lack of education to gnaw away at the edge of the forest and turn it to worthless fallow behind him (*sic*: such myths often carry a curious gender bias). Thus, with characteristic tunnel vision, forestry staff are reported to lament illegal cutting by people 'who do not understand the ecological consequences of forest destruction' (Soerianegara 1982, p. 82).

However, between apparent irrationality and greed lies a whole world of decision-making where farming households, and forest pioneers, actually live (Townsend *et al.* 1995). Institutional and economic factors are the primary influences on frontier agricultural settlers in Amazonia, affecting both the productivity and the sustainability of agriculture (Richards 1997). The decisions of forest farmers are entirely rational at the household scale, even where they are significant agents of forest loss, as in south-east Nigeria (Ite 1996). Forest farmers have a very clear understanding of the ecology of fallow plots in the forest: they leave trees in cleared plots (they have many recognised uses, but are also hard to remove), and are aware of the economic and ecological options in cropping systems (cassava in intensive cleared land near the house, and banana and plantain in regenerating forest further away), and aware of the implications of the total loss of forest cover. Forest farms tend to be small, diverse, and integrated in space and time with the forest around them (Plate 9.2). Patterns of forest clearance, and post-clearance field management and forest regeneration, are therefore the result of both necessity and choice at household scale. Forest farmers are intelligent and active agents of forest management, far from the unwitting agents of degradation so often portrayed in accounts of deforestation. Tropical forests are often profoundly managed ecosystems, even when they are rich in biodiversity and to the untutored eye look 'natural'.

The biggest impact of forest clearance is often on forest people themselves. There is increasing awareness of the impacts of deforestation on indigenous people and other forest dwellers, for example in Amazonia (Vickers 1984, Arvelo-Jiménez 1984, Survival International 1985). The World Bank's policy paper *Tribal Peoples and Economic Development* stated the Bank's policy not to develop areas presently occupied by tribal people, with the proviso that if this was unavoidable, the Bank would ensure that 'best efforts have been made to obtain the voluntary, full, and conscionable agreement (i.e. under prevailing circumstances and customary laws)' of the tribal people or that of their advo-

Plate 9.2 Land cleared for a farm in the dry forests of the Madhupur Tract, Bangladesh. Fertility is high on newly cleared land, and a mixture of crops is grown without extensive cultivation, achieving rapid coverage of bare soil.

cates, and that project design and implementation 'are appropriate to meet the special needs and wishes of such peoples' (World Bank 1984a, p. 1). Many observers remain unconvinced by policy statements of this kind, but Maybury-Lewis argued that 'we have to come up with strategies of development that offer reasonable and equitable opportunities to participate to all members of society, both Indians and non-Indians' (1984, p. 134); it should not be assumed that Indian societies are doomed just because of the strength of the ideologies behind the second conquest.

On the ground, forest 'development' has destroyed the environment and economy of forest-dwelling people in many parts of the world, their small but sustained flow of revenue from the forest swept away in the pursuit of larger short-term benefits for the state or (more usually) private business interests. Thus in Médio Amazona, large-scale ranching on *terra firme* had a serious impact on small peasant producers (Bunker 1980). Peasant producers often lacked title to land, and were unable to sustain production in the face of competing demands from ranchers for accessible land as land prices rose. Ranching therefore was socially regressive as well as environmentally damaging. Indigenous people in Rondônia (Brazil) lost land to gold prospectors (whose mercury polluted rivers far from their scene of operation), ranchers, loggers and land speculators (Barraclough and Ghimire 1995). Riverine cultivators, established Portuguese-speaking farmers of the seasonally flooded *varzeas*, suffered from ecological changes due to pollution as well as dam construction (see Chapter 8).

The close link between forest people and their land, and their interest in defending their way of life have led to various kinds of resistance (Colchester 1994; see also Chapter 12). The Amazonian boom in rubber collapsed in the 1940s, but production from wild trees persisted (Coomes and Barham 1994). However, as land colonisation, deforestation and mining gathered pace, those living in the forest collecting rubber and Brazil nuts steadily lost land to speculators and ranchers. The rubber tappers began to organise and protest, and met with violent oppression from powerful economic interests. The plight of the rubber-tappers became part of international protest at Amazonian 'development' following the murder of the head of the recently formed National Council of Rubber Tappers, Chico Mendez, in 1988 (Barraclough and Ghimire 1995). The rubber tappers' protest eventually led to the establishment of extractive reserves in 1990 (Campbell *et al.* 1996). By 1990, indigenous reserves covering 270,000 square kilometres had been set aside in Amazonia, and the rights of indigenous people were recognised in international law, although in practice and on the ground these rights were still widely abused (Barraclough and Ghimire 1995).

While First World environmentalists worry about the loss of tropical forest biodiversity associated with forest clearance, the impacts of that clearance on the livelihoods of the poor are often more alarming. Clearly, there can be a large measure of common interest between those advocating forest conservation because of its diversity, and those concerned for forest dwellers. Both deplore unsustainable logging and privatisation of public assets. In other circumstances, however, conservation is itself a direct challenge to the immediate livelihood needs of the rural poor. There is a political ecology of conservation, as will be discussed in the next section.

The political ecology of conservation

Conservation, like forestry, is a complex issue in the context of sustainability. As has been argued earlier in the book, concern for the conservation of species and ecosystems was one of the most important roots of sustainable development (see Chapters 2 and 3). While the first document of the mainstream, the *World Conservation Strategy* (IUCN 1980), addressed the question of development, poverty alleviation and wider environmental management, it grew from a determination on the part of wildlife conservationists in IUCN and the WWF to search for ways to redirect development that are more benign for nature conservation.

As Chapter 2 argued, a significant shift in conservation thinking was begun in the *World Conservation Strategy*. This involved both an assault on conventional environmentally destructive development, in order to better preserve threatened ecosystems and species, and a parallel attack on those within conservation who saw people as the enemies of wildlife. It was neatly argued that conservation could (indeed must) sustain development, and also that development could be reconfigured to promote conservation. It was with the concept of sustainable development that conservationists began to claim that these objectives could be achieved at all levels: globally, nationally and locally. Therefore it came to be argued that conservation could meet the true interests of poor people, and particularly the rural poor, who were themselves often the victims of development. Conservation was portrayed as something which, if properly organised, could (almost by definition) be made to meet the needs of, and hence appeal to, local people.

This transition in conservation ideology was a considerable one, and did not happen without planning and effort, and did not happen instantaneously (nor indeed is it everywhere complete; Adams and Hulme 1998, Hulme and Murphree 1999). The rise of this new ideology or narrative of community conservation will be discussed in Chapter 12. Here I want to concentrate on the ideas that it sought to replace, and on what might be called the narrative of 'fortress conservation' (Adams and Hulme 1998, cf. Leach and Mearns 1996).

The history of conservation in the Third World is not one of happily shared interests between rural people and state conservation bodies, but one of exclusion and latent or actual conflict. The historical evolution of conservation in sub-Saharan Africa has been described in Chapter 2, for it forms an important part of the heritage of thinking in mainstream sustainable development. The conventional approach to conservation, in Africa as in the rest of the Third World, has followed the experience of industrialised countries in establishing protected areas, land set aside for 'nature' or 'wildlife', where human use is either prevented or severely constrained. This 'fortress conservation' approach, also referred to as the 'fences and fines' approach (Wells and Brandon 1992), placed conservation in direct conflict with many people with rights to, or need for, resources in the protected area.

In the industrialised world, protected areas have been based on two models. The dominant model in terms of its influence on the global ideas about conservation is that developed in the USA, where national parks were created in remote and sparsely populated areas, where the human costs of eviction have been small and the voice of those excluded faint. Such land has often been regarded as 'wilderness', and evidence of former occupation and ecosystem modification (by indigenous people) has been forgotten or downplayed. Of course, a huge industry has developed in such parks, ranging from the automobile-based mountain viewing from newly made roads to the specialised lightweight 'back-country' campers, but these people, and the staff who serve them and clean up after them, are permitted presences in the wilderness; indeed, they are as invisible in their Lycra-clad glory as the ghosts of Indian sites in the 'wild' place they visit (A. Wilson 1992).

In Europe, conservation has also been built around systems of protected areas; however, they have been less extensive and less exclusive than the North American model. In the UK, for example, while a small number of national nature reserves have been established by government agencies since 1949, conservation has also been pursued on much larger areas of private land. A myriad of small pieces of habitat within densely settled and intensively used economic landscapes have been designated, and their management to favour wildlife has been promoted by financial incentives and grants, and conservation advice (Adams 1996). British national parks are very different from the US and international (IUCN) model. Their purpose is to preserve scenic beauty, and it is recognised that much of this depends on continued human management of land, primarily in agriculture. British national parks are large areas of privately owned and mostly farmed land with their own planning framework, rather than tracts of empty (and emptied) state-owned land believed to be wilderness (Evans 1992, Bunce 1994, Adams 1996).

The British model of national parks, and indeed the wider experience in Europe of integrating people and wildlife in a landscape context, has had little influence on conservation in the Third World, although the IUCN Commission on National Parks and Protected Areas now recognises the category of 'protected landscape', and there is interest in the applicability of the UK model in the non-industrialised world. Internationally, however, it is the exclusionary concept of protected areas that has been dominant, even in areas such as anglophone Africa where conservation regulations were developed under British colonial rule. In the Third World, many government conservation departments have their roots in agencies established to prevent hunting (or 'poaching') by local people and to set aside areas of land for 'game' or wildlife, and if necessary to remove people from them (for in a 'natural' area, people – particularly black people – were almost universally seen as an 'unnatural' presence, a disturbance to ecological equilibrium with their grazing animals, fires, spears and snares). Thus attempts to stop hunting and tackle 'the menace of shifting cultivation' in forest reserves, key elements in a conservation policy for a country such as Nigeria for example (Anadu 1987, p. 249), have been made against intense local opposition.

In the conventional fortress conservation model, landscapes were deemed to be 'natural', and humans were excluded. At least, certain humans were excluded, for while farmers, hunters and other uneducated resource users were unacceptable, tourists with cameras, hotel proprietors and tour operators, scientists and sometimes big game hunters were allowed (Plate 9.3). Neumann (1998) describes the ways in which the Anglo-American nature aesthetic was applied to the national parks of Africa, both at their inception (in the colonial period) and today. Unlike America, Africa was not emptied of people upon annexation and settlement, yet for the purposes of conservation large tracts were adjudged to be empty, or empty enough to be treated conceptually as 'wilderness'. The idea of nature that dominated the establishment and management of national parks, for example at Serengeti and Arusha in Tanzania, involved a denial of African history – and of use rights. Nature was allotted its fixed place in the development plans of colonial (and subsequently independent) governments, and people had no right there. Wildlife and people were to be kept apart, the animals confined to reserves and shot as 'problem animals' when they transgressed invisible administrative boundaries and raided crops, and people kept at bay by the policing of protected-area boundaries and the control of incursions through paramilitary anti-poaching patrols. The plight of people evicted from protected areas is directly comparable to that of reservoir evacuees (see Chapter 8), and the economic impacts of such evictions can be considerable (Brockington and Homewood 1996, Infield and Adams 1999). The threat of state coercion and violence is ever-present. 'Shoot-to-kill' policies against poachers (like that in Kenya in the early 1990s) may seem a sensible strategy to deal with organised and well-armed elephant or rhinoceros poaching gangs, although clearly a desperate one. However, 'enforcement' more often consists of taking action against unarmed women collecting firewood or herders straying over an unmarked park boundary. The people most effectively excluded from the fortress were the rural poor. As Neumann argues, 'parks and protected areas are historically implicated in the conditions of poverty and underdevelopment that surround them' (1998, p. 9).

Fortress conservation therefore involved the suppression of resource use by local people. Conservation's priorities have tended to be remote from local day-to-day economic reality, and are indeed from this perspective irrational, since they involve forced abandonment of rights and resource use patterns that were in place. Regulations (and boundaries) are set from outside the community, imposed in plans that are never seen by government officials who have power, but are faceless and nameless to local people, advised by scientists whose expertise is untestable, remote, and not always sound. Fortress conservation reflects the priorities of national conservation agencies and international organ-

Plate 9.3 The gate of the Mgahinga Gorilla National Park in south-west Uganda
symbolises the separation of the 'wild' and the 'tame', and separates
intensively managed farmland from both residual forest and land cleared for
agriculture, and then cleared of farmers to create the national park, within
the past forty years. The park is small, consisting of the northern slopes of
three volcanoes forming the international border with Rwanda, and contains
giant heather forests and bamboo thickets visited by groups of gorillas
moving between Rwanda, Uganda and the Democratic Republic of Congo.
Treks to see habituated gorilla groups attract international tourists and
generate significant revenue, much of which is used for conservation in the
park and through 'revenue-sharing' in infrastructure development for local
communities in the form of school buildings.

isations. Work is often funded from overseas by specialist conservation aid agencies who employ their own visiting experts. They carry their own assumptions,
stay for short periods, and suffer from the same seasonal, urban, tarmac biases
as their economic and engineering counterparts (Chambers 1983). Repeatedly,
such expert missions identify the current actions of local people as a threat to
the survival of some feature of conservation interest. The policy proposals of
these conservationists are likely to be as alien to local people as any proposed
by conventional development planners (dams or irrigation schemes, for example;
see Chapter 8), and potentially as adverse to their interests.

Conservationists' plans can also be highly coercive. There are many examples that could be chosen to demonstrate the destructive nature of conservation
as development. Most stem, as do their agricultural counterparts, from the
strength and impact of the ideologies that underpin planning. The curious

psychology and sociology of hunting in colonial Africa was described in Chapter 2 (J.M. MacKenzie 1987, 1989). The rise of the hunt as an acceptable ritualistic pastime for whites was matched by the proscription of hunting by Africans. The latter was poaching – it was quite different, and was to be prevented. When ecology developed as a tool of land use decision-making, it was marshalled in the service of proving the hypothesis created by this ideology: that African hunting damaged wildlife populations. Collett (1987), for example, describes a continuity in government attitudes to the Maasai based on precisely this supposition. He finds no evidence in the archaeological record that Maasai pastoralists damaged wildlife populations. Early colonial administrators developed the view that the Maasai were 'predators terrorising neighbouring groups', accumulating stock and refusing to trade (ibid., p. 144), and a threat to wildlife. As a result, it seemed axiomatic that they must be excluded from conservation areas (Homewood and Rodgers 1991).

The Maasai also face entrenched policy narratives about overgrazing (see Chapter 7). Brockington and Homewood (1996) describe the power of these ideas in legitimising the eviction of pastoralists from the Mkomazi Game Reserve in northern Tanzania. They were cleared from the reserve, making it 'wilderness' for the first time, because of conservation planners' fears of the people, and their present and unknown future impact (ibid., p. 104). From the viewpoint of the Maasai, conservation policy at Mkomazi and elsewhere has primarily been experienced as the arbitrary (and unfair) imposition of state power to negate their rights and take away their resources. Its effects are not readily distinguishable from those of other evictions by the state, for example that suffered by the pastoral Barabaig further south in Tanzania, evicted to make way for a Canadian-funded state wheat farm (Lane 1992). More generally, local people's experience of conservation has often been as a form of imposed development, the more or less unwelcome work of remote planners and the state. Conservation has often imposed significant costs on local people in its attempt to secure benefits from species preservation at the wider national or international scale.

Turton (1987) discussed the impact of state conservation on the Mursi of the Omo Valley in Ethiopia, and described the very different cultural values of the Wildlife Conservation Department and the Mursi themselves. This area was perceived by conservationists, quite incorrectly, as 'wilderness'. In fact it is an anthropogenic ecosystem created by the Mursi. Their economy is based on cattle-herding, dry-season cultivation and flood-retreat farming along the Omo River. All three have to be combined to achieve subsistence. The Mursi also hunt in the hungry season before harvest, and trade ivory, leopard skins and other products. The Omo National Park was established in 1966 west of the River Omo, and a second park to the east, the Mago National Park, was planned in the 1970s by a Japanese team. Over this period Mursi had been driven south into the Mago area by drought, and some land uses had intensified. The 1978 report saw the Mursi as a threat to conservation, although without defining in what way, and proposed resettlement. Turton (1987) ridiculed the policy of exclusive conservation, and drew a sharp contrast between attitudes to the conservation of wildlife and those towards the survival of the Mursi themselves. He comments that if the integrity of the Omo and Mago Park boundaries had been successfully defended against human use, there would simply have been no Mursi economy left.

Schoepf (1984) reviewed the impact of the Man and the Biosphere (MAB) programme established in the Lufira Valley in Zaire (now the Democratic Republic of Congo) in 1979. The Lufira Valley MAB reserve covered 50,000 ha in savanna Miombo woodland in south-eastern Shaba. The Lemba people used to integrate shifting cultivation with more intensive practices, including mounding and ridging using mulch, and hand irrigation of streamside gardens. During the colonial period labour emigration was significant, and villages and fields were relocated along roads 'to facilitate administrative surveillance' (ibid., p. 272). The Lufira reserve was selected by an expatriate ecologist, and by 1981 had yet to be visited by the chairman of the committee at Lubumbashi designated to oversee it. MAB reserves are zoned, with a core where all productive activities are prohibited, surrounded by a buffer zone where existing activities may continue, but innovations causing environmental change are banned. The Lufira Valley MAB reserve became caught up in efforts by the state to increase agricultural production through obligatory cultivation of cassava and maize, which was to be enforced in the experimental zone. Moreover, while the reserve boundary was adjusted to exclude the charcoal-cutting areas of parastatal and private firms, and certain large farms, the central zone was said to be uninhabited. The chiefs of the Upper Lufira Valley complained in 1980 about the establishment of the reserve, and listed 2,000 people who lived or cultivated there. The chiefs were dismissed as uncooperative, obstructionist and anti-state.

Conservation can have significant impacts, even where local people are tolerated in protected areas. Gordon (1985) describes the impact of conservation on the 'Bushmen' of Namibia. Game reserves were first declared in Namibia in 1907, and involved bans on the hunting of giraffe, buffalo, and female eland and kudu. Various game reserves were declared, and in some (notably what became the Etosha Game Park) Bushmen were tolerated, and used as trackers and piece labourers. Elsewhere, they were moved out. The Gemsbok National Park was declared, in 1931, on land occupied by Bushmen. The National Parks Board banned hunting and prosecuted 'poachers'. In 1941 land was set aside for Bushmen to hunt adjacent to the park, but there and elsewhere cultural contact and economic integration brought cultural change, and attempts to move Bushmen who were no longer 'true' Bushmen living in a traditional manner, out of the parks and reserves. In 1955 the Department of Nature Conservation was established, and in the 1960s white tourism became important. In 1980 the Department began to open up new areas for tourism, including a 562,000-ha game park in Bushmanland, in the Kalahari on the Botswana border. Gordon argues that this park makes sense only by making the Bushmen themselves the subject of tourists' quests, 'the objects of the leisure rituals of the affluent whites' (ibid., p. 40). The people of Nyae-Nyae will be able to hunt 'traditionally', but not develop their economy to embrace subsistence pastoralism. Gordon titles his piece 'conserving the Bushmen to extinction' (ibid., p. 28).

The impact of conservation, and particularly protected areas, on local people has been widely discussed (e.g. McNeely and Pitt 1987, Neumann 1998). The problem is effectively global, and recognition of the political and economic costs of conservation has helped stimulate a massive change in conservation policy, leading to the rise of 'community conservation'. This will be described in detail in Chapter 12.

However, political ecology is also relevant to conservation policy at the international scale. The biologist E.O. Wilson observed the 'awful symmetry' of economic wealth and biodiversity, 'whereby the richest nations preside over the smallest and least interesting biotas, whilst the poorest nations, burdened by exploding populations and little scientific knowledge, are stewards of the largest' (1992, p. 260). Since the end of colonial rule, conservation in Third World countries has been part of indigenous government agencies. Conservation can no longer be considered an alien ideology, since like democracy and Coca-Cola it has been incorporated into modern states in almost every country of the world. Indeed, some countries, for example Tanzania, have adopted and expanded on the international model of conservation with enormous enthusiasm. Tanzania has designated more than 20 per cent of the country as protected areas, making the establishment of protected areas part of its thrust for modernity and its apparatus of nation-creation, as well as a vital element in its foreign affairs strategy.

However, it is somewhat disingenuous to argue that conservation in the Third World is not profoundly influenced by the interests of governments and citizens in industrialised countries. There is a standardised global conservation ideology, and it is created and disseminated by visionaries, thinkers and the media in ways that strongly reflect the interests of a global literate urban elite that is predominantly in industrialised countries. The ideologies that dominate media coverage of the environment in the North (indeed, the very concept of 'the environment' as a separate cognitive category) are very different from those in the South (Chapman *et al.* 1997). The strength and spread of global conservation ideology are reflected in the high membership of environmental organisations in industrialised countries, and the limited (and purely elite membership) of such organisations in the South. It is articulated and turned into policy by international NGOs and First World aid donors. To the Third World, they offer ideas and values, they suggest technical and bureaucratic structures. They seek partners in the South and sometimes help them, and they also attempt to aid (and to an extent capture) grassroots environmental movements.

First World governments and NGOs establish international fora (of which the Rio Conference in 1992 was the type example) that Third World governments are invited to attend and at which they are urged to sign agreements that will further conservation. The power of a small number of Northern (mostly US-based) environmental NGOs was revealed at Rio. These NGOs (the Sierra Club, the National Audubon Society, the National Parks and Conservation Association, the Izaak Walton League, the Wilderness Society, the National Wildlife Federation, Defenders of Wildlife, the Environmental Defence Fund, Friends of the Earth, the Natural Resources Defence Council, IUCN, the WWF and the World Resources Institute; Chatterjee and Finger 1994) have acquired a global corporate culture. Their leaders wear business suits and seek to talk the power-language of bankers and world leaders. At Rio they had the resources, the experience and the expertise to lobby to some effect, and their voices were far louder than those of most of the other 4,200 accredited lobbyists, who were disorientated, confused and disorganised (ibid.).

The international conservation movement also seeks more direct influences on policies within Third World countries. NGOs provide capital and recurrent funding for projects, and they obtain permission to run pilot projects to show how conservation can be done. They also influence First World donors, and push them to disburse money in particular ways, establishing environmental conditionality on aid. International NGOs also seek to secure effective environmental gains within particular countries through 'debt-for-nature' swaps (whereby part of a country's debt to a foreign bank is bought by an NGO at a discounted price in return for agreed expenditure on conservation in the country; Gullison and Losos 1993).

All of these things may on balance be desirable, and the argument here is not that NGOs should not seek to have influence on development. Nor is it true that conservation necessarily involves actions against the interest of people (particularly poorer and less powerful people). It is simply that it often has done so, and sometimes still does. At all scales from the local to the global, decisions about conservation are highly political. Actions aimed at the conservation of nature are also, by definition, actions that engage with society. However convincing the technical analysis of conservation biologists, and however pressing their conclusions about the need for drastic conservation action, conservation policy is profoundly political. It is therefore within the framework of political ecology that conservation policy needs to be understood.

The political ecology of famine

It was argued at the beginning of this chapter that the understanding of the human dimensions of environmental change demanded a political ecology approach. An area where this is well illustrated is in the study of famine in the context of ideas about human-induced environmental degradation and desertification, discussed in detail in Chapter 7. As that chapter argued, environmentalists have learned enough human ecology since the 1970s to reject simple climatic determinism as a cause of the nexus of poverty and degradation in the Sahel. Yet environmentalist analysis of the human impacts of drought, social dimensions of desertification, and famine often remains constrained by a very limited view of social relations. As the discussion in Chapter 7 showed, debates about environmental degradation tend to be carried forwards in isolation from broader questions concerning society and political economy. Even those who recognise that there is a relationship between what people do to respond to environmental change, and the environmental change that takes place, tend to think about these issues in a theoretical vacuum. As Michael Watts memorably observed, social perspectives on drought and desertification too often degenerate into 'a pluralist grab-bag of ideas embracing everything from land tenure to international political organisation' (1987, p. 188).

An increasing volume of work on the intersection of drought, environmental degradation and hunger (much of it broadly falling within a political ecology perspective) makes an explicit attempt to place an understanding of environmental change within an understanding of political economy. Classically, Piers Blaikie (1985) argued that soil degradation and erosion were the result of sets of decisions about land use made over time by land-users that could not be isolated from their political-economic context. In the words of Jean Copans, 'we can no longer separate the natural phenomenon and the necessary political translation of its effects' (1983, p. 94). Mortimore criticises the 'doomsday' scenario so often painted for African drylands, asking, 'what are the effects on environmental management of world markets, global economic recession and the impoverishment of African governments through debt repayments, diminishing revenues, inefficiency, corruption and war?' (1998, p. 4). These are exogenous political-economic challenges to the livelihood security of the rural poor, just as important as drought or climate change – and moreover, they interact with environmental change in complex and often insidious ways.

The individual household facing a shortfall in rain is linked through the market and the state to the international economy. Historically, colonialism has been a critically important process mediating this linkage, but local communities and household economies can be disrupted by the demands of the state and capital in many other ways (for example 'taxation, land-grabbing or amassment, demands for cash crop production and the extraction of labour and surplus'; Blaikie and Brookfield 1987, p. 121).

Academic debate about environmental degradation, poverty and hunger in the last three decades of the twentieth century was stimulated by the two Sahel famines of the 1970s and 1980s. These were both famines of the television age, with the suffering of refugees relayed nightly to living rooms in the North by television, stimulating a surge of humanitarian concern (most notably, perhaps, in the Live Aid Concert, beamed live around the world, organised by Bob Geldof ultimately in response to an item on the UK evening news; Geldof 1986).

The outrage of televised famine demanded a response; the prevention of famine obviously demanded an explanation of what caused it. The simplest, and perhaps most obvious, explanation was that famine was caused by a decline in food availability; and in semi-arid regions the most obvious cause of food availability decline (FAD in the literature) was drought. In a dry area, such as the Sahel, it was easy to picture farmers at risk from fickle rains in every year, and in years when the rains were delayed or too small, being tipped over into hunger; failure of the rains on a large scale would lead to famine. At one level this association between rainfall decline and food production is valid enough. Drought prefaced the famines on the 1970s and 1980s, and, as Chapter 7 described, scientific work was undertaken that demonstrated that the variability of rainfall between years was considerable, and far greater than had previously been realised by scientists. (It is likely that this variability caused little surprise to Sahelian farmers and pastoralists.)

Furthermore, the powerful neo-Malthusian ideas about desertification provided a further reason for explaining food availability decline: excessive numbers of people caused land to be overused, reducing soil cover and soil fertility, vegetation cover and productivity, and setting the preconditions for catastrophic failure of production when the next drought arrived (Chapter 7). This explanation of the 'drought–degradation nexus' was widely believed in the governments of Sahelian countries, in the international agencies that provided capital for relief and rehabilitation, and in the universities that claimed to understand how the Sahel worked (Mortimore 1989, 1998). These ideas proved convenient to many parties. In Ethiopia, for example, the government and Western aid donors found that a neo-Malthusian explanation of the causes of famine (too many people causing a degraded environment) could be expressed in technical terms (free of political ideology) and provided a rationale for a large food-aid programme (Hoben 1995).

Even in the 1970s, however, several bodies of work suggested that the explanation of famine, and its putative link with environmental degradation, was somewhat more complicated. Field research on how farming families coped with drought and hunger gave a very different picture from that so readily adopted by well-meaning outsiders. In Nigeria, for example, work by Michael Mortimore (1989) on villages in the Sahelian north of the country demonstrated the enormous depth of indigenous capacity to plan and cope with variable rainfall. The variability of rainfall both within and between years was considerable, and had great significance for production, but people had developed strategies to cope with it, spreading risk, diversifying (within farming into livestock and other economic activities) and through mobility (for example seasonal labour migration). Furthermore, farmers were well aware of the importance of soil fertility, and had (in areas such as the Kano close-settled zone) long since established permanent and sustainable systems of agriculture. Sahelian farmers were far from the passive victims of fate, driven to a marginal and famine-stalked exposure to famine by their own fecundity.

A second, and highly influential, approach to famine was that taken by Amartya Sen, in his book *Poverty and Famines* (1981). Here, and in the subsequent *Hunger and Public Action* (Drèze and Sen 1989), Sen argued that famine and hunger were caused by a collapse in entitlements to food, and not in food availability. Entitlements derive from trade, production, labour, and inheritance or transfer. In a series of case studies, including accounts of the Great Bengal Famine of 1943 (when people starved outside the doors of grain warehouses), and of famine in the Sahel and in Ethiopia in the 1970s, Sen argued that it was a breakdown in the ability of people to obtain food that led to widespread starvation, not a physical shortage of food as such. Sen's work on famine has been the subject of intensive academic debate (Watts and Bohle 1993). It has provided an important framework for many studies, for example de Waal's account of famine in Darfur in the Sudan (1989), although he has also been criticised, in particular for his interpretation of individual famines, and the evidence of price changes and food availability.

The entitlements approach views famine 'as economic disasters, not just as food crises' (Sen 1981, p. 162), and demonstrates that diverse forces can lead to the predicament of famine. While many starve in a famine, not all will do so:

> it is by no means clear that there has ever occurred a famine in which all groups in a country have suffered from starvation, since different groups typically do have very different commanding powers over food, and an over-all shortage brings out the contrasting powers in stark clarity.
>
> (ibid., p. 43)

What the entitlements approach is less good at is explaining the 'long-term structural and historical processes by which specific patterns of entitlements and property rights come to be distributed – in other words political economy' (Watts and Bohle 1993, p. 48). Political ecology provides a more complete interpretation of environmental degradation, and of food shortage and famine. Copans (1983), writing about the social and political context of the 1970s drought in the Sahel, argued that the phenomenon of famine must be understood in the context of a series of scales: first, the conditions of production (e.g. rainfall); second, the social organisation of production (for example arrangements for agriculture); third, the national political economy (for example issues of inequality of both class and region); and fourth, the international political economy (dependency and neocolonialism). He was concerned to demonstrate that real explanations of rural poverty and exposure to famine often lie outside the rural community itself, just as they lie outside ecology and climatology.

Copans suggested that Africanists failed to understand the economic and political causes of the 'actual ecological imbalances' that occurred during the 1970s Sahel drought (1983, p. 83). He criticised the research 'boom' that followed the drought for its conventional nature and disciplinary constraints, and highlighted the lack of work on the links between social, economic and political history and related changes in the natural environment which these produced. He called for a new focus on the impact of modernisation on both the social and natural environment.

Several authors have done this, notably Michael Watts (1983a, b, 1984, 1987) in his studies of the village of Kaita in Katsina Emirate in northern Nigeria, and Franke and Chasin in their book *Seeds of Famine* (1980). Franke and Chasin describe the impact on the Sahel of the expansion of European economic power and trading influence in the seventeenth century, and latterly, at the end of the nineteenth century, by colonial expansion. The French colonial governments in French West Africa brought about the expansion of groundnut (peanut) cultivation through a mixture of incentives and coercion (particularly taxation and forced labour), and the commoditisation and monetisation of the economy. Regular shipments of groundnuts to Marseilles began in 1884, and they were exempted from duty in 1892. Railways and roads, and after 1913 agricultural research, credit and extension, were all marshalled to promote groundnut production. Production rose through the colonial period and after independence. Production in Senegal was 45,000 tons in 1884–5 and over 1 million tons in 1965–6. By 1964, groundnuts represented 58 per cent of exports by value in Mauritania, 59 per cent in Mali, 63 per cent in Niger and a staggering 79 per cent in Senegal.

Franke and Chasin argued that individual farmers were caught in a production trap as groundnut yields declined, demanding the expansion of production into fallow areas and into areas used for growing staple food crops of millet. As a result, cash was required to purchase foodgrains, demanding further groundnut production.

Furthermore, intensification of groundnut cultivation led to abandonment of former practices: rotating groundnuts and millet, and following three years of groundnut production with a fallow period of six years to allow nutrients to recover in the soil (Franke and Chasin 1979, 1980). Continuous sole cropping with groundnuts and the violation of fallow led to soil deterioration in agricultural regions. The expansion of groundnut cultivation northwards, and the cultivation of fallow lands further south, in turn disrupted the pastoral economy, creating conflicts over dry-season pastures and additional pressure on seasonal northern grasslands. Thus cash cropping, driven initially by colonial policy and latterly by the demands of the national economy and by international political economy, created both degraded croplands and desertified rangelands in West Africa. In this lay the seeds of famine, while soil exhaustion caused further declines in yields.

A similar intensification trap worked at the level of national economies. Declining terms of trade told against these one-crop economies, while competition in the post-war period for the European market from other oilseed producers, particularly American soya oil, reduced prices. The expansion of groundnut production became a national goal, with production doubling in Senegal for example between 1954 and 1957, but it carried serious implications of soil exhaustion and the erosion of food self-sufficiency. It was this that created conditions in which Sahelian producers were exposed to famine during the drought of the early 1970s.

The sweep of this argument is undoubtedly attractive, although its history is rather simplistic. There is for example no analysis of factions within either the colonial state or the African producers, no consideration of winners and losers within the fundamental system. There is a need for detailed case studies of the ecological implications of continuous agricultural production, or the nature of change in pastoral and agricultural ecosystems. Nonetheless, this account, which reflects widely held views about the nature of the colonial impact in Africa, is probably basically sound. The existence of important links between political economy and environment is clear. The impacts of the actions of colonial and post-colonial states on the actions of farmers and pastoralists, and the effects of their actions in turn on the ground, must be the centre-piece of accounts of environmental degradation.

In his work in Nigeria, Michael Watts offered a detailed account of the relations between producers and non-producers, and the social mechanisms for surplus extraction. He argued that they are the means of understanding the 'connections between material circumstances and ecological conditions' (1987, p. 190). His analysis started therefore from the peasant household, locked into the local economy and ecology. He discussed the impact of capitalism on production systems in rural Nigeria, looking in particular at the articulation of the pre-capitalist mode of production with the global capitalist system through the agency of the colonial state (Watts 1983b). With colonial rule came taxation, cash cropping (in the north of Nigeria chiefly cotton and groundnuts) and railways. The railway reached Kano in 1913 and generated a rapid rise in groundnut exports and (one must assume) a parallel shift in land use. While some may have profited from this trade, it is argued that these changes reduced the 'margin of security' which poorer Hausa farmers had previously enjoyed (ibid., p. 251), both from the highly adaptive ecology of their production system (based on intercropping and the skilled exploitation of the diversity of upland and wetland environments open to them) and from the range of sources of livelihood outside farming (crafts, farm labour, livestock, seasonal migration and sale of land).

Hausa farmers became progressively more involved in cash cropping. Falls in groundnut prices caused a 'reproduction squeeze' where farmers either increased production or reduced consumption. This squeeze was felt unequally by the poor in the differentiated Hausa society of northern Nigeria, and it promoted (among other things) the decline of the moral economy – the patterns of reciprocity between households, particularly between rich and poor, that had provided a safety net in times of drought or disease. Poor farmers, 'shackled by their own poverty, are largely powerless to effect the sorts of changes which might mitigate the debilitating consequences of environmental hazards' (Watts 1983b, p. 256). With drought, in the twentieth century as before, came famine (von Apeldoorn 1980). Watts sees drought 'refracted through the prism of community inequality' (1983b, p. 256), and the environmental crisis of drought as something that reveals the structure of the social system reworked by colonialism and maintained by international political economy.

It is now accepted that famine is a phenomenon with complex causes, and that many of these are political. Thus in an analysis of famine in the Sudan, Olsson (1993) concludes that the idea that environmental degradation was a significant cause of famine in 1984–5 should be abandoned. Famine was the result of drought and market failure. Severe reductions in rainfall in 1984 triggered widespread speculation in food, which pushed prices beyond the reach of ordinary rural people. At a national scale there was sufficient food, but poor distribution allowed famine to develop. Colonialism reduced the autonomy of West African rural communities (Mortimore 1989); decades of development have done little to increase their capacity to withstand drought and have in many instances exposed them to new risks that have eroded their flexibility and adaptability (Mortimore and Adams 1999).

Risk and vulnerability are critical concepts in understanding who is threatened by environmental change (Blaikie *et al.* 1994). Whether the issue is deforestation, conservation or famine, political ecology offers a coherent and interdisciplinary framework for analysis. The prevention of famine, hunger and environmental degradation must start from an understanding of political ecology, and an approach that addresses both environmental variability and political economy. Röling and Wagemakers describe sustainability as 'the outcome of collective decision-making that arises from the interaction among stakeholders' (1998, p. 7). The search for sustainability is an essentially political process.

Summary

- Poverty and environment change are linked in a close and complex way. Political ecology offers a challenge to established approaches to understanding social action and environmental change. Writing in political ecology is diverse, embracing the links between the logics of capitalist growth and environmental change, the politics of social action for the environment, and the discursive power of social constructions of nature (including scientific explanations of environmental change).

- Tropical deforestation is a classic issue that demands a rich explanation drawing on political ecology. Data on tropical forest loss are poor, although the fact of rapid reduction in area is undisputed. Conventional explanations include population growth, economics (chiefly market failure) and poor governance. However, forest clearance needs to be understood in the light of political economic structures and decisions by actors at a range of levels.

- Environmental myths or narratives can be enormously important in justifying policies towards forests and forest clearance, and can be profoundly erroneous. Forest policy has serious implications for forest dwellers.

- Conservation is intensely political in its impacts. The establishment of protected areas in particular can impose significant costs on local people. The question of who bears the costs and who derives the benefits of the preservation of species is important at a range of scales from local to global.

- The causes of famine are complex and deeply political, demanding an explanation that addresses both the environmental and the political-economic context of rural production and consumption.

Further reading

Barraclough, S.L. and Ghimire, K.B. (1995) *Forests and Livelihoods: the social dynamics of deforestation in developing countries*, Macmillan, London.

Blaikie, P. (1985) *The Political Economy of Soil Erosion in Developing Countries*, Longman, London.

Blaikie, P. and Brookfield, H. (1987) *Land Degradation and Society*, Methuen, London.

Blaikie, P., Cannon, T., Davis, I. and Wisner, B. (1994) *At Risk: natural hazards, people's vulnerability and disasters*, Routledge, London.

Bryant, R. and Bailey, S. (1997) *Third World Political Ecology*, Routledge, London.

Hecht, S. and Cockburn, A. (1990) *The Fate of the Forest: developers and defenders of the Amazon*, Penguin Books, Harmondsworth.

Morse, S. and Stocking, M. (eds) (1995) *People and the Environment*, UCL Press, London.

Mortimore, M. and Adams, W.M. (1999) *Working the Sahel: environment and society in northern Nigeria*, Routledge, London.

Neumann, R.P. (1998) *Imposing Wilderness: struggles over livelihood and nature preservation in Africa*, University of California Press, Berkeley.

Parnwell, M.J.G. and Bryant, R.L. (eds) *Environmental Change in South-East Asia: people, politics and sustainable development*, Routledge, London.

Peet, R. and Watts, M. (eds) (1996) *Liberation Ecologies: environment, development, social movements*, Routledge, London.

Rocheleau, D., Thomas-Slayter, B. and Wangari, E. (eds) (1996) *Feminist Political Ecology: global issues and local experiences*, Routledge, London.

Stott, P. and Sullivan, S. (eds) (2000) *Political Ecology: science, myth and power*, Arnold, London.

Web sources

<*http://www.survival.org.uk/index2.htm*>   Survival International is a major international campaigner against the threats to indigenous people.

<*http://www.wrm.org.uy/*>   The World Rainforest Movement is a global network of non-governmental organisations, founded in 1986, and campaigns on threats to tropical rainforests; the site has information on deforestation, mining, colonisation and forest people.

<*http://www.cs.org/specialprojects/maasai/maasai.htm*>   The Maasai Environmental Resource Coalition was founded in 1987 to address the illegal appropriation and destruction of the natural environment in Maasai people's traditional lands in Kenya and Tanzania.

<*http://www.unep-wcmc.org/forest/homepage.htm*>   UNEP-World Conservation Monitoring Centre Forests and Drylands Programme home page for data on forests and deforestation.

<*http://www.fao.org.forestry/*>   The website of the Food and Agriculture Forestry Organisation programme, including material on the state of the world's forests.

<*http://www.fews.net/*>   USAID Famine Early Warning System network: reports, monthly updates, information on vulnerability and hazard monitoring (climate, prices, vegetation).

10 Sustainability and Risk Society

Not only do the rich occupy privileged niches in the habitat while the poor tend to work and live in the more toxic or hazardous zones ... but the very design of the transformed ecosystem is redolent of its social relations.

(David Harvey 1996, p. 185)

Risk Society

In 1986 the German sociologist Ulrich Beck published *Risikogesellschaft* (published in English in 1992 as *Risk Society*). It had a dramatic impact on thinking both within social science and on environmentalism, transforming understanding of the relations between science, technological hazard and social action. For sociologists, the reason for the excitement was the way in which he addressed the nature of modernity, and offered an alternative to the fragmentation and relativism of postmodernism. For environmentalists, Beck addressed the issue of technological hazard, taking seriously their concerns that the scientific-technical complex was producing toxic materials and life-threatening processes that were novel in their longevity, spatial scope and complexity.

Beck is often taken by environmentalists to argue against modernity: against the whole package of science, industrialisation and social and economic organisation of the West. In fact he by no means speaks for a return to some simpler and less destructive or less technological past. Nor does he argue, as postmodernists do, against the very possibility of scientific realism. Beck suggests that there is a break within modernity, and a transition from classical industrial society towards a new (but still industrial) form of 'Risk Society'. He writes, 'this concept designates a developmental phase of modern society in which the social, political, economic and individual risks increasingly tend to escape the institutions for monitoring and protection in industrial society' (1994, p. 5). Global ecological crisis no longer appears as an 'environmental problem' in the world outside, but 'a profound institutional crisis of industrial society itself' (ibid., p. 8). Just as modernisation transformed traditional society (for example the industrialisation and urbanisation of feudal society in Europe), so industrial modernity is being forced to change by a process of 'reflexive modernisation'. Beck believes that the way forwards that is already emerging is not a retreat to the past, in opposition to modernity, but 'more modernity, a modernity radicalised *against* the paths and categories of the classical industrial setting' (Beck 1992, p. 14, emphasis in the original). The universal principles of modernity are valid (civil rights, equality, functional differentiation, methods of argumentation and scepticism), but they are now in conflict with existing industrial society.

The development of Risk Society may be seen in the changes taking place in the North in social organisation (the emancipation and employment of women, the decline of the nuclear family, the decline of marriage) and in economic organisation (globalisation, flexibilisation of labour, changing patterns of work and non-work, the rise of computers and remote working, unemployment and systematic underemployment). He writes, 'the system of coordinates in which life and thinking are fastened in industrial modernity – the axes of gender, family and occupation, the belief in science and progress – begins to shake, and a new twilight of opportunities and hazards comes into existence – the contours of the Risk Society' (Beck 1992, p. 15). The advent of Risk Society may be seen also in the inadequacy of systems of democratic governance to cope with increasingly technical and globalised systems of production: democratic institutions are ill-designed to cope with the challenges raised by science and technology and business. The conventional modernist goal of eliminating scarcity is replaced with the task of eliminating risk, and this reveals 'a vacuum of institutionalised political competence' (ibid., p. 48); he discusses whether intangible and common risks can be made the subject of political action.

The techno-economic system has produced risks, just as it has produced wealth. Beck argues that at early stages of industrialisation these could be dismissed as 'side-effects', but that this argument becomes increasingly unconvincing. Risks might once have been confined to particular factories or sets of industrial processes (as with the US Radium Corporation workers in the 1920s who contracted mouth cancer licking the brushes with which they put luminescent dots of radioactive paint on watch faces; Caufield 1989). However, now they have a tendency to be global, supra-national and often not class specific. These 'global hazards' present new threats with which the existing systems of governance are ill-suited to cope: to the problems of distribution in a society of scarcity have been added 'the problems and conflicts that arise from the production, definition and distribution of techno-scientifically-produced risks' (Beck 1992, p. 19) The risks of modernisation are 'a *wholesale product* of industrialisation and are systematically intensified as it becomes global' (ibid., p. 21, emphasis in the original).

Of course, risk is not a new phenomenon of later twentieth-century industrialisation. The attention given to the problem of 'natural hazards' was discussed in Chapter 9, as was the role of political economy in determining who suffers their impacts (Hewitt 1983, Blaikie *et al.* 1994). However, Beck argues that there is something distinctive in techno-scientifically produced risk. First, this arises directly from the processes of wealth creation (introduced by modernisation itself). Second, hazards are not local or personal but global (they are not tied to point of production). Third, contemporary technological risks escape straightforward perception: they are knowable only through science, and are peculiarly open to social construction and interpretation (whether risk is magnified in lay people's fear, or minimised by complacent governments or corporations). The social, cultural and political risks of modernisation remain hidden.

Beck argues that the threat of late modern risks is so enormous that they cut across class lines, and new social movements form that are not constrained by class division. In this analysis he is imagining West Germany's lack of economic scarcity and its dissolving social classes. Such conditions were not universal (and indeed have not persisted in Germany); it is not global but local technological hazards that have been most important politically. Such hazards reinforce existing class divisions rather than transcending them (Marshall 1999). While technological assessments of risk focus on the average (average dosages, people undifferentiated by age or patterns of work or residence, on regional averages of pollution loads), not everybody experiences the risks of modernisation in the same way, or to the same extent. Some classes of people are more exposed to risk than others.

Risk and environmental class

Beck argues that the risks of modernisation will eventually rebound even on those who profit from them (the risks of a major nuclear accident being hard for even the rich and powerful to avoid, for example), and are therefore, potentially at least, universal. Nonetheless, there is a distributional logic to risk, both between rich and poor in industrialised countries, and internationally between First and Third World states. Beck comments that the risks of modernisation seem to strengthen and not abolish class society, for the wealthy 'can purchase safety and freedom from risk' (1992, p. 35). In Third World countries, determination to overcome poverty and lax environmental regulations attract hazardous industries 'like magnets', and create an explosive mixture: 'The devil of hunger is fought with the Beelzebub of multiplying risks', and to poverty is added the 'destructive powers of the developed risk industry' (ibid., p. 43). That industry is increasingly globalised in its organisation. In the face of globalised corporate power, the nation state (particularly in the Third World) is not necessarily a particularly powerful player in regulating environmental risk. There is therefore a danger that the search for development is accompanied by neglect of environmental justice (Low and Gleeson 1998).

Environmental justice has been an important element in wider concern about environmental health, a key element in post-war environmentalism, particularly in the USA (Hays 1987). Debate in the USA in particular about 'environmental racism' has drawn attention to the differential exposure of minorities, native people and people of colour to environmental risks, particularly in the location of toxic waste dumps and other polluting facilities. The protest about Love Canal in Buffalo, New York, in 1977 (where houses were built on a polluted landfill site) was followed by wider protests in the 1980s about inequities in the location of land uses that threatened human health and well-being. By the 1990s the 'environmental justice' movement was well established (Harvey 1996). In response, the US Environmental Protection Agency established an Office of Environmental Equity in 1992, and in 1996 an Executive Order required every federal agency to consider the effects of its programmes on health and well-being of minority communities (Low and Gleeson 1998).

One problem with a pragmatic focus on environmental hazard as a planning and land use issue is that it can reduce the entire question of industrial risk to a locational problem, a dilemma over where to put noxious facilities. In economic terms, one logical outcome of such a focus is the movement of pollution to low-cost locations, where the poor are willing to accept low levels of compensation; thus community leaders are faced with the dilemma of 'trading their people's environmental health in return for basic material security' (Low and Gleeson 1998, p. 119). To extend this argument, a logical place to locate dumps and other noxious sites is not on the doorstep of the First World poor, but at the feet of the poor of the Third World. And precisely that has at times been proposed, not least within the World Bank in 1992 (Rich 1994).

Many Third World countries have weak or poorly resourced regimes of environmental regulation, and hence effectively trade off hopes of future economic security against present risk. The movement of polluting industries to the Third World is seen as important industrial investment on all sides: by Third World governments, by international banks and financial houses, and by global corporations. Lack of facilities for identifying or measuring environmental hazards and costs means they are underestimated or ignored. The resulting 'traffic in risk' (Low and Gleeson 1998) is the result of corporate strategic planning, whereby firms seek low-cost environments with weak environmental regulation and high potential for evasion, but may also be attractive to industrialised country governments, which are able to externalise industrial risks by moving hazardous elements in the production process outside their borders. 'Sustainability' within one jurisdiction is being achieved by exporting risk and hazard elsewhere.

The export of toxic wastes is a good case in point, and has become an increasingly urgent international environmental issue since the 1980s. Awareness of the problem was aided by particular events such as the voyage of the *Karin B* around the ports of Europe in search of a place to unload a cargo of European toxic waste found dumped in leaking oil drums on farmland in the Niger Delta in Nigeria. While the evasion of proper practices was particularly blatant in that case, and the 1991 Bamako Convention subsequently banned waste imports to Africa, international hazardous waste transfers remain standard. The trade is covered by the 1989 Basel Convention on the Control of Transboundary Movements of Hazardous Wastes, and was addressed in *Agenda 21*. The door to international traffic in waste has therefore closed somewhat. The World Trade Organisation's determination to remove restrictions on free trade offers little hope for tighter regulation on the global market in pollution and environmental hazard.

The problems of hazardous production processes, with their implications both for workers and for the pollution of the wider environment, are probably worst in those Third World countries that are industrialising fastest. Here the economic opportunities for investing firms are greatest (with more chance of a trained and accustomed workforce, and a burgeoning local market). Here too, domestic industry (whether owned by the state itself, or by private industrialists, cronies of the head of state or, in the case of China, by organs of the state such as the army) is likely to be an enthusiastic adopter of bad but profitable practice. The poor are the losers when, for whatever reason, unregulated industrialisation is organised.

In the case of India, Gadgil and Guha (1995) argue that the pattern of industrialisation followed since independence in 1947 has concentrated government investment in larger centres, which have received subsidised water, power, transport and communication facilities. Here it pays industry 'to manipulate and bribe politicians and bureaucrats rather than worry about technological innovation, efficient resource use or pollution control' (ibid., p. 31). As Beck comments, 'safety and protection regulations are insufficiently developed, and where they do exist, they are often so much paper' (1992, p. 42).

Gadgil and Guha divide India's people into three categories: omnivores, ecosystem people and ecological refugees. The majority of India's rural people (half of the total population) they describe as ecosystem people, dependent on fields, forests and rivers for their subsistence. A third of Indian people live as environmental refugees, displaced by poverty, landlessness or development projects such as dams, and living on the margins of rural or urban life. The remaining sixth of Indians are the omnivores, entrepreneurs, larger landowners, professionals and formal-sector workers. Four decades of planned development have created urban islands of relative prosperity in a sea of poverty. Gadgil and Guha write:

> The omnivores inhabiting these islands are securely on firm ground. The bulk of India's ecosystem people are submerged in the sea of poverty. The ecological refugees are hangers on at the end of the islands of prosperity, somewhat like mud-skipper fishes hopping around on the muddy beaches fringing mangrove islands. From time to time the tide swallows them; they can manage to clamber back on to the mud, but can never make it to dry land.
>
> (1995, p. 34)

Gadgil and Guha describe an 'iron triangle' established by India's omnivores, at the expense of the subsistence sector and the environment (Figure 10.1). This triangle is an alliance between the omnivores and those who decide the size and scale of the favours of the state (the politicians) and those who implement their delivery (technocrats and bureaucrats).

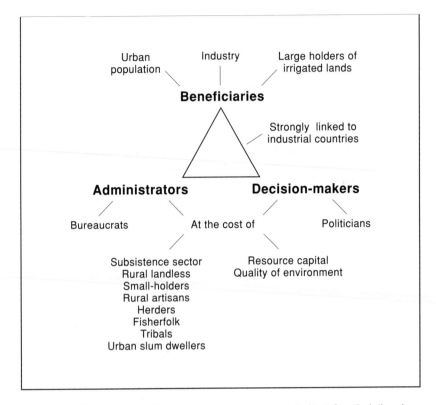

Figure 10.1 The 'iron triangle' governing resource use in India (after Gadgil and
Guha 1995).

Development and environmental pollution

Pollution was one of the concerns that gave rise to the UN Conference on
the Human Environment in Stockholm in 1972, but although at the time it
was seen to be a problem of affluence and an industrialised economy, it is clear
that it is also very much a problem of the Third World. The Founex Meeting
that preceded the Stockholm Conference (Chapter 3) adopted a broad definition of 'environment' (one subsequently adopted by UNEP). This recognised
that it was the immediate need to tackle the problems of malnutrition, disease,
infant mortality and illiteracy that set priorities for environmental action in
Third World countries (Walter and Ugelow 1979).

Certainly Third World countries lack the environmental and anti-pollution
safeguards that have become standard in the industrialised world. Walter and
Ugelow (1979) present comparative data on environmental standards derived
from a survey of 145 countries conducted in 1976 by the UN Conference on
Trade and Development (UNCTAD) showing an inverse relationship between
level of development and rigour of environmental policies. In national compar-

isons, indicators of environmental stress such as carbon dioxide emissions and municipal waste tend to rise with income (for example from a rise in pollution from motor vehicles). Phenomena such as acid rain are serious in many industrialising and urbanising parts of the Third World where soils are susceptible (McCormick 1997; and see Figure 10.2).

Problems of industrial pollution exist on a particularly grand scale and acute form in those countries in the Third World that have most successfully industrialised such as India, Malaysia and Brazil. It is ironic that it is these 'newly industrialised countries' that have been most successful in achieving the 'development' for which less industrialised countries strive. Pollution seems an inescapable part of the development process, part of the price of the right to enter the development race. It is a price disproportionately borne by the poor.

Industrialisation is conventionally taken to be a prerequisite of long-term economic growth, and such growth as a necessary precursor to tackling mass poverty. In this paradigm, only economic growth can allow escape from the downward spiral of poverty and environmental degradation (Broad 1994). Clearly, this is unfortunate, if industrialisation is indeed inevitably associated with increased environmental risk (Lipton 1991). The World Bank argues that industrialised countries have begun to manage to delink economic growth and pollution through control of emissions of sulphur dioxide, nitrogen dioxide and particulates and lead, and they imply from this that industrialisation and environmental improvement do not need to be incompatible (World Bank 1992, see Figure 5.3 on p.109). The importance of this relationship between economic growth and environmental quality (the so-called 'environmental kuznets curve') has been discussed in Chapter 5. Grübler (1994) shows that, over time, industry is tending to reduce the amount of energy used and the amount of carbon produced per unit of value. However, this shift in industrial ecology is not sufficient to balance out growth in volume of output. Consumption is therefore still driving industrial pollution up. Huq (1994) raises the question of whether the transition of all countries to full industrialisation can be sustained, even assuming it can be achieved.

Auty (1997) reviews the evidence on changing pollution patterns during the transition from a rural non-industrial economy to a manufacturing industrial economy and thence to a service economy. Historically, the pattern has been for pollution to increase with GDP, and then fall, giving an S-shaped curve. The experience of industrialising countries in South-East Asia suggests a shift over time as industry develops, from water-borne organic pollution through solid waste and airborne pollution towards toxic pollution. Information and service economies, which have lost much of their heavy industry, tend to have reduced pollution levels.

However, as Auty points out, there are a number of potential problems with all analyses of this sort. First, they use historical data on already-industrialised countries as a model for contemporary industrialisation, and technology and regulatory environments change over time. The ability of a country such as the UK to reduce air pollution, for example, is a measure of the rapid decline of manufacturing industry, and may not be a good guide to the future in India

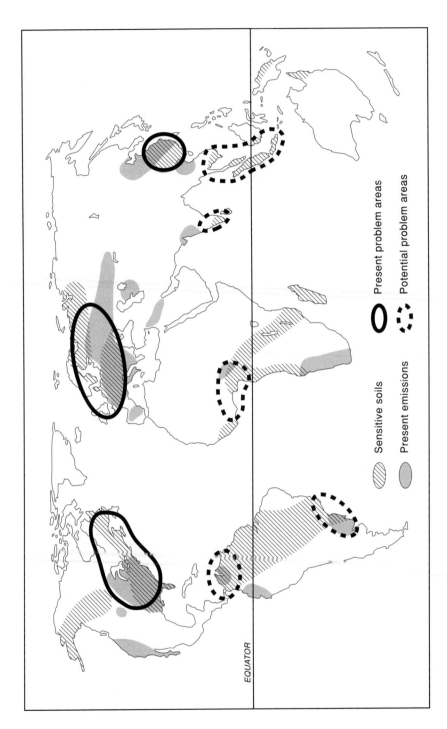

Present problem areas

Potential problem areas

Sensitive soils

Present emissions

EQUATOR

Figure 10.2 Regions at risk from acidification (after Rodhe *et al.* 1988).

or Malaysia. It is no guide at all to the problems of toxic waste that might develop in countries far from any sustained 'industrial transition', such as Papua New Guinea or Mauritania. Second, aggregate data mask differences in national policy and resource endowment. Third, growth in population and consumption is causing increases in overall global industrial pollution whatever happens locally.

Beck's Risk Society is not confined to the deindustrialising North; it is a strong feature of the South, too. Technological hazard and riskiness are rapidly growing problems in most Third World countries, particularly those where industrialisation and modernity have most strongly taken root. As in the North, technological risk is accompanied by changes in culture (family, marriages, informal institutions), in economic organisation (the form of labour and employment) and in governance (the system of patterns of trust, delegation and fairness), all carried along in a torrent of change.

For the poor in rural areas, problems of resource depletion are often more serious than those of pollution (Lipton 1991), but for the urban poor this is not the case. It is in cities that the most serious and intractable pollution problems in Third World countries arise (Hardoy *et al.* 1992). Cities such as São Paulo in Brazil or Mexico City contain some of the most heavily polluted environments in the world. Air pollution, for example, is a major issue, causing both acute and chronic (low-level) health effects. There are several major sources of air pollution (Hardoy *et al.* 1992). Coal- and woodfuel-burning produce smoke or suspended particulates, sulphuric acid and polycyclic aromatic carbons (the classic 'London smog' complex, still a problem in many Third World cities). Vehicle traffic and other hydrocarbon combustion produce photochemical pollutants (including hydrocarbons from evaporating petrol or other sources, nitric oxide, nitrogen dioxide, ozone and aldehydes and other oxidation products; the classic 'Los Angeles smog' complex; Plate 10.1). Other common air pollutants associated with vehicle use include carbon monoxide and lead. Respiratory disease is a major killer in Third World cities: in Bangkok there are estimated to be 1,400 deaths a year due to airborne particulates (Hardoy *et al.* 1992).

Problems of air quality do not originate in urban areas alone. The forest fires of 1997–8 in Sumatra and Kalimantan (associated with low rainfall due to an El Niño event) caused massive palls of smoke and haze across Indonesia, Malaysia and Singapore (Byron and Shepherd 1998). The health of 20 million people was adversely affected (especially that of the old and the young) through asthma and upper respiratory tract infections, and there were serious economic impacts through business shutdowns, airport delays, accidents and depressed tourist revenues. These were not 'natural' fires, but were often started deliberately, not least as a cheap (and government-sanctioned) way to clear forest for oil palm plantations. The hazard of forest fires in Indonesia was as much a product of modernity, its system of governance and regulation and its capitalist economy, as any more obviously technological pollution hazard.

The disposal of solid wastes and the by-products of industrial processes is also a major source of pollution and hazard, particularly in large urban centres,

Plate 10.1 Traffic in Mexico City. Mexico City, built on a dry lake bed surrounded
by hills, has legendary air pollution problems. Vehicle traffic is a major
culprit, with hydrocarbons, nitrogen dioxide, ozone and other compounds
creating a heavy smog. Traffic levels continue to rise, despite the creation
of an underground metro system.

for example in China (Zhao and Sun 1986) and Thailand (Tuntamiroon 1985).
Heavy metals (for example lead, mercury and cadmium) in various forms, poly-
chlorinated biphenyls (PCBs), hydrocarbons, organic solvents, asbestos, cyanide
and arsenic are all recognised problems. The threat to health posed by many
of these chemicals is invisible, their effects cumulative – and exacerbated in the
bodies of children, and people weakened by hunger and disease.

Although in theory, relatively dense settlement makes it cheaper to supply
services such as health or water to the urban poor, rates of mortality and sickness are high. Risks exist in the residential environment (due to water-borne
disease, limited water availability and poor water quality, and to lack of sewage
and waste disposal), in the workplace (due to unsafe industrial practices, pollution, low wages and lack of sickness or disability provision) and in the general
environment. Children are particularly exposed to hazards, including the hazards
of workplaces (Hardoy *et al.* 1992). In Metro Manila, for example, only 15
per cent of the population is served by sewers or septic tanks, and 1.8 million
people lack adequate water supply or sanitation (Hardoy *et al.* 1992). In
Guayaquil in Ecuador, almost half of urban residents (1 million people) lack
reliable sources of potable water; those without piped supply pay high prices
to water vendors for water of dubious quality (Swyngedouw 1997). Within
the growing city over the past century, the supply of water has been the site

of intense social struggle; the distribution of water reflects the distribution of power (ibid.).

Risk Society is as ubiquitous in the South as in the North. Techno-scientifically produced risk arises from processes of wealth creation in the South as it does in the global industrial core. Indeed, its power and scope are less confined and its proliferation less moderated because technologies and systems of industrial organisation are applied without the institutions of understanding, regulation and governance developed in the North to control them. The global industrial frontier is a wild and open place – sometimes a profitable place to invest capital, but a dangerous place to live and work.

Manufacturing pollution

Industrial operations are the source of some of the most toxic pollutants and the most seriously polluted environments. Minerals extraction is particularly problematic. Mines can have complex impacts on environments over long distances (through water or air pollution), and because they are often in remote locations can have drastic effects on local people that are both unrecognised and uncompensated. An Australian corporation, BHP, developed an opencast gold and copper mining operation in the western mountains of Papua New Guinea in the mid-1980s (Low and Gleeson 1998). The government of Papua New Guinea owned 30 per cent of the shares of Ok Tedi Mining. The mine employs 1,700 people, and supplies over 16 per cent of national export earnings. However, it also releases about 80,000 tonnes of mine tailings daily into the Tedi River (which drains into the Fly River system). The tailings contain a suite of heavy metals. The original plan was to hold these within a tailings dam, but this failed in 1984 in a massive landslide, and they have since flowed directly into the river (Hyndman 1994). Over the life of the mine, some 250 billion tonnes of waste will have been dumped, and despite sanguine predictions of the potential for containing the downstream effects (Petr 1979), floodplain forest and river fisheries have been drastically affected; the bed of the upper river has been raised by several metres.

In search of economic compensation, the downstream villages, excluded from the original mining agreement, sued the company for compensation in the Australian courts (Low and Gleeson 1998). An out-of-court settlement was eventually reached, involving resettlement, clean-up and a trust fund for compensation. However, the mine continued to operate. In 1999 it was announced that up to 900 km^2 of forest would be destroyed if mining continued for a further ten years, until the ore body was exhausted.

The environmental impacts of the Ok Tedi mine are not the result of some unfortunate accident. They are integral to the ore extraction process, and are inevitable unless pollution control facilities are built. Such facilities cost money, and their construction cuts profits. Once the original tailings dam had collapsed, it was not replaced. A major attraction of mining in the Third World is that capital and operating costs can be kept low. Limited statutory protection for workers and weak requirements for the control of pollution (whether in the environment or in workers' bodies) often mean that costs are lower than for an equivalent First World mine, and profits correspondingly greater (although, interestingly, Ok Tedi has lost money steadily, and the government has been the chief enthusiast for keeping it open).

Many Third World mineral extraction projects have serious environmental impacts. Baluyut (1985) describes the impacts of three gold and copper mines on the Agno River in the Philippines. These use cyanidation and flotation to obtain ores, and produce large volumes of tailings which are impounded in tailing ponds. Failure of the retaining dams of these ponds has resulted in the release of tailings comprising fine sediments and toxic materials into the Agno River. Downstream ecosystems, particularly coastal mangroves, have been adversely affected by sedimentation, and heavy metals (mercury, cadmium and lead) have accumulated in fish and shellfish downstream. Other mineral extraction and purification processes also cause pollution. The treatment of bauxite with caustic soda to produce alumina creates similar toxic waste disposal problems; for example in Jamaica, 13 million tonnes of caustic alkaline slurry is produced every year by the transnational-dominated aluminium industry (Bell 1986).

Oil and gas exploration and extraction bring with them well-known environmental problems. These have been prominently reported in a number of areas, particularly in the Niger Delta in Nigeria. Ikporukpo (1983) describes the problem of crude oil pollution in the Niger Delta of Nigeria; its impacts include loss of crops, sterilisation of soils and the death of fish. In the decade 1970–80 Nigeria experienced eighteen major oil spillages involving over 1 million barrels of oil. The problem was not addressed by either the Nigerian government or the oil corporations (of which the most prominent was Shell). Environmental degradation associated with oil extraction and gas-burning persisted through the 1980s and 1990s. There were 300 spills per year in Delta and Rivers States in 1991 and 1993, in addition to massive gas-flaring, and repeated disputes over land rights with local communities (Frynas 2000). Protests against pollution, and infringements of human and civil rights, particularly by the Ogoni people, met with heavy-handed treatment by government forces. Demands for a separate Ogoni state (and the share of oil revenue this would bring) were met with violence (Simonsen 1995, Rowell 1996). The judicial murder of the Nigerian playwright Ken Saro-Wiwa and other activists in 1995 brought the environmental and political performance of oil companies, and particularly Shell, into the international spotlight, although it by no means ended the disruption to environment and economy in the Niger Delta.

Environmentalist and human rights NGOs were vociferous in their opposition to the more destructive activities of mineral companies in the 1980s and 1990s. Friends of the Earth ran a major campaign against Rio Tinto's plans to exploit mineral sands in the forests of south-east Madagascar in 1995. In common with other large corporations facing pressure from shareholders and public opinion in the North, most such companies have taken steps to become more 'green', or at least to present themselves as doing so. Some companies in the minerals and energy sector such as Rio Tinto, Placer Dome and BP, have adopted a policy of seeking out and engaging their critics in the search for legitimacy (e.g. Mulligan 1999). There is increasing interest by both corporate and NGO strategists about the best ways to engage and the degree to which common aims can be defined. Most analysts welcome the new initiatives, but wait to make judgement on the significance of their benefits in terms of sustainability (Mulligan 1999).

Manufacturing plants in the Third World can also generate significant risks. The explosion at the Union Carbide factory at Bhopal in Madhya Pradesh, India, in December 1984 is perhaps now the epitome of the hazards of industrialisation in a poorly regulated environment. The plant manufactured chemicals needed to make pesticides, and an explosion in a large tank of methyl isocyanate spread a cloud of pollution over the city of Bhopal. Up to 10,000 people may have died as a direct result of the tragedy, and up to 300,000 were injured, including perhaps 20,000 severely disabled (Low and Gleeson 1998). Responsibility for the tragedy remains a moot point: the explosion might have been averted had the automatic controls standard in such Union Carbide plants in the USA been installed, had the Indian government not insisted that manual controls be fitted, or had the Indian subsidiary of Union Carbide followed plans more closely. Wherever blame is shifted, the risk is clear, as is the identity of those who bore it. The Indian government made an out-of-court settlement with Union Carbide in 1989 for US$470 million.

While the Bhopal tragedy was (thankfully) exceptional, the conditions for such tragedies are not exceptional. They represent part of the risk of modernity that Beck describes. The Bhopal incident could have occurred in many places. Indeed, similar poisonings, on a smaller and more piecemeal basis, are occurring everywhere all the time, around manufacturing plants (for example over the US border in Mexico) or in the effluent from tanneries, smelters or other plants. Castleman (1981), for example, describes piles of asbestos-cement waste and the discharge of untreated waste water outside a factory in Ahmedabad in India making building materials. There are many such factories, innumerable instances of hazardous production processes and inadequate regulation or control of pollution.

Pollution can occur for reasons that relate to the poverty of experience and expertise, for example because the technology is lacking to clean up discharges, to the inappropriateness of certain Western technologies or to a failure of regulation. Fundamental to many incidents of chronic pollution is economics. The costs of pollution control tend to be passed either forwards to product prices (thus making manufactured goods more expensive), or backwards, reducing returns on capital. Pollution control is therefore unattractive. It can therefore be argued that Third World countries without strict pollution legislation can achieve a competitive advantage over industrialised countries (Walter and Ugelow 1979). Lack of pollution controls (like cheap labour) cuts the cost of production, or rather transfers costs to the host environment and community.

In a sense, therefore, pollution is a hidden subsidy for Third World industry. The hazards and costs of industrialisation without pollution control can therefore seem to have a certain attractiveness. When this is combined with the inertia, inefficiency and corruption with which Third World government bureaucracies are plagued, the result can be – as in the case of Bhopal – disastrous.

Low pollution control standards in the Third World are also greatly advantageous to First World companies faced with rising production costs at home. Transnational companies in particular have moved industrial plants that are highly polluting to locations in the Third World as a direct result of environmental protection policies in industrialised countries. Thus Suckcharoen *et al.* (1978) discussed the location of a caustic soda factory in Thailand by a Japanese company in 1966 in response to weak local environmental protection legislation. The risk was of pollution by methyl mercury in aquatic ecosystems, and especially in fish, which form half the animal protein intake in the average Thai diet. Mercury poisoning among the fishing community at Minimata in Japan (causing extensive birth defects in children) had been a major factor in raising consciousness of the dangers of industrial pollution in industrialised countries (D'Itri and D'Itri 1977). Suckcharoen *et al.* (1978) showed that mercury levels in fish near the caustic soda factory are on average 1.48 ppm compared to low background levels of 0.07 ppm.

There is evidence to support theoretical arguments about the attractions of lax pollution controls in the Third World. Walter and Ugelow (1979) presented data on the percentage of capital expenditure spent on pollution control by US-based multinationals in the US and overseas. In primary metals industries pollution control comprised 21.1 per cent of domestic capital outlays, but only 9.8 per cent of overseas capital, and proportions were similar for other industries, for example chemicals (8.9 per cent in the USA, 5.3 per cent abroad). Industrial plants converting raw materials into more complex products can be particularly polluting, because the costs of pollution control can be high compared to the value of the product. The manufacture of pulp and paper is an important case in point. Christiansson and Ashuvud (1985) discussed the potential environmental effects of a paper mill in Mufundi District in Tanzania. Standards do vary between the pollution risk from plants in the First and Third Worlds. Walter and Ugelow (1979) found that US-based multinationals in the USA spent 21.9 per cent of capital expenditure on pollution control in US plants, and 11.8 per cent abroad.

Murphy (1994) argues that manufacturers have a good idea of the dangers implicit in their operations, although they seek to shield this knowledge from environmental movements and the public, and sometimes from the state: 'Transnational companies, often in complicity with state organisations, have overestimated the safety of their factories in order to convince the public to allow them to pursue their search for profit' (ibid., p. 137). Accidents, therefore, are in a sense not unexpected, but events whose likelihood of occurrence is carefully calculated by corporate planners. Techno-scientific risk emerges from modernity, an integral element in the evolution of industrial society.

The underestimation of risk (and attempts to estimate risks where no estimate is possible) is a strategy useful to corporate interests, since it serves 'to legitimate the imposition of risks on the population' (Murphy 1994, p. 142). Beck (1992) points out that science has a monopoly on the definition of hazard: science *determines* risks, while the population simply *perceives* risks (and, of course, suffers from their effects). Risk analysis is therefore something carried out solely by scientific-technical experts, either within industry or government. In theory such experts are independent, but in complex technological areas (which means an increasing number of areas), regulation depends heavily on the expertise and the data of the private sector, and government and business operate a 'revolving door' for 'experts'. Inevitably, commercial factors have influence.

The risks of pesticide production at Bhopal were part of 'a bargain struck by the Indian state on behalf of its people'; in many ways,

> the complex installations of the chemical factories with their imposing pipes and tanks are expensive symbols of success. The death threat they contain, by contrast, remains largely invisible. For them, the fertilisers, pesticides and herbicides they produce signify above all emancipation from material need.
>
> (Beck 1992, p. 42)

The pesticides being produced in Bhopal were part of the package of technical inputs of the 'Green Revolution', intended to revolutionise agricultural productivity in India as elsewhere, and push back hunger and poverty; as Beck comments, 'in the competition between the visible threat from hunger and the invisible threat of death from toxic chemicals, the evident fight against misery is victorious' (ibid.).

The problem of pesticides

The disaster at Bhopal has, of course, a particular irony in that the chemicals being produced at this factory were intended to be directly useful in terms of human welfare, as agricultural pesticides. These compounds, used both in agriculture and in the control of disease vectors, provide an interesting example of the delicate balance between benefit and risk that the techno-science of Beck's Risk Society offers. Pesticides (the word embraces insecticides, herbicides, fungicides, etc.) are central elements in the modernist strategy for Third World agriculture, increasing food and commodity production to deal with hunger and poverty. They have also been enormously effective in reducing the incidence of disease spread by insects or other vectors. The effectiveness of pesticides in controlling crop pests and disease vectors is one of the most commonly cited pieces of evidence in support of a technology-driven strategy of development: do they not allow the hungry world to be fed, and the sick to be protected from disease?

The scale of pesticide use, and the speed with which that use has developed, are staggering. In Africa, sleeping sickness, and an equivalent disease of livestock, *nagana*, was a major scourge over 10 million square kilometres of Africa, in thirty-four countries. It is spread by the hefty biting tsetse fly (*Glossina*). From the 1940s the fly began to be controlled in Central Africa by aerial dusting with the organochlorine pesticides DDT and gamma-BHC (lindane). In due course this was replaced with spraying of these pesticides and dieldrin, both indiscriminately and from the ground using manual or vehicle sprayers on tsetse resting sites, and more recently with fine aerosol spraying of synthetic pyrethroids from aircraft over larger areas (Ford 1971, Matthiessen and Douthwaite 1985, Ormerod 1986). Fortunately, these methods have now given way to a much more targeted technique using traps coated with pesticides baited with chemicals attractive to tsetse (cheaper, more effective and less risky).

Pesticides are one of the linchpins of the Green Revolution, which depends on the spread of high-yielding hybrid crops capable of much greater yields than local crops, but only if supported by inorganic fertilisers (and often irrigation), and protected from insect and fungal attack, and weed competition, by pesticides. Crop losses through pest attack are a major problem in the Third World. In the 1970s, in-field crop losses were estimated to be 42 per cent in Africa: 13 per cent from insects, 13 per cent from disease and 16 per cent from weeds (Ghatak and Turner 1978). In Asia, in-field losses were 43 per cent, in Latin America 33 per cent. Food losses in storage are also considered a major factor in hunger and poverty. 'Improved' crops need protection from pests to a much greater extent than indigenous crops: the technology of the Green Revolution, designed to reduce the risk of hunger, is itself a generator of risk. Furthermore, the distribution and sale of pesticides is almost entirely in the hands of transnational agribusiness companies, which eagerly promote the advantages of chemical pest control, and have a clear vested interest in doing so. The technologies of risk avoidance and creation are an important market.

In conventional analyses, no let-up in the defence of technology-dependent food production systems may be countenanced. Populations of locusts boomed in Africa in the mid-1980s following the temporary reduction in chemical control measures and the disbanding of the Office International contre la Criquet Migrateur Africain in 1986, providing graphic evidence of the need for continued chemical control (Jago 1987). The larger grain borer, a beetle pest of stored maize in particular, was accidentally introduced to Tanzania in the late 1970s and has spread rapidly through both East and West Africa despite international control efforts. Control demands fumigation with methyl bromide (a serious ozone-depleting chemical) or phosphine (Taylor and Harris 1994).

The Green Revolution has been the subject of endless academic debate. Work on crop-breeding was primarily done within the public sector in the International Agricultural Research Centres (IARCs), for example IITA at Ibadan in Nigeria, IRRI at Los Baños in the Philippines and ICRISAT in Pakistan, although increasingly new seed development (now drawing on genetic manipulation technologies) is becoming privatised. Higher-yielding and more fertiliser-responsive varieties of several major crops (particularly rice, wheat and maize) have transformed food production in parts of the Third World, particularly in East and

South Asia, although much less has been achieved in Africa (Richards 1985, Lipton and Longhurst 1989). The impact of the package has been fiercely criticised, for example because it helps rich farmers more than poor farmers or landless labourers, although others argue that even the poor can gain from increases in demands for labour on richer neighbours' fields and crops (Lipton and Longhurst 1989).

Yapa (1996) argues that the green revolution embodies a particular epistemology of development, of understanding food, technology, nature and culture. He suggests that 'miracle seeds' need to be understood not as objects – things that do or do not produce a solution to poverty – but as the product of interlocking relations. These relations are technical, but also social, cultural, political and environmental, and even academic. The green revolution created a technology that required farmers to purchase inputs, replacing other technologies and other options, 'devaluing the "reproductive power" of nature by substituting the "productive power" of industrial inputs' (Yapa 1996, p. 82). It produced an academic discourse that castigated 'backward farmers' and indigenous methods, supported the 'progressive farmers' that adopted the new technology, and minimised its risks (ibid.).

The economic costs and benefits of pesticide are somewhat complex to calculate. It is relatively easy to measure the direct costs of pesticide use, but much harder to estimate the benefits (and hence the costs of damage avoided, particularly by prophylactic use before pest outbreaks occur). It is also difficult to obtain data on pollution effects, including problems of the impact of pesticides on non-target organisms and the development of resistance of target organisms.

At a national scale Ghatak and Turner (1978) stressed the need to consider the foreign exchange costs of importing pesticides (for few countries in the Third World have indigenous production capacity), and point out that pesticides often serve as substitutes for labour inputs (for example in weeding), and labour is neither expensive nor in short supply in much of the Third World. Certainly the substitution of pesticides purchased for cash for family labour is unlikely to make sense in many peasant households. Cox (1985) studied the economics of the widely used pesticides DDT and inorganic copper fungicides in Tanzania. These are both persistent in the environment, but the implications of this – for example on the use of drainage water for drinking downstream – are not known, and long-term social costs cannot therefore be calculated. He discussed the problems of using conventional cost–benefit analysis, and suggested that pesticides should be used only when private benefits (that is, to the user) exceed private costs and all immediate social costs. This approach fails to take account of the longer-term environmental and social costs. It externalises the risk of the technology, and legitimises the application of technology whose impacts are mostly unknown.

There are two particular risks associated with pesticide use. The first is the development of resistance to pesticides under continuous use, and the second is the threat to human health (Conway and Pretty 1991, Pretty 1995). The development of resistance to pesticides is well documented. The problem of the rapid development of resistance to pesticides is combined with the impact of broad-spectrum insecticides and herbicides on non-target species, particularly natural predators. A classic example is the Gezira Scheme in the Sudan, where cotton losses to whitefly kept pace with increasing use of pesticides: the area of cotton sprayed increased rapidly from 600 ha in 1946 (0.7 per cent of the total area) to 98,000 ha (100 per cent) in 1954, and the number of sprays then increased from one per year in the late 1950s to nine per year in the late 1970s. By 1976, 2,500 tonnes of insecticide was being used per year at Gezira, yet despite this effort, and the massive rise in production costs it represented, pest losses continued (Bull 1982). In Malaysia the number of rice pests resistant to at least one pesticide rose from eight in 1965 to fourteen in 1975, the brown plant hopper (*Niliparvata ingens*), for example, bouncing back in the late 1970s in the face of growing and intensifying pesticide use (Bull 1982; see also Figure 10.3). Pest resurgence following pesticide use is an acknowledged problem, and can affect adjacent unsprayed fields because of the spatial dynamics of recolonisation. Decisions about the benefits and costs of pesticide use in terms of pest numbers are therefore more complex than is often assumed (Trumper and Holt 1998).

The problem of pesticide resistance, and the escalating cost of multiple pesticide applications, have led to new strategies to try to reduce levels of pesticide use and control crop losses instead by integrated pest management (IPM). In the Philippines, for example, a national IPM programme was launched in 1993 to train farmers in pest management strategies that minimise chemical pesticide use (Bolido 1998). It is now clear that the brown plant hopper is controlled by a wide guild of predators, and rapid resurgence follows accidental control by pesticides of these organisms. Strategies against the brown plant hopper include breeding rice varieties with pest-resistant attributes. There is increasing interest in the possibility of IPM for cotton, which is one of the most heavily treated with pesticides (FAO 1994a). Other IPM strategies include the release of natural enemies of pests, the use of a bio-pesticides such as the bacterium *Bacillus thuringiensis*, and the integrated management of fields and surrounding habitats to minimise pest damage and maximise natural predation (Pretty 1995).

IPM, and related approaches such as integrated plant nutrition, are part of the increasingly important field of sustainable agriculture (Pretty 1995). While some new technologies may be beyond the reach of poor Third World farmers, many of the modifications of agribusiness practice called for by environmentalists in industrialised economies (including organic cropping, without inorganic pesticides and fertilisers, or low-input farming) are perforce normal practice in the Third World. Indeed, ironically, they are among the kinds of 'backward' practices that agricultural modernisers from governments and agribusiness companies spent a great deal of time and money in the twentieth century trying to persuade farmers to abandon.

The most commonly quoted example of pesticide resistance is not in agriculture, but in public health: the attempt by the World Health Organisation

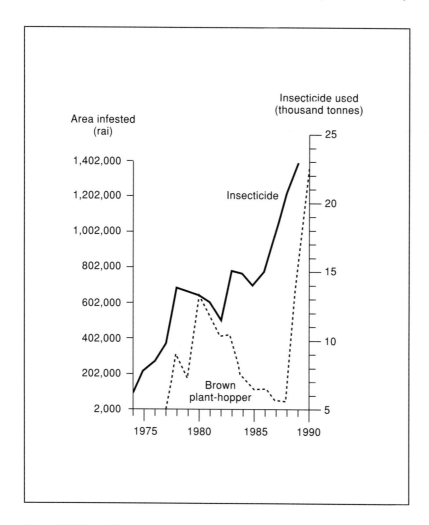

Figure 10.3 Pesticide use on rice and pest resistance, Thailand (after Hadfield 1993).

(WHO) to eradicate malaria using DDT (an organochlorine compound) and drug treatments on sufferers in the 1950s. Initially this met with enormous success, the number of cases being reduced from an estimated 300 million pre-1946 to 120 million in the late 1960s, a period when world population had greatly increased. However, from the mid-1960s mosquitoes resistant to organochlorine pesticides began to appear, and the number of new cases of malaria rose once more (Bull 1982). Resistance to the prophylactic drugs taken against the malaria parasites is also now a major problem in many areas of the Third World where such drugs are widely used, particularly those popular with global tourists such as the Kenyan coast.

There is also evidence that the use of organophosphorus and carbamate insecticides in agriculture is implicated in some instances in the development of pesticide resistance in mosquito vectors of malaria. Resistance can appear quickly in a target population. The Onchocerciasis Control Programme in West Africa began spraying the organophosphorus insecticide temephos to control the biting fly *Simulium damnosum*, the vector for river blindness, in 1974. The project sprayed 18,000 km of river per week in the wet season and 7,500 km per week in the dry season in Upper Volta, Mali, the Ivory Coast and Ghana (Walsh 1985). Resistance was first recorded on the lower Bandama River in the Ivory Coast in March 1980, but spread rapidly to surrounding areas. The project switched to chlophoxim (another organophosphorus compound), but resistance to this appeared by mid-1981 and this too was withdrawn. It appears that resistance to one organophosphorus pesticide promoted resistance to others (Walsh 1985).

Pesticides also have impacts on other aspects of the ecosystem (Conway and Pretty 1991). The organochlorines, such as DDT and dieldrin, were the subject of much controversy in Britain and other industrialised countries in the 1950s and 1960s because they do not metabolise, but become stored in fatty tissue. There were a number of direct poisoning incidents on farmland birds in the 1950s, and subsequent research showed that in top predators such as birds of prey, concentrations were reached which caused physiological problems such as eggshell thinning. There were serious population declines in some species (Sheail 1985). In Nigeria, Perfect (1980) found that DDT accumulated in the soil over the four-year experimental period, although at lower levels than those in temperate areas (2 per cent of that applied) because of more rapid volatilisation. Soil organisms showed shifts in numbers and species complement, and there were both qualitative and quantitative changes in the pathways and rates of litter breakdown. Although yields were higher in the short term under pesticide use, contamination in soils following pesticide use reduced yields. Perfect suggested that DDT may impair the system's capacity to regenerate fertility in fallow periods.

The second, and in many ways more acute problem of pesticide use is the threat pesticides pose to human health. Accidental poisoning by pesticide use is a major problem (Conway and Pretty 1991). Third World countries account for only 15 per cent of global pesticide use, but their consumption is rising rapidly, much of it sponsored by First World development aid. Over half the cases of pesticide poisoning occur in the Third World (Bull 1982). It is estimated that there are 375,000 cases of poisoning by pesticides in the Third World each year, 10,000 of them fatal (Caufield 1984). In Vavuniya in northeast Sri Lanka, for example, there were 938 deaths from pesticide poisoning in 1977, more than from malaria, tetanus, diphtheria, whooping cough and polio put together (Bull 1982, p. 44). Pretty (1995) reports WHO data for 1990 suggesting that between 3 and 25 million agricultural workers may be poisoned by pesticides annually, with perhaps 20,000 deaths.

In the Philippines, sales of pesticides increased by 70 per cent between 1988 and 1992, and between 1980 and 1987 there were over 4,000 cases of pesticide poisoning and over 600 deaths (Pretty 1995). A rise in pesticide purchases in the early 1970s was accompanied by a 27 per cent rise in the death rates of men of working age (Pearce 1987). Deaths up until 1976 peaked in August, the height of the spraying season. In that year double cropping began on the local irrigation scheme, and spraying in February was matched by a second peak in deaths in that month. Protective clothing was not worn, and the backpack sprayers used put 40 mg of active ingredients of pesticide onto operators per hour (ibid.).

The causes of this toll in the Third World are fairly clear. The danger that those people applying pesticide, whether in agriculture or disease control, will be poisoned has become more acute with the transition from organochlorine pesticides to organophosphorus compounds. The big advantage of organochlorine insecticides (e.g. DDT, dieldrin and lindane), apart from their cheapness, is that they are relatively safe to human users. Their replacements, organophosphorus pesticides such as pirimphos-ethyl, endosulfan or disulfoton, are hugely toxic to humans, and (unlike organochlorines) can poison through skin contact. Disulfoton, for example, inhibits the production of cholinesterase, an enzyme important in nerve function, lack of which causes convulsions. Product labelling is often poor and inappropriate, often not in a relevant language, and often without any provision for use by illiterate people, who are of course the majority. Farmers are therefore unable to read warnings about toxicity, and lack the knowledge to interpret dangers, or the equipment to apply pesticides safely. It is quite unrealistic to expect rules of application devised in developed countries, for example using clothing proof against sprays and protective masks, or washing before eating, to be practicable in the rural Third World. Medical support in the case of poisoning is rarely obtainable.

Ulrich Beck comments specifically on pesticides, noting the way in which technological risk is embedded within corporate decision-making:

> The 'industrial naiveté' of the rural population, which often can neither read nor write, much less afford protective clothing, provides management with un-imagined opportunities to legitimise the ways of dealing with risks that would be unthinkable in the more risk-conscious milieus of the industrial states. Management can issue strict safety instructions, knowing they will be unenforceable, and insist that they be obeyed. This way they keep their hands clean, and can shift responsibility for accidents and death to the people's cultural blindness to hazards, cheaply and in good conscience.
>
> (1992, p. 42)

Pesticides are a brilliant product of the modern techno-scientific system, long regarded by conventional development thinkers as an essential element in strategies to promote economic growth and freedom from hunger in the Third World. They also offer serious technological hazards that are inseparable from their intended function. As controls on pesticide use tighten in the industrialised world, Third World markets are increasingly attractive to producers, particularly when pesticides banned at home can be exported to countries where environmental controls are more lax. In 1979 25 per cent of the pesticides exported by US companies were either banned or unregistered in the USA (Caufield 1984).

Pesticide use can be intensive in the Third World, particularly where non-traditional crops are sold to urban markets. Guatemala, for example, is one of a number of countries that have become suppliers of out-of-season vegetables and fruit for Northern markets. The US Food and Drug Administration regularly detains shipments of fruit and vegetables from Guatemala because of excessive pesticide residues (costing the country $18 million between 1984 and 1994), but there are no equivalent controls for domestic consumers (Arbona 1998). In the highlands of western Guatemala most farmers apply pesticides in greater doses and more frequently than is recommended, they rarely purchase protective clothing available in local agrochemicals stores, and they regularly wash out containers in irrigation canals. There are about 1,200 cases of acute reaction to pesticides a year, in addition to a greater (but unrecorded) incidence of chronic low-level exposure (particularly problematic for children, and those with immune systems suppressed by malnourishment; ibid.).

Pesticides also represent a serious problem of hazardous waste disposal, or when used wholly outside their intended context. In Guyana the rodenticide thallium sulphate was imported in 1981 to kill rats in sugar cane plantations. It was misused as a fertiliser and in other ways, and contaminated milk and grain caused a number of deaths (D. MacKenzie 1987). Pesticide use rose in the Pacific, as elsewhere, in the 1970s, and there have been a number of reported cases of the death of fish from pesticide spills, for example of lindane and DDT in the lagoon of Tokelau, or the leakage of endrin and sodium arsenate into streams feeding the lagoon on Yap (Brodie and Morrison 1984).

There has been growing international concern about the spread and longevity in organisms and the environment of persistent organic pollutants (POPs), including a number of synthesised pesticides (aldrin, dieldrin, DDT, chlordane and endrin), as well as dioxins and PCBs. There is now a process led by UNEP to negotiate an international agreement banning such chemicals (McGinn 2000).

The expansion of pesticide use in the Third World is an integral part of the development process. It is sanctioned and promoted by development agencies, and financed by First World loans. The industry is run from the industrialised world, and the expansion of pesticide use in the name of development is good business. There are both ecological and economic costs to pesticide use, although these are difficult to identify, and it is clear that at certain times and in certain places these costs will outweigh the benefits of increased yields and disease control. There are also human costs, particularly from those who apply poisons unprotected and in ignorance of their toxicity, and these are not always borne by those who stand to benefit. Higher productivity may mean cheaper food for urban consumers, but it may also mean lower income and perhaps an early death for a peasant farmer. Pesticides may raise yields, but if they also raise the cost of production, and take the farmer onto a treadmill where pest resistance demands new and larger pesticide applications, their benefits may be a cruel illusion. Pesticides, like other aspects of development, are not magic. Like industrial pollution, they are just one element in the complex equation linking poverty and environmental quality.

Conventional development thinking emphasises the importance of both pesticide use in agriculture and public health, and industrialisation, as part of a strategy of development. Both carry with them a high degree of environmental hazard, as do many other aspects of conventional development paths. Beck's concept of Risk Society captures the way in which risk emerges from the 'business as usual' of conventional development. While he did not have the South in mind in describing Risk Society, it fits only too well. As many of the critics of conventional development have pointed out, 'development' is a two-edged sword, promising to hack away at the choking creepers of poverty, but at the same time bringing with it unrecognised, unregulated and often deeply hazardous change. Furthermore, the risks development creates are not distributed uniformly, but are concentrated in space and time. The logic of industry relocating in search of cheap labour and resources and freedom to externalise the costs of production is inherent to capitalism, but the profitability of relocation to the Third World is a relatively recent phenomenon (Marshall 1999). It is a function of increasing globalisation, a revolution in communications and the mobility of capital. In the second half of the twentieth century the First World began to evade the risks it conjured up with its industrial technology by moving them elsewhere. Now 'the world capitalist system thrives on passing on the costs of environmental degradation to the ecosystem people of the Third World' (Gadgil and Guha 1995, p. 122).

Risk society may be a product of late modernity but it exists in a world with more and less risky places. Where does this leave those who advocate sustainable development as an alternative to conventional development? Both mainstream and countercurrent thinkers about sustainability recognise and deplore the problem of pollution and indiscriminate pesticide use, and they would like a world with a lot less of both. They differ in their explanation of the problem, in the significance they attach to global and local inequities in hazard, and in their ideas about possible solutions. To what extent can environmental risk be taken out of the context of Risk Society, and dealt with inside the existing paradigm of development? In other words, can the problem of risk be handled *within* the mainstream of sustainable development? Alternatively, are the challenges of sustainability and risk such that a more profound or more radical approach, drawing on the countercurrents to mainstream sustainable development, is required? These issues are addressed in the next two chapters, 'Mainstreaming risk' (Chapter 11) and 'Sustainable development from below' (Chapter 12).

Summary

- Ulrich Beck's Risk Society offers a challenging analysis of modernity and environmental hazard that is directly relevant to debates about sustainable development. Risk Society is the product of change forced on modernity by global ecological crisis. Techno-scientific risk is ubiquitous, arising from processes of wealth creation, yet hidden from straightforward perception.
- Environmental risk is unevenly distributed and unequally shared. Environmental justice, environmental racism and environmental class are critical issues for sustainable development, and relevant at both the global scale (for example in the trade in hazardous wastes) and the more local scale (for example the location of polluting industrial plants).
- Industrial and urban pollution are major problems in many parts of the South. While in theory it can be argued that industrial pollution declines over time with economic growth, environmental hazards of development are both real and persistent.
- Mineral extraction and manufacturing are both forms of industrial investment where pollution is endemic, and where poor environmental and employment regulation effectively offer an incentive to pollution and hazard creation.
- The integration of hazard and potential development benefit is well demonstrated by the case of pesticides. Accidental pesticide poisoning, environmental impacts and pest resistance to biocides all represent environmental or social hazards that can balance or outweigh the planned benefits from reduced disease incidence of food losses to pests.
- The existence of environmental hazards from pesticides or industrial pollution is integral to the modern project of development through industrialisation. Sustainable development challenges the assumptions and methods of conventional developmentalism, but there are important questions as to strategy between the confusion of mainstream sustainable development and the radicalism of countercurrents to that mainstream.

Further reading

Beck, U. (1992) *Risk Society: towards a new modernity*, Sage, London.
Beck, U. (1995) *Ecological Politics in an Age of Risk*, Polity Press, Cambridge.
Conway, G.C. and Pretty, J.N. (1991) *Unwelcome Harvest: agriculture and pollution*, Earthscan, London.
Gadgil, M. and Guha, R. (1995) *Ecology and Equity: the use and abuse of nature in contemporary India*, Routledge, London.
McCormick, J. (1997) *Acid Earth*, Earthscan, London.
Murphy, R. (1994) *Rationality and Nature: a sociological inquiry into a changing relationship*, Westview Press, Boulder, CO.
Pretty, J.N. (1995) *Sustainable Agriculture: policies and practice for sustainability and self-reliance*, Earthscan, London.

Web sources

<*http://www.pan-uk.org/Reviews/ar99p6–8.htm*> Information on international campaigns against pesticides from the Pesticide Action Network UK.

<*http://www.iied.org/agri/index.html*> The International Institute for Environment and Development Sustainable Agriculture Programme, which sets pesticides in context. Documents that can be downloaded are listed at <*http://www.iied.org/pdf/index.html*>.

<*http://www.fao.org/organicag/*> The Food and Agriculture Organisation on organic agriculture. The FAO's Pesticide Management Unit is at <*http://www.fao.org/WAICENT/FAOINFO/AGRICULT/agp/agpp/Pesticid/Default.HTM*>. This site contains information on recommendations for the distribution and use of pesticides, and maximum pesticide residual levels in food.

<*http://www.bhopal.org/*> Links to the latest on the Bhopal story.

<*http://www.mosopcanada.org/index1.html*> The Movement for Survival of the Ogoni People (MOSOP), formed in 1990, an umbrella organisation for a number of Ogoni organisations.

<*http://www.worldwildlife.org/toxics/globaltoxics/index.htm*> The World Wide Fund for Nature's Global Toxics Initiative Programme, a site reflecting campaigns on toxic chemicals worldwide, including persistent organic pollutants (POPs) and agricultural pollution.

<*http://www.worldbank.org/html/fpd/mining/index.html*> A website on mining and development (e.g. 'mining and environment' and 'mining and poverty') of the International Finance Corporation (IFC), part of the World Bank Group. The website of Ok Tedi Mining Limited (which operates Ok Tedi in Papua New Guinea), <*http://www.oktedi.com/*> has interesting entries on 'people' and 'environment'; the website of Placer Dome, the fifth largest gold-mining company in the world, is interesting on the subject of sustainability and mining (<*http://www.placerdome.com/sustainability/index.asp*>).

<*http://www.bp.com*> See the thinking behind the logo: what a major oil company (arguably an energy company) has to say about issues such as climate change, cleaner fuels, energy conservation and human rights. An interesting contrasting view of the importance of these issues is presented by Exxon (<*http://www.exxon.com*>).

11 Mainstreaming risk

There is no expert on risk.

(Ulrich Beck 1992, p. 29)

Risk and sustainability

Ulrich Beck defined risk as 'the probabilities of physical harm due to given technological or other processes' (Beck 1992, p. 4). He imagined primarily techno-
scientifically induced risks – of pollution as a product of industrial processes for
example. As Chapter 10 demonstrated, such risks extend to the South; indeed, in
many ways they are attracted there, and proliferate relatively unregarded and
unchecked. Risk in this sense is distributed geographically and socially. While
global in scope, such risk offers a significant threat to the Third World poor.

However, this is but a part of the risk that challenges sustainable livelihoods
in the South. Environmental change and degradation is also a significant source
of risk (see Chapters 7 and 9), as is development planning itself (see Chapter
8). The exploitation of resources such as forests, just as the conservation of
resources such as wildlife, involves the distribution of costs and benefits, risks
and rewards (Chapter 9). Risk is not only the product of late modern techno-
industrialism, but a fruit of modernity. Conventional 'developmentalism' (Ekins
1992), discussed in Chapter 6, is a prominent fruit of that modernity – an
attempt to engage rationality in the service of human betterment (Murphy
1994, Cowen and Shenton 1995, 1996).

In the context of the South, therefore, the risks of modernity have a somewhat different complexion from those in Beck's account of Risk Society. His
notion of risk reflects the concerns of First World environmentalism, particularly concerns about health, pollution and technology (e.g. Hays 1987).
Southern environmentalism is broader, including issues of livelihoods security
and poverty (a conventional 'development' agenda), rather than focusing on
a discrete category of 'the environment' (Gadgil and Guha 1995, Chapman
et al. 1997). To Beck's hazards of techno-industrialism must be added others,
more mundane perhaps, but just as dangerous, intractable and alarming.

Development projects and programmes themselves are very often a response to
certain of these forms of risk, particularly of scarcity of environmental goods (e.g.

responses to drought and famine; see Chapters 7 and 9), perceived degradation (e.g. 'desertification'; see Chapter 7), shortage of power, water or food (e.g. dams and irrigation; see Chapter 8), or poverty more generally. Development planning has also itself been a major cause of risk, creating its own chains of environmental and socio-economic impacts (see Chapter 8). Such impacts were a major stimulus to ideas of sustainable development (see Chapter 2). Indeed, mainstream sustainable development (see Chapter 5) represents the dominant political strategy for tackling the risks of modernity that Ulrich Beck described.

The idea of risk is fundamental to mainstream sustainable development. Ecological Modernisation involves a commitment to the adoption of cleaner technologies, and a shift from 'end-of-pipe' solutions to problems of industrial waste towards upstream strategies that address industrial processes, product design and research and development (e.g. Hajer 1996, Christoff 1996; see Chapter 5). There is a large scientific literature on risk assessment, attempting to adapt techniques devised for particular technological or industrial processes (dam or nuclear reactor safety, for example) for wider application. However, as Wynne (1992) demonstrates, there are critical distinctions to be drawn between different kinds of lack of knowledge or certainty about the outcomes of human action. He distinguishes four: first, risk (where the chances of something happening are known); second, uncertainty (where the chances are not known, at least in detail); third, ignorance (where it is not even known what the problems are, let alone the chances of their happening); and fourth, indeterminacy (where the outcomes are inherently not predictable). Wynne's argument is that techniques of 'risk assessment' alone will not be able to address all dimensions of risk. That task needs a far more open engagement with social and cultural values. Nonetheless, it is precisely at this technical level that mainstream sustainable development *does* seek to address the risks of technology and industrialisation, and of formal development projects and programmes. What can mainstream sustainable development offer as a challenge to the risks engendered by conventional development?

In 1995 Beck suggested that there might be two phases of Risk Society. In the first, the risks of global self-destruction are present, but the social response draws on established strategies of industrial society (scientific, legal, economic and political). This is a 'technocratic' variant of risk society, placing reliance on experts, on the bureaucracy and regulation of risk control. The second phase develops as Risk Society comes into existence and is acknowledged, once 'ecological hazards are recognised as forms of social self-jeopardisation, and are cut back and contained by institutional innovations' (Beck 1992, pp. 67–8). Risk Society challenges industrialism, the social hegemony of science, and the legal system. Beck asks what good is a legal system 'that prosecutes technically manageable small risks, but legalises large-scale hazards on the strength of its authority, foisting them on everyone, including even those multitudes who resist them?' (1995, p. 69).

Mainstream sustainable development lies within the first of these phases, the attempt to control risk and steer economic change without transforming patterns of social or economic organisation. Critical in this project is the assessment of the environmental impact of human action. In this chapter, attention is focused on the techniques for considering environmental and social impacts in planning, and particularly the political economy of what Beck rather nicely calls 'the bureaucracy of knowledge' (1992, p. 54).

The assessment of environmental and social impacts

Environmental impact assessment (EIA) is the standard technique for inserting environmental concerns into project and programme planning (O'Riordan 1976a, Munn 1979, Wathern 1988, Barrow 1997). Its origins were described briefly in Chapter 2. Barrow suggests that EIA should be treated as 'a generic term for a process that seeks to blend administration, planning, analysis and public involvement in assessment prior to the taking of a decision' (1997, p. 1), thus embracing approaches such as social impact assessment (SIA), discussed on p. 324. EIA is frequently much less than this – a short technical study carried out as an afterthought to project design to satisfy some bureaucratic criterion laid down in the vague hope of promoting sustainability. My purpose here is not to describe how to carry out an EIA (or some broader environmental assessment, EA), but to explore the limitations of such processes in eradicating environmental degradation and environmental risk.

One obvious problem with EA methods concerns quantification. Even at the stage of assessing the severity of environmental impacts, the science of ecological survey starts to give way to informed value judgement of the analyst. These problems are compounded at later stages of the assessment in the interpretation of impacts, although there are a series of methodologies for determining weights or scores in a quasi-independent manner (Barrow 1997). Quantified evaluations, like those of cost–benefit analysis (CBA), may be criticised because they make evaluation the preserve of the 'expert', remote from public comprehension and accountability. Whether cloaked in quantification or not, such procedures are essentially qualitative, and therefore highly dependent on the skills, prejudices and perceptions of the analyst. Quantified methods may appear to confer enhanced legitimacy, particularly among a scientific audience, and may also make the task of making decisions on the basis of the EA easier by effectively prejudging the issue. They are therefore attractive both to scientists and consultants wanting to present a clear case, and to politicians faced with difficult decisions. At the same time, however, they place excessive reliance on the skills of 'experts' often far removed from appointed arenas of decision-making. Quantified methods can also make the communication of results harder, particularly to affected groups (Lee 1983). O'Riordan (1976b) argues that there is a tendency to use EA procedures to collate these unquantifiable factors together in a systematic way, but without integrating them into the cost/benefit calculation. Inasmuch as EA procedures satisfy environmental concern without allowing that concern to be reflected in adapted project design, they are counter-productive: a waste of resources and a negation of their integrative and potentially transforming role.

Clearly, EAs are only as good as the policy frameworks within which they are carried out. The sequencing of tasks and the nature of the players among whom the EA will be created, assessed and acted on are particularly important (Munn 1979, Barrow 1997). Procedures work through the definition of national development goals and the establishment of policy and programme activities before particular projects are considered. The next step is to determine whether particular developments will in fact have a significant impact, before formal EA procedures begin. The tasks do not end until the EA has been reviewed by the competent body, the project implemented and a post-project audit carried out. Barrow (1997) identifies an ideal pattern of project planning as a 'helix', with environmental assessment taking place at both programme and project levels, its results being fed into ongoing cycles of planning (Figure 11.1).

It need hardly be said that many of these steps are sensitive to critical decisions and pressures within and on bureaucratic systems, and also that in many cases some or all of these procedures are skipped. It is, for example, one thing to have an EA commissioned and carried out, but quite another to integrate it into decision-making. The size of EA documents can be out of all proportion to the capacity of agencies to digest data and reach informed conclusions. Indeed, the pursuit of gigantism in an EA is one strategy open to the promoter of a major project who is hopeful of drowning objectors in a flood of indigestible data. In the USA, for example, early EA documents were too long, their language was too technical and complex, and they failed to summarise issues vital to decision-making (Lee 1983).

Environmental assessment in the Third World

The importance of carrying out assessments of both the long- and short-term environmental impacts of projects in the Third World is clear (Lohani and Thanh 1980). The EA methodologies open to investigators are broadly similar to those in developed countries (Ahmed and Sammy 1985); indeed, USAID projects in the Third World should (in theory) be subject to the same EA procedures as federal actions in the domestic arena (Horberry 1988). However, the constraints on the effectiveness of EA procedures in developed countries are greatly heightened in the Third World. A series of factors make Third World EA a great deal more problematic.

One problem affecting Third World EA relates generally to the nature of environmental impacts themselves. In practice it is often far from easy to identify the nature of environmental impacts at a particular point in space and time, let alone predict them. Significant impacts are often remote geographically from a project, for example the impacts of logging or dam construction on sedimentation or river degradation downstream. They may therefore be beyond the boundary of the specific development project, and therefore be unperceived by project developers. Such problems are compounded if, as in the case of downstream floodplain wetland environments, the place where the impacts

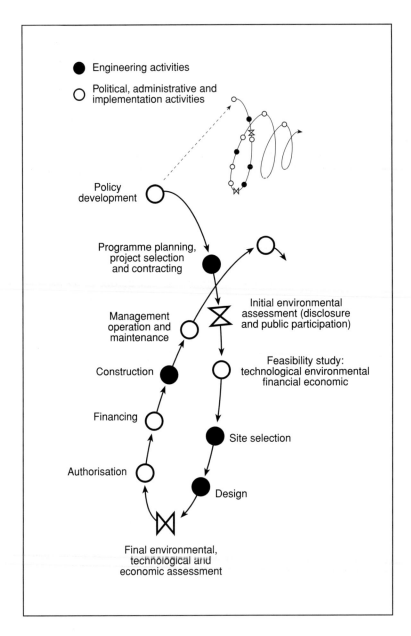

Figure 11.1 Environmental assessment and the project planning helix (after Barrow 1997).

occur is physically remote from centres of planning and decision-making, difficult to access and of marginal importance politically (see Chapter 5).

Environmental impacts can also be delayed, occurring some time after project development, and can involve complex sets of knock-on impacts, for example ecological change in a floodplain following hydrological and geomorphological change. Indeed, relatively few geomorphological or ecological processes have instant responses, and most relate to each other in dynamic and complex ways. As was described in Chapter 8, the possible impacts of upstream dams on the fertility of floodplain land downstream or on patterns of erosion and deposition down to the delta are obviously lagged in this way. Environmental impacts can also be increased by other developments or natural changes that have synergistic effects on ecosystems. A good example here is the impact of drought on the discharge of dammed rivers in the Sahel (Hollis *et al.* 1994).

There can also be problems knowing where to set the boundary for assessment of environmental impacts. The problem of boundaries has been discussed in the context of the problem of trade-offs in the assessment of sustainability in Chapter 5. The spatial and temporal scope chosen, and the range of development initiatives considered, can have significant effects on the outcome of an EA, as indeed on other aspects of project appraisal. For example, the controversial Sardar Sarovar Dam on the Narmada River in India made technical sense only if it were built to operate with the Narmada Sagar Dam upstream (to control inflows) and the smaller Omkareshwar and Maheshwar dams (Rich 1994). Without Narmada Sagar, the hydro-power generation of Sardar Sarovar would be reduced by 25 per cent, and irrigation by 30 per cent. However, Narmada Sagar was expensive, would flood a large area of tropical forest, and would force resettlement of an additional 100,000 people. The World Bank decided to appraise Sardar Sarovar as a discrete project and did not take account of its links with the other schemes (ibid.). The project's full costs were therefore not taken into consideration. Dam projects are particularly vulnerable to artificially narrow assessment in this way; for example, assessment of the viability of the Bakolori Dam and irrigation scheme in Nigeria specifically excluded any consideration of the considerable downstream impacts (Adams 1985a), and assessment of the Pangue Dam on the Bíobío River in Chile in the early 1990s ignored its dependence on other dam projects (Usher 1997b).

The technical challenge of assessing the nature and extent of environmental impacts is exacerbated by the enormous data-hunger of EA procedures. EA requires a great deal of environmental, and particularly ecological, data. Assessment of impacts demands large data sets and long time-series, and such resources are unusual. The problem is simply another version of that which faces most forms of planning in the Third World, the task of 'planning without facts' (Stolper 1966). The problem is not simply the lack of data (although this is real enough), but a deeper ignorance. In development economics 'all too often it is quite unclear precisely what question should be asked; sometimes the question is asked wrongly, and it is by no means certain that answers always exist' (ibid., p. 9).

There are of course various technical tricks to extend short data sets, for example the work of hydrologists who correlate a short run of river discharge figures with a longer run of rainfall data and generate a synthesised data set offering a long time-series. Such methods have been devised for temperate environments, and assumptions about the variability and lack of secular change in conditions valid there may be misleading and even dangerous in the tropics. For example, the enormous spatial and temporal variability of rainfall in areas such as the Sahel within and between years is now becoming clear (Hulme 1987). There are severe risks in extrapolating from short data sets to derive longer-term averages, particularly if they relate to periods such as the 1960s (which were unusually wet compared to the long-term record; Hulme 1996).

The lack of environmental data in the Third World is matched by a lack of ecologists and other scientists. Impact assessment also demands a high degree of knowledge of the dynamics of ecosystems, and an ability to make realistic predictions about their response to stresses of various kinds. Long-term ecological monitoring is unusual, and experienced ecologists are few in the Third World. Indeed, there are few textbooks on tropical ecology directed at Third World audiences, although there are some creditable exceptions such as that by Deshmukh (1986). Ecologists or other environmental scientists with extensive knowledge and expertise exist, and their numbers are increasing, but in practice EAs do suffer from the lack of indigenous skills.

The established response to the lack of indigenous expertise is to bring in outside 'experts' in the form of consultants. The importance of such outside influences on planning and development is considerable (Chambers 1983). There may be many reasons for calling outside experts: the department in the host country may be empire-building; the civil servant may want a 'fall-guy', more time on his farm or access to foreign travel; or the experts may be the only way to obtain resources in the form of aid, coming as part of a tied aid package. Whatever their justification, foreign consultants and planners are ubiquitous in the Third World. Information and the power to act flow outwards and downwards from the rich, centralised modern core of the world's developed countries to the periphery of the Third World, and the employment of expatriate consultants and technical companies is a significant feature of this flow. The insights to be derived from what Chambers (1983) calls 'rural development tourism' are limited.

Foreign experts may command a legitimacy denied prophets in their own country derived from their presentation of technology and experience, but this expertise may be less valuable than it at first appears. Thus Winid (1981) argued that one of the reasons why developments in the Awash Valley of Ethiopia have not been more successful was the lack of knowledge on the part of the French consultants of the political, ecological, social, cultural, scientific and technical aspects of the region. This in turn was related to the fact that those in the field had contact only with the local elite, lacked supervision and failed to transfer technology. There are therefore serious pitfalls in a planning system built on the backs of foreign experts. Foreign companies offering to carry out impact assessments without suitable in-house ecologists may well fail to produce an integrated understanding of environmental impacts. Even if temperate-zone ecologists are brought in as consultants, they may yet prove a poor (and costly) substitute for local expertise, perhaps glibly repeating dominant preconceptions and myths about tropical ecosystems (for example about overgrazing or desertification) which have misdirected development so often in the past (Leach and Mearns 1996).

The lack of ecological expertise matters more because of the disciplinary bias inherent in the planning process (Chambers 1983). Project appraisal is dominated by the 'hard' technical disciplines such as engineering, hydrology and agronomy, and by the most technocratic of the social sciences, economics. Most project feasibility studies are undertaken by companies dominated by these disciplines, or by consortia typically dominated by engineering companies. These are not, however, the disciplines relevant to EA. The 'soft' disciplines such as ecology, geography, anthropology or sociology central to EA (and vital to social cost–benefit analysis) tend to be marginal to the planning process. Increasingly they have a place in project planning, but it is a small one. Sociology and ecology are typically slotted in with perhaps a one- or two-person-month input on a project where the total planning input is several person-years. Both environmental studies of development projects and investigations of social impacts are often begun too late. Hamnet (1970) laments the way in which as a project sociologist he could not initiate ideas, but could simply react to problems created by technical proposals. Jaundiced practitioners sometimes see sociology as the task of selling unattractive projects to unwilling participants, and in the same way an EA can seem to become part of the process of justifying a project rather than a genuine means of appraising it. EAs based on whistle-stop tours by visiting experts are unlikely to be of great value.

Ecologists and other experts from 'soft' disciplines are rarely employed on the full-time payroll of engineering companies. They are rarely, if ever, project managers. Furthermore, those who do lead appraisal teams usually lack the knowledge and experience necessary to know how to feed environmental data effectively into the planning process. It is clear that environmentally sound development demands that environmental factors be taken into account early in planning, for EA is useful only if it influences the nature, speed and extent of development. This demands teams strong in relevant skills; however, project appraisal is still often done by teams rich in only a limited range of technical skills, and they give a low priority to environmental expertise. Furthermore, such teams are rarely if ever led by those from 'soft' disciplines. Those who do lead them rarely even have the knowledge and experience to know what those disciplines require and can offer. They do not appreciate how EAs need to be integrated into the planning process.

EA procedures can be costly, particularly if taken seriously. In the 1970s, EA under US legislation in California for example absorbed roughly 1 per cent of project costs, and more if there were delays over decisions (Sandbach 1980). EA costs are sunk before it is known whether the project is technically, environmentally and economically viable, and will not be recouped if the project does not go ahead. Appraisal investigations, whether they involve EA or not, demand work in a number of technical disciplines, and frequently inputs from costly overseas consultants who need to be paid in foreign exchange, often financed by loans that become part of national debt. Furthermore, completion of an EA can therefore hold back the project appraisal process, and delay the moment when economic benefits can be realised. Tight project development schedules can prevent the adoption of effective EA procedures. EA therefore not only raises the cost of sunk investment in project appraisal, but postpones the commencement of streams of possible benefits. There are obviously strong economic and financial incentives to truncate or speed up EA procedures, or even omit them altogether.

Environmental assessment in project appraisal

The limited account taken of environmental impacts in project planning reflects not simply the many problems and limitations of environmental assessment in the South, but also the low priority attached to the environmental compared to economic assessment. (Similarly, in developed countries, EA is too often seen as a way of paying lip-service to environmental problems without solving them (O'Riordan 1976b).) Inevitably, Third World EA is something grafted in various often loose ways onto social cost–benefit analysis (Abel and Stocking 1981), or is seen as a discrete and often superfluous element within the project planning process.

It is CBA that is the established procedure by which economic decisions about project development can be made. Environmental sustainability was not recognised as an issue when the standard manuals of Third World project appraisal, using CBA, were written (Van Pelt *et al.* 1990). CBA techniques are widely known and reviewed (e.g. Sugden and Williams 1978, Barrow 1997). CBA is systematic, based on principles that are widely comprehended (people's choices as revealed by surveys and markets), and it produces a neat result that can feed directly into the planning process. Debate about the shortcomings of CBA is considerable, focusing particularly on the problem of determining social values, and the impact on public participation of the procedure's technical complexity (Barrow 1997). Nonetheless, CBA remains a central element in mainstream sustainable development. Arguably it 'meets more of the requirements of the evaluation process and more nearly satisfies the other criteria than does any other evaluation method' (Abelson 1979, p. 57).

CBA can be used to assess projects, programmes or policies. It is a decision tool that compares options in terms of present and future economic costs and benefits. There are many manuals of project assessment, setting out methods for CBA and warning against undue haste and narrow thinking (Bridger and Winpenny 1987, Brent 1990, OECD 1995). Key issues are the need to compare with- and without-project benefits and costs (doing nothing can sometimes prove a unexpectedly beneficial strategy), and the need to consider the opportunity costs of investments (that is, what the same investment could yield if spent on something different). A critical problem is time, since both benefits and costs tend to change as proposed projects develop and mature. Some projects can have high initial costs and slow payback (for example pollution control technologies), or delayed payback (for example afforestation). Time is conventionally dealt with by discounting future costs and benefits to calculate net present value. The 'discount rate' used is a critical figure, since it tends to cause projects to be favoured that yield benefits early and demand gradual investment. Environmental projects are often the reverse – expensive at the outset and yielding benefits slowly – and CBA tends to make them look unattractive. Furthermore, typical discount rates tend to give any cost or benefit that occurs more than thirty years in the future such a low value as to make it effectively irrelevant to the analysis. Thirty years may be a long time to a politician or a government planner, but it is little more than one human generation, less than a tenth of the lifespan of many trees, and an eye-blink in terms of evolution (Maser 1990). The discount rates conventionally used in CBA obviously raise significant issues as a basis for the discussion of sustainability.

There are also serious questions about the extent to which it is possible to express social and environmental values in monetary terms. Intrinsic values of nature, and cultural and religious values, are particularly problematic. However, monetary valuation of non-human nature can be done, and is being done, as a contribution to project and programme appraisal, using a variety of techniques (Chapter 5), although many of these remain somewhat experimental and the whole field is the subject of intense debate (Gouldner and Kennedy 1997).

For obvious reasons, environmentalists and human rights groups tend to dislike and distrust CBA. However, if applied carefully, it has several useful things to offer. First, CBA provides a valuable discipline for gung-ho development planners. It can halt environmentally and socially destructive development that lacks even simple economic justification. Second, it can also provide a written record of the calculations that justified a project that is, at least in theory, available for subsequent scrutiny (without which any mistakes cannot be understood). Third, CBA should ensure that all externalities of a project are taken into account, including environmental and social impacts. Of course, the possibility of doing this is constrained by the problems of discount rates and the quantification of unquantifiable values discussed above; nonetheless, the structured approach of a CBA has some value. A clear distinction is drawn between economic and financial rates of return. Financial analysis measures the money costs and returns on a project (as a private corporation would do); economic analysis tries to measure the real or resource costs to the economy, which should include externalities (for example health or environmental impacts), secondary and tertiary impacts, and questions of subsidy. Taking these into account, of course, makes project appraisal both harder and more expensive than it would otherwise be, but in theory at least, a careful CBA will consider these things (or its failure to do so should be clear from the record).

Enthusiasts for CBA (mostly economists) argue that environmentalists reject the technique too intemperately, and fail to see its potential. They argue that opponents of CBA criticise the technique of appraisal, when what is really at fault is the planning or political process that wields it. In this, CBA and EA techniques share the same weakness, as will be discussed in the next section.

Why projects fail

Shortcomings exist with EA procedures in most contexts. In the Third World in particular, they are greatly exacerbated by the complex political milieu within which project appraisal takes place. Political pressures on national politicians combine with the sometimes counter-intuitive functioning of bureaucracies and the commercial interests of companies to create an arena of decision-making where the technical merits of a project are just one factor among many. In theory, development projects undergo a strict and technically sophisticated appraisal procedure from the project identification, pre-feasibility and feasibility studies, and design, before implementation begins.

This multi-stage process should allow all problems to be ironed out, successive consultants obtaining contracts on the basis of experience and competitive bidding, and checking their predecessors' findings. In practice, the project appraisal procedure is far from perfect. First, badly framed terms of reference can constrain the range of options the consultant is prepared to investigate, perhaps meaning that viable alternative schemes are not considered. Thus, if asked to investigate the potential for large-scale irrigation in a river basin, a consultant will get small thanks for a study investigating the benefits of small-scale alternatives. Since success in bidding for the next job depends on contacts made during this, there are strong reasons for not rocking the boat by presenting unpalatable appraisals. Second, competition has the effect of paring bids for jobs to a minimum, with the result that 'extras' get cut out. Such extras can often include EAs, or at least the resources necessary to make them effective. Third, client organisations in the host country can steer the consultant in various ways, for example by demanding changes in a report at draft stage, or simply suppressing it. Professional and commercial considerations may push in very different directions.

Pressures of this kind distort the technical processes of appraisal. Perhaps for this reason, projects often achieve a self-sustaining life of their own that carries them through the supposed hurdles of appraisal unchanged. Pitt suggests that, once created, plans 'achieve their own reality and ritual inviolability' (1976, p. 41). The Volta River Project was first conceived of when Sir Albert Kitson, Director of the Gold Coast Geological Survey, travelled down the Volta (David Hart 1980). He saw a possible dam site, and having discovered bauxite the previous year, he conceived the idea of building a dam for hydroelectric power generation to produce aluminium. He made proposals to this effect in the 1920s, but they were not taken up. The idea was brought up again in the 1930s by a South African engineer who formed the African Aluminium Syndicate, and proposed a dam 40 metres high. The idea was pursued in various quarters through the 1940s, with increasing interest from international aluminium interests, before the decision by the colonial government to commission a survey of the basin in the 1950s, and the eventual decision to build the project (under very different terms) in the 1960s (ibid.). Many other development projects share this kind of history, living as a blueprint in a limbo of experts' minds in the depths of planning bureaucracies until awakened to re-emerge in the light of a more auspicious dawn.

Even without this kind of analysis, it is clear that the interests of the state can be far from those assumed by the economic basis of technical appraisal. Keith Hart points out that a small economic revenue which is predictable may be highly attractive to the state, and even a project with a sharply negative cost/benefit ratio may be attractive if it is 'a rural manifestation of the state's active presence' (Hart 1982, p. 89). Certainly, strategic factors can be important influences on development policy, for example in the case of Brazilian settlement in Amazonia, and the construction of an irrigation scheme at Bura on the Tana River in north-east Kenya (Adams 1992). More prosaically, perhaps, there is a political justification for settlement projects of this kind that is quite separate from the question of their economic performance.

Corruption is a significant factor in decision-making by Third World bureaucracies, although it is hard to trace and substantiate (Usher 1997c). Wellings (1983) describes what he calls 'capital leakage' in the public accounts of Lesotho of such a scale that the report of the Auditor-General for 1975–8 was suppressed. Wade (1982) has made it clear that corruption in canal irrigation in India is not some kind of curious aberration, but an integral part of the structure of decision-making and performance. The same may well be true in other contexts. Hart comments:

> Today, international construction firms and development bureaucracies vie for shares of what usually turn out to be very lucrative projects for all concerned (including government officials) – lucrative that is for everyone save the farmers whose work on the land can never possibly generate a surplus large enough to pay for all these overheads.
>
> (1982, p. 94)

Large projects mean large pay-offs, and it is clear that in some circumstances this can be an important influence on decision-making. Furthermore, commercial competition between consultants, and commercial independence between consultants appraising projects and contractors building them, may be more apparent than real (Usher 1997c). Commercial pressures also influence aid donor decision-makers, sensitive to the need for domestic companies to win contracts on aid projects: the dam construction and turbine industry of Norway and Sweden, for example, is a major influence on Nordic donor decisions (Usher 1997c).

The process of conception, design and approval of major development projects is in practice highly complex, bureaucratic and often political. The increased visibility of environmental aspects of project development in the 1980s and 1990s means that it is also frequently controversial. Usher (1997b) describes the anatomy of decision-making in the approval of Swedish and Norwegian aid to allow the Nordic multinational Kvaerner to supply turbines for the Pangue Dam on the Bíobío River in Chile in the early 1990s. This was a complex case, not least because it was being built by a private company (the privatised electricity utility ENDESA), and under OECD rules not eligible for aid, and because although Pangue was one of an integrated suite of dams for the Bíobío River, its environmental impacts were assessed as if the other dams, with which it formed an integrated package, were not being planned. Reports were prepared under tight deadlines, and decision-making by the Swedish and Norwegian governments, and other donors, was highly political (Usher 1997b).

Chile itself developed an EA process following the end of dictatorship, but it is inadequate, presuming the possibility of social compensation and environmental mitigation (and hence making the rejection of a project almost impossible), and dependent on excessively close relationships between consultants and the dam construction industry (Silva 1997). In the case of Pangue, the EA did not consider downstream impacts and water release patterns of Pangue, and failed to provide a full picture of Pangue and Ralco dams on the environment of the Bíobío valley (Silva 1997). Usher and Ryder (1997) argue that such failures on environmental assessment are characteristic of dam projects. The impact of the Theun Hinboun Dam in Laos (on a tributary of the Mekong) on subsistence fisheries and the food security of floodplain communities was systematically ignored or underplayed in 'expert' environmental impact assessments. In the case of the Pak Mun Dam on the Mekong, an environmental assessment was carried out, but not released. When opponents saw a leaked copy in the USA in the early 1990s they attacked it fiercely for its inaccuracy. The shortcomings of the environmental assessment of this dam (which among other problems lies within the Koeng Tana National Park) were a significant factor in criticisms of the World Bank's environmental record (Rich 1994). Usher and Ryder conclude that 'the failure to identify such crucial issues raises questions about both the experts and about the aid agencies that take these claims at face value' (1997, p. 99). Usher concludes that the process for reviewing the environmental impacts of Third World dams is 'rigged' (Usher 1997c, p. 59).

The extent to which the EA of a development project identifies critical risks, weaknesses and impacts is obviously open to a wide range of commercial and political influences (the latter reflecting both international geopolitics and a myriad of political forces at smaller spatial scales and within state, parastatal and private-sector organisations). However, there are obviously also technical problems that can be addressed. First, the lack of data needs to be tackled, and this requires long-term monitoring and baseline research. Much of this can be done by national institutions, although pump-priming and support funding may be necessary in the form of foreign aid to support education, professional training and sharing of expertise between institutions, particularly internationally within the Third World. There can also obviously be a series of progressive reforms in the way EAs are fitted into the project planning process: they need to be better resourced, begun sooner and given time for completion, and to be truly interdisciplinary. They also need to be led by people with appropriate skills. Perhaps this will only be achieved when it becomes routine for the management of appraisal projects to be done by experts from disciplines other than economics and engineering. Another approach is to cease to try to put together multidisciplinary teams by simply aggregating diverse skills, but to seek to train people in such a way that they acquire multi-disciplinary skills and experience.

Above all, perhaps, there is a need for greater openness and for a willing-ness to learn from mistakes. No one in development is willing to be proved wrong. Chambers suggests that intelligence is too often defined as 'avoiding being demonstrably wrong about anything' (1977, p. 411). Pork-barrel poli-tics is the norm and not the exception in much of the Third World, particularly for heads of state. Both politicians and bureaucrats share a fear of association with failure. Retrospective studies of environmental impacts are vital (Ashby 1980), not simply as an end in themselves, but as a step towards impact management. EA procedures should be seen as the start of a process and not the end-point. Few development projects evolve without substantial shifts of design, extent, or speed of implementation; indeed, the failure of appraisal processes generally to cope with the unpredictable elements in development is a major problem.

Barnett (1980) suggests that development should not be seen as a kind of black box process, where known inputs create entirely predictable outputs, but as a kind of Pandora's box. Investment produces change, but it is not possible to predict the extent and direction of that change. In other words, development is a probabilistic process, full of uncertainties and full of risks. If development projects are not to fail, they must be allowed to evolve as circumstances change. Environmental appraisals need to evolve with them as they change. Development is a continuing process, and development planning is a form of permanent crisis and risk management. Both development planning and environmental appraisal need to aim for the same flexible and interactive relationship with change.

Reforming impact assessment

Mainstream ideas about sustainability have penetrated conventional develop-ment planning disciplines in various ways. One is the incorporation of environmental concerns into national economic planning in the form of taxa-tion and monetary policy (e.g. Munasinghe 1993b), new approaches to national accounting and the measurement of welfare (e.g. Daly and Cobb 1990), and integrated environmental and economic accounting (Bartlemus 1994). These were reviewed in Chapter 5.

A second way in which ideas from mainstream sustainable development have influenced development planning is more directly through initiatives to broaden and strengthen techniques of project appraisal (Van Pelt *et al.* 1990). There are alternatives to CBA as a form of project appraisal (OECD 1995). These include risk–benefit analysis, multicriteria analysis and decision analysis. Risk–benefit analysis effectively inverts CBA by looking at the risk of not taking action, for example the risk of pollution if investment is not made in safety measures. Multicriteria analysis includes criteria in addition to quantifiable economic rate of return, for example costs per beneficiary, or the number or characteristics of beneficiaries. Decision analysis drops the assumption in CBA that decision-makers are risk-neutral, and seeks to assess their preferences, judgements and trade-offs. None of these will necessarily lead to development projects with smaller environmental and social impacts, but they do represent attempts to diversify and strengthen the range of tools available.

Two other approaches address sustainability more explicitly, strategic environmental assessment (SEA), and social impact assessment. SEA pushes the assessment of impacts upstream from the project, to the programme and policy arenas. Arguably, unless the environment is taken fully into account at the strategic level of planning, it will not be surprising if individual projects generate unforeseen and unwelcome impacts. This notion is not novel – it was, for example, a key recommendation of the *World Conservation Strategy* in 1980 (Chapter 3) – but it remains important, because it is still not standard practice.

Social impact assessment (SIA) is a term that implies a significant broadening of conventional EA processes (Vanclay and Bronstein 1995). SIA is a methodology that seeks to assess and manage planned intervention, although at another level it is also a recognition of the need for sustainability in development and a commitment to achieving it (Vanclay 1999). The aim of SIA is to achieve better outcomes from development in terms of both ecological and social sustainability. This is likely to involve a participatory approach (working with communities and building their own capacities to plan), a recognition of the connections between social and biophysical impacts, and an explicit consideration of second- and later-order impacts (ibid.).

Unlike EIA, SIA seeks to address not only social impacts arising from biophysical impacts, but also those generated through social change that can both follow from and be created by planned development interventions. Thus people forced to resettle as a result of dam construction may be affected by a complex chain of social and environmental impacts associated with loss of resources, changed disease incidence, conditions in a resettlement location, and social change resulting from translocation (loss of social cohesion or loss of respect for elders, loss of traditions or religious belief). SIA attempts to recognise and consider all of these. It could be used as part of a state's regulatory process and address a single project, but it could also be used by communities or other actors as a means not only to appraise, but also to steer development. In that sense, SIA can be seen as a philosophy or paradigm rather than simply a methodology (Vanclay 1999).

The ability of improved project and programme appraisal methodologies to contribute to sustainability depends crucially on institutional capacity. A key issue here is governance and regulation in Southern countries themselves, and their capacity to understand and solve environmental problems. This capacity has been increasing, particularly since the Rio Conference in 1992, with the support of the GEF and the imposition of 'environmental conditionality', for example the World Bank's insistence on National Environmental Action Plans (NEAPs). By 1982 over 100 Third World countries had established procedures for assessing environmental impacts. This trend was bolstered by the Rio Conference, and such provisions had become almost universal by the mid-1990s (Barrow 1997), although in many countries capacity to implement them still remained weak. In the absence of effective governance and environmental regulation, the effectiveness with which proper account is taken of sustainability depends to a considerable extent on the capacity of the aid donor organisations that fund so much development, as discussed in the next section.

Aid agencies and environmental policy

The environmental policies of aid agencies have a disproportionate importance in the Third World because much of the Third World is dependent on aid donations to finance major projects. Aid agencies are particularly important sources of funding for the investigative stages of development planning, for example project definition and appraisal. Aid agencies often pick up those aspects of development projects without a direct contribution to economic return, among which the assessment of environmental impacts is notable. The attitude of those First World institutions therefore has a considerable impact on the nature of EA in development planning. Furthermore, aid is often tied to the involvement of First World consultancy companies, and the aid agencies are thus common routes through which 'experts' are channelled into development project planning. The performance of both multilateral and bilateral aid agencies began to receive attention in the 1970s and 1980s (Stein and Johnson 1979, Johnson and Blake 1980, Kennedy 1988, Horberry 1988). Since that time there has been growing pressure by environmental groups for reform of aid agency environmental policy, particularly in the USA, aimed at both USAID itself and the World Bank, whose income is dominated by the size of the US contribution.

The need for tight procedures and clear guidelines for aid projects is great. Hughes (1983) reviewed the role of environmental assessment (or the lack of it) in the development of projects on the Tana River in Kenya. She argued that although it was known from an early stage that a series of hydroelectric schemes and an irrigation scheme would have potentially serious environmental effects on downstream floodplain environments, funding agencies (including the World Bank, the EC and the UK's Overseas Development Administration (ODA) consistently failed to carry out the studies necessary to determine their significance. ODA did insist that a proposed study by UNEP be carried out.

This study suggested that the construction of the Masinga Dam could go ahead, but it recommended that a series of further, more detailed studies should be a condition of loan, and that an environmental monitoring and control group should be set up. After confrontation between donors and the Kenyan government this idea was scrapped (ibid.).

The other critical issue relating to development in the Tana basin, of the impact of fuelwood demand from the irrigation scheme on riparian woodland, was raised in successive project reports, but the necessary detailed research into impacts was not funded (Hughes 1983). At no stage was a formal EA carried out on developments in the Tana. Hughes was left wondering if the current concern for EA on the part of aid donors is a smokescreen, and felt that in practice 'little pressure is put on governments receiving aid to carry out high-quality environmental studies and monitoring' (ibid., p. 181).

The World Bank (or more properly the International Bank for Reconstruction and Development and the International Development Association, which are separate arms of the Bretton Woods family of organisations) is the largest multi-lateral aid donor. The sheer size of its lending makes its environmental policy of considerable importance, and of consuming interest to environmentalists (Fox and Brown 1998b). Bank lending policy has undergone a number of shifts, the most notable being the move in the 1960s away from exclusive concentration on large-scale capital-intensive infrastructural projects to projects designed to aid the poor (Rich 1994). The addition of environmental considerations to the cocktail of poverty-related concerns was potentially problematic (Stein and Johnson 1979). In the 1970s most institutions lacked clear procedures and criteria for EIA, and have had to develop new forms of accountancy or analysis to bring longer-term environmental effects of development projects into consideration in project appraisal. Rich speaks of the 'fatal hubris about the Bank's ability to know, plan and direct the evolution of human societies and the natural systems they depend on' (1994, p. 105).

At the centre of the World Bank's evaluation procedures in the 1970s was the Office of Environmental and Health Affairs (OEHA). In practice this had little influence on project design. It was represented on few World Bank project identification missions, and mission reports suffered from the constraint of 'paying almost unique attention to financial and economic considerations', rarely discussing environmental issues in enough detail (Stein and Johnson 1979, p. 17). The OEHA received the project brief, and thus advanced notice of the general nature of a project. Project preparation was in theory the recipient country's job, but in practice it was often assisted (or ghosted) by Bank staff. The adequacy of consideration of environmental matters in the preparation phase depended on the experience of project planners, but it could be cursory, particularly when done by less aware regional office staff. While in theory project appraisal should have included environmental aspects of the project, in practice environmental data were less clear-cut and operationally more workable, and few projects were rejected on environmental grounds alone. Although projects were seen by the OEHA before going to the Loan Committee of vice-presidents of the Bank and then the executive directors, and in theory projects could be sent back for reappraisal, by the approval stage it was effectively impossible to redesign projects, and difficult to build in environmental safeguards. The pressure to keep projects on schedule is intense, and delays most unwelcome.

The key to the actual level of consideration of environmental aspects of project development in any aid bureaucracy (or consultancy company) is the attitudes and perception of particular individuals within the normal planning structure. There is often a wide gap between 'the increasingly alert concern of individuals and the official response of most institutions' (Stein and Johnson 1979, p. 133). Bureaucracies such as aid agencies are extremely difficult to reform from within. In the 1970s and early 1980s this was a problem for the Bank, despite its surprising record of employing staff popularly identified with the environmentalist cause such as Robert Goodland and the economist Herman Daly (Holden 1988).

There is a danger that such 'mavericks' are 'good for the image but mouse-sized in impact' (Watson 1986, p. 275). As an ecologist working for the Bank, Watson found her ideas rejected as unrealistic and impractical every time they contained implications for practice. Environmental advisers with NORAD (the Norwegian aid agency) who were critical of dam projects on environmental grounds found their advice ignored. Usher suggests that environmental and social issues are obstructions to the process within the aid bureaucracy to 'push development projects through the "pipeline"' (1997c, p. 69). Watson's experience was similar: she commented, 'by and large we were viewed by project and technical staff as an unessential office which at best put useless icing on the cake and at worst could halt or slow projects' (1986, p. 269).

However, the World Bank has moved some way down the road of 'greening' its corporate banking mentality. In 1984 it adopted its first official statement on environmental aspects of its work, an Operational Manual Statement (OMS; World Bank 1984c). With the exception of a specific mention of environmental audits of completed projects, this effectively set out the status quo on environmental procedures through the project cycle. The Bank at this time treated projects individually, in the context of local setting. The OMS drew attention to guidelines to be followed in Bank operations unless the borrowing country's standards were more strict (ibid., p. 3). These principles include for example the provision that the Bank would not finance projects that 'cause severe or irreversible environmental degradation', would affect the environment of neighbouring countries, or that would 'significantly modify' biosphere reserves, national parks or other protected areas (ibid., p. 4). It was also stated that 'project designers, consulting firms and Bank project officers are expected to provide effective and thorough environmental input into project design construction and operation' (Goodland 1984, p. 13). With the World Environment Centre, the Bank organised a series of seminars for major consulting firms in the early 1980s to raise awareness of the need for proper environmental appraisal of projects (Goodland 1990).

Environmental groups maintained their critical pressure on the Bank, particularly US environmental NGOs, especially the National Wildlife Federation, the Sierra Club and the Environmental Defence Fund (e.g. Goldsmith 1987, Horowitz 1987, Fox and Brown 1998a). The US Congress held oversight hearings on the multilateral development banks in 1983, hearing in particular severe criticism of the World Bank's funding of Brazil's Polonoroeste Programme (Northwest Region Development programme), and the construction of Highway 364 into the heart of the remote rainforest region (Rich 1994). The barrage of environmental criticism had some effect. In 1987 the Bank created an Environment Department with forty new staff, and new scientific and technical staff in regional offices (Holden 1987). Behind these organisational changes lay some softening of the rigid doctrines of Bank economics, with the appointment to the Latin American office of the zero-growth economist Herman Daly (Holden 1988). New systems of national accounting were developed to reflect the depletion of non-renewable resources and unsustainable exploitation of renewable resources, and to revise the discount rate which biased appraisals in favour of projects with short-term pay-offs against longer-term cost/benefit considerations (ibid.).

The Bank also produced a series of policy papers in the 1980s addressing issues central to sustainability, including an overall environmental policy statement and papers on involuntary resettlement, wildlands conservation, pollution control and pesticides, and tribal people. These codified existing best practice, and had already existed in draft for several years. Such policy statements were valuable, but they did not change corporate culture. Goodland noted that 'their full implementation by the Bank's disparate staff of 6000, with many urgent priorities, cannot be achieved overnight, only through time' (1990, p. 151).

By the early 1990s, environmental critique of the Bank had developed from specific issues (often concerning rainforests, roads or dams) into a broader attempt to make the Bank accountable to civil societies in donor and borrowing countries (Fox and Brown 1998b). The Bank's need for donor government contributions to the International Development Association every three years provided a cyclic window of opportunity for NGO pressure. The Bank instituted a range of reforms in the early 1990s, strengthening its policies on EA, involuntary resettlement and indigenous peoples. However, practice was publicly shown to be behind policy, particularly in the case of the Sardar Sarovar Dam on the Narmada River in India. The Bank's board of directors set up a review of this project in 1991 under Bradford Morse, former director of UNDP (Fox 1998). The Morse Commission concluded in 1992 that the Bank had flouted its own environmental and resettlement policies, and recommended that it 'step back' from the project, which it eventually did (to no avail, since construction continued despite a storm of protest; see Chapter 13). The Bank also cancelled a proposed loan for Nepal's Arun II Dam (Usher 1997c). A review of resettlement in Bank projects followed directly from the Morse Report, recommending changes already proposed internally by Michael Cernea (Fox 1998). In 1993 a new water resources management policy was finally approved, including the stipulation that 'environmental protection and mitigation' would be integral parts of a comprehensive approach to water development (Moore and Sklar 1998).

The World Bank transformed its policies on the environment and sustainability in the 1990s, even creating a vice-presidency of environmentally sustainable development, and leading the development of economic ways to factor the environment into economic thinking (e.g. Munasinge 1993b). Nonetheless, the Bank's practice still falls short of its promises of reform (Fox and Brown 1998b).

Other multilateral and bilateral donors have not received the same level of environmental criticism as the World Bank, but most have felt the same pressures for reform. Most have adopted environmental appraisal procedures of some kind. The Asian Development Bank established an environmental unit in the early 1980s, and the UK's Department for International Development (DFID, formerly the ODA) produced a manual of environmental appraisal in 1989 (Barrow 1997). The 1997 UK White Paper on international development explicitly addressed the challenge of sustainable development, committing the DFID to the promotion of 'sustainable livelihoods' and protection and improvement of the 'natural and physical environment' (Carney 1988a).

The World Bank and other multilateral and bilateral aid donors changed their tune on sustainability in the 1980s and 1990s. To a variable (but mostly much lesser) extent they have also changed their practices. The degree of this change is critical, because of the enormous influence of donor organisations on planning in many Southern countries. However, donor organisations are large bureaucracies: hierarchical, autonomous, populated by people jealously aware of their professional skills and their established procedures. How can such organisations change?

Fox and Brown (1998b) distinguish between institutional adaptation and learning. If external political pressure is sufficiently strong, changes in organisational behaviour (adaptation) may take place without learning (changes in the way problems are perceived and explained). At the same time, learning can take place without adaptation, if staff lack the power to change organisational behaviour. Organisational learning that threatens dominant paradigms (for example the hegemony of neoclassical economics) is likely to provoke resistance. However, over time, learning will take place, as a result of recruitment and staff training, the penetration of new ideas (for example environmental and ecological economics) and changing internal institutional incentives. External pressure can also lead to the release into positions of influence of innovators who *have* learned. It is an interesting question, therefore, whether aid donor 'greening' is the fruit of internal learning or simply adaptation to external pressure from NGOs. So far, Fox and Brown (1998b) believe that World Bank staff have done more adapting than learning.

Learning is not something confined to the bureaucracies of aid organisations and governments. The evolution of thinking within the professional disciplines involved in development planning is also very important. New approaches to environmental and social planning can form the basis for new disciplines, as

SIA does (Vanclay 1999). The infiltration of environmental (and even ecological) economics into the economics faculties of less conservative universities represents a similar evolution (and in the case of ecological economics a potential revolution) within disciplinary boundaries.

A critical discipline in terms of sustainability is undoubtedly engineering, and here there has certainly been considerable institutional learning and reform 'from within', for example in the work of the International Commission on Large Dams (1981) and the development of EIA (see Chapter 2). This engagement with environmental problems, which began with the Conservation Foundation's conference on 'the ecological aspects of international development' in Virginia in 1968 and the resulting publications *The Careless Technology* (Farvar and Milton 1973) and *Ecological Principles for Economic Development* (Dasmann *et al.* 1973), has continued. In 1995, for example, the FAO published *Environmental Impact Assessment of Irrigation and Drainage Projects*, setting out a methodology to predict environmental impacts and provide an opportunity to mitigate those that are negative and enhance any that are positive. EA should become 'as familiar and important as economic analysis in project evaluation' (Dougherty and Hall 1995, p. 1), and should 'facilitate sustainable development' (ibid., p. 2).

The International Commission on Irrigation and Drainage (ICID), the NGO linking relevant irrigation, flood control and drainage professionals, established an international working group on environmental aspects of irrigation, drainage and flood control projects in 1986, and in 1993 it published an environmental checklist to provide a simple approach to environmental assessment. Its aim was to educate specialists and non-specialists concerned with irrigation and drainage about environmental change, to allow assessments to take place where it was impossible to assemble a broad, multidisciplinary EA team, to clarify the tasks that can be undertaken by specialists and non-specialists, and to bring expertise together in a practical form (Mock and Bolton 1993). The range of impacts included was wide (hydrology, pollution, soils, sediments, ecology, society/economy, health), and the strength of interlinkages between them was also specified. The checklist offered not just a textbook but a methodology to assist irrigation and drainage planners (the majority of whom are engineers) to acquire the skills necessary to start avoiding adverse environmental impacts. As the ICID checklist notes,

> a growing sensitivity to environmental effects will be needed if the increasing manipulation of the natural environment for irrigated food production, land reclamation and flood protection, for the benefit of an expanding human population, is to be sustained into the future.
>
> (Mock and Bolton 1993, p. 1)

This is a pragmatic approach in a field dominated by civil engineers to the task of rebuilding development planning procedures to take account of ideas of sustainability, from within the system.

The most extensive engagement between water resource development professionals and their environmental and social critics was through the work of the World Commission on Dams (WCD) between 1998 and 2000. The 1990s saw increasingly vocal and coherent protest against dam construction (for example in the 1994 Manibeli Declaration, calling for a moratorium on all World Bank funding for large dams, and the Bank's own independent review of the Sardar Sarovar Dam in 1992, the Morse Report (World Commission on Dams 2000; see also Chapter 13)). A workshop organised in Switzerland in April 1997 by the IUCN and the World Bank brought together a wide range of parties, including governments, funding organisations, engineering companies and protest groups, to debate the benefits and costs of dam construction.

The WCD eventually began work in May 1998, first to review the development effectiveness of large dams and assess alternatives for water resources and energy development, and second to develop internationally accepted criteria, guidelines and standards for the planning, design, appraisal, construction, operation, monitoring and decommissioning of dams (WCD 2000). It took evidence for two years, holding four regional consultations and taking 900 submissions from individuals and organisations, commissioning seventeen 'thematic reviews' and eight detailed case studies of individual dams, as well as carrying out a cross-check survey of dams in fifty-two countries.

The WCD's report, launched in London in November 2000, is clearly expressed and ambitious. It offers a clear and new basis for planning water and energy resources that embraces participatory decision-making and an explicit engagement with both rights and risks; it emphasises the centrality of social and environmental issues in dam planning (WCD 2000). The WCD report is mainstream sustainable development at its best: carefully professional, yet challenging to the status quo, building on best practice and yet pushing (and pushing hard) at the outside of the envelope of normal planning practice. For those living upstream or downstream of dams, and those caring about riverine ecosystems, it is undoubtedly good news. It remains to be seen whether it will prove effective in achieving reform of planning in practice worldwide, and whether it can survive the inevitable rearguard action of those who imagine their interests threatened by its recommendations.

Neither a general intention to improve environmental appraisal nor a superficial awareness that environmental impacts may occur is enough to guarantee that development will become sustainable, or that aid agencies will take the steps necessary to control the impacts of the developments they fund. Large bureaucracies are inherently conservative, and the 'greening' of development is bizarre theoretically in the context of the established disciplines of development planning, troublesome in terms of policy, and highly inconvenient administratively. Reforming the practice of development bureaucracies has to go a great deal deeper than the superficial transformation of rhetoric and terminology. Arguably, sustainability can only to a limited extent be driven 'from above', through reform of planning procedures. The potential for promoting sustainable development in a different way, from below, is the subject of the next chapter.

Summary

- Mainstream sustainable development is an approach to tackling the risks of modernity. In the South these risks extend beyond the realm of techno-scientific risk. Development is itself a source of risk where it generates significant adverse social and environmental impacts. These are the fruit of the same late modernity that creates the techno-scientific risks.
- The assessment of social and environmental impacts has become an important element in project appraisal, involving specific techniques such as environmental impact assessment. In the Third World the environmental assessment of projects faces numerous problems relating to the complexity of environmental responses to human action (for example timescales, spatial boundaries), the difficulty of knowing about those impacts (for example lack of data or researchers), the nature of the organisations carrying out assessments (for example their dependence on foreign expertise), haste and cost.
- One reason for the poor account taken of environmental and social issues in project planning is the entrenched power of economic considerations, particularly in cost–benefit analysis (CBA). Some failings of CBA relate to the planning process rather than the limitations technique as such.
- In practice, project planning is often rushed, incomplete, and exposed to the problems of poor governance, corruption and the prioritisation of political considerations over those of local welfare or environment.
- Broader project appraisal methodologies have been developed, including strategic environmental assessment (SEA) and social impact assessment (SIA). SIA addresses all social impacts, including direct impacts, those resulting from environmental impacts, and the chain of impacts resulting from them.
- Aid donor organisations have a particular role in transforming the process of project appraisal. Most have 'greened' to some extent. The World Bank came under sustained pressure from environmental groups, and in the early 1990s adopted a series of new policies, for example on the environment, indigenous people and resettlement. The process of institutional change, in both aid organisations and the academic disciplines from which their staff are recruited, is slow.

Further reading

Barrow, C.J. (1997) *Environmental and Social Impact Assessment: an introduction*, Arnold, London.
Barrow, C.J. (2000) *Social Impact Assessment: an introduction*, Arnold, London.
Beck, U. (1992) *Risk Society: towards a new modernity*, Sage, London.
Fox, J.A. and Brown, L.D. (eds) (1998) *The Struggle for Accountability: the World Bank, NGOs and grassroots movements*, MIT Press, Cambridge, MA.
Petts, J. (ed.) (1999) *Handbook of Environmental Impact Assessment*, Blackwell Scientific, Oxford.

Rich, B. (1994) *Mortgaging the Earth: the World Bank, environmental impoverishment and the crisis of development*, Earthscan, London.

Usher, A.D. (ed.) (1997) *Dams as Aid: a political anatomy of Nordic development thinking*, Routledge, London.

Vanclay, F. and Bronstein, D.A. (eds) (1995) *Environmental and Social Impact Assessment*, Wiley, Chichester.

Wathern, P. (ed.) (1988) *Environmental Impact Assessment: theory and practice*, Unwin Hyman, London.

World Commission on Dams (2000) *Dams and Development: a new framework for decision-making*, Earthscan, London.

Web sources

<*http://www.worldbank.org/html/extdr/projects.htm*>   This site gives information on the World Bank's operations and policies, for example on resettlement and indigenous people.

<*http://www.iaia.org/*>   The International Association for Impact Assessment exists to advance innovation and development and communication of best practice in impact assessment. The site describes activities and meetings, and lists key references on impact assessment.

<*http://www.wbcsd.ch/*>   The World Business Council for Sustainable Development, a coalition of 140 international companies formed in January 1995 through a merger between the Business Council for Sustainable Development (BCSD) in Geneva and the World Industry Council for the Environment, in Paris. The WBCSD claims to be 'the pre-eminent business voice on sustainable development'.

<*http://www.dams.org*>   The World Commission on Dams offers some carefully considered approaches to the planning and implementation and monitoring of controversial large projects such as dams.

12 Sustainable development from below

Surely, if decades of failed international development efforts have taught anything, it is the folly of induced, uniform, top-down projects. Such schemes ignore and often destroy the local knowledge and social organisation on which sound stewardship of ecosystems as well as equitable economic development depend.

(Rich 1994, p. 273)

Development from below

The conventional model of development, development 'from above', became increasingly battered in the last two decades of the twentieth century. The presumption of a monolithic value system and a uniform basis for human happiness (which 'automatically or by policy intervention will spread over the entire world'; Stöhr 1981, p. 41) began to be quite widely challenged in development studies. The debates about sustainable development reviewed in this book were an integral element in the academic and policy reassessment of conventional thinking.

Part of that critique concerned the question of *how* development was to be done, as much as about *what* was done. From the 1970s, the failure of high-technology and large-scale development projects which were imposed on rural communities by outside agents began to be contrasted with the merits of alternative technologies, small-scale projects developed by or with local communities, in versions of the vision that 'small is beautiful' (Chapter 5; cf. Schumacher 1973). It began to be believed that small-scale projects could minimise adverse environmental impacts, maximise economic benefits, reduce waste on vast management bureaucracies and make best use of the energies and talents of participants.

Criticism of individual projects and development outcomes grew into (and in a cycle of criticism and affirmation fed upon) a profound shift in the dominant discourses of development during the 1970s. 'Top-down', 'technocratic', 'blueprint' approaches to development came under increasing scrutiny as they failed to deliver the economic growth and social benefits that had been promised (Turner and Hulme 1997). An alternative agenda emerged, associated in particular with the work of Robert Chambers (1983, 1988b, 1997). It began to be widely argued that development goals could only be achieved by 'bottom-up planning', 'decentralisation', and 'participation' and 'community development' (Agrawal and Gibson 1999). By the early 1990s, aid donors and development planners were heavily committed to participatory approaches.

This 'development from below' demanded a reversal of conventional development thinking, working from the 'bottom up' and the 'periphery inwards' (Stöhr 1981, p. 39). It was an approach, not a package; an idea, not a set of rules. It suggested that for success, developments must be not only innovative and research based, but locally conceived and initiated, flexible, participatory and based on a clear understanding of local economics and politics. There was an alternative to large-scale centralised development, one 'characterised by small-scale activities, improved technology, local control of resources, widespread economic and social participation and environmental conservation' (Ghai and Vivian 1992b, p. 15). Development had to begin to 'put people first' (Cernea 1991).

As was discussed in Chapter 5, this neo-populist vision of development was integral to mainstream sustainable development thinking, although it was by no means confined to it. The idea that development should come 'from below', from the community and not the state, in the Third World reflected a curious interlocking of opposing political ideologies in Europe and North America in the 1980s. On the one hand, political acceptance of neo-classical economic analysis in the 1980s led to a view of development as something driven by the market, not the state: that the market alone delivers development. This had a profound influence on ideas about national development strategies (Toye 1993). It was argued that to achieve public policy goals (sustainable development, for example) the economic incentives for all of the main actors must be set correctly, and that that was best done by the market mechanism. Right-wing political thinkers also argued that the state had become bloated and inefficient, and that the state (and other non-market actors) tended to distort markets and impede efficient delivery of economic growth. The dominant political view was that state power had become too great and too centralised, generating a tidal wave of privatisation of state-run services. The notion that 'the community' should take an active part in development previously handled 'from above' by the state was therefore reasonably palatable to neo-classical economists in the World Bank and other international institutions, and to industrial donor governments.

On the other hand, the notion of empowering people at the 'grassroots' was also attractive to political thinkers on the left and to communities themselves, threatened by the globalisation of businesses and the social costs of rapid economic restructuring. New social movements of various kinds had become a political force, both in the former state socialist countries of Eastern Europe and in the West (notable among them, of course, the environmental movement), and governments faced challenges to their policies (including those of privatisation and market dominance), and demands for new openness and improved local democracy.

Participatory development was therefore the hybrid fruit of two contrasting sets of ideas. The first sought to expose more of public life to the discipline of the market. This meant a reduced role for the state and created spaces for 'communities' (villagers, private individuals, companies, groups of companies) to be more involved in development. These ideas about market, state and civil society formed the basis of a 'New Policy Agenda' for foreign assistance developed in the USA in the early 1990s (M. Robinson 1993, Moore 1993). The second sought to move power down from the state to more local levels, and emphasised the capacity of communities to organise themselves to manage development. This emphasised an enhanced role for civil society and democracy.

As might be expected from its hybrid political roots, the notion that sustainable development could and should be pursued 'from below' is common to both mainstream and countercurrent thinking. To those in the mainstream, it provides the counterbalance to ecological modernisation and environmental-economic managerialism, a release of energy and enterprise at the grassroots to drive the green economy towards sustainability. To countercurrent thinkers, it represents the possibility of challenge to the dual hegemony of the bureaucratic state and corporate power, a solidarity of citizens engaging with each other and with non-human nature for mutual benefit. This chapter does not rehearse the diversity of either mainstream or countercurrent thinking about sustainable development (for this see Chapters 5 and 6), but it is worth noting that despite the eclecticism of environmental ideas, the notion that sustainable development should be sought in whole or in part through social action 'from below' is a common thread from anarchism or ecofeminism through to the economistic hardheads of the World Trade Organisation. Interestingly, Ulrich Beck (1992) argued that the second phase in the development of Risk Society that would challenge the hegemony of technology and science would emerge from contact between citizens and communities, not through the action of the state regulation of market enterprise.

This chapter will discuss the potential for the delivery of sustainability 'from below', through community action and through grassroots transformation of thinking among industrialists and planners. Two things need to be noted. First, while the chapter discusses a range of examples (conservation, rainforests, industry and river basin planning), it does not seek to be encyclopedic. Second, while in a sense these case studies are examples of good practice, and are intended to provide an element of hope after the rather depressing litany of failure and underperformance in development discussed elsewhere in this book, this chapter should not be read as a recipe book of 'solutions'.

Indeed, the whole neo-populist project of 'participatory' development or 'development from below', which has so commended itself to development thinkers in the North, is itself problematic (Kitching 1982). It invites naïve, simplistic and idealistic analyses of society, social engagement with nature, and the political economy of development. The idea of 'the community' as a source of legitimacy and a means of achieving effective and lasting developmental change touches on deeply wired Western romantic notions of communities as 'natural' organic social entities. These draw heavily on European works of fiction (for example Tolkien's 'Middle Earth' and other fantastic versions of a rural European past), and were important elements in the opposition to modernity (industrialisation, urbanisation, pollution, specialisation) in Northern environmentalism (Hays 1987, Veldman 1994). When reflected through lenses of paternalistic colonialism or idealistic post-colonial guilt, these ideas make it seem self-evident that rural people in the Third World live together in discrete 'villages', share common (and fixed) 'tribal' identities, and are committed to each other by co-residence, kinship and shared poverty in a way that people in the urbanised, industrialised, 'developed' West have lost. This romanticism about 'community', and the associated vagueness about political conflict at local level (be it district, village or household), is characteristic of many of the documents of the sustainable development mainstream. This slightly odd romantic heritage by no means invalidates ideas about 'development from below', although it may explain why local people may have different ideas about the desirability of 'community action' from those held by development workers.

It is also worth noting that there is nothing magical in sprinkling the word 'participation' across the development process. White (1996) calls participation a 'Hurrah' word, whose use 'brings a warm glow to its users and hearers', but she points out that this very quality prevents its detailed examination, and masks the fact that participation can take many forms and serve many different interests (ibid., p. 7). Participation is a highly political process, both within 'the community' and in the relations of the community to other agents (not least the eager development worker). However 'participatory' a development projects seeks to be, White argues that it cannot escape the grip of wider power relations within society. Development 'from below' provides no escape from hard distributional questions.

Development planning and indigenous knowledge

The rise of calls for 'development from below' reflects the recognition of the value of indigenous knowledge, both to the livelihood security of many rural people in the Third World and as an important source of innovation and tested solutions to environment and development challenges. Indigenous knowledge has gone from being denigrated and dismissed by development planners to being recognised as a vital knowledge resource. Just as government and development agency bureaucrats and their 'expert' consultants have failed to take adequate account of the environmental impacts of their actions, they have also frequently been blind to the needs and capacities of those people they seek to 'develop'.

In contrast, it is now commonplace in some quarters to celebrate the skills and understanding of peasant farmers in the Third World. Such praise forms one of the buttresses of environmentalist critiques of the practice of Third World rural development, and a plank of sustainable development thought in works such as the *World Conservation Strategy*. Enthusiastic celebration of indigenous skills can form the foundation of suggested alternative strategies of development (e.g. Harrison 1987). Huijsman and Savenije argue that the basis for building strong community-based environmental management systems and decision-making structures is to 'respect and make use of native wisdom and indigenous knowledge and experience, and to accept local decision making' (1991, p. 25); local specificity is a vital component of such planning.

This new orthodoxy of peasant rationality and skill is of relatively recent date. Formerly, attitudes to peasant production tended to be less generous and less benign, as well as less perceptive. As was described in Chapter 2, when colonial administrators came to tropical Africa, for example, they mostly failed to see order or skill in rural production systems. Practices such as mixed cropping or intercropping presented an image of confusion and poor husbandry, and the cautious risk avoidance strategies of peasant farmers were dismissed as the result of a stultified conservatism. There were, however, commentators with greater vision. In West Africa, for example, Jones (1936) celebrated the diversity of native farming, Faulkner and Mackie commented that farmers would generally 'obtain a maximum of return for a minimum of labour' (Faulkner and Mackie 1933, p. 6), and Stamp (1938) praised the soil conservation practices of Nigerian farmers. In time, studies of strategies such as shifting cultivation began to demonstrate their logic, and the high degree of environmental knowledge of practitioners such as Allan's 'African husbandman' (1965). It began to be appreciated that Western science might not have the monopoly of answers: in describing the shifting cultivation of the Zande in Sudan, de Schlippe wrote, 'the teacher of a culture is its environment, and agriculture is its classroom' (1956, p. xiv).

Ideas of this kind re-emerged in the 1970s as part of a liberal and populist reaction against the unsuccessful technological triumphalism of rural development practice, for example in Belshaw's call to 'take indigenous technology seriously' (1974), and in the book *Indigenous Systems of Knowledge and Development* (Brokensha *et al.* 1980). In the West African context of Sierra Leone, Richards (1985, 1986) developed the compelling notion of 'indigenous agricultural revolution', and drew a sharp contrast between the high degree of ecological adaptation in Mende swamp rice production systems and the grim comedy of repeated attempts by the colonial and post-colonial developers to transform them. Like other swamp and floodplain farmers in West Africa, the swamp rice farmers of central Sierra Leone have knowledge of a range of rice varieties with different ecological requirements, and a cultivating environment of considerable variability in small stream-fed inland swamps. Upland and swamp farming are integrated to optimise labour use. Farmers show considerable skill in selecting rice varieties suited to particular flooding conditions, and are also highly innovative. Faced with new seed varieties, they will deliberately experiment with them by planting them alongside varieties of known requirements and performance, planting across the down-slope catena to assess performance in different conditions, and carrying out input–output trials (Richards 1985, 1986).

The skills and resourcefulness of these farmers of Sierra Leone are shared widely by others elsewhere in the Third World. Thus, for example, Barker and

Spence (1988) describe the 'rich and functional environmental knowledge' of Maroon farmers in Jamaica which underpins their 'food forests'. These are multi-tier crop complexes involving tree crops (such as coconut), shrubs (coffee or cocoa) and smaller plants, sometimes of up to fifty species. There has been similar interest in the skills of agricultural management in rainforest areas, both indigenous and of more recent origin (e.g. Denevan *et al.* 1984, Gliessman 1984, Rambo 1982).

Scientific research has tended to vindicate those who recognised the skill of indigenous cultivators. When agronomists and soil scientists finally carried out scientific research on indigenous farming systems, numerous advantages over monoculture in tropical environments were identified. Jodha (1980) showed that the complexity and diversity of traditional dryland farming systems enabled farmers to meet multiple objectives simultaneously. Therefore, rather than trying to identify alternatives to traditional polycultural systems, scientists should concentrate on developing more and better options for the farmer, and leaving the choice between them up to him or her. Crops such as millet and sorghum, once dismissed by agricultural extension workers, are amazingly well adapted to low rainfall, short growing seasons, and desiccation during growth. They are the linchpin of sustainable agriculture in the Sahel (Kowal and Kassam 1978, Mortimore and Adams 1999). Studies of comparative farming systems (e.g. Turner and Brush 1987) now focus attention on indigenous techniques and their adaptation to environment.

The root of colonial attitudes to indigenous farming practice was a confidence in the superiority of Western civilisation, and particularly its science. The cause of the failure of many innovations in development was the inadequacy of that science to elucidate tropical environments. De Schlippe argued that 'if the question is asked why modern civilisation has failed to improve African agriculture, the answer must be that it has proved so far incompatible with the environment of the wet tropics' (1956, p. xv). Western science has been just part of the wider impact of Western planning. Haugerud commented, 'practitioners of positivist, empiricist science are the new missionaries who would convert developing countries to Western bureaucratic norms and values' (1986, p. 4). The trauma of development lies in the alien nature of those norms and values, inherent in the modernist attributes of Western notions of development. The impact of such imposed ideas extends far beyond agricultural change, just as they are carried over from the colonial period to the work of the modern independent state.

Indigenous knowledge has become a recognised element in development planning, acknowledged both for its cost-effectiveness (in the narrow sense of helping development planners make desired outcomes happen), and also for the wider set of values and ideas it represents (a reminder of the integrity and moral value of cultural practices perhaps very different from those of a globalised, modernising, monetised and urbanised world economy).

There are, however, no easy boundaries between 'indigenous' and 'modern' knowledge, the latter being open to continuous appropriation and incorporation into existing environmental management practices. The changes created by formal 'development' over the second half of the twentieth century have been profound and in many cases novel, and might seem well labelled as 'modern', as contrasted with an unchanging 'traditional' past. However, the 'modern' innovations of the twentieth century have followed innumerable others. Rural societies have not waited, locked immobile in some unchanging cultural time-warp, for the onset of the siege engines of modernity: in the past, as now, they have innovated. Weiskel (1988) describes how a wide range of domestic plants were introduced to the Ivory Coast in West Africa from the fifteenth century onwards, a backwash from the burgeoning trade in slaves: from the New World came cocoa, cassava, groundnut, tomato, maize, sweet potato, cocoyam, pineapple, papaya, avocado, hot peppers, tobacco and New World cottons, while from Asia came Asian rice, taro, sweet banana, sugar cane, citrus fruits and mango. The critical point here is that by the 1890s, when colonial rule began in the Ivory Coast, these crops 'became so fully integrated into African agricultural systems that virtually all of them were referred to as "traditional" crops by Europeans' (ibid., p. 162). Not only new crops, but composting techniques, bicycles, veterinary drugs or even (conceivably) the Internet can be 'indigenised' in this way.

It is therefore a mistake to speak of 'indigenous' knowledge as something unmodern, and there are real dangers in an over-romantic view of local knowledge. Bebbington (1996) criticises the proposal that agricultural programmes should build primarily, or only, on farmers' own existing techniques and innovations. Outsiders may have a commitment to 'native' or 'traditional' agro-ecological techniques, but this enthusiasm is often not shared by indigenous peoples' organisations' (ibid., p. 87). Instead, he finds local organisations in the Andes in Ecuador pursuing agrarian development not by standing against modernisation, but by reforming, managing and adapting it. Specifically, Indian federations in the Andes incorporate 'green revolution' technologies in programmes that seek to promote development while reinforcing Indian culture and society. Thus Bebbington argues that 'what gives a strategy its alternative, indigenous orientation is not its *content* (i.e. that it uses indigenous technologies, etc.) but rather its *goal* (i.e. that it aims to increase local control of social change)' (ibid., p. 88; emphasis in the original).

Indigenous knowledge is important for development planners, and was for too long unrecognised. However, its importance lies less in the technical superiority of existing over new ideas (although this may hold true), than in issues of ownership of ideas and control of change. Theorists need to be less idealistic about the distinction between 'indigenous' and 'modern', and more open to the importance of the political process whereby different technologies and ideas may be used by local people to build sustainable livelihoods, and to secure acceptable paths to the future. One way of approaching this is to use the notion of 'primary environmental care', and the idea of promoting integrated natural resource management by 'building on local skills, local resources, local forms of co-operation, flexible planning and participation' (Pretty and Guijt

1992, p. 36). Ghai and Vivian comment that 'sustainable development requires that local communities enjoy genuine autonomy, have control over adequate resources, and, in some cases, that they be provided with financial and technical assistance to restore their resource base and re-establish their control over resources' (1992b, p. 19).

Community conservation

The ideology of development through participation has been particularly influential in the field of biodiversity conservation. The dominant approach to conservation for most of the twentieth century, in both First and Third World countries, was the exclusion of people from 'natural' places, the ideology of 'fortress conservation' already discussed in Chapter 9. Ideas about the place of people in conservation strategies changed rapidly in the 1980s and 1990s, with emphasis being placed on the needs of local people (Western and Wright 1994). Indeed, the involvement of local people in conservation has become a major feature of conservation policy in many countries. Debates at the Third and Fourth World Congresses on National Parks and Protected Areas in Bali in 1982 and Caracas in 1992 mark this change (McNeely and Miller 1984, McNeely 1993). At these meetings the problem of adverse impacts of parks on local people was widely recognised and discussed, not least by delegates from indigenous groups who came and spoke. A much higher priority was accorded to the need both to integrate parks into economic development planning frameworks and to prioritise the interests (and recognise the rights) of local people (McNeely 1993, Kemf 1993). There is now a strong consensus that local people must be involved in management decisions about protected areas, and that they should benefit from them economically. The support of local people is seen to be essential to their long-term survival and integrity (Hannah 1992). Furthermore, it is argued that the old confrontational approach between park managers and local people needs to change, and park managers needed to start to address the needs of local communities by providing services such as education and healthcare, and by allowing local people to participate in park management and allowing consumptive and non-consumptive resource use (for example hunting and gathering, agriculture, religious practices, pastoralism; Western and Wright 1994, Hulme and Murphree 1999).

Such a change in emphasis demands considerable ideological change on the part of conservation management professionals, restructuring of conservation organisations and effective responses to the obvious institutional weaknesses of both protected area systems and the government wildlife administrations that operate them. Legislation is often weak or confusing, and conservation or protected area policies are frequently divorced from policy development elsewhere in government (for example in agriculture, forestry, river basin planning or tourism). Conservation staff are often undertrained and badly paid, and government bureaucracies are frequently immobilised for lack of operating expenses or means of transport. More seriously, it is asking a lot of game guards and wardens trained in a military/policing role, armed and uniformed, to begin to see people they have been harrying (however unsuccessfully) for their law-breaking poaching or other activities as 'partners'. The reassignment of rangers from anti-poaching patrols to be 'community rangers' can be problematic (particularly if they still sometimes bear weapons and join anti-poaching patrols). The result can be highly confusing for both staff and local people. Of course, institutional weakness can be tackled by training and targeted external support, and institutional overlap and inertia through innovative approaches to planning. However, the challenges are considerable.

The slogan of the Fourth World Congress on National Parks, held in 1992 in Caracas, Venezuela, was 'Parks for Life'. It emphasised the idea of expanding partnerships for conservation, and McNeely (1996) offers some suggestions as to how this might be done (Table 12.1). Among these are a series of points that emphasise directly the importance now given to local communities around protected areas. First, he argues that protected areas must provide direct benefits to local people, that therefore its benefit/cost ratio to local people (before, or as well as, nationally) must be positive if it is to prosper, and that those people must be involved in its planning. Second, he suggests that protected areas must be planned and managed in such a way as to meet local needs as well as biodiversity conservation goals. Third, protected areas must be planned in a way that is integrated with surrounding human uses. Fourth, while protected areas must be planned as a system that addresses both national and international objectives, they must also (fifth) be managed individually in collaboration with local people (Table 12.1; McNeely 1996).

McNeely argues that while national parks are as important as ever, and should be as carefully protected as ever, they must be supplemented by other kinds of protected areas 'to meet the social and economic development needs of modern society' (1984, p. 1). In fact, the IUCN Commission on National Parks and Protected Areas recognises eight different categories of protected area, ranging from the scientific reserves or strict nature reserves through the national parks and protected landscapes to resource reserves and multiple-use

Table 12.1 Ten principles for successful partnerships between protected-area managers and local people

1	Provide benefits to local people.
2	Meet local needs.
3	Plan holistically.
4	Plan protected areas as a system.
5	Plan site management individually, with linkages to the system.
6	Define objectives for management.
7	Manage adaptively.
8	Foster scientific research.
9	Form networks of supporting institutions.
10	Build public support.

Source: McNeely (1996)

management areas. This last category provides for sustained production of natural resources (for example water, timber, wildlife, pasture, fish or outdoor recreation). This concept of creating reserves to allow for sustainable production was integral to the biosphere reserves introduced by the Man and the Biosphere (MAB) Programme in the 1970s (see Chapter 2). The notion of biosphere reserves was discussed early in MAB's development in 1969. Criteria for selection and objectives were set out by the Expert Panel in 1973 (UNESCO 1973). Biosphere reserves were to be located in representative areas, and to be of large enough size 'to accommodate different uses without conflict', including ecological research (Batisse 1982, p. 102).

The importance of taking the needs, ideas and aspirations of local people seriously in conservation planning was for too long perhaps unrecognised by conservationists, but is now part of the language of conservation planning, and methods such as participatory rural appraisal (PRA; Nichols 1991, Chambers 1994) have now been integrated into the professional conservationist's portfolio. Caldecott, for example, emphasises the need to ensure that 'all parts of a project are both socially and environmentally durable', and that reliance on local participation can help ensure that projects 'put down mental, social, financial and institutional "roots" among the people of the project area' (1996, p. 261). 'Co-management' is now an accepted concept in national parks in North America and Australia, and is spreading (e.g. Hill and Press 1994).

In some cases, a community's cultural values and practices can provide a solid basis for an effective conservation programme (Kleymeyer 1994), as in the case of institutions for coastal-zone management in Indonesia (Zerner 1994). However, other commentators warn against excessive optimism about the supposed 'new paradigm' of community conservation, pointing out that 'the troubling question of whether communities actually can resolve resource conflicts and slow environmental degradation better than centralised authority remains' (Western and Wright, 1994). Furthermore, invoking 'the community' as a concept may well not provide a solution to conservation problems, particularly if the terms of local people's involvement in decisions about land and resources are set by outsiders. As Murphree (1994) comments, 'imposed community-based conservation is a contradiction in terms, and implies an exercise in futility' (p. 404).

Conservation with development

The concept of sustainable development is central to the community approach to conservation. Sustainable development, from the *World Conservation Strategy* onwards, has been put forward as a means of avoiding adverse environmental impacts, destructive socio-economic impacts, and the systematic land degradation created by poverty. To what extent does sustainable development provide a means of integrating conservation and development objectives in conservation areas?

The most often-cited example of the integration of conservation and development based on conservation areas is Amboseli National Park in Kenya. This was declared in 1974, covering 488 km² (subsequently reduced to 390 km²). The park was created from land originally set aside as a game reserve in southern Kenya in 1899, the Amboseli National Reserve being created in 1952 as one of a series of smaller reserves within this area. Amboseli lies in land grazed by the Maasai in the dry season, with cattle comprising up to 60 per cent of large-animal biomass (Lindsay 1987). In 1961 the District Council assumed control of the area, but although it received some entrance fee revenue and hunting licence fees, conflict developed because of Maasai suspicions that further grazing rights would be lost. To the Maasai of Amboseli the District Council was 'only another level of bureaucratic authority' (ibid., p. 153). Large game animals began to be killed. The Maasai demanded formal ownership of the area, the conservationists demanded a national park.

When the national park was declared, a complex agreement was made whereby Maasai gave up the right to graze within the park in return for joint owner-ship of surrounding land in group ranches, piped water supplies, compensation for lost production through wildlife grazing, economic opportunities in the form of tourist lodges, and developments such as a school and a dispensary. The District Council retained control of lodges on 160 ha in the heart of the park. Initial reviews suggested that the Amboseli programme was a great success (Western 1982). A school and cattle dip were built, wood and gravel were sold to the park, and campsite fees were paid. However, the borehole and pipeline system malfunctioned, wildlife utilisation fees were paid irregularly after 1981, no new viewing circuits or lodges were created, and park entrance fees were retained by the Treasury (Lindsay 1987). Tourist numbers levelled off in Kenya in the 1970s, and Maasai incomes did not rise as predicted. Killings of rhinoceros and elephant began again in the early 1980s, and poaching increased. The Amboseli Park Plan, like previous programmes, 'failed to provide the Maasai community with continuous appreciable benefit in return for compro-mises in their use of land' (ibid., p. 161). Income to the Maasai was too little and too unpredictable, and continuing cultural, social and economic change among the Maasai undermined static assumptions about the long-term accept-ability and sustainability of the group ranching system.

Projects that attempt to combine both conservation and development under a single project umbrella are often labelled 'integrated conservation–develop-ment projects' (ICDPs; Wells and Brandon 1992, Barrett and Arcese 1995), or 'conservation-with-development projects' (Stocking and Perkin 1992). The first generation of ICDPs have had mixed success (Brandon and Wells 1992, Stocking and Perkin 1992, Wells and Brandon 1992). Wells and Brandon note that 'linking conservation and development objectives is in fact extremely diffi-cult, even at a conceptual level' (1992, p. 567). Conservationists may have been naïve in assuming that a commitment on paper to sustainability and partic-ipation or 'bottom-up' planning would yield successful projects where more conventional development projects have a poor record. Stocking and Perkin (1992) provide a case study of ICDPs in action in the East Usambaras Agricultural Development and Environmental Conservation Project in Tanzania.

The East Usambaras reach an altitude of 1,500 m and support submontane forests with a very high level of endemic species. The IUCN project began in 1987 with three aims: to improve the living standards of the people; to protect the functions of the forest (particularly its role as a catchment for downstream water supply); and to preserve biological diversity. Traditional conservation objectives were deliberately de-emphasised to stress revenue generation and development. After four years, achievements were modest. A vast range of project activities had been begun, from agricultural extension to attempts to control illegal pit-sawing, most with limited success.

The problems of the East Usambaras Project included lack of funds, leading in turn to a lack of breadth in technical expertise, and the way in which capital and energy were dissipated in too wide a range of activities. Behind many of these problems lay the lack of proper feasibility study, a common failing in conservation projects (Caldecott 1996). Conservation organisations are discovering (like developers before them) that development plans are hard to transfer from paper to reality (Stocking and Perkin 1992). As projects, ICDPs are inherently highly complex and demand high levels of skill on the part of project staff. They also demand substantial funds and a realistic (i.e. slow) timescale. Their chances of success depend on local perceptions of the project, and these are vulnerable to the public failure of particular components. Clear and precise objectives, careful evaluation of the costs and benefits of project components at the level of the individual household, long-term commitment to funding, and strong local participatory linkages are essential. Projects of this sort will not be cheap to implement, and will not yield results quickly. Furthermore, there is a real risk that positive impacts of the project on the local economy will be transient and dependent on the maintenance of flows of project revenues (Infield and Adams 1999). Barrett and Arcese conclude that ICDPs are 'no more than short-term palliatives' (1995, p. 1081).

There can be dangers from a conservation perspective if emphasis on development leads to such a profound de-emphasis of conservation goals that they are no longer seriously addressed. Oates (1995) argues that precisely this has happened in the case of the Okumu Forest Reserve in south-west Nigeria. He is very critical of the sustainable development rhetoric in *Caring for the Earth*, and blames this for new conservation programmes that have accelerated forest loss to small farmers. There may be a need to distinguish between the merits of development interventions of this kind as a contribution to local livelihoods and the merits of their contribution to conservation. While at a rhetorical level it may be desirable to argue that conservation and development can go hand in hand through a joint programme, development expenditure for conservation purposes may not give results that are cost-effective in either livelihood or conservation terms.

There are potentially common interests between conservation of nature and the protection of the interests of local people, both at the level of opposition to destructive change (clear-cut logging, for example, or a dam) and at the pragmatic level that the wildlife of a peopled ecosystem depends on the way those people work the environment, and hence on the ways they are them-selves influenced by government development policy. However, nature conservation and the interests of disadvantaged groups are distinct issues. Thus in the Brazilian Amazon, the way native peoples relate to their environment changes over time as populations grow and inevitable cultural changes take place (Seeger 1982). The romantic belief that such peoples are a 'natural' part of the ecosystem has led to proposals for the establishment of multiple-use reserves, creating national parks for native peoples and allowing native peoples in national parks. Seeger argues that this is unworkable, and that 'where resources are limited, the conflict between Indians and forest management has no real solution' (1982, p. 188), if only because of the proven capacity of others to use Amerindians to exploit park resources.

It is likely that conservation and the indigenously controlled development of rural peoples are not wholly compatible. Seeger argues that although land designation is a valuable strategy for both, separate territories are needed. In this he is supported by biologists concerned about the preservation of wildlife. For example, while Johns discusses at length the possibility of achieving wildlife objectives in areas of rainforest cut by selection felling, he stresses that this 'should not be regarded as an alternative to maintaining primary forest areas' (Johns 1985, p. 370), for these alone will contain the full diversity of species. Oates (1999) draws similar conclusions about the success of attempts to inte-grate conservation and development in the West African rainforest. He believes that such projects, far from creating a 'win–win' outcome, end up satisfying neither human needs nor conservation objectives. From a narrow biodiversity conservation perspective, multiple land use is therefore seen as a pragmatic second best, valuable primarily where there is no political will to preserve the rainforest without human presence.

The primary objective of conservationists is the preservation of biodiversity. Development, even if packaged as 'sustainable development', is attractive chiefly as a secondary strategy where it promotes their primary objective. Where con-servationists have little chance of commanding the kind of resources they desire, they are willing to push those kinds of development (which they call sustainable development) least damaging to wildlife. In doing so, they may well align them-selves with the needs and aspirations of indigenous groups. Both interests have much to gain from collaboration that increases power to withstand certain kinds of development action, whether from corporate capital or the state.

Nature conservationists and local people may make effective political part-ners, locally, nationally, and in the international field where environmental pressure groups have valuable experience and resources. However, such part-nerships may be short-lived, unless a way can be found to make wildlife 'pay its way' (Eltringham, 1994). This can be done through non-consumptive use, through wildlife tourism. However, there are environmental costs to the tourist industry (not least resource consumption and pollution from air flights), and opportunities may be limited, particularly in those protected areas that do not share the attractiveness of the grassland savannas of East Africa or the gorillas of the Congo basin. As Eltringham points out, tourism is a fickle industry, 'subject to the vagaries of the political climate' (1994, p. 164). It is a market highly subject to fashion, and one that has traditionally developed, exploited and abandoned destinations on a short cycle. Although much is written about ecotourism, and great hopes are hung upon the concept for both long-term business profitability and developmental benefit, the possibility of sustainability of tourism in many Third World tourist locations remains highly debatable.

More controversially to some, wildlife can be made to pay its way through 'consumptive use', meaning killing or harvesting. This may take the form of hunting by local people (for example for bushmeat), killing in return for a licence fee by big game hunters, or through the collection of marketable or consumable natural products (for example rainforest rattans or turtle eggs). This approach to conservation is built on the idea of wildlife simply as an economic resource that should be exploited in an effective and sustainable way. In 1990 IUCN established a Specialist Group on 'sustainable use', and began to develop guidelines for utilisation of wild species. However, the resulting debate saw polarisation within IUCN, with members from Third World countries speaking out against the traditional protectionist approaches to conservation advocated by First World organisations, and calling for new approaches to conservation that explicitly recognised the dependence of rural communities on wild species as resources (Allen and Edwards 1995).

ICDPs based on the consumptive use of wild species depend on the feasibility with which a 'sustainable harvest' can be defined (a complex scientific task conventionally requiring good data over long periods and regular monitoring, things often not available in practice). They also require effective institutions to enforce that harvest (rules, agreement by potential hunters that these are fair and reasonable rules, and measures to deal with those who break them). There are both monetary and non-monetary reasons why people harvest illegally (Barrett and Arcese 1995). It is often not 'local' people who hunt, and even if it is, they often do so to supply an organised national trading network and an urban market in bushmeat. It may therefore be hard for ICDPs (or other conservation-driven engagements with local people) to provide sufficient incentives to decouple rural livelihoods from unsustainable patterns of wildlife harvest.

There are use-based conservation projects and programmes in a number of Third World countries, particularly in Africa. In Zambia, for example, controlled hunting was an element of the Lwangwa Integrated Resource Development Project (LIRDP), begun in 1987. Revenue from safari hunting in the Lupande Game Reserve was used for development projects in the local area, as well as to finance the cost of game guards in the South Lwangwa National Park. Game guards were locally recruited, and both safari hunting and community game-harvesting (particularly of hippopotamus) take place legally (Wells and Brandon 1992). The LIRDP approach formed the basis for the ADMADE programme in 1987 in ten other game management areas in Zambia. Revenue from safari and other hunting fees is used to meet wildlife management costs (40 per cent to these activities within the game management area itself plus 15 per cent to the national parks system and 10 per cent to the Zambian Tourist Bureau), and to generate revenue for local community projects (35 per cent). ADMADE is often reported as a success, and its economic (consumptive use) and benefits-sharing approach held to be a valuable model for conservation elsewhere (Swanson and Barbier 1992). However, research suggests that neither the LIRDP nor ADMADE has been effective in changing the level of hunting by local people. Gibson and Marks (1995) argue that the community benefits generated by ADMADE fail to compensate for the economic, social and political benefits of hunting: hunters change tactics, but they keep hunting. Wainwright and Wehrmeyer (1998) argue that LIRDP has failed to achieve its conservation goals and has generated few community benefits. ADMADE is inflexible, and does not direct economic benefits through democratic institutions. It does not provide a strong model for situations where human populations are denser and wildlife numbers less abundant than in the extended savanna woodlands of Zambia.

The other much-reported approach to community-based conservation is the CAMPFIRE programme in Zimbabwe. Wildlife ranching has become important economically on large commercial farms in the driest areas of Zimbabwe, and is also important in conservation terms since populations of species such as black rhinoceros and cheetah within Zimbabwe are mostly on private land. There has also been development of communal wildlife utilisation projects under the Communal Areas Management Programme for Indigenous Resources (CAMPFIRE). This is similar to Zambia's ADMADE programme, but places greater emphasis on communal initiation and control of the programme and hunting activity. About 12.7 per cent of Zimbabwe lies within national parks and equivalent areas, and a further 2.4 per cent lies within forest reserves (Child 1995). Wildlife policy in Zimbabwe was reviewed in the late 1960s, and the Parks and Wildlife Act 1975 allowed authority over wildlife resources to be devolved to the district level (Metcalfe 1994). The revolutionary step of CAMPFIRE was the de facto granting of power to local authorities, such that they can make decisions about hunting and conservation, and gain from revenues so generated. Outside the protected area, wildlife is considered a renewable resource to be managed to create maximum human welfare.

Like ADMADE, the CAMPFIRE model is seen internationally by conservation policy-makers to offer a form of conservation that is both popular and affordable (Barbier 1992, Olthof 1994, Child 1995). However, here too problems are emerging, particularly because authority (and hence revenues) are devolved only to district level, not to communities themselves (Murombedzi 1999). While CAMPFIRE seems to work well and harmoniously in some areas, in others, particularly those less rich in high-value trophy species such as elephant, lacking in good tourist infrastructure and a ready supply of wealthy safari hunters, and with rapid rates of immigration, the story is less clear-cut. It is also not clear to what extent CAMPFIRE can provide a model for other countries, in areas with a biogeography or political history very different from those of Zimbabwe.

The most controversial dimension of the debate about the consumptive use of wildlife is the harvesting of elephants. It has been widely argued that the high value of ivory means that it is economically desirable to treat elephants as a resource, and that it is scientifically possible to set sustainable harvest levels (e.g. Barbier *et al.* 1990a). This approach has been pursued strongly by the countries of Southern Africa such as Zimbabwe (Hill 1995), although opponents of elephant hunting (chiefly motivated by ethical considerations, or fears that institutional failure in measures to control hunting would allow indiscriminate and illegal killing to continue, particularly in East and West Africa) won their case for a total ban on international trade in ivory from African elephants at the meeting of contracting parties to the Convention on International Trade in Endangered Species (CITES) in 1989 (Princen 1994). The ban was subsequently partially lifted for certain elephant-range states (Namibia, Botswana, Zimbabwe) at an emotional CITES meeting in Harare in 1998. However, as Allen and Edwards (1995) point out, the debate about consumptive use cuts deeply into established patterns of conservation thinking. While to many observers in the Third World it looks like pragmatic common sense, to others (including many conservationists in Western urban societies), sustainable use is 'a very threatening concept because it challenges our perceptions of what conservation is about' (ibid., p. 97).

Top-down conservation has been transformed in a number of ways by participatory approaches, and the attempt to integrate development and conservation aims. The idea of sustainable development has been fundamental to this transition, and conservation projects are now at the forefront of experiments in achieving sustainability in practice. Such projects reveal the confused diversity of thinking inherent to sustainable development, and may not create win–win solutions, but instead reveal divided interests and awaken latent controversy. However, local people are not the only 'community' that can be addressed in pursuit of 'sustainable development from below'. A similar approach may be used in quite different circumstances to bring about change in resource management by addressing global consumers. This will be discussed in the next section.

Sustainability through consumption

The history of mainstream sustainable development, as discussed in Chapters 4 and 5, has essentially been one of attempting to convince opinion leaders in governments, NGOs and (to a lesser but growing extent) in business to deal with the environment differently. The mainstream documents, from the *World Conservation Strategy* to *Agenda 21*, have a rhetorical tone, admonishing, advising and encouraging those perceived to have power and influence in the global economic and political system. Environmental NGOs have long sought to develop campaigns that tie environmental issues to particular products and companies, to harness commercial fear of the power of consumer backlash in order to persuade conservative corporations to change their practices. The Friends of the Earth campaign against Schweppes' decision to use non-returnable glass bottles, or Greenpeace's action against Shell's attempt to dispose of the *Brent Spar* oil platform at sea, are clear examples of this.

In the 1980s and 1990s Northern NGOs attempted to extend this approach to global campaigns. In particular, they sought to link unsustainable harvesting and trade practices in the South (particularly in rainforest products) to purchases in Northern high streets. The Friends of the Earth (FoE) Rainforest Campaign, launched in 1985, involved a boycott on high street outlets using tropical timber obtained through clear-felling or non-replacement selective logging. Obviously the importance of logging as a cause of forest clearance varies in different places (see Chapter 6), but the importance of First World markets for timber means that such pressure could in places have significant impacts (Oldfield 1988).

One problem with this as a campaigning strategy was that there was no immediately available alternative to the unacceptable and unsustainably produced tropical timber products – no carrot to combine with the stick of consumer outrage. An attempt was made to provide this through certification of the origins of products and the conditions under which they were produced. The idea was that if Northern consumers could distinguish products produced sustainably from those that were not, their purchasing power could drive production towards more sustainable methods.

NGOs therefore campaigned for the adoption of a code of conduct by UK and EC timber traders, under which they would stock only timber from concessions that have a government-approved management plan that stipulates post-logging management, where annual timber extraction does not exceed the concession's sustainable yield and where logging impacts are minimised by sympathetic extraction methods. Under the code of conduct, traders should not stock wood from plantations established in virgin forest areas, and they must label country and concession of origin of wood products. Furthermore, traders would be asked to devote 1 per cent of profits towards a fund to promote the sustainable use of rainforest.

The FoE campaign has continued, and has been echoed in the campaigns of many other environmental organisations. Campaigning strategies have evolved from defining a global 'environmental problem' (such as 'rainforest destruction') and linking it to specific events around which public protest can focused, into wider campaigns involving consumer boycotts of specific products (drawing explicit links between environmental issues and specific products on sale in the industrialised world, such as hardwood garden furniture). In the early 1990s various environmental groups (Friends of the Earth, rainforest action groups and eventually Earth First!) began direct-action protests against the biggest six DIY superstore chains in the UK. These tactics in turn developed into 'solutions-led' campaigning such as the Greenpeace development of propane refrigeration technology as an alternative to the use of CFCs or HCFCs (Rose 1993, Rawcliffe 1998). Such campaigning has borne fruit in some industrialised countries (notably in Western Europe, less so in the USA and barely at all in Asia), and 'green consumerism' has become a potent idea in the minds of corporate planners, sales staff and consumers.

The 1990s saw a breakdown in the simple oppositionism between environmental groups and business, and the exploration of the possibility that there might be shared agendas. *Agenda 21* called for partnerships between environmental groups and businesses, and the development of such relationships has been an important element in the post-Rio agenda. The Business Council for Sustainable Development (BCSD) and the International Chamber of Commerce were present at the Rio Conference in 1992, representing the view of international business. In 1992 the World Business Council for Sustainable Development (a coalition of 120 international companies) published its response to the Rio Conference in 1992, *Changing Course* (Murphy and Bendell 1997). Business response to the challenge of sustainable development can be thought of as having passed through three phases: pollution prevention (around 1970), self-regulation (1980s) and sustainability (in the 1990s) (ibid.).

There is much written about the phenomenon of 'green capitalism' in the 1990s (e.g. Welford and Starkey 1996). Undoubtedly there has been much hype, both about the possibility of a society buying its way to sustainability, and the specific commitments of individual manufacturers or traders. Nonetheless, there have been company directors and boards who have been persuaded of the moral imperative to take environmental aspects of their business seriously, and rather more who have seen marketing and growth opportunities in realigning their business to meet the new demands of consumers and the new opportunities in 'green' products. One such company is the UK hardware chain B&Q, which responded to campaigns on hardwoods and other products positively, taking a very public lead. In 1995 B&Q's managing director wrote, 'we recognised that our stakeholders, that is our customers, staff, local communities, young people, government and shareholders, believed that the environment did matter' (B&Q 1995, p. 3): in some instances improved environmental performance brought cost savings, and it certainly enabled the company to claim an enhanced corporate image as a responsible retailer.

There have also been some significant partnerships between environmental groups and business. One is the World Wide Fund for Nature UK initiative, the '1995 Group'. WWF challenged the timber industry to make the world's forest production sustainable by 1995. In 1991 ten companies committed themselves to meeting this target, and formed the '1995 Group' (Murphy and Bendell 1997). The 1995 Group is effectively an informal environmental management system accreditation process with set targets. WWF-UK did not verify company policy, and companies did not gain use of the WWF panda logo (ibid.). Members of the group committed themselves to the Forest Stewardship Council (FSC) as a source of certification and labelling of 'sustainable production', and to moving towards phasing out wood not accredited in this way. They committed themselves to identifying the source of wood and wood products, and they agreed to appoint a named senior manager to implement this commitment, and to monitor progress very six months. In return they could use the FSC logo on appropriate products (WWF 1996). The FSC, which came into existence formally in 1993, is a membership organisation,

Table 12.2 Forest Stewardship Council (FSC) principles of forest management

---

1 *Compliance with laws and FSC Principles:* Forest management shall respect all available laws of the country in which they occur and international treaties and agreements to which the country is a signatory, and comply with all FSC Principles and Criteria.

2 *Tenure and use rights and responsibilities:* Long-term tenure and use rights to the land and forest resources shall be clearly defined, documented and legally established.

3 *Indigenous peoples' rights:* The legal and customary rights of indigenous peoples to own, use and manage their lands, territories and resources shall be recognised and respected.

4 *Community relations and workers' rights:* Forest management operations shall maintain or enhance the long-term social and economic well-being of forest workers and local communities.

5 *Benefits from the forest:* Forest management operations shall encourage the efficient use of the forest's multiple products and services to ensure economic viability and a wide range of environmental and social benefits.

6 *Environmental impacts:* Forest management shall conserve biological diversity and its associated values, water resources, soils and unique and fragile ecosystems and landscapes, and, by so doing, maintain the ecological functions and integrity of the forest.

7 *Management plan:* A management plan – appropriate to the scale and intensity of the operations – shall be written, implemented and kept up to date. The long-term objectives of management, and the means of achieving them, shall be clearly stated.

8 *Monitoring and assessment:* Monitoring shall be conducted – appropriate to the scale and intensity of forest management – to assess the condition of the forest, yields of forest products, chain of custody, management activities and their social and environmental impacts.

9 *Maintenance of natural forests:* Primary forests, well-developed secondary forests and sites of major environmental, social or cultural significance shall be conserved. Such areas shall not be replaced by tree plantations or other land uses.

10 *Plantations:* Plantations shall be planned and managed in accordance with Principles and Criteria 1–9, and Principle 10 and its Criteria. While plantations can provide an array of social and economic benefits, and can contribute to satisfying the world's needs for forest products, they should complement the management of, reduce pressure on, and promote the restoration and conserva-tion of natural forests.

Source: WWF (1996)

with both industry and environmental interests represented, in two chambers. It adopted ten principles of forest management (Table 12.2).

WWF-UK has claimed considerable success for this partnership strategy. It sug-gests that the 1995 Group companies had made demonstrable progress in areas of certification, sourcing and open reporting, and, perhaps more importantly, they had opened up the debate within the industry, and pressed environmental concerns up the supply chain (towards the source in rainforest countries). Nonetheless, only 4 per cent of the timber sold by members of the group was certified, and only twenty-three of the forty-seven companies in the group had purchased some certified wood or wood product (although fourteen of the remaining twenty-four companies used paper, for which there was no certified source at this time; WWF 1996). WWF sought to extend the work of the 1995 Group, forming a '1995 Plus Group', members of which would commit themselves to use only certified timber. However, in the mid-1990s there was an industry backlash against the partnership and certification process, driven by organisations such as the Canadian Pulp and Paper Association. In 1996 the timber industry organisation the Timber Trade Federation opposed this arrangement as contrary to the principles of free competition and free movement of goods under UK and EU law, and WWF-UK was forced to water down the group's membership requirements (Murphy and Bendell 1997). This was one dimension of a wider and more systematic assault on environmentalism in the 1990s, as industries and right-wing political activists in a number of countries (but particularly in the USA) attempted to reverse the influence of several decades of populist environmentalism, and challenge the growing hegemony of mainstream sustainable development (Rowell 1996).

Despite this setback, and the broader anti-environment backlash that it reflected, a number of environmental organisations, including WWF, have been actively seeking to extend the partnership approach with business. Murphy and Bendell (1997) cite a number of examples, including the sometimes rocky relationship between McDonald's and the environmental movement. The drive for 'free trade' and the increasing awareness of the global ramifications of business organisation have made it clear that national government regulation is of limited power, and that businesses themselves must be central to any significant move towards sustainability. Murphy and Bendell suggest that 'the mythic power of sustainable development provides a platform for exploring new models of society' (ibid., p. 245), and these can include new ways of working for both environmental groups and businesses.

Such initiatives can bear fruit, but the harvest is as yet slow in coming and pitifully small. In many parts of the Third World, for example in Cambodia, rainforests are still being logged illegally (but with the full knowledge of senior government figures) without even a nod in the direction of sustainability. The timber extracted is used to meet First World demand for such vital products as hardwood garden furniture (Alley 1999). The economic values of the standing forest to local people are liquidated and privatised. Because of its illegality, the benefits of such trade are lost to the public purse and recycled abroad or though elite consumption and the black economy.

Rainforest management reform

Consumer campaigns by environmental pressure groups, and environmentalist–business partnerships, seek to bring about sustainability 'from below' by working *within* the global economic system. They are wholly dependent on the success with which sustainable resource management strategies can be defined and made technically feasible. The notion of 'sustainable management' of tropical forests has become an important element in debates between First (and Third) World environmental pressure groups and the global timber industry. Unsurprisingly, there has been a great deal of research to establish what 'sustainable tropical forest management' would look like, and there have been a succession of guidelines for development in rainforest regions (e.g. Poore 1976, Poore and Sayer 1987, Schreckenberg and Hadley 1991, Gómez-Pompa *et al*. 1991, Dykstra and Heinrich 1996). The *FAO Model of Code of Forest Harvesting Practice* (Dykstra and Heinrich 1996) attempts to contribute to the expectations of the Rio Conference about sustainable forestry, and particularly the Forest Principles. It provides a model for national forest harvesting codes that sets out best practice in the areas of harvest planning, forest road engineering, extraction, landing operations, transport operations, harvesting assessment and the forest harvesting workforce.

One repeated theme of such works is their emphasis on the multiple functions of forests, and the need to look beyond the timber resource. Thus, for example, the FAO's *Mangrove Forest Management Guidelines* argued that 'the traditional "*management paradigm*" implying that if forests are well managed then, *ipso facto*, the non-wood ecosystem components will remain stable, is notionally flawed' (FAO 1994b, p. xxiii). The products of mangroves are legion, including capture and culture of fish, salt production, honey, agriculture, charcoal, firewood, poles, tannin, palm and wildlife. Mangroves are also vital in many instances to coast protection. Social and economic benefits are maximised through multiple-use management, involving both integration with management of the wider coastal zone, and an approach to planning that allows for the participation of the rural poor (FAO 1994b).

There is now a growing literature on attempts to transform forest management policies in Third World countries to include the designation of forest reserves, the creation of sustainable agriculture in rainforest regions, the managed exploitation of natural forests, and the restoration of logged and degraded forest lands (see for example Gradwohl and Greenberg 1988), and/or the creation of systems of incentives for sustainable production systems (Repetto 1987).

The Tropical Forestry Action Plan was released at the time of the World Forestry Congress in Mexico (Poore and Sayer 1987). It was the work of the World Resources Institute (a powerful and wealthy environmental think-tank based in Washington, DC), IUCN, the FAO, the UNDP and the World Bank. It argued for increased investment in the forestry sector, proposing to make sustainable forestry in the humid tropics a reality by doubling spending on timber and fuelwood plantations. Its target, however, was unreservedly industrial forestry, and environmentalists argued that it would neither promote the protection of remaining rainforest areas nor protect the interest of rainforest people. It was, they said, a 'top-down' approach to preserving forests (Caufield 1987).

An even more 'top-down' organisation is the International Tropical Timber Organization (ITTO). This was formed in 1983 following the UN Conference on Tropical Timber held under the auspices of the UN Conference on Trade and Development (UNCTAD), and a huge amount of behind-the-scenes wrangling (Johnson 1985). The ITTO is a joint organisation between producing and consuming countries, with votes on the Tropical Timber Council divided equally. It has been dogged by financial shortages because member countries have been slow to pay their dues. There were complaints of underfunding at its first meeting, in Yokahama in March 1987. The ITTO is different from other commodity agreements because it specifically promotes reforestation and national polices of sustainable utilisation and conservation. Its aims are broad, embracing the improvement of tropical forest management, the improvement of marketing and distribution of tropical timber, and the promotion of wood processing in the producing countries. In 1991 the ITTO eventually set the goal to make the global timber trade sustainable by the year 2000, through the establishment of demonstration plots, the development and dissemination of guidelines, and the promotion of the case for sustainable management to member governments. Some observers suggested that the ITTO offered an opportunity for conservationists to promote conservation 'in the maximum degree of harmony with the tropical timber trade that is consonant with sound conservation objectives' (Johnson 1985, p. 44). However, its progress has been slow, and its achievements very modest (WWF 1991). While the word 'sustainable' is attached to many tropical hardwood products, and claimed for many production systems, the majority of tropical forestry fails to sustain the full range of values of forests, and much fails even to sustain the timber resource itself.

The institutional challenge of creating a forestry industry capable of something other than timber mining is considerable. There has been some progress: the broad technical requirements for sustainable forest management have been identified and are now well established (e.g. Gómez-Pompa *et al.* 1991, Schreckenberg and Hadley 1991). The environmental impacts of logging vary considerably with the logging regime adopted. Typical problems of most large-scale commercial forms of timber extraction include soil compaction and loss of internal structure, effects of leaching and runoff on soil nutrients and microorganisms, the loss of nutrients in timber, and changes to the internal microclimate of the forest (Shelton 1985).

The most critical factor in determining the ecological impacts of logging is subsequent land use. This can range from abandonment following timber clear-felling through replacement with a pasture for cattle-raising, establishment of a plantation of tree crops, to an agroforestry system. In many cases forest can regenerate, and with it, eventually, timber value, non-timber forest products and biodiversity. For example, a local operating company of the Japanese company Honshu Paper, part of the Mitsui group, began felling for wood chips and saw logs in the Madang and Naru Valleys in Papua New Guinea in 1973 on a logging concession of 68,000 ha (Lamb 1980, Seddon 1984). By 1983 more than half had been logged. Logging had caused the removal of humus and topsoil, leading to high rates of soil erosion and leaching, deposition of sediments, and waterlogging on valley floors and flats, with associated soil acidity and the loss of phosphorus. However, by 1984 regeneration of secondary forest was taking place. There were significant effects on the diversity of the forest which regenerated, partly because of the loss of soil seed sources. However, where logged land was cultivated or disturbed (for example in log loading areas), tree regeneration was prevented, and grassland took over (Saulei 1984).

This is not an untypical story. It can be read from the standard idealist conservation perspective (of biodiversity lost as natural forest is logged). However, the 'naturalness' of the forest is not in fact so straightforward. Despite appearances, the forest cleared was itself a secondary growth resulting from drought and fires in the 1930s and during 1944–5 (Saulei 1984). Many areas of tropical forests have a long history of human use. In Marovo Lagoon in the Solomon Islands, the apparently 'natural' is in fact rich in cultural sites of fields, settlements and economic trees (Hviding and Bayliss-Smith 2000). The diversity of such forests may make them a legitimate target of concern for conservationists, but their apparently pristine nature is a product of the romanticism and ignorance of outside observers. Arguably, a sustainable logging regime might be perfectly possible, and perfectly acceptable, in such a location.

The most acute impacts are associated with clear-felling, followed by permanent conversion (Plate 12.1). Even in this situation the ecological impacts are complex. Studies of the ecology of forest fragments in Amazonia are increasing knowledge of exactly what may survive in refugia left after widespread deforestation (Lovejoy *et al.* 1983). Inappropriate management of clear-felled land can lead to widespread environmental impacts such as the large-scale fires that afflicted large areas of Indonesia in the late 1990s. On clear-felled land, ground vegetation is lost through desiccation, and animal and bird species dependent on shaded forest conditions are also lost. Wildlife species such as primates are unable to survive in heavily logged forests, although a limited number of species can survive selective logging (Johns 1985). Territorial species such as gibbons are tenacious in surviving forest fragments (Shelton 1985). Long-term responses of primate populations (and by implication other species) depend on the availability of refugia, the nature of hunting pressure (a massive problem in Africa in particular because of the huge urban market for 'bushmeat'), and the willingness of foresters to adhere to silvicultural rules designed to achieve sustainability (Johns and Johns 1995).

The extent of forest transformation in many countries is such as to present considerable challenges in terms of biodiversity conservation. The problems facing conservation in West African forests suggest to Oates (1995, 1999) that anything other than strict protection will fail to preserve biodiversity (certainly for easily hunted primate species); arguably, conservation projects have to be designed around existing patterns of logging and land clearance. Thus in Bendel State, Nigeria, forest reserves are already penetrated by farmers and hunters, and their exclusion in the interests of wildlife conservation (even if desirable) would be politically impossible (Osemiebo 1988). As a result, conservation objectives will have to be achieved through multiple land use (ibid.). A similar approach is to seek to modify regional planning in rainforest areas such as the Amazon to foster development that is both 'environmentally adaptive and

Plate 12.1 Bwindi Impenetrable National Park, Uganda. The mountain gorilla only survives in the wild in four remaining areas of forest in central Africa, Bwindi Impenetrable National Park in Uganda, with Mgahinga Gorilla National Park (Plate 9.2), the Parc Nacional des Volcans in Rwanda and the Parc National des Virungas. These parks are islands set in a sea of recently cleared and now intensively managed farmland. Gorillas and farms do not mix, presenting a sharply focused choice between retaining the forests and perhaps keeping the gorillas, and clearing the remaining forest for farms.

economically profitable' (Eden 1978, p. 401). Data on birds and plants from Puerto Rico suggest that models based on island biogeography theory tend to overestimate the rate of species extinction associated with deforestation. There may be potential to maintain and enhance biodiversity through appropriate management (Lugo *et al.* 1993). On the other hand, Shankar Raman *et al.* (1998) demonstrate that shifting cultivation with a typical five- to ten-year cycle in north-east India has significant implications for birds and plant species richness; a fallow interval of twenty five years for birds and fifty to seventy-five years for plants would be necessary to maintain community composition. The extent to which 'sustainable environmental management' can maintain particular elements of biota (such as birds, or still less primates) remains debatable.

Selective commercial logging is generally much less environmentally damaging than clear-felling, although it is not negligible. Many non-target trees can be damaged by falling timber trees and drag paths or other works. In East Kalimantan, intensive selective felling of fourteen stems per hectare by mechanised methods caused damage to 41 per cent of residual trees; skid tracks, haul roads and log yards occupied 30 per cent of the logged area (Kartawinata *et al.* 1981). The recovery of these compacted and scraped areas is slow: water infiltration rates are low and erosion is high. Selective logging also involves the loss of the best specimens of commercial species, and sometimes the extinction of species, forms of 'genetic erosion' that have significant impacts on the physical and floristic structure of the forest (ibid.).

Although claims are now widely made that logging practices in rainforest countries have made substantial shifts towards sustainability, much doubt remains about whether that is the case. In Sarawak, for example, selective logging and an annual cut below estimated maximum sustainable yield are used to justify a claim of sustainable forestry. However, the state forest output (15.8 million cubic metres in the mid-1990s) was 70 per cent over the estimated sustainable cut, and there are serious problems of collateral damage to non-target trees, soil erosion, and slow revegetation of skid trails; the planned cutting cycle (twenty-five years) is too short for dipterocarp trees. Furthermore, logging takes place without regard to the needs of forest dwellers, with the exception of one ethnic group, the Penan. In Sarawak, as in many other places, sustainability is but a 'thin veneer' (Pearce 1994, p. 30).

Traditional selective logging, of the kind still practised in the late 1970s along the Urumbamba River in eastern Peru (White 1978), has died out in most areas of tropical moist forest. Logging of natural forest can broadly be classified as selective or clear-felling, but within these categories there are distinctions. Selective logging intensities can vary from two to three stems per hectare taken out in some African forests to twenty or more stems per hectare in parts of South-East Asia. This difference is caused less by the attitude of foresters to conservation than by the density of timber of suitable size and quality. Both selective felling and clear-felling have extensive impacts on forest ecology, although of the two, clear-felling is obviously the more destructive. The two may be operated together in different zones of the same forest, and worked forests may exhibit a complex and fragmented structure with some areas logged at near 100 per cent and others with a canopy reduced by 30–50 per cent. Selective logging can take place on a monocyclic basis (a single operation to remove all saleable trees) or in a polycyclic system where trees are removed in a series of felling cycles as they reach suitable sizes (Johns 1985).

Conventional forestry usually involves a company (often a transnational) paying government for a licence to extract timber, in the form of chips or sawn logs. It may employ a second company as felling contractor. The benefit of logging to the local economy is usually minimal (some low-paid work, but loss of non-timber forest products). Benefit to the national economy is restricted, because while value is added to the timber when it is sawn and made into products, this typically takes place elsewhere. Moves to promote sustainable forest operations therefore not only address the ecological impacts of timber extraction but also seek to capture more of the value of wood extracted in the local economy. This might involve increases in local involvement in (or even ownership of) commercial forestry operations, and increases in the extent to which timber is processed locally.

One early example of an alternative approach to logging is the Gogol Valley development in Madang Province, Papua New Guinea. From 1975 to 1978 the Gogol Project was the subject of MAB-funded research, and was seen in some quarters as a showcase of forestry management (Lamb 1980). Initially the project was intended to integrate sawn timber, veneers and woodchips for pulpwood. The plan was for 48 per cent of the land to be clear-felled, 22 per cent selectively felled and 30 per cent unlogged (Lamb 1980), and the project was to include an element of reforestation. In practice, reforestation fell far short of the 800 ha per year target, only 1,000 ha being planted in the first five years, mostly with species of *Eucalyptus, Terminalia* and *Acacia*. The rate of planting subsequently increased, but the extent of plantation remained too little to sustain the pulp mill after the fifteen- to twenty-five-year life of the project (Seddon 1984). The potential sustainability of the Gogol Valley model depended not only on the profitability of the timber operation, but also on the economic benefits and costs to residents of the valley (ibid.). The Gogol Valley is home to 2,000–4,000 indigenous people who lived by shifting agriculture, and for whom forms of permanent agriculture have been sought. Royalties paid to the indigenous people have certainly injected cash into their economy. Experience of other promised benefits (access, education, jobs) has been mixed. For example, logging roads have not survived the extraction operation. Against this must be balanced the trauma of the loss of traditional forests (although areas around villages and river corridors were spared), and the economic fact that the holding company JANT failed to show a profit, and thus escaped taxation. Profits were written off against capital borrowing costs and (it is alleged) transfer pricing allowed profits to be repatriated (ibid.).

Timber certification provides a valuable opportunity to maximise local capture of timber values (Upton and Bass 1995). The price paid for certified eco-timber in the Solomon Islands in 1991 was 3.7 times that for unprocessed round logs (Hviding and Bayliss-Smith 2000). Moreover, relatively simple and affordable technology (chainsaw and milling frame) goes some way to making micro-scale logging feasible. A project called SWIFT (Solomon Western Islands Fair Trade) was established in 1994 by a Dutch group associated with the Solomon Islands United Church's Integrated Human Development Programme to produce Forest Stewardship Council-certified 'eco-timber', and make the vital connection to European consumers. Hviding and Bayliss-Smith are dubious about the financial feasibility of the small-scale timber model being developed by SWIFT, and suggest that it is too dependent on top-down management and Dutch expertise and money. Nonetheless, it and many other projects like it do represent genuine and hopeful attempts to develop an alternative strategy for forest management to the 'strip and run' of so much conventional tropical forestry.

The replacement of forest with farms leads to the loss of much of the existing biodiversity. Critical though this problem is from a conservation perspective, it is compounded when the agricultural 'development' in whose name the forest is replaced cannot generate sustained yields and sustainable livelihoods. Then indeed the loss of the forest leaves no lasting benefit (beyond, perhaps, the private profits of the timber concessionaire, any private payments to bureaucrats and politicians to acquire the logging rights, and, perhaps, the benefits flowing from the government's use of royalties on timber extracted. Clearly, much rainforest agriculture by recent settlers has not been sustainable (for example in the Amazon), just as, probably, much indigenous shifting cultivation has been. Much research has been done on how to create sustainable agro-ecosystems in rainforest areas (e.g. Altieri *et al.* 1984), and on the problems and prospects of agroforestry (Winterbottom and Hazelwood 1987). Agroforestry involves the integration of trees and crop plants, and under the right conditions can provide fuelwood, increase agricultural productivity (for example through enhanced soil fertility in alley cropping) and generate employment and income (Winterbottom and Hazelwood 1987). However, initial enthusiasm for novel alley cropping combinations of crops and leguminous trees, developed in part through work by the International Centre for Research on Agroforestry (ICRAF), has been moderated by the recognition of problems of labour demand. It is also now more widely appreciated that while the term 'agroforestry' is novel, many rainforest communities have practised farming systems that integrate crops and trees in space and time in what is essentially an agroforestry system (Hviding and Bayliss-Smith 2000).

Multiple-use forest management that takes account of the needs of existing local populations approaches is commonly labelled 'social forestry', although that label has been applied loosely to a great many different kinds of projects and initiatives. It has been incorporated into the rubric of most international agencies involved in forestry, particularly the FAO and ITTO (Barraclough and Ghimire 1995). In a number of countries, for example in India, conventional forestry strategies based on narrowly defined commercial timber production have been broadened to embrace a wider range of non-timber forest products. Moves to relate the needs of tribal people to forestry management, for example in Gujarat (Murdia 1982), have seen institutional innovations such as Joint Forest Management in India, where to some extent at least technical forestry bureaucracies begin to share power with local communities.

However, it should not be expected that communities will necessarily share the opposition to commercial clear-cutting expressed by Northern environmentalists. Opposition by local people (backed by Northern environmentalists) to logging in northern New Georgia (Solomon Islands) by Lever Pacific Timbers (part of Lever Brothers) in the 1980s caused the company to withdraw. Despite this apparent victory for conservation, and explicit local demands for sustainable development, a new contract was soon awarded to another company, which promptly moved back into the old Lever camp and carried on much as before. The reasons for this are complex (Hviding and Bayliss-Smith 2000), but it serves as a reminder that sweeping assumptions should not be made that 'the community' will choose a particular interpretation of sustainability.

Forest products are of enormous importance to both forest dwellers (hunters and farmers) and those living on the forest margin. There are important distinctions to be drawn between those who are dependent on the forest and those whose use is the subject of choice (Byron and Arnold 1999). Both can be adversely affected by the arrogation of rights in timber by the state, and the closure of the forest to alternative uses (grazing, fuelwood and poles, charcoal-making, collection of medicinal plants and hunting). The removal of forest cover can create new frontiers for settlement, but at great cost to existing communities at the forest edge, and most particularly to groups economically dependent on the forest itself. Patterns of forest product demand, use and supply are changing almost everywhere, particularly as forests are cleared and product supply is limited to bush fallow and farm trees. The balance of individual and communal tenure of forest resources tends to change, and the social arrangements surrounding forest product extraction are complex and fluid (Coomes and Barham 1997).

The state has a vital role in supporting institutional arrangements that maximise sustainable livelihoods, and the matching of those institutions to social and economic change is a key challenge for the future. In Costa Rica rates of deforestation have been very rapid (over 7 per cent of forest area per year was lost in the 1980s; Gottfried *et al.* 1994). The Golfo Dulce Forest Reserve (613 km²) lies on the Osa peninsula, which contains the largest area of rainforest on the Pacific coast of Central America. Some 8,000 people live within the reserve, where forest loss has run at 5 per cent per year. The Costa Rican government has sought to maintain forest cover and enhance sustainable livelihoods through natural forest management and small-scale timber enterprises. However, this strategy faces the major problem of insecure land tenure (forcing farmers to clear land to prove effective possession) and lack of credit (forcing farmers to seek a quick return through cattle-raising, even though this is ecologically unsustainable). There are also tax incentives and government loans for land clearance (Gottfried *et al.* 1994). The pursuit of sustainability in forest environments therefore depends crucially on the institutions governing social, economic and environmental change.

Joined-up thinking: the example of river basin planning

This chapter has argued that if sustainable development is to be delivered, in whole or in part, 'from below', it will require three things. First, it will need a new respect for indigenous knowledge, and new forms of engagement (and power-sharing) between development agents and local people. Second, it will require new alliances, new ways of cutting through the conservatism of industry's ingrained destructive practices and environmentalism's knee-jerk oppositionism. Third, it will require new and deeper understanding of how ecosystems work, how they respond to and recover from management, and how that recovery can be enhanced by restoration.

However, sustainable development 'from below' will also require something else, a significant loosening up of constrained ways of thinking about development and environment. It is not so much that new knowledge is needed (although it may be, since technology and the world economy keep creating new risks and new extractive and absorptive demands on the biosphere), as a new openness to ways of fitting that knowledge together: a commitment to 'joined-up thinking'. The challenge is to link sophisticated understandings of ecosystem dynamics and human impacts with innovative institutional models for planning change. There have been some interesting experiments in addressing this challenge in the field of river basin planning.

The adverse environmental impacts of river control on downstream fisheries and riparian agriculture discussed in Chapter 8 are now reasonably widely recognised. While the development of more sophisticated techniques for assessing environmental and social impacts allows economists to calculate net benefits and costs more accurately, and allows dam designers to assess the relative merits of different dam sites on a wider range of criteria, conventional approaches to river basin planning are still locked into a rather sterile technocentrist 'control and transform' mind-set in response to the natural functions of rivers and flood-plains and the existing uses people make of them (Chapter 6). It is possible to start thinking outside this straitjacket. Thayer Scudder (1980, 1988, 1991b) calls for a quite different approach to dam design and operation. In a number of African river basins, artificial floods have been proposed and in some cases released to re-establish ecological function and sustain economic activity in huge downstream floodplains (e.g. Scudder 1991a, Horowitz and Salem-Murdock 1991, Acreman and Howard 1996). This reflects developments elsewhere, for example in the USA (Hecht 1996, National Research Council 1992) and Europe (Petts 1996). By the end of the 1990s, the need to integrate dam releases and downstream environments had become widely recognised, and adopted by the World Commission on Dams (WCD 2000).

Scudder suggests that the simulation of the seasonal flood peak would make downstream production possible and also allow cultivation in the drawdown zone of the reservoir. In the case of a hydroelectric dam, management in this way would offset the many costs to downstream producers that need to be taken into account, and also open up a new resource for reservoir evacuees. In the case of dams built for flood control and irrigation, this form of management would reduce downstream flooding losses and allow further development to be piecemeal and locally instigated and managed, thus avoiding the high costs of centrally planned large-scale irrigation. Gross benefits might be smaller, for example through reduced power generation or slower expansion of irrigation, but the cost/benefit ratio would improve considerably. There would, of course, be problems, for example in management. However, as Scudder points out, it is time that attention was paid to the management of tropical river basins rather than simply their 'development', seen as a one-off process.

There have now been experiments with controlled flood releases in several places in Africa, for example on the Phongolo River in Natal in South Africa (Scudder 1991b, Bruwer *et al.* 1996), on the Senegal River between 1988 and 1990 (Hollis 1996, Salem-Murdock and Horowitz 1991) and on the Waza-Logone floodplain in Cameroon (Ngantou 1994, Wesseling *et al.* 1996). In each of these places, dam construction or other engineering works have created serious downstream ecological and economic impacts. On the River Senegal, related cropping is practised on floodplain *waalo* land (Lericollais and Schmitz 1984). The area cultivated has varied from about 150,000–200,000 ha in the 1960s (when rainfall was good) to about 20,000 ha in the drought years of the 1970s. Studies showed that the value of lost downstream production following construction of the Manantali Dam outweighed marginal benefits of hydroelectric power generation. Between 1904 and 1984 there was sufficient water both to generate 74 MW of power and to release an artificial flood large enough to inundate some 50,000 ha of land downstream in every year except the most severe drought years (1913, 1977 and 1979–84). An experimental release of an artificial flood has been carried out, although with mixed success.

The use of controlled floods to convert single-purpose dams (e.g. for hydro-electric power generation) into a tool for multi-purpose multi-environment management – indeed, the wider notion of using dams to work with the natural patterns in the rivers of Africa – is attractive, but presents problems for river managers. In particular, the idea is extremely demanding technically, since it requires knowledge of a number of complex variables (Table 12.3), and planning and project development that is based on effective real-time monitoring and decision-making. African river basin planning agencies attempting such integrated planning will require extensive training, technical support and institutional strengthening.

The task of integrating the releases of water from upstream dams and the needs of people and ecosystems in downstream floodplains cannot simply be seen as a technical one. The diversity of downstream needs makes it effectively impossible to devise a single solution that automatically takes account of all interests. One approach to the complex planning required is to involve flood-plain communities in the planning and management of releases. This has been done in the Phongolo floodplain in South Africa (Bruwer *et al.* 1996). Here some 70,000 people depend on wetland resources sustained by the flooding

Table 12.3 Knowledge required to integrate dams and downstream environments

- The topography of the downstream floodplain.
- Predictive models of probable flood volumes and durations at all points downstream of the release site.
- The implications of floodplain morphology for flood depth and duration.
- The depth and duration of flooding in past years.
- The nature of aquatic and riparian vegetation in past years.
- Changes in aquatic and riparian ecology since changes in flooding, and the implications of reversing (or further changing) those flooding patterns on ecology.
- The social and economic impacts of changes in past flooding patterns and assessment of the implications of reversing (or further changing) those flooding patterns
- Cost–benefit analyses of present and future management regimes.
- Monitoring of the hydrology of released floods.
- Monitoring of the ecological impacts of the released floods.
- Monitoring of the social and economic impacts of the released floods.

of the river. The decision to build the Pongolapoort Dam was taken in the 1950s for political reasons, but it was filled only to 30 per cent of capacity to avoid inundation of part of Swaziland. Surplus water was released from the dam to serve downstream communities, but the restructured floods up to 1984 were smaller and unpredictable in timing, and created a risky environment for floodplain resource use. In 1984 the reservoir was filled to capacity by flood-waters from Cyclone Dominoa, and larger releases began to be possible. This enabled ecological conditions in the floodplain to be restored, but it did nothing to reduce uncertainty for floodplain people. From 1983, downstream villages began to organise themselves to present their needs and interests, and gradually 'combined water committees' were set up. In 1988 a 'liaison committee' met to hear the views of all stakeholders. There are now carefully agreed proce-dures for ward water committees to communicate their needs for floods to the Department of Water Affairs. The result is reported to be positive local attitudes to the possibility of managing floodplain water effectively, and a move towards sustainable utilisation of the floodplain (ibid.).

The idea of integrating the management of upstream dams and downstream environments through controlled flood releases essentially seeks to transform river basin development from the conventional closely directed and externally imposed blueprint of future development based on large-scale projects to a more open-ended, flexible and diverse picture of locally initiated smaller-scale projects. Scudder seeks to offer dam builders and river basin planners a prac-tical alternative development model that can be implemented using existing planning frameworks. However, the implications of his suggestions are more fundamental. They start to challenge the whole established 'development from above' model of development planning (Stöhr 1981). They also challenge the Promethean arrogance of conventional approaches to development, and start to demonstrate how development planning can 'design with nature' (McHarg 1969), tackling development problems holistically and intelligently.

That development should be made to come 'from below' has become almost axiomatic in the 1980s and 1990s. However, while this is a rhetorical flourish that is easily made, it is harder to make it mean very much. There are few studies of development written from the perspective of the developed, giving the view 'from below'. Academics, particularly anthropologists, sometimes attempt to speak for those without voices in the remote literature of devel-opment, but most researchers and almost all development professionals are condemned to some form of 'rural development tourism' (Chambers 1983). They turn up in a village on a tight timetable with a tight budget, seeking something. Whether that something is information, agreement or votes, village people might well regard with considerable caution the stranger in a jeep with city clothes, a bagful of papers and a clipboard.

Researchers are particularly suspect. In her 'letter to a young researcher', Adrian Adams (1979) tried to explain the experience of the village of Jamaane on the River Senegal of Europeans. The researcher had come from the develop-ment agency (the Organisation pour la Mise en Valeur du Fleuve Sénégal,

OMVS) filled with their presumptions and philosophy. He was but one in a long chain of visitors, mostly white, mostly staying only short periods, who transformed *un développement paysan* into *un développement administratif*. A peasant-run project based on communal work groups was gradually taken over by the Société d'État with the introduction of mechanised irrigation exclusively for rice, funded by USAID.

To the technician through whom the alienation of indigenous change began, 'develop' was 'an intransitive verb, and "development" was one and indivisible' (Adams 1979, p. 473). The changes he and his successors introduced were 'presented as neutral steps, objectively required on technical grounds as part of the development process'. The technician was an 'expert', someone with 'a halo of impartial prestige' lent by his skills, able to neutralise conflicts and package political issues as technical ones. Such an expert embodies 'modernity, progress, efficiency', and, as Adrian Adams points out, 'no-one may be an expert in his own country: it's an expatriates' role'. Ulrich Beck recognises the problematic role of the 'expert' in risk society, and calls for a 'demonopolisation of expertise'. He says that 'people must say farewell to the notion that administrations and experts always know exactly, or at least better, what is right and good for everyone' (1994, p. 29).

The technicians visiting Jamaane village were such 'experts', full of knowledge and ideas, but they could not communicate and they would not listen. They failed to see that Jamaane village was the home of thinking people, with memories of past visitors, problems of the present and ideas for the future (Adams 1979). Croll and Parkin suggest that development itself can be thought of as 'a form of self-conscious or planned construction, mapping and charting both landscapes and mindscapes' (1992, p. 31). Mainstream sustainable development is pre-eminently part of that mapping and charting process. The idea of 'development from below' challenges dominant notions of who should influence, or control, the making of those charts.

Summary

- Development 'from below' has become an increasingly important theme in development planning since the 1970s. Its roots are complex, but it is fundamental to both mainstream and countercurrent ideas about sustainable development because of the inherent limitations of technocratic planning 'from above' as a means of delivering sustainable improvements in human welfare or environmental quality.
- Indigenous knowledge is an important element in 'development from below', but its importance lies as much in what it says about the power of people to control the changes they experience as it does in the details of indigenous technology itself.
- Wildlife conservation is a prime example of a significant shift in planning 'from above' to planning 'from below', with the community. Integrated conservation and development projects are now numerous, but have

experienced a number of problems. Attempts to make wildlife 'pay its way', particularly through sustainable consumptive use (for example hunting revenues), are being vigorously pursued as a way of achieving both conservation and community development goals.

- A different form of sustainable development 'from below' is the attempt to use Northern consumer pressure and eco-labelling or certification schemes to make fields such as tropical forest management more sustainable.
- Tropical forests offer numerous examples of attempts to change systems of management, including management guidelines or good-practice guides for commercial timber extraction and community-run timber businesses.
- Sustainable development 'from below' demands innovativeness and open thinking. There are a number of examples in the field of river basin planning, for example the idea of 'planned releases' from dams to supply and maintain downstream floodplain ecosystems and economies.

Further reading

Barraclough, S.L. and Ghimire, K.B. (1995) *Forests and Livelihoods: the social dynamics of deforestation in developing countries*, Macmillan, London.

Baviskar, A. (1995) *In the Belly of the River: tribal conflicts over development in the Narmada Valley*, Oxford University Press, Delhi.

Cernea, M. (ed.) (1991) *Putting People First: sociological variables in rural development*, Oxford University Press, Oxford, for the World Bank.

Chambers, R. (1997) *Whose Reality Counts? Putting the first last*, Intermediate Technology Publications, London.

Ghai, D. and Vivian, J.M. (eds) (1992) *Grassroots Environmental Action: people's participation in sustainable development*, Routledge, London.

Hviding, E. and Bayliss-Smith, T. (2000) *Islands of Rainforest: agroforestry, logging and ecotourism in Solomon Islands*, Ashgate, Aldershot.

McCully, P. (1996) *Silenced Rivers: the ecology and politics of large dams*, Zed Press, London.

Murphy, D.F. and Bendell, J. (1997) *In the Company of Partners: business, environmental groups and sustainable development post-Rio*, The Policy Press, Bristol.

Richards, P. (1985) *Indigenous Agricultural Revolution: ecology and food production in West Africa*, Longman, London.

Usher, A.D. (ed.) (1997) *Dams as Aid: a political anatomy of Nordic development thinking*, Routledge, London.

Western, D., White, R.M. and Strum, S.C. (eds) (1994) *Natural Connections: perspectives in community-based conservation*, Island Press, Washington, DC.

World Commission on Dams (2000) *Dams and Development: a new framework for decision-making*, Earthscan, London.

Web sources

<http://www.iucn.org/2000/communities/content/index.html>   Information on the policies of the World Conservation Union (IUCN) on people and conservation.

<http://www.iucn.org/themes/sui/index.html>   The World Conservation Union (IUCN)

Sustainable Use Initiative; describes efforts to promote the sustainable use of wild renewable resources.

<http://www.campfire-zimbabwe.org/> The website of the Zimbabwean CAMPFIRE Association. CAMPFIRE is an innovative approach to rural communities in the driest regions of Zimbabwe to obtain sustained incomes from safari hunters, effectively making wildlife 'pay its way'.

<http://www.fsc-uk.demon.co.uk/> The Forest Stewardship Council (FSC) is an international non-governmental organisation founded in 1993 to promote good forest management worldwide through certification of forest products. This FSC UK site explains how this is done in relation to UK timber producers and timber product traders.

<http://wedo.org/> The Women's Environment and Development Organization is an international advocacy network seeking social, political, economic and environmental justice through the empowerment of women.

<http://www.livelihoods.org/index.html> A site that offers guidance and distance learning materials to those interested in adopting the UK Department for International Development (DFID) 'sustainable livelihoods' approach: thought-provoking good sense.

<http://iclei.org> International Council for Local Environmental Initiatives, an international environmental agency for local governments aimed at promoting globally the conditions for sustainable development and global environmental improvement through local actions; it has more than 350 towns, cities and local governments as members.

<http://www.un.org/esa/sustdev/success.htm> Sustainable development success stories from the United Nations.

13 Green development: reformism or radicalism?

> But be ye doers of the word, and not hearers only, deceiving your own selves.
> (James 1:22, Holy Bible, Authorised Version)

Claiming sustainability

There is a tension that has run through this book, and that also runs through a great deal of writing about the environment and development, both South and North. The argument set out here has drawn a distinction between 'mainstream' approaches to sustainability, and various more radical countercurrents. This distinction is not an absolute one, as even a cursory closer analysis of the kaleidoscope of ideas about sustainability soon demonstrates. Nonetheless, sustainable development thinking contains both technocentric and ecocentric responses to the threats of development to environment and people. The former tend to be pragmatic, involving technical and implementable steps towards the reform of development practice. The latter tend to be radical, and demand fundamental change in political economic structures.

It has not been my intention to argue that one end of this reformist–radical continuum is right and the other wrong. Nor have I tried to identify some pure essence of thinking within which 'real' sustainability is to be found, a secret policy formula with which the entire stew of muddled and well-meaning thinking of recent decades can be clarified and made both wholesome and palatable. There is no simple and single recipe for sustainability, and no easy answers for those who want to continue the job of global 'development', in the classic normative sense as progress towards the goal of universal human improvement (Cowen and Shenton 1996). It follows from this that there are hard decisions to be made. These are multiple, and contingent on endlessly repeated dilemmas in different places and at different times. Sustainability is not something that can be delivered formulaically through the adoption of new and improved analytical structures (even those as effective as environmental economics), or new planning procedures (even those that seek to achieve 'development from below'). Pitfalls await those who expect a quick and unproblematic switch from 'unsustainable' to 'sustainable' development projects simply by altering the style of development planning, the nature of consultation with affected people and the scale of projects (Conroy and Litvinoff 1988).

The end of the twentieth century has seen the collapse of confidence in the myth of development as continuous human improvement. While economists (and aid donors) worry most about the measurable economic records of projects, even economic successes can involve substantial costs, and these are often borne most unequally. Development in practice too often holds little comfort for the poor. Furthermore, development everywhere tends to transform, homogenise and degrade biological diversity. Development offers mixed prospects to environmentalists, whether wealthy nature lovers in industrialised countries or poor 'ecosystem people' dependent on ecosystem functions and resources for survival (Gadgil and Guha 1995).

Clearly, 'development' is not necessarily good; it depends who you are, it depends how the structures in society expose you to its hazards or open to you its fruits. It depends how you value the changes created around you by others, and whether your own voice can gain purchase on the behemoths of state planning and business profit-seeking. Development planning involves choices, and tough decisions. Very often in the past those decisions have been taken by 'experts', trained to see the world through clever but reductionist lenses, and insulated by wealth, culture and place of residence from the consequences of their decisions. However, even where planning is brought down to earth, dragged out of the tangle of government bureaucracy and politics, extracted from the spreadsheets of experts and freed from the stranglehold of consultancy contracts, the hard decisions do not go away.

We know the limitations and failures of development, of course. They have been key elements in the litany of sustainable development since the Stockholm Conference in 1972. Indeed, 'sustainable' development has been one of the ideas through which we have sought to recapture a sense of moral trajectory, and a means of measuring our success in driving economies and societies forwards. Through the 1980s and 1990s, more and more has been added to the concept, until it groans under the weight of ideas not only about the environment, but about equity, democracy, openness and freedom. As the economic system has become increasingly globalised, with power leaking from nation states towards transnational corporations linked in a highly interconnected global order (Lash and Urry 1994), the attraction of the moral agenda apparently offered by sustainable development has grown. However, 'sustainable development' offers no escape from the dilemmas of development. The huge achievement of the debate about sustainability has been that it has expanded the horizons of development thinking to embrace the environment. However, it offers no resolution of the moral ambiguities inherent in development. It offers no route around development's hard choices.

Development does create victims, such as those displaced by dams, those whose subsistence is taken and sold by logging companies, or those whose livelihoods or health are destroyed by factory effluent. Even if these development projects generate benefits – creating employment, paying taxes for government to recycle in improved health services or water supply systems, or generating cheap power for small enterprises – these benefits are reaped at a larger spatial scale by others, elsewhere, later. If politicians prefer soldiers to teachers and limousines to schools, these benefits are dissipated. If environmental regulation is poor and the costs of pollution are successfully externalised by industry, the benefits may never outweigh costs, and the victims of development may keep appearing for generations to come.

To an extent, sustainability can be sought through the classic strategies of the mainstream, improving governance and regulation and planning, revising economic assessments to internalise costs, reforming industrial processes to minimise risk, and seeking to mitigate and compensate for environmental and socio-economic costs. Much can be achieved by a reformist approach to sustainability. Yet this leaves much that is undone. Arguments about the costs and benefits of development that aggregate across place and time may well not satisfy those whose livelihoods and lives are shattered by development, or who suffer environmental degradation today. For them, a reformist approach may not yield sustainability or environmental justice on a meaningful timescale. Therefore, sustainability must not only be planned for, but also be *claimed*.

This is the real contribution that the Western environmental movement has made to debates about environment and development in industrialised countries. It is easy to dismiss Western environmentalism as selfish, the concern about obscure rare species and the money spent on their protection, and the apparently selfish obsessions with marginal aspects of environmental quality (access to the countryside, scenic beauty, organic food), as the special pleading of an effete and selfish class of global hedonists. And yet behind these obsessions lies the solid core of challenge to the emerging pattern of economy and society of the late twentieth century, the recognition that 'it doesn't have to be like this'. Most successful environmental protests of the past three decades have been built on a shared recognition that things need to be made to happen differently, whether the issue be the disposal of oil platforms or of non-returnable bottles, the thinning of the ozone layer or the clearance of old-growth forests (Rowell 1996, Rawcliffe 1998). From the Monkey Wrench Gang to the Sierra Club, Friends of the Earth to Earth First!, environmentalists have recognised that sustainability (whatever they called it) was not something that could safely be left to industry and government, to expert technical planners and politicians (Abbey 1975, Bookchin and Foreman 1991, Devall 1988).

In the South, too, sustainability has to be claimed. However complete the transformation of the mind-sets of development planners, and the adoption of the sustainable development mainstream as orthodox development thinking, it will be necessary for individual people and communities to lay claim to the prerequisites for a sustainable livelihood. Beyond a certain point, sustainability is not something that can be administered from above; it has to be seized from below. The poor, in particular, need to become environmental activists, both against their own degradation of the environment on which they depend, and against the environmental impacts of development. Three kinds of responses to environmental risk and degradation, and to the unwanted impacts of development, can be distinguished: adaptation, resistance and protest.

Adapting for sustainability

The most basic response to environmental degradation or risk is for people to adapt their lives and systems of production to cope with it. The livelihoods of people in high-risk or highly variable environments tend to exhibit considerable flexibility. Thus, for example, pastoralist groups in drought-prone savanna environments are remarkably flexible in their use of terrain in the face of seasonal and inter-annual rainfall variability. Flexibility is built into the composition of their herds, their relations with adjacent agricultural groups and their awareness of rich kinship networks (Behnke *et al.* 1993; see Chapter 7).

Farmers in drought-prone dryland environments such as the Sahel have also developed livelihood systems that allow them to adapt to environmental and economic conditions between and within years (Mortimore and Adams 1999). In agriculture, these include diverse crop varieties, diverse cropping systems, and integrated management of crops and livestock. Farm households choose crops, particular varieties of crops and cropping mixtures to suit the soils of their fields, their observations and expectations of rainfall, and the availability of labour to manage them. However, agricultural activity is also balanced against off-farm income from buying and selling food or petty products, from craft activities (making palm leaf mats, for example) and seasonal and even longer-term migration in search of work. The effort invested at the household level by different people into these different activities is a direct response to the opportunities each offers, and particularly to the amount and timing of rain. For any one household, the portfolio of human and technical resources available varies through each season and from one season to the next, and so too does their decision about what they do. There is, of course, considerable variation between one household and another in both their endowment of social and environmental resources and their decisions about how they should be allocated (ibid.).

The same organising principles of livelihood adaptation prove useful when environmental or socio-economic change is endogenous (from within local society), driven by the development process. In the face of deforestation, peasant farmers respond to shrinking forest and land resources defensively, trying to maintain traditional systems of resource management, to intensify crop, livestock and forest production, and squeezing consumption (Barraclough and Ghimire 1995). Thus in Lusotho District in the Usambara Mountains of Tanzania, growing population pressure was contained with minimal deforestation by adaptation of consumption and production systems. Poorer families used less fuelwood and other forest products and more bricks and other non-forest materials. Livestock were fewer in number but of better quality, and were stall-fed and not open-grazed; higher-yielding crops such as cassava, maize and Irish potatoes were grown, even supplanting coffee; vegetables were grown for urban markets, and farmers invested in irrigation, contour-bunding and

tree-planting (ibid.). These adaptations of agricultural husbandry, echoed elsewhere such as at Machakos in Kenya (Tiffen *et al.* 1994), were strategies to
maintain environmental productivity. Their relative success was related to the
supportive institutional environment created by the Tanzanian state. Similar
pressures in different circumstances in Totonicapan in Guatemala (particularly
with regard to land tenure) prevented Mayan farmers from making a success
of such adaptations (Barraclough and Ghimire 1995).

Sometimes production systems can be adapted to cope with even traumatic
development-induced environmental change. The example of the adaptation of
farmers in the wetlands of the Hadejia-Jama'are Valley of north-eastern Nigeria
to desiccation caused by upstream dam construction and several years of low
rainfall was described in Chapter 8. Production of wet-season flood rice and
flood-recession crops such as cowpeas declined through much of the area, as
did the fishery (Thomas and Adams 1999). However, over time most communities were able to adapt their agriculture and re-establish or even enhance
their livelihoods. This was possible because of new agricultural technology that
fortuitously became available in the area, in the form of imported petrol pumps
and shallow tubewells. It would be a mistake to expect this kind of luck to
hold more generally, but the principle that people can and do adapt successfully to regain and retain sustainable livelihoods, and that they will work to
make and keep those livelihoods sustainable, is fairly universally valid. The institutional context (of state, market and civil society) is enormously important to
their capacity to adapt successfully, but in favourable circumstances they will
be able to do so.

Of course, in this Nigerian example by no means everybody was able to
profit by the newly available technology and regain their lost livelihoods. The
patterns of impact and response were socially, spatially and temporally quite
complex. Some parts of the floodplain remained dry, and not all areas were
suitable for irrigation. Also, the desiccation of the floodplain led to disagree-
ments about fishing rights between long-established and recently arrived fish
catchers, while the new technology of irrigation pumps allowed dry-season
farming in places previously available for Fulani cattle to graze, resulting in a
series of violent local conflicts between farmers and graziers.

A more universal discriminator of ability to adapt successfully is wealth.
Barraclough and Ghimire describe how in the hill districts of Rasuwa and
Nuwakot in Nepal, poorer households used less fuelwood and no longer kept
fires burning overnight, while richer households with sufficient land planted
trees for fodder and fuelwood; poor households cut even mango trees for wood
and to make room for crops (Barraclough and Ghimire 1995). Not everybody
can adapt to achieve sustainability, even when circumstances are kind.

Resistance to development

A second response strategy to environmental risk and the environmental and
social impacts of development is resistance. The transformation of rural

economies by colonial and post-colonial states has often involved direct state coercion of rural producers in the name of development (Williams 1981, Crummey 1986). The state's ability to 'capture' the peasantry politically and economically (or its failure to do so) has been widely discussed (Hyden 1980). Many Third World governments expected peasant farmers to contribute to development by providing the resources for others to develop the urban indus- trial economy (Williams 1976). With an agricultural policy aimed at taxing producers to pay for urban infrastructure, and price controls to keep urban food prices down, the construction of projects such as hydroelectric dams in rural areas may well make the state and its 'development' a highly unattrac- tive actor on the rural scene (Good 1986).

Peasant farmers rejected colonial 'development', as they reject development today, for many reasons. One is its alien nature, and the drastic cultural changes it demands. Another is simply its technical blindness and incompetence. For several years in the late 1940s, expatriate agriculturalists tried to grow ground- nuts on the Niger Agricultural Project at Mokwa in Nigeria, with conspicuous lack of success. Their plans for mechanised production by smallholders simply did not work. Nigerian farmers were reluctant to accept contracts as settlers; they did not think that they were being offered 'development', but that they were being called to the rescue (Baldwin 1957). The inadequacy of such blun- dering projects in post-war Africa, most notoriously the Groundnut Scheme at Kongwa in Tanganyika, makes the reluctance of peasants to become involved seem entirely understandable. As was pointed out in Chapter 11, the myth of the conservative peasant ran deep in the colonial mind, even though contem- porary commentators such as Faulkner and Mackie pointed out that 'the prevalent idea that the native farmer is excessively conservative is largely due to the mistakes of Europeans in the past' (Faulkner and Mackie 1933, p. 7). The failure of development projects was blamed on this non-participation, the 'intransigence and primitiveness of peasants' (Hill 1978). Developers saw – and to some extent still see – a problem in the recalcitrance of peasants to outsiders' conceptions of progress, and their reluctance to join in as required (Williams 1981). Such attitudes persist to a greater extent than might be expected given the almost universal protestations about indigenous knowledge and participa- tory development, for example in attitudes to rainforest people, or people reluctant to be resettled from behind a hydroelectric dam.

Reluctant participation in the projects of the state forms part of wider peasant resistance. Commonly, subordinate classes resist impositions and demands made upon them (whether by the state or richer neighbours) silently, subtly, passively and without overt organisation. They meet the demands for food, labour, taxes, rents and interest on loans with what Scott calls '*everyday* forms of peasant resistance' (p. xvi, emphasis in the original): 'the ordinary weapons of relatively powerless groups: foot dragging, dissimulation, desertion, false compliance, pilfering, feigned ignorance, slander, arson, sabotage and so on' (1985, p. xvi).

Spittler (1979) describes the defensive strategies of peasants faced with the unreasonable demands of the colonial state in Niger. One strategy was evasion:

lookouts outside villages reported the arrival of soldiers and policemen, and people hid in the bush to evade tax collection, requisitioning or labour demands, and hid cattle and children during censuses. Another strategy was silent disobedience, for example ignoring orders about groundnut smuggling or voting; another to misuse material or money or livestock given for a specific purpose; another to end difficult interviews by agreeing with everything. The state in turn responded with a paradoxical mix of 'laissez-faire and force' (ibid., p. 33), often alternating the two, for example rounding people up who refused to participate in a development project, but eventually dropping it.

Scott's own study *Weapons of the Weak* concerns conflict within a village on the Muda Irrigation Scheme in Peninsular Malaysia, where mechanisation of the rice harvest in the second half of the 1970s had drastically reduced field labour opportunities for poor households. He is concerned to understand both the acts of resistance and their symbolic context. In the theatrical metaphor he uses, he is interested in both the 'on-stage' behaviour and the 'offstage' language that people use to contextualise it. The rich are effectively immune to material sanctions, but not to what Scott calls 'symbolic sanctions', such as slander, gossip and character assassination. He comments that

> those with power in the village are not, however, in total control of the stage. They may write the basic script of the play but, within its confines, truculent and disaffected actors find sufficient room for manoeuvre to suggest subtly their disdain for the proceedings. The necessary lines may be spoken, the gesture made, but it is clear that many of the actors are just going through the motions and do not have their hearts in the performance.
>
> (Scott 1985, p. 26)

Protest for sustainability

The third form of response to environmental degradation and imposed development is open protest. Resistance to coercion may be typified by 'silence and stealth' (Crummey 1986, p. 10), but many protests are more vocal or active. In the context of colonial Africa, for example, Bates links the commercialisation of agriculture with the rise of political protest in the rural areas of colonial Africa (1983, p. 104), and the importance of forced terracing campaigns in focusing political mobilisation in countries such as Kenya and South Africa is now widely recognised (Throup 1987, Beinart 1984, Beinart and Coates 1995). However, peasant movements are typically limited in aims and achievements, and deficient in organisation and execution, as Beinart and Bundy (1980) commented in the context of the Transkei.

While peasant revolts have fascinated Western scholars excited by Marxism or the Vietnam War, peasants rarely engage in open revolt. When they do, the revolt is rarely successful. Hildermeier (1979) distinguishes between agrarian revolts, often short-lived and violent, and longer-term and more peaceful agrarian movements. The latter are less dangerous and more common. The land occupancy protests that followed the alienation of common lands in Latin America by *haciendados* in the 1920s were 'more or less spontaneous and localised affairs, involving peaceable squatting in the first place, but almost inevitably developing later, as the troops moved in, into violent confrontations, at great cost in peasant lives' (Rudé 1980, p. 69). Crummey refers to this as the waking of 'the other beast' of state violence (1986, p. 21).

Rural protest at state action is important in the dynamics of rural change in many Third World countries. It occurs in particular in response to the expanding centralised bureaucracy of development, to the central role of 'outsiders' and 'experts' in the development process and to the impacts of transnational agribusiness enterprise. Despite numerous examples of failure in development, many schemes are, as Hill (1978, p. 25) points out, 'little better planned than their more spectacularly misbegotten predecessors', and many government development plans have been the stimulus for organised grassroots protests.

One much-quoted example is the Chipko movement in the Garhwal Himalaya in India. The historical roots of Chipko are long and complex (Guha 1989). The adoption of 'scientific' forestry, in both the traditionally governed Tehri Garhwal and the colonially governed Kumaon, and the reservation of forests for timber production towards the end of the nineteenth century caused considerable disruption of production systems, and considerable hardship (Guha 1989). In both areas, in different ways and with different effects, peasants protested about their exclusion from the forests throughout the twentieth century. Tactics included strikes and go-slows in providing porters for visiting officials, refusals to pay fines and incendiarism (the standard technique of burning litter on the forest floor to open it up for grazing was banned in order to improve tree regeneration). By 1921, Guha reports 'a near total rupture between the colonial state and its subject population' (ibid., p. 137).

Chipko therefore built on a long history of challenges to the demands of commercial forestry. Such protests had been openly rebellious, and had been met with state violence (notably in Tehri Garhwal in the 1940s). Protests, and government attempts to dissipate or appease them, continued. Conflict over access to forests and allocation of felling rights in trees, as well as problems of landslides and floods, led to a series of actions to prevent contractors from logging forests in the 1970s. The action of women to prevent the cutting of trees in the village of Reni in 1974 (Guha 1989) has become the centre of the global environmentalist myth of Chipko. Guha concludes *The Unquiet Woods* by suggesting that 'peasant movements like Chipko are not merely a defence of a little community and its values, but also an affirmation of a way of life more harmoniously adjusted with natural processes' (1989, p. 196).

Chipko is in fact a very diverse movement politically. It has also been remarkably effective, achieving an effective ban on green felling in the Himalaya above 1,000 m in 1981. However, it is the wider ramifications of Chipko both within India and globally, and more particularly of beliefs about it, that are particularly important. Bandyopadhyay (1992) describes the way in which Chipko metamorphosed from a peasant movement to a global campaign for the sustainable management of forests in general, and those of the Himalaya in particular. He suggests that Chipko is 'no longer a hill people's movement against forest fellings', but has become a philosophy, an extension of Gandhian thought.

This may well be true, but other observers are critical of the romanticism of some of those enthralled by Chipko. Rangan criticises environmentalists inside and outside India for being 'rapt and slavish in their adoration and assiduous pursuit of romance with Chipko's ecological reincarnation' (1996, p. 222). He suggests that Chipko's leaders are in fact reactionary, their allegiance to the myth of Chipko preventing the protest and arguments of village leaders and activists from being heard outside the region. There are militant local calls for more development, for tree-felling rather than tree-hugging, and for the establishment of a new state of Uttaranchal. Rangan argues that both Chipko in its early days and the Uttaranchal movement have addressed access to resources, but that while the latter's message of secular development and social justice commands widespread political support locally, Chipko has won the support of government and environmentalists, and the wider movement's voice is stifled. It has also influenced protest against another kind of development, the construction of the Tehri Dam Project in Tehri Garhwal (Bandyopadhyay 1992).

The construction of large dams has been one of the most controversial areas of large-scale development (see Chapter 8), and it is one that has occasioned considerable protest, both locally among reservoir evacuees and internationally, within the environmental movement. Anti-dam protest within the environmental movement prior to the 1970s involved essentially preservationist opposition to the inundation of wilderness areas. Gradually, however, protest extended to include directly threatened communities, and to address the plight of evacuees and the wider issues of unsustainable development (McCully 1996). One of the formative moments in the early history of the environmental movement in the USA was the battle between John Muir (and the Sierra Club) and the city of San Francisco over the proposal to flood the Hetch Hetchy Valley in Yosemite National Park in the early years of the twentieth century. That campaign was self-consciously re-echoed in the 1950s in the ultimately successful campaign against the US Bureau of Reclamation's proposal to build construction of the Echo Park Dam on the Green River in Utah in the 1950s, and the unsuccessful campaign against Glen Canyon Dam on the Colorado a few years later (ibid.). That dam was closed in 1963, but protests by conservationists continued, and by the end of the 1980s the Bureau of Reclamation had more or less abandoned large dams as a formal and official policy. From 1976 a Dam Fighters' Conference was held annually in the USA.

Elsewhere in the industrialised world, similar protest campaigns took place. Indeed, Usher (1997d) argues that opposition to dams exists in almost every country where there is democratic space to express dissent. In Sweden, for example, there was fierce debate about dam construction on the Vindel River in the 1960s, and opposition to dams was strong through the 1970s. An inter-basin River Savers Association was established in 1974, and in 1987 a Natural Resources Act halted all dam construction on the last four large free-flowing rivers (Lövgren 1997). In Arctic Norway a proposal to dam the Alta River in the 1970s led to intense protest (including civil disobedience, protest camps on-site and hunger strikes by Sami protesters). Protest was stopped by the police (with the army in reserve), and the dam was finally finished in 1987 (Dalland 1997). In Australia the Tasmanian Wilderness Society led protest against the proposal to dam the Gordon River in the early 1980s (McCully 1996). In India the successful campaign by the Kerala Shastra Sahitya Parishad against the flooding of the Silent Valley in Kerala was essentially similar: the land threatened was tropical forest, rich in wildlife, but containing few human evacuees (Singh *et al.* 1984, Singh 1997).

Protest against dams elsewhere in the world is less easily classified within the confines of the conservation movement. Protest about the construction of dams formed the focus for broad social movements in a number of countries in Eastern Europe during the 1980s, before the collapse of the Iron Curtain, for example in Bulgaria and Latvia, and in Hungary against dams on the Danube at Nagymaros and Gabčíkovo (McCully 1996). In the Third World, dams have triggered powerful social movements of social and environmental protest, for example in Brazil in the 1980s.

One example of such protest is the Bakolori Project in northern Nigeria. This consists of a dam on the River Sokoto to store water to supply an irrigation scheme of 30,000 ha, one of a series of large-scale irrigation projects developed in the 1970s that have been the subject of considerable criticism for their high costs, poor planning and adverse socio-economic impacts (Wallace 1981, Adams 1992). Land clearance on the area to be irrigated took place on land already farmed with rain-fed crops. Farmers complained about the land expropriation and reallocation process, about planting bans and damage to growing crops, and above all about compensation for economically important trees and other improvements. From 1978 to 1980 there were repeated blockades of parts of the project area by aggrieved farmers (Adams 1988b). These culminated in a violent police action in 1980 in which a substantial number of people died. While this did not solve the grievances, it effectively ended overt public protest about the development. Beckman says of these events, 'the Bakolori peasant rising and the extreme violence by which it was repressed will continue to cast long shadows over the rural scene in Nigeria, providing inspiration to some and warnings to others' (1986, p. 154).

Studies of Bakolori that embrace this protest have placed it firmly in the context of national political economy, and particularly the concentration and centralisation of state power over the rural sector in the 1970s. Oculi stresses the powerful position of the banks and transnational corporations at the time of project appraisal at Bakolori (which came before the oil boom), and the undermining of local planning through the co-option of local state officials (in the Federal River Basin Authority) via 'bribery or ideological collaboration' (1981, p. 204). Beckman (1986) develops a similar argument, focusing in the case of Bakolori on attempts by the state to control land and labour, and the common concerns of international capital and the Nigerian elite. Andrae and Beckman (1985) focus on the 'wheat trap', placing the failure to develop large-scale irrigation for foodgrain production in the context of increasing dependence on imported wheat, and the power of milling and baking interests. The nature of the planning process at Bakolori is also extremely important in understanding the nature of the protests (Adams 1988a). Policy-making is not mechanistic or simple, rather 'the whole life of policy is a chaos of purposes and accidents (Clay and Schaffer 1985, p. 192). The question of what actually happens in project planning is seriously under-researched.

The outstanding example of dam protest, however, is that of Narmada Bachao Andolan against the Sardar Sarovar Dam on the Narmada River (Baviskar 1995). The notion of damming the Narmada was first considered in 1946, although planning waited until the resolution of disputes between Maharashtra, Gujarat and Madhya Pradesh on water-sharing in 1978. The Sardar Sarovar Project forms only one element in a vast programme of river engineering, the Narmada Valley Project, which includes 135 medium and 3,000 minor dams. The Sardar Sarovar is the second largest, and construction began in earnest in 1985 with World Bank funding. The dam is 139 m high and is intended to supply drinking and irrigation water, and to generate hydroelectricity. It will flood 37,000 ha of land and displace 152,000 people in 245 villages in the three states (ibid.). These include hill *adivasis* and other groups. In addition, substantially greater numbers of people (up to 1 million) will be affected who live outside the reservoir area, for example in linked forestry schemes or in irrigation infrastructure (ibid.).

There was a short-lived movement against the project in the 1970s which died when the politician who began it was elected and dropped the issue. Anti-dam mobilisation began again in 1985; its initial focus was a demand for adequate rehabilitation for evacuees, but the Andolan subsequently developed its position to outright opposition. The campaign has been based on popular mobilisation within the Narmada valley. Since 1988, people threatened by the reservoir have removed survey markers, held demonstrations both in the reservoir area and outside, and staged hunger strikes (Baviskar 1995). In 1989 for example the Andolan organised a National Rally against Destructive Development, and in 1990 they marched on the dam site and declared a programme of non-cooperation with the state. Many of these protests were met with violence by the police and Indian Administrative Service.

However, the work of the Narmada Bachao Andolan has not been confined to localised protest. It has been extensively supported by other groups, both within India and internationally. Urban environmental NGOs within India have raised funds and lobbied, and rural mass organisations from elsewhere in the country have joined protests. The movement has carried out research and launched legal challenges within India, and has proposed alternative development strategies (Baviskar 1995). Internationally, lobbying of the US Congress by NGOs put pressure on the World Bank to stop funding the dam, and an independent review was carried out. Its report, in 1992, was highly critical of the project, and the Bank pulled out of the project in 1993. At the same time, the Japanese government suspended aid to Sardar Sarovar in response to lobbying by Friends of the Earth Japan.

The response of the Indian government to the World Bank's withdrawal of funding was an escalation of violence against protesters, and an attempt to complete the project using Indian finance. Flooding began as a result of the part-completed works in 1993. In 1994 the Narmada Bachao Andolan filed a case against the project in the Supreme Court in Delhi, which called for a detailed review of the project; further hunger strikes were begun (McCully 1996).

Protests against dams have become increasingly internationalised, and increasingly linked with wider social movements. For example, the Narmada Bachao Andolan helped establish the National Alliance of People's Movements in 1996 (McCully 1996). Internationally, *The Ecologist* magazine began to campaign on large dams in the 1980s (Goldsmith and Hildyard 1984), while in the USA, publication of the *International Dams Newsletter* began in 1984. From this grew the International Rivers Network and its newsletter *World Rivers Review* (McCully 1996). In 1988, activists from round the world met in San Francisco and signed a 'San Francisco Declaration' which demanded a moratorium on all new large dams that failed to meet criteria for participation by those affected, access to project information, and environmental, social health, safety and economic performance (ibid.). In 1994 the Manibeli declaration was presented to the World Bank's president on its fiftieth anniversary, signed by 326 groups and coalitions in forty-four countries calling for a moratorium on all loans for large dams until certain conditions were met (ibid.).

Sustainability and civil society

Clearly, people threatened by development both adapt and resist. Sometimes that resistance is informal, small scale and hidden. At other times it is organised and overt, sometimes even involving illegal acts, and certainly often awaking violence by the state or its allies. Increasingly, protesters link internationally, and try to exert pressure not only locally but also at different points through the structures that drive or regulate change. Sometimes improved development planning is sufficient to move towards sustainability, and 'win–win' solutions are possible. At others the hard choices inherent in development still have to be made, and when they are made, the decision comes down against the poor, the marginal, the uneducated and the powerless.

Faced with such challenges, people can sometimes adapt successfully, and failing that they can flee, either down networks of kinship to cities, or to some resource frontier. Alternatively, they can stay in place and resist passively. Under some circumstances they organise, making their resistance directed and active. Under what circumstances can those who need to claim livelihood sustainability become activists? Broad (1994) identifies three necessary conditions for activism. First, the natural resource base on which people depend has to be threatened; second, people have to have lived in the area for some time or have some sense of permanence there; third, civil society must already be somewhat politicised and organised. On Mindanao in the Philippines, for example, people acted 'once environmental degradation began to transform poor people who lived in a stable ecosystem into marginal people living in vulnerable and fragile ecosystems' (ibid., p. 814).

It might seem obvious from the perspective of Western environmentalism that it is a good idea to organise against developments that challenge sustainability and in favour of those that foster it. However, there are strong reasons why people do not organise in this way. Very often, people fear the consequences of collective action, whether from the state or from other actors. In the language of institutional economics, the anticipated costs of deviation from established (and often imposed) norms are not balanced by anticipated benefits. People dare not rock the boat. Linked with that legitimate fear is powerlessness: without education to make a case in the formal terms required by a court of law, without the money to pay lawyers or transport to reach a politician to petition, without the knowledge of the development planned or of the bureaucratic process of project decision-making, without enough time to do anything significant, how can people organise themselves to protest, or to suggest what should be done?

Many other factors stifle voices of protest or alternative futures (Scott 1985). Changes that are sudden, sharp and unexpected (for example the introduction of combine harvesters in the village of Sedaka on the Muda Irrigation Scheme) may stimulate people to organise, but many changes are gradual and piecemeal. People cope with them progressively, and often alone. Changes that create complex (and especially not visible) categories of winner and loser are also less likely to lead to collective responses than those that create a large and self-aware group of losers. Other obstacles to collective action lie in class structure, and in the cleavages and alliances that cut across it such as kinship, friendship, faction, patronage and ritual ties (Scott 1985).

The prospects for grassroots demands for sustainable management of the environment, and for sustainable livelihoods, depend a great deal on the nature of civil society. One way of looking at this is to use the slightly confused term 'social capital' (Harriss and de Renzio 1997). This entered the language of development studies through Robert Putnam's book on Italy, *Making Democracy Work* (1993). He argued that networks of trust and shared norms in society facilitate cooperation for mutual benefit, and hence provide a secure base for effective government and economic development (Putzel 1997). Through the 1990s, social capital has been the centre of enormous attention in development studies, especially because of enthusiasm for it within the World Bank, although its meaning has become progressively obscured, and the possibility of a 'dark side' has been recognised (Harriss and de Renzio 1997, Putzel 1997).

Where actors are able to organise themselves, and are allowed to organise themselves, collective social action can be remarkably effective. In the Mexican city of Monterrey, housewives meeting together organised effectively to force an improvement in water supply (Bennett 1998). By 1980, water supplies to over half of Monterrey were rationed. Meeting at the collective tap or at a street or neighbourhood meeting, housewives identified a common problem and began negotiating with government authorities. If that did not work, they began to organise direct meetings with government officials and meetings and protest rallies. The next step was to bypass the water authority and contact the state governor or the mayor. The final stage was direct action, either blocking roads or kidnapping water service vehicles. In the short term these protests usually persuaded hard-pressed engineers to find water for the protesting neighbourhood. In the longer term they helped demonstrate the scale of the water supply problem and forced the centralised government planning process to move towards investment in improved water supplies. People can, and do, organise to address the problems of poverty and environmental degradation, and the organisations and structures they see as responsible. Sometimes it works.

Green development: reformism or radicalism?

There is no magic formula for sustainable development. Despite the enthusiastic rhetoric, the technical guidelines and the oft-lauded 'greening' of development agencies, there is no easy reformist solution to the dilemma and tragedy of poverty. Behind the slogans about environment and development lies the 'hard process' of development itself, wherein choices 'are indeed cruel' (Goulet 1971, p. 326). Development ought to be what human communities do to themselves. In practice, however, it is what is done to them by states and their bankers and 'expert' agents, in the name of modernity, national integration, economic growth or a thousand other slogans.

Fundamentally, it is this reality of development – imposed, centralising and often unwelcome – that the 'greening' of development challenges. It throws attention back on the ethical questions that underlie the idea of development itself. It recognises that societies are 'developing' whether or not they are the target of some specific government 'development' scheme. In practice, in the Third World (as elsewhere) ideas, culture and the nature of society are in flux. Farming practice, production system, economy, are all sucked into the whirlpool of the world economy to some extent, moving in response to the pull of capital. There is no 'real development' to be reached for that escapes this pull, and sustainable development is no magic bridge by which it can be attained.

Part of the limitation of the sustainable development thinking and the reformist technical guidelines discussed in this book is their failure to address political economy. Without a theory of how the world economy works, and without theories about the relations between people, capital and state power, sustainable development thinking – and most conservation action – is locked within a limited compass. In practice, power and the initiative for development planning are highly centralised, globally and nationally. Development planning is centralised and imposed, as is much conventional conservation, although both seek to involve (and co-opt) local interests and both believe they are operating in the interests of some notional wider constituency. Development initiatives and the context of aid-giving and project formulation have to be understood in the way the world economy functions. Pollution and environmental degradation in the Third World have to be understood in terms of their relations to the urban, industrial cores of the First World (Chambers 1988b, p. 15).

The Green alternative in development is not simply about reforming environmental policy, it also issues a challenge to the very structures and assumptions of development. It is also, first and foremost, about poverty and human need, about sustainable livelihood security (Chambers 1988b). Debates about the mechanisms and dynamics of development have tended to obscure its ethical basis, but Chambers argues strongly that there are moral as well as practical reasons for putting poor people first in development planning. Goulet suggests that the 'shock of underdevelopment' can be overcome only by creating 'conditions favourable to reciprocity', in which 'stronger partners ... offset the structural vulnerability of weaker interlocuters by being themselves rendered politically, economically and culturally vulnerable' (1971, p. 328). The feasibility of such a vision can be debated, but the extent of its challenge to reformist tinkering with environmental aspects of development policy is clear.

Green development focuses on the rights of the individual to choose and control his or her own course for change, rather than having it imposed. The Green agenda is therefore certainly radical, but it is also open-ended, flexible, diverse. Green development is almost a contradiction in terms, not something for which blueprints can be drawn, not something easily absorbed into structures of financial planning, or readily co-opted by the state. It shares the 'politically treacherous' characteristics of O'Riordan's concept of sustainability (O'Riordan 1988, p. 30), and is subject to the very real tensions within competing ideologies of technocentric and ecocentric environmentalism (O'Riordan 1988, Turner 1988b). It is something that very often emerges in spite of, rather than as a direct result of, the actions of development bureaucracies. Green development programmes must start from the needs, understanding and aspirations of individual people, and must work to build and enhance their capacity to help themselves. As Chambers comments, 'The poor are not the problem, they are the solution' (1988b, p. 3)

This view of the desirable characteristics of development is counter-intuitive, even subversive. It is most often found among those who are recipients rather than creators of development projects. It is very much the view of the development process 'from below', looking up at what is descending from the modernising state. In her 'letter to a young researcher', described in Chapter 12, Adrian Adams presented this view with a terrible clarity. Radical dependency theorists were no more able than Senegalese development bureaucrats to acknowledge what she called the 'existence' of the village of Jamanne. But Jamanne did (and does) exist, 'here and now, having a past and a future because it is alive today' (Adams 1979, p. 477). Furthermore, there are people there with a legitimate and informed view of the development process.

Tragically, Adrian Adams (no relation to the author) was killed in Senegal in 2000, her life of commitment to the people of the Senegal Valley, in the face of outsiders' plans for 'development', cut short. But what she wrote in 1979 is still true: the village of Jamanne exists, as do countless places like it. Their people deserve and demand to be heard. Culture, society, economy and environment are complex and dynamic, changing continually, sometimes in subtle ways and sometimes dramatically and fast. 'Development' that is based on programmes and policies that are conceived and imposed within institutions distanced from those they affect is unlikely to be able to cope with these changes effectively, or to meet human needs. Better environmental and developmental planning is both needed and possible, and is at the core of mainstream sustainable development. But this is just the beginning of the challenge of greening development.

'Green' development is not about the way the environment is managed, but about who has the power to decide how it is managed. Its focus is the capacity of the poor to exist on their own terms. At its heart, therefore, 'greening' development involves not just a pursuit of new forms of economic accounting or ecological guidelines or new planning structures, but an attempt to redirect change to maintain or enhance the power of the poor to survive without hindrance and to direct their own lives. 'Sustainable development' is the beginning of a process, not the end. It is a statement of intent, not a route-map.

References

Abbey, E. (1975) *The Monkey Wrench Gang*, J.B. Lippincott, New York.

Abel, N. and Stocking, M. (1981) 'The experience of underdeveloped countries', pp. 253–95 in T. O'Riordan and W.R.D. Sewell (eds) *Project Appraisal and Policy Review*, Wiley, Chichester.

Abelson, P. (1979) *Cost Benefit Analysis and Environmental Problems*, Saxon House, Farnborough.

Ackerman, W.C., White, G.F. and Worthington, E.B. (eds) (1973) *Man-Made Lakes: their problems and environmental effects*, Geophysical Monograph 17, American Geophysical Union, Washington, DC.

Acreman, M. and Howard, G. (1996) 'The use of artificial floods for floodplain restoration and management in sub-Saharan Africa', *IUCN Wetlands Programme Newsletter* 12, 20–5.

Adams, A. (1979) 'An open letter to a young researcher', *African Affairs* 38(313): 451–79.

Adams, J. (1992) 'Horse and rabbit stew', pp. 65–73 in A. Coker and C. Richards (eds) *Valuing the Environment: economic approaches to environmental valuation*, Belhaven, London.

Adams, W.M. (1985a) 'The downstream impacts of dam construction: a case study from Nigeria', *Transactions of the Institute of British Geographers* N.S. 10: 292–302.

Adams, W.M. (1985b) 'River basin planning in Nigeria', *Applied Geography* 5: 292–308.

Adams, W.M. (1987) 'Approaches to water resource development, Sokoto Valley, Nigeria: the problem of sustainability', pp. 307–325 in D.M. Anderson and R.H. Grove (eds) *Conservation in Africa: people, policies and practice.* Cambridge University Press, Cambridge.

Adams, W.M. (1988a) 'Rural protest, land policy and the planning process on the Bakolori Project, Nigeria', *Africa* 58. 315 36.

Adams, W.M. (1988b) 'Irrigation and innovation: small farmers in the Sokoto Valley, Nigeria', pp. 83–94 in J. Hirst, J. Overton, B. Allen and Y. Byron (eds) *Small Scale Agriculture*, Commonwealth Geographical Bureau and the Department of Human Geography, Research School of Pacific Studies, Australian National University, Canberra.

Adams, W.M. (1991) 'Large scale irrigation in northern Nigeria: performance and ideology', *Transactions of the Institute of British Geographers* N.S. 16: 287–300.

Adams, W.M. (1992) *Wasting the Rain: rivers, people and planning in Africa*, Earthscan, London.

Adams, W.M. (1995) 'Sustainable development?', pp. 354–73 in R.J. Johnston,

P.J. Taylor and M.J. Watts (eds) *Geographies of Global Change: remapping the world in the late twentieth century*, Blackwell, Oxford.

Adams, W.M. (1996) *Future Nature: a vision for conservation*, Earthscan, London.

Adams, W.M. and Hughes, F.M.R. (1990) 'Irrigation development in desert environments', pp. 135–60 in A.S. Goudie (ed.) *Techniques for Desert Reclamation*, Wiley, Chichester.

Adams, W.M. and Hulme, D. (1998) *Conservation and Communities: changing narratives, policies and practices in African conservation*, Community Conservation Discussion Paper, Institute of Development Policy and Management, University of Manchester.

Adams, W.M. and Mortimore, M.J. (1997) 'Agricultural intensification and flexibility in the Nigerian Sahel', *Geographical Journal* 163: 150–60.

Adamson, R.S. (1938) *The Vegetation of South Africa*, British Empire Vegetation Committee, London.

Addo, H., Amin, S., Aseniero, G., Frank, A.G., Friberg, M., Frobel, F., Heinrichs, J., Hettne, B., Kreye, O. and Seki, H. (1985) *Development as Social Transformation: reflections on the global problematique*, Hodder and Stoughton, London, for the United Nations University.

Adeniyi, E.O. (1973) 'The impact of change in river regime on economic activities below Kainji Dam', pp. 169–77 in A.L. Mabogunje (ed.) *Kainji Lake Studies*, vol. 2, Nigerian Institute of Social and Economic Research, Ibadan.

Agnew, C. (1990) 'Spatial aspects of drought in the Sahel', *Journal of Arid Environments* 18: 279–93.

Agnew, I. (1995) 'Desertification, drought and development in the Sahel', pp. 137–49 in T. Binns (ed.) *People and Environment in Africa*, Wiley, Chichester.

Agrawal, A. and Gibson, C.C. (1999) 'Enchantment and disenchantment: the role of community in natural resource conservation', *World Development* 27: 629–49.

Ahmed, Y.J. and Sammy, G.K. (1985) *Guidelines to Environmental Impact Assessment in Developing Countries*, Hodder and Stoughton, Sevenoaks.

Allan, W. (1965) *The African Husbandman*, Oliver and Boyd, Edinburgh.

Allen, C.M. and Edwards, S.R. (1995) 'The sustainable-use debate: observation from IUCN', *Oryx* 29: 92–8.

Allen, D.E. (1976) *The Naturalist in Britain*, Penguin, Harmondsworth.

Allen, F. (1983) 'Natural resources as natural fantasies', *Geoforum* 14: 243–7.

Allen, J.C. and Barnes, D.F. (1985) 'The causes of deforestation in developing countries', *Annals of the Association of American Geographers* 75: 163–84.

Allen, R. (1980) *How to Save the World: strategy for world conservation*, Kogan Page, London.

Allen, R.N. (1973) 'The Anchicayá Hydroelectric Project in Colombia: design and sedimentation problems', pp. 318–42 in M.T. Farvar and J.P. Milton (eds) *The Careless Technology: ecology and international development*, Stacey, London.

Alley, P. (1999) From the dragon's tail to suburbia', *Earth Matters* 41: 24–6.

Altieri, M.A., Letourneau, D.K. and Davis, J.R. (1984) 'The requirements of sustainable agro-ecosystems', pp. 175–89 in G.K. Douglass (ed.) *Agricultural Sustainability in a Changing World Order*, Westview, Boulder, CO.

Ambio (1979) 'Review of environmental development', *Ambio* 8: 114–15.

Amin, M.A. (1977) 'Problems and effects of schistosomiasis on irrigation schemes in the Sudan', pp. 407–11 in E.B.Worthington (ed.) *Arid Land Irrigation in Developing Countries: environmental problems and effects*, Pergamon Press, Oxford.

Amin, S. (1985) 'Apropos the "green" movements', in H. Addo, S. Amin, G. Aseniero, A.G. Frank, M. Friberg, F. Frobel, J. Heinrichs, B. Hettne, O. Kreye and H. Seki (eds) *Development as Social Transformation: reflections on the global problematique*, Hodder and Stoughton, Sevenoaks, for the United Nations University.

Anadu, P.A. (1987) 'Progress in the conservation of Nigeria's wildlife', *Biological Conservation* 41: 137–251.

Anderson, C.W. (1912) *Forests of British Guiana* (2 vols), Department of Lands and Mines, Georgetown, British Guiana.

Anderson, D.M. (1984) 'Depression, dust bowl, demography and drought: the colonial state and soil conservation in East Africa during the 1930s', *African Affairs* 83: 321–44.

Andrae, G. and Beckman, B. (1985) *The Wheat Trap: bread and underdevelopment in Nigeria*, Zed Press, London.

Arbona, S.I. (1998) 'Commercial agriculture and agrochemicals in Almolonga, Guatemala', *Geographical Review* 88: 47–63.

Arvelo-Jiménez, N. (1984) 'The politics of cultural survival in Venezuela: beyond indigismo', pp. 105–26 in M. Schmick and C.H. Wood (eds) *Frontier Expansion in Amazonia*, University of Florida Press, Gainesville, FL.

Aseniero, G. (1985) 'A reflection on developmentalism: from development to transformation', pp. 48–85 in H. Addo, S. Amin, G. Aseniero, A.G. Frank, M. Friberg, F. Frobel, J. Heinrichs, B. Heltne, O. Kreye and H. Seki, *Development as Social Transformation: reflections on the global problematique*, Hodder and Stoughton, Sevenoaks for the United Nations University.

Ashby, E. (1980) 'Retrospective environmental impact assessment', *Nature* 288: 28–9.

Atkinson, G., Dubourg, R., Hamilton, K., Munasinghe, M., Pearce, D. and Young, C. (1997) *Measuring Sustainable Development: macroeconomics and the environment*, Edward Elgar, Aldershot.

Atkinson, T.R. (1973) 'Resettlement programmes in the Kainji Lake Region', in A.L. Mabogunje (ed.) *Kainji: a Nigerian man-made lake*; Kainji Lake Studies vol. 2: *Socio-economic conditions*, Nigerian Institute for Social and Economic Research, Ibadan.

Attwell, R.I.G. (1970) 'Some effects of Lake Kariba on the ecology of a floodplain of the mid-Zambezi Valley of Rhodesia', *Biological Conservation* 2: 189–96.

Aubréville, A. (1949) *Climats, Forêts et Désertification de l'Afrique Tropicale*, Société d'Édition Géographiques, Maritimes et Coloniales, Paris.

Auty, R.M. (1997) 'Pollution patterns during the industrial transition', *Geographical Journal* 163: 216–24.

B&Q (1995) *How Green is my Front Door? B&Q's second environmental review, February 1993–July 1995*, B&Q plc, Eastleigh, Hants.

Bahro, R. (1978) *The Alternative in Eastern Europe*, New Left Books, London (republished 1981, Verso, London).

Bahro, R. (1982) *Socialism and Survival*, Heretic, London.

Bahro, R. (1984) *From Red to Green: interviews with 'New Left Review'*, Verso, London.

Baker, K.M. (1995) 'Drought, agriculture and environment: a case study from the Gambia, West Africa', *African Affairs* 94: 67–86.

Baldwin, K.D.S. (1957). *The Niger Agricultural Project: an experiment in African Development*, Blackwell, Oxford.

Balon, E.K. and Coche, A.G. (eds) (1974) *Lake Kariba: a man-made tropical ecosystem in Central Africa*, Monographiae Biologicae 24, Junk, The Hague.

Baluyut, E.A. (1985) 'The Agno River Basin (The Philippines)', pp. 15–54 in T. Petr (ed.) *Inland Fisheries in Multi-purpose River Basin Planning and Development in Tropical Asian Countries: three case studies*, FAO Fisheries Technical Paper 265, Rome.

Bandarage, A. (1984) 'Women in development: liberalism, Marxism and Marxist-feminism', *Development and Change* 15: 495–515.

Bandyopadhyay, J. (1992) 'Sustainability and survival in the mountain context', *Ambio* 21: 297–302.

Barbier, E.B. (1992) 'Community-based development in Africa', pp. 103–135 in T.M. Swanson and E.B. Barbier (eds) *Economics for the Wilds: wildlife, wildlands, diversity and development*, Earthscan, London.

Barbier, E.B. (1994) 'Natural capital and the economics of environment and development', pp. 290–322 in A. Jansson, M. Hammer, C. Folke and R. Costanza (eds) *Investing in Natural Capital: the ecological economics approach to sustainability*, Island Press, Washington, DC.

Barbier, E.B. (1998) 'Valuing environmental functions: tropical wetlands', pp. 344–369 in E.B. Barbier, *The Economics of Environment and Development: selected essays*, Edward Elgar, Cheltenham.

Barbier, E.B., Burgess, J.C., Swanson, T.M. and Pearce, D.W. (1990a) *Elephants, Economics and Ivory*, Earthscan, London.

Barbier, E., Markandya, A. and Pearce, D. (1990b) 'Environmental sustainability and cost–benefit analysis', *Environment and Planning A* 22: 1259–66 (reprinted as pp. 54–64 in E.B. Barbier, *The Economics of Environment and Development: selected essays*, Edward Elgar, Cheltenham).

Barbier, E.B., Burgess, J.C. and Folke, C. (1994) *Paradise Lost? The ecological economics of biodiversity*, Earthscan, London.

Barbier, E.B., Burgess, J.C., Bishop, J. and Aylward, B. (1995) *The Economics of the Tropical Timber Trade*, Earthscan, London.

Barbier, E.B., Adams W.M. and Kimmage, K. (1998) 'An economic valuation of wetland benefits', pp. 322–343 in E.B. Barbier, *The Economics of Environment and Development: selected essays*, Edward Elgar, Cheltenham.

Bardach, J.F. (1973) 'Some ecological consequences of Mekong River development plans', in M.T. Farvar and J.P. Milton (eds) *The Careless Technology: ecology and international development*, Stacey, London.

Barker, D. and Spence, B. (1988) 'Afro-Caribbean agriculture: a Jamaican Maroon community in transition', *Geographical Journal* 154: 198–208.

Barnett, T. (1978) *Gezira: an illusion of development*, Frank Cass, London.

Barnett, T. (1980) 'Development: black box or Pandora's box?', pp. 306–24 in J. Heyer, P. Roberts and G. Williams (eds) *Rural Development in Tropical Africa*, Macmillan, London.

Barraclough, S.L. and Ghimire, K.B. (1995) *Forests and Livelihoods: the social dynamics of deforestation in developing countries*, Macmillan, London.

Barrett, C.B. and Arcese, P. (1995) 'Are integrated conservation–development projects (ICDPs) sustainable? On the conservation of large mammals in sub-Saharan Africa', *World Development* 23: 1073–84.

Barrow, C.J. (1981) 'Health and resettlement consequences and opportunities created as a result of river impoundment in developing countries', *Water Supply and Management* 5: 135–50.

Barrow, C.J. (1987) 'The environmental effects of the Tucuruí Dam on the middle and lower Tocatins River Basin, Brazil', *Regulated Rivers* 1: 44–60.

Barrow, C.J. (1997) *Environmental and Social Impact Assessment: an introduction*, Arnold, London.

Bartelmus, P. (1994) *Environment, Growth and Development: the concepts and strategies of sustainability*, Routledge, London.

Bass, S.M.J. (1988) 'National Conservation Strategy, Zambia', pp. 186–91 in C. Conroy and M. Litvinoff (eds) *The Greening of Aid: sustainable livelihoods in practice*, Earthscan, London.

Bassett, T.J. (1993) 'Introduction: the land question and agrarian transformation in sub-Saharan Africa', pp. 3–31 in T.J. Bassett and D.E. Crummey (eds) *Land in African Agrarian Systems*, University of Wisconsin Press, Madison.

Bate, J. (1991) *Romantic Ecology: Wordsworth and the romantic tradition*, Routledge, London.

Bates, M. (1953) *Where Winter Never Comes: a study of man and nature in the tropics*, Victor Gollancz, London.

Bates, R.H. (1983) *Essays in the Political Economy of Rural Africa*, Cambridge University Press, Cambridge.

Batisse, M. (1975) 'Man and the biosphere', *Nature* 256: 156–8.

Batisse, M. (1982) 'The biosphere reserve: a tool for environmental conservation and management', *Environmental Conservation* 9: 101–11.

Baviskar, A. (1995) *In the Belly of the River: tribal conflicts over development in the Narmada Valley*, Oxford University Press, Delhi.

Baxter, R.M. (1977) 'Environmental effects of dams and impoundments', *Annual Review of Ecology and Systematics* 8: 255–84.

Bayliss-Smith, T.P., Bedford, R., Brookfield, H. and Latham, M. (1988) *Islands, Islanders and the World: the colonial and post-colonial experience of eastern Fiji*, Cambridge University Press, Cambridge.

Bebbington, A.J. (1996) 'Movements, modernisations and markets: indigenous organisations and agrarian strategies in Ecuador', pp. 86–109 in R. Peet and M. Watts (eds) *Liberation Ecologies: environment, development, social movements*, Routledge, London.

Beck, U. (1992) *Risk Society: towards a new modernity*, Sage, London (originally published in German in 1986).

Beck, U. (1994) 'The reinvention of politics: towards a theory of reflexive modernization', pp. 1–55 in U. Beck, A. Giddens and S. Lash, S. (eds) *Reflexive Modernization: politics, traditions and aesthetics in the modern social order*, Polity Press, Cambridge.

Beck, U. (1995) *Ecological Politics in an Age of Risk*, Polity Press, Cambridge (originally published in German in 1988 by Suhrkamp Verlag, Frankfurt am Main).

Beckerman, W. (1974) *In Defence of Economic Growth*, Jonathan Cape, London.

Beckerman, W. (1994) 'Sustainable development: is it a useful concept?', *Environmental Values* 3: 191–209.

Beckerman, W. (1995a) 'How would you like your "sustainability" sir? Weak or strong? A reply to my critics', *Environmental Values* 4: 169–79.

Beckerman, W. (1995b) *Small is Stupid: blowing the whistle on the Greens*, Duckworth, London.

Beckman, B. (1986) 'Bakolori: peasants versus the state in Nigeria', pp. 140–55 in G. Goldsmith and N. Hildyard (eds) *The Social and Environmental Effects of Large Dams*, vol. 2: *Case Studies*, Wadebrige Ecological Centre, Wadebridge, Cornwall.

Behnke, R.H. and Scoones, I. (1991) *Rethinking Range Ecology: implications for range management in Africa*, ODI/IIED, London.

Behnke, R.H. Jr, Scoones, I. and Kerven, C. (1993) *Range Ecology at Disequilibrium: new models of natural variability and pastoral adaptation in African savannas*, Overseas Development Institute, London.

Beinart, W. (1984) 'Soil erosion, conservation and ideas about development: a Southern African exploration 1900–1960', *Journal of Southern African Studies* 11: 52–84.

Beinart, W. and Bundy, C. (1980) 'State intervention and rural resistance: the Transkei 1900–1965', pp. 272–315 in M. Klein (ed.) *Peasants in Africa: historical and contemporary perspectives*, Sage Publishing, Beverly Hills, CA, and London.

Beinart, W. and Coates, P. (1995) *Environment and History: the taming of nature in the USA and South Africa*, Routledge, London.

Belshaw, D. (1974) 'Taking indigenous technology seriously: the case of intercropping techniques in East Africa', *Institute of Development Studies Bulletin* 10: 24–7.

Benech, V. (1992) 'The northern Cameroon floodplain: influence of hydrology on fish production', pp. 155–64 in E. Maltby, P.J. Dugan and J.C. Lefeuve (eds) *Conservation and Development: the sustainable use of wetland resources*, IUCN, Gland, Switzerland.

Bennett, V. (1998) 'Housewives, urban protest and water policy in Monterrey, Mexico', *Water Resources Development* 14: 481–97.

Berkes, F. (ed.) (1989) *Common Property Resources: ecology and community-based sustainable developmnt*, Belhaven, London.

Berkes, F. and Folke, C. (1994) 'Investing in natural capital for the sustainable use of natural capital', pp. 128–49 in A. Jansson, M. Hammer, C. Folke and R. Costanza (eds) *Investing in Natural Capital: the ecological economics approach to sustainability*, Island Press, Washington, DC.

Berkes, F. and Folke, C. (eds) (1998) *Linking Social and Ecological Systems: management practices and social mechanisms for building resilience*, Cambridge University Press, Cambridge.

Berry, L. (1984) 'Desertification in the Sudan-Sahelian region (1977–1984)', *UNEP Desertification Control Bulletin* 10: 23–8.

Bews, J.W. (1916) 'An account of the chief types of vegetation in South Africa, and notes on the plant succession', *Journal of Ecology* 4: 129–59.

Bideleux, R. (1987) *Communism and Development*, Methuen, London.

Bigg, T. (1995) 'The UN Commission on sustainable development: a non-governmental perspective', *Global Environmental Change* 5: 251–3.

Biswas, A.K. (ed.) (1980) 'The Nile and its environment', *Water Supply and Management* 4: 1–113.

Biswas, A.K. and Biswas, M.R. (1976) 'Hydropower and the environment', *Water Power and Dam Construction* 28: 40–3.

Biswas, M.R. and Biswas, A.K. (1984) 'Complementarity between environment and development processes', *Environmental Conservation* 11: 35–44.

Blaikie, P. (1981) 'Class, land use and soil erosion', *ODI Review* 2: 57–77.

Blaikie, P. (1985) *The Political Economy of Soil Erosion in Developing Countries*, Longman, London.

Blaikie, P. (1988) 'The explanation of land degradation in Nepal', pp. 132–58 in J. Ives and D.C. Pitt (eds) *Deforestation: social dynamics in watersheds and montane ecosystems*, Routledge, London.

Blaikie, P. (1995) 'Understanding environmental issues', pp. 1–30 in S. Morse and M. Stocking (eds) *People and the Environment*, UCL Press, London.

Blaikie, P. and Brookfield, H. (1987) *Land Degradation and Society*, Methuen, London.

Blaikie, P., Cannon, T., Davis, I. and Wisner, B. (1994) *At Risk: natural hazards, people's vulnerability and disasters*, Routledge, London.

Bledsoe, C. (1994) '"Children are like young bamboo trees": potentiality and reproduction in sub-Saharan Africa', pp. 105–38 in K.L. Kiessling and H. Landberg (eds) *Population, Economic Development and the Environment: the making of our common future*, Clarendon Press, Oxford.

Bliss-Guest, P.A. and Keckes, S. (1982) 'The regional seas programme of UNEP', *Environmental Conservation* 9: 43–9.

Boardman, R. (1981) *International Organisations and the Conservation of Nature*, Indiana University Press, Bloomington.

Boffey, P.M. (1976) 'International Biological Programme: was it worth the cost and effort?', *Science* 193: 866–8.

Bolido, L. (1998) 'Helping farmers fight pests', *People and the Planet* 7(1): 18–19.

Bookchin, M. (1971) *Post-scarcity Anarchism*, Ramparts, Berkeley, CA.

Bookchin, M. (1979) 'Ecology and revolutionary thought', *Antipode* 10(3)/11(1): 21–32.

Bookchin, M. (1982) *The Ecology of Freedom: the emergence and dissolution of hierarchy*, Cheshire, Palo Alto, CA.

Bookchin, M. and Foreman, D. (1991) *Defending the Earth: a dialogue between Murray Bookchin and Dave Foreman*, ed. S. Chase, South End Press, Boston, MA.

Boserup, E. (1965) *The Conditions of Agricultural Growth: the economics of agrarian change under population pressure*, Allen and Unwin, London.

Botkin, D.B. (1990) *Discordant Harmonies: a new ecology for the twenty-first century*, Oxford University Press, New York.

Boulding, K.E. (1966) 'The economics of the coming Spaceship Earth', in *Environmental Quality in a Growing Economy* (reprinted pp. 121–32 in H.E. Daly (ed.) (1973) *Towards a Steady-State Economy*, W.H. Freeman, New York).

Bourn, D. (1978) 'Cattle, rainfall and tsetse in Africa', *Journal of Arid Environments* 1: 9–61.

Bovill, E.W. (1921) 'The encroachment of the Sahara on the Sudan', *Journal of the African Society* 20: 174–85.

Boyd-Orr, Sir J. (1953) *The White Man's Dilemma: food and the future*, Allen and Unwin, London.

Bragdon, S. (1996) 'The Convention on Biological Diversity', *Global Environmental Change* 6: 177–9.

Brandon, K.E. and Wells, M. (1992) 'Planning for people and parks: design dilemmas', *World Development* 20: 557–70.

Brandt, W. (1980) *North–South: a programme for survival*, Pan, London.

Brandt, W. (1983) *Common Crisis North–South: cooperation for world recovery*, Pan, London.

Breitbart, M. (1981) 'Peter Kropotkin: the anarchist geographer', in D.R. Stoddart (ed.) *Geography, Ideology and Social Concern*, Blackwell, Oxford.

Bremen, H. and de Wit, C.T. (1983) 'Rangeland productivity and exploitation in the Sahel', *Science* 221: 1341–7.

Brent, R.J. (1990) *Project Appraisal for Developing Countries*, Harvester Wheatsheaf, New York.

Bridger, G.A. and Winpenny, J.T. (1987) *Planning Development Projects: a practical guide to the choice and appraisal of public sector developments*, HMSO, London.

Broad, R. (1994) 'The poor and the environment: friends or foes?', *World Development* 22: 811–23.

Brockington, D. and Homewood, K. (1996) 'Wildlife, pastoralists and science: debates concerning Mkomazi Game Reserve, Tanzania', pp. 91–104 in M. Leach and R. Mearns (eds) *The Lie of the Land*, James Currey, Oxford.

Brodie, J. and Morrison, J. (1984) 'The management and disposal of hazardous wastes in the Pacific Islands', *Ambio* 13: 331–3.

Brokensha, D.W. and Scudder, T. (1968) 'Resettlement', pp. 22–62 in N. Rubin and W.M. Warren (eds) *Dams in Africa*, Frank Cass, London.

Brokensha, D.W., Warren, D.M. and Warner, O. (eds) (1980) *Indigenous Systems of Knowledge and Development*, University Press of America, Washington, DC.

Bromley, D.W. (1989) *Economic Interests and Institutions: the conceptual foundations of public policy*, Blackwell, Oxford.

Brookfield, H. (1975) *Interdependent Development*, Methuen, London.

Brooks, F.T. (ed.) (1925) *Imperial Botanical Conference, London 1924: report of proceedings*, Cambridge University Press, Cambridge.

Brown, K. (1997) 'The road from Rio', *Journal of International Development* 9: 383–9.

Brown, K., Adger, N.W. and Turner, R.K. (1993) 'Global environmental change and mechanisms for North–South resource transfers', *Journal of International Development* 5: 571–89.

Brown, L. and Wolf, E.C. (1984) *Soil Erosion: quiet crisis in the world economy*, Worldwatch Paper 60, Washington, DC.

Brundtland, H. (1987) *Our Common Future*, Oxford University Press, Oxford, for the World Commission on Environment and Development.

Bruwer, C., Poultney, C. and Nyathi, Z. (1996) 'Community-based hydrological management of the Phongolo floodplain', pp. 199–211 in M.C. Acreman and G.E. Hollis (eds) *Water Management and Wetlands in Sub-Saharan Africa*, IUCN, Gland, Switzerland.

Bryant, R. (1998) 'Power, knowledge and political ecology in the Third World', *Progress in Physical Geography* 22: 79–94.

Bryant, R. and Bailey, S. (1997) *Third World Political Ecology*, Routledge, London.

Buckley, G.P. (ed.) (1989) *Biological Habitat Reconstruction*, Belhaven, London.

Buckley, P. (1995) 'Critical natural capital: operational flaws in a valid concept', *Ecos: A Review of Conservation* 16(3/4): 13–18.

Budiardjo, C. (1986) 'The politics of transmigration', *The Ecologist* 16: 57–116.

Bull, D. (1982) *A Growing Problem: pesticides and the Third World*, Oxfam Books, Oxford.

Bunce, M. (1994) *The Countryside Ideal: Anglo-American images of landscape*, Routledge, London.

Bunker, S.G. (1980) 'The impact of deforestation on peasant communities in the Medio Amazonas of Brazil', pp. 45–60 in V.H. Sutlive, N. Altshuler and M.D. Zamora (eds) *Where Have All the Flowers Gone? Deforestation in the Third World*, Studies in Third World Societies No. 13, College of William and Mary, Williamsburg, VA.

Burgess, R. (1978) 'The concept of nature in geography and Marxism', *Antipode* 10: 1–11.

Bush, E.J. and Harvey, L.D.D. (1997) 'Joint implementation and the ultimate objective of the United Nations Framework Convention on Climate Change', *Global Environmental Change* 7: 265–85.

Butcher, D.A. (1967) *An Operational Manual for Resettlement: a systematic approach to the resettlement problem created by man-made lakes, with special relevance for West Africa*, FAO, Rome.

Buxton, P.A. (1935) 'Seasonal changes in vegetation in the north of Nigeria,' *Journal of Ecology* 23: 134–9.

Byres, T. (1979) 'Of neo-populist pipe-dreams', *Journal of Peasant Studies* 4: 210–44.

Byron, N. and Arnold, M. (1999) 'What future for the people of the tropical forests?', *World Development* 27: 789–805.

Byron, N. and Shepherd, G. (1998) *Indonesia and the 1997–8 El Niño: fire problems and long-term solutions*, ODI Natural Resource Perspectives 28, Overseas Development Institute, London.

Cairns, J. Jr (1991) 'The status of the theoretical and applied science of restoration ecology', *The Environmental Professional* 13: 186–94.

Caldecott, J. (1996) *Designing Conservation Projects*, Cambridge University Press, Cambridge.

Caldwell, L.K. (1984) 'Political aspects of ecologically sustainable development', *Enviromental Conservation* 11: 299–308.

Campbell, C. in collaboration with the Women's group of Xapuri (1996) 'Out on the front line but still struggling for voice: women in the rubber tappers' defense of the forest in Xapuri, Acre, Brazil', pp. 27–61 in D. Rocheleau, B. Thomas-Slayter and E. Wangari (eds) *Feminist Political Ecology: global issues and local experiences*, Routledge, London.

Capra, F. and Spretnak, C. (1984) *Green Politics*, E.P. Dutton, New York.

Carney, D. (ed.) (1998a) *Sustainable Rural Livelihoods: what contribution can we make?*, Department for International Development, London.

Carney, D. (1998b) 'Implementing the sustainable rural livelihoods approach', pp. 3–23 in D. Carney (ed.) *Sustainable Rural Livelihoods: what contribution can we make?*, Department for International Development, London.

Carney, J. (1992) 'Peasant women and economic transformation in the Gambia', *Development and Change* 23: 67–85.

Carney, J. (1993) 'Converting the wetlands, engendering the environment: the intersection of gender with agrarian change in the Gambia', *Economic Geography* 69: 329–48.

Carney, J. and Watts, M. (1990) 'Manufacturing dissent: work, gender and the politics of meaning in peasant society', *Africa* 60: 207–30.

Carr, S. and Mpande, R. (1996) 'Does the definition of the issue matter? NGO influence and the International Convention to Combat Desertification in Africa', pp. 143–66 in D. Potter (ed.) *NGOs in Africa and Asia*, Frank Cass, London.

Carroll, L. (1872) *Through the Looking Glass and What Alice Found There*, Puffin Books, London (1984 edition).

Carr-Saunders, A.M. (1922) *The Population Problem: a study in human evolution*, Clarendon Press, Oxford.

Carr-Saunders, A.M. (1936) *World Population: past growth and present trends*, Clarendon Press, Oxford.

Carter, A. (1993) 'Towards a green political theory', pp. 39–62 in A. Dobson and P. Lucardie (eds) *The Politics of Nature: explorations in green political theory*, Routledge, London.

Castleman, B. (1981) 'Double standards: asbestos in India', *New Scientist* 26 February: 522–3.

Caufield, C. (1982) *Tropical Moist Forests: the resource, the people, the threat*, Earthscan/IIED, London.

Caufield, C. (1984) 'Pesticides: exporting death', *New Scientist* 16 August: 15–17.

Caufield, C. (1985) *In the Rainforest*, Pan Books, London.

Caufield, C. (1987) 'Conservationists scorn plans to save tropical forests', *New Scientist* 25 June: 33.

Caufield, C. (1989) *Innocent Exposures: chronicles of the radiation age*, Secker and Warburg, London.

Cernea, M. (1988) *Involuntary Resettlement in Development Projects: policy guidelines in World Bank-financed projects*, Technical Paper 80, World Bank, Washington, DC.

Cernea, M. (ed.) (1991) *Putting People First: sociological variables in rural development*, Oxford University Press, Oxford, for the World Bank.

Chambers, R. (ed.) (1970) *The Volta Resettlement Experience*, Pall Mall, London.

Chambers, R. (1977) 'Challenges for rural research and development', pp. 398–412 in B.H. Farmer (ed.) *Green Revolution? Technology and change in rice-growing areas of Tamil Nadu and Sri Lanka*, Macmillan, London.

Chambers, R. (1983) *Rural Development: putting the last first*, Longman, London.

Chambers, R. (1988a) *Managing Canal Irrigation: practical analysis from South Asia*, Cambridge University Press, Cambridge.

Chambers, R. (1988b) 'Sustainable rural livelihoods: a key strategy for people, environment and development', pp. 1–17 in C. Conroy and M. Litvinoff (eds) *The Greening of Aid: sustainable livelihoods in practice*, Earthscan, London.

Chambers, R. (1994) 'The origins and practice of participatory rural appraisal', *World Development* 22: 953–69.

Chambers, R. (1997) *Whose Reality Counts? Putting the first last*, Intermediate Technology Publications, London.

Chapman, G.P., Kumar, K., Fraser, C. and Gaber, I. (1997) *Environmentalism and the Mass Media: the North–South divide*, Routledge, London.

Charney, J. (1975) 'Dynamics of deserts and drought in the Sahara', *Quarterly Journal of the Royal Meteorological Society* 101: 193–202.

Charney, J., Stone, P.H. and Quirk, W.J. (1975) 'Drought in the Sahara: a biogeophysical feedback mechanism', *Science* 187: 434–5.

Chatterjee, P. and Finger, M. (1994) *The Earth Brokers: power, politics and world development*, Routledge, London.

Chien, N. (1985) 'Changes in river regime after the construction of an upstream reservoir', *Earth Surface Processes and Landforms* 10: 143–59.

Chilcote, R.H. (1984) *Theories of Development and Underdevelopment*, Westview, Boulder, CO.

Child, G. (1995) Managing wildlife in Zimbabwe', *Oryx* 29: 171–7.

Chisholm, A. (1972) *Philosophers of the Earth: conversations with ecologists*, Sidgwick and Jackson, London.

Chisholm, N. and Grove, J.M. (1995) 'The lower Volta', pp. 229–50 in A.T. Grove (ed.) *The Niger and its Neighbours: environmental history and hydrobiology, human use and health hazards of the major West African rivers*. Balkema, Rotterdam.

Christiansson, C. and Ashuvud, J. (1985) 'Heavy industry in a rural tropical ecosystem', *Ambio* 14: 122–33.

Christoff, P. (1996) 'Ecological modernisation, ecological modernities', *Environmental Politics* 5: 476–99.

Ciriacy-Wantrup, S.V. (1952) *Resource Conservation, Economics and Policies*, University of California Press, Berkeley.

Clark, W.C. and Munn, R.E. (eds) (1986) *Sustainable Development of the Biosphere*, Cambridge University Press, Cambridge.

Clarke, R. and Timberlake, L. (1982) *Stockholm Plus Ten: promises promises? The decade since the 1972 UN Environment Conference*, Earthscan, London.

Clay, E.J. and Schaffer, B.B. (eds) (1984) *Room for Manoeuvre: an exploration of public policy in agriculture and rural development*, Heinemann, London.

Cleaver, F. (2000) 'Moral ecological rationality, institutions and the management of common property resources', *Development and Change* 31: 361–83.

Cleaver, K. (1994) 'Deforestation in the Western and Central African forest: the agricultural and demographic causes, and some solutions', pp. 65–78 in Cleaver, K. (ed.) *Conservation of West and Central African Rainforests*, World Bank Environmental Paper No. 1, World Bank, Washington, DC.

Coe, M.J., Cummings, D.H. and Phillipson, J. (1976) 'Biomass and production of large African herbivores in relation to rainfall and primary production', *Oecologia* 22: 341–54.

Cohen, S., Demeritt, D., Robinson, J. and Rothman, D. (1998) 'Climate change and sustainable development: towards dialogue', *Global Environmental Change* 8: 341–71.

Colchester, M. (1985) 'An end to laughter? Hydropower projects in central India', in *An End to Laughter?*, Review 44, Survival International, London.

Colchester, M. (1994) 'Sustaining the forests: the community-based approach in South and South East Asia', *Development and Change* 25: 69–93.

Cole, S. (1978) 'The global futures debate 1965–1976', in C. Freeman and M. Jahoda (eds) *World Futures: the great debate*, Martin Robinson, London.

Coleman, J.S. (1990) *Foundations of Social Theory*, Harvard University Press, Cambridge, MA.

Collett, D. (1987) 'Pastoralists and wildlife: image and reality in Kenyan Masailand', pp. 129–48 in D.M. Anderson and R.H. Grove (eds) *Conservation in Africa: people, policies and practice*, Cambridge University Press, Cambridge.

Collins, R.O. (1990) *The Waters of the Nile: hydropolitics and the Jonglei Canal 1900–1988*, Clarendon Press, Oxford.

Colson, E. (1971) *The Social Consequences of Resettlement*, Kariba Studies iv, University of Manchester Press, Manchester.

Commoner, B. (1972) *The Closing Circle: confronting the environmental crisis*, Jonathan Cape, London.

Conroy, C. (1988) 'Introduction', pp. xi–xiv in C. Conroy and M. Litvinoff (eds) *The Greening of Aid: sustainable livelihoods in practice*, Earthscan, London.

Conroy, C. and Litvinoff, M. (eds) (1988) *The Greening of Aid: sustainable livelihoods in practice*, Earthscan, London.

Conway, G.R. (1985) 'Agroecosystem analysis', *Agricultural Administration* 20: 31–55.

Conway, G.R. and Pretty, J.N. (1991) *Unwelcome Harvest: agriculture and pollution*, Earthscan, London.

Conwentz, H. (1914) 'On national and international protection of nature', *Journal of Ecology* 2: 109–22.

Coomes, O.T. and Barham, B.L. (1994) 'The Amazon rubber boom: labor control, resistance, and failed plantation development revisited', *Hispanic American History Review* 74: 231–57.

Coomes, O.T. and Barham, B.L. (1997) 'Rain forest extraction and conservation in Amazonia', *Geographical Journal* 163: 180–8.

Copans, J. (1983) 'The Sahelian drought: social sciences and the political economy of underdevelopment', pp. 83–97 in K. Hewitt (ed.) *Interpretations of Calamity: from the viewpoint of human ecology*, Allen and Unwin, Hemel Hempstead.

Coppock, D.L., Ellis, J.E. and Swift, D.M. (1986) 'Livestock feeding ecology and resource utilisation in a nomadic pastoral ecosystem', *Journal of Applied Ecology* 23: 573–85.

Corbridge, S.E. (1982) 'Interdependent development? Problems of aggregation and implementation in the Brandt Report', *Applied Geography* 2: 253–65.

Corbridge, S.E. (1993) 'Marxisms, modernities, and moralities: development praxis and the claims of distant strangers', *Environment and Planning D: Society and Space* 11: 449–72.

Costanza, R. (ed.) (1991) *Ecological Economics: the science and management of sustainability*, Columbia University Press, New York.

Costanza, R. and Daly, H.E. (1992) 'Natural capital and sustainable development', *Conservation Biology* 6: 37–46.

Costanza, R., d'Arge, R., de Groot, R., Farber, S., Grasso, M., Hanon, B., Limburg, K., Naeem, S., O'Neill, R.V., Paruelo, J., Raskin, R.G., Sutton, P. and van den Belt, M. (1997) 'The value of the world's ecosystem services and natural capital', *Nature* 387: 253–60.

Cotgrove, S. (1982) *Catastrophe or Cornucopia: the environment, politics and the future*, Wiley, Chichester.

Cotgrove, S. and Duff, A. (1980) 'Environmentalism, middle class radicalism and politics', *Sociology Review* 28: 235–351.

Coughenour, M.B., Ellis, J.E., Swift, D.M., Coppock, D.L., Galvin, K., McCabe, J.T. and Hart, T.C. (1995) 'Energy extraction and use in a nomadic pastoral ecosystem', *Science* 230(4726): 619–25.

Courel, M.F., Kandel, R.S. and Rasool, S.I. (1984) 'Surface albedo and the Sahel drought', *Nature* 307: 528–31.

Cowen, M.P. and Shenton, R.W. (1995) 'The invention of development', pp. 27–43 in J. Crush (ed.) *Power of Development*, Routledge, London.

Cowen, M.P. and Shenton, R.W. (1996) *Doctrines of Development*, Routledge, London.

Cox, P. (1985) *Pesticide Use in Tanzania*, ODI Economic Research Bureau, Dar es Salaam.

Croll, E. and Parkin, D. (1992) 'Cultural understandings of the environment', pp. 11–36 in E. Croll and D. Parkin (eds) *Bush Base: Forest Farm; culture, environment and development*, Routledge, London.

Crosby, A.W. (1986) *Ecological Imperialism: the ecological expansion of Europe, 1600–1900*, Cambridge University Press, Cambridge.

Crummey, D. (ed.) (1986) *Banditry, Rebellion and Social Protest in Africa*, James Currey/Heinemann, London.

Crush, J.C. (1995) 'Imagining development', pp. 1–23 in J.C. Crush (ed.) *Power of Development*, Routledge, London.

Culwick, A.T. (1943) 'New beginning', *Tanganyika Notes and Records* 15: 1–6.

Curry-Lindahl, K. (1986) 'The conflict between development and nature conservation with special reference to desertification', pp. 106–130 in N. Polunin (ed.) *Ecosystem Theory and Application*, Wiley, Chichester.

Cushing, D.H. (1988) *The Provident Sea*, Cambridge University Press, Cambridge.

Dahlberg, K.A., Soroos, M.S., Ferau, A.T., Harf, J.E. and Trout, B.T. (1985) *Environment and the Global Arena: actors, values, policies and futures*, Duke University Press, Durham, NC.

Dalland, Ø. (1997) 'The last big dam in Norway: whose victory?', pp. 41–56 in A.D. Usher (ed.) *Dams as Aid: a political anatomy of Nordic development thinking*, Routledge, London.

Daly, H.E. (ed.) (1973) *Towards a Steady-State Economy*, W.H. Freeman, New York.

Daly, H.E. (1977) *Steady-State Economics: the economics of biophysical equilibrium and moral growth*, W.H. Freeman, New York.

Daly, H.E. (1990) 'Toward some operational principles of sustainable development', *Ecological Economics* 2: 1–6.

Daly, H.E. (1994) 'Operationalising sustainable development by investing in natural capital', pp. 22–37 in A. Jansson, M. Hammer, C. Folke and R. Costanza (eds) (1994) *Investing in Natural Capital: the ecological economics approach to sustainability*, Island Press, Washington DC.

Daly, H.E. and Cobb, W. (1990) *For the Common Good: redirecting the economy towards community, the environment and a sustainable future*, Beacon Press, Boston.

Dasmann, R.F. (1980) 'Ecodevelopment: an ecological perspective', pp. 1331–5 in J.I. Furtado (ed.) *Tropical Ecology and Development*, International Society of Tropical Ecology, Kuala Lumpur.

Dasmann, R.F., Milton, J.P. and Freeman, P.H. (1973) *Ecological Principles for Economic Development*, Wiley, Chichester.

Davies, B.R. (1979) 'Stream regulation in Africa: a review', pp. 113–42 in J.V. Ward and J.A. Stanford (eds) *The Ecology of Regulated Streams*, Plenum Press, New York.

Davies, B.R. and Walker, K.F. (eds) (1984) *The Ecology of River Systems*, Monographiae Biologicae 60, Junk, The Hague.

Davies, B.R., Hall, A. and Jackson, P.B.N. (1972) 'Some ecological aspects of the Cabora Bassa Dam', *Biological Conservation* 8: 189–201.

Davis, T.A.W. and Richards, P.W. (1933) 'The vegetation of Moraballi Creek, British Guiana: an ecological study of a limited area of tropical rain forest, Part I', *Journal of Ecology* 21: 350–84.

Davy, J.B. (1925) 'Correlation of taxonomic work in the Dominions and Colonies with work at home', pp. 214–34 in F.T. Brooks (ed.) *Imperial Botanical Conference, London 1924, Report of Proceedings*, Cambridge University Press, Cambridge.

de Schlippe, P. (1956) *Shifting Cultivation in Africa: the Zande system of agriculture*, Routledge and Kegan Paul, London.

de Waal, A. (1989) *Famine that Kills: Darfur, Sudan, 1984–1985*, Clarendon Press, Oxford.

Denevan, W.M. (1973) 'Development and the imminent demise of the Amazonian forest', *Professional Geographer* 25: 130–5.

Denevan, W.M. (1980) 'Swiddens and cattle versus forest: the imminent demise of the Amazonian rain forest re-examined', pp. 225–44 in V.H. Sutlive, N. Altshuler and M.D. Zamora (eds) *Where Have All the Flowers Gone? Deforestation in the Third World*, Studies in Third World Societies No. 13, College of William and Mary, Williamsburg, Virginia.

Denevan, W.M., Treacy, J.M., Alcorn, J.B., Padock, C., Denslow, J. and Paitan, S.F. (1984) 'Indigenous agroforestry in the Peruvian Amazon: Bora Indian management of swidden fallows', *Interciencia* 9: 346–57.

Dennett, M.D., Elston, J. and Rodgers, J.A. (1985) 'A reappraisal of rainfall trends in the Sahel', *Journal of Climatology* 5: 353–61.

Department of the Environment (1988) *Our Common Future: a perspective by the UK on the Report of the World Commission on Environment and Development*, Department of the Environment, London.

Derman, W. (1984) 'USAID in the Sahel: development and poverty', in J. Barker (ed.) *The Politics of Agriculture in Tropical Africa*, Sage, Beverly Hills, CA.

Deshmukh, I. (1986) *Ecology and Tropical Biology*, Blackwell, Oxford.

Devall, B. (1988) *Simple in Means, Rich in Ends: practicing deep ecology*, Gibbs Smith, Layton, UT.

Devall, B. and Sessions, G. (1985) *Deep Ecology: living as if nature mattered*, Peregrine Smith Books, Salt Lake City.

DFID (2000) *Departmental Report 2000*, Department for International Development, London.

di Castri, F. (1986) 'Interdisciplinary research for the ecological development of mountain and island areas', pp. 301–16 in N. Polunin (ed.) *Ecosystem Theory and Application*, Wiley, Chichester.

Dietz, T. (1996) *Entitlements to Natural Resources: contours of political environmental geography*, International Books, Utrecht.

D'Itri, P.A. and D'Itri, F.M.D. (1977) *Mercury Contamination: a human tragedy*, Wiley, Chichester.

Dixon, J.A. and Fallon, L.A. (1989) 'The concept of sustainability: origins, extensions, and usefulness for policy', *Society and Natural Resources* 2: 73–84.

Dobson, A. (1990) *Green Political Thought*, HarperCollins, London.

Dougherty, T.C. and Hall, A.W. (1995) *Environmental Impact Assessment of Irrigation and Drainage Projects*, FAO Irrigation and Drainage Paper 53, Food and Agriculture Organisation, Rome.

Dove, M.I. and Noguiera, J.M. (1994) 'The Amazon rain forest, sustainable development and the biodiversity convention: a political economy perspective', *Ambio* 23: 491–5.

Dovers, S.R. and Handmer, J.W. (1992) 'Uncertainty, sustainability and change', *Global Environmental Change* December 1992: 262–76.

Dregne, H.E. (1984) 'Combating desertification: evaluation of progress', *Environmental Conservation* 11: 115–21.

Dròze, J. and Sen, A. (1989) *Hunger and Public Action*, Clarendon Press, Oxford.

Dryzek, J.S. (1995) 'Toward an ecological modernity', *Policy Sciences* 28: 231–41.

Dykstra, D.P. and Heinrich, R. (1996) *FAO Model Code of Forest Harvesting Practice*, Food and Agriculture Organisation, Rome.

Eckersley, R. (1992) *Environmentalism and Political Theory: toward an ecocentric approach*, UCL Books, London.

Eckholm, E.P. (1982) *Down to Earth: environment and human needs*, IIED and Earthscan, London.

Ehrlich, P.R. (1972) *The Population Bomb*, Ballantine, London.

Ehrlich, P.R. and Ehrlich, A.H. (1970) *Population, Resources and Environment: issues in human ecology*, W.H. Freeman, New York.

Ekins, P. (1992) *A New World Order: grassroots movements and global change*, Routledge, London.

Ekins, P. and Jacobs, M. (1995) 'Environmental sustainability and the growth of GDP: conditions for compatibility', pp. 9–46 in V. Bhaskar and A. Glyn (eds) *The North, the South and the Environment: ecological constraints and the global economy*, United Nations University Press and Earthscan, London.

El Serafy, S. (1996) 'In defence of weak sustainability: a response to Beckerman', *Environmental Values* 5: 75–81.

Elliot, R. (1997) *Faking Nature: the ethics of environmental restoration*, Routledge, London.

Elliott, J.A. (1999) *An Introduction to Sustainable Development*, Routledge, London (2nd edition).

Elmhirst, R. (1998) 'Reconciling feminist theory and gendered resource management in Indonesia', *Area* 30: 225–35.

Elton, C. (1927) *Animal Ecology*, Sidgwick and Jackson, London.

Eltringham, S.K. (1994) 'Can wildlife pay its way?', *Oryx* 28: 163–8.

Enzensberger, H.M. (1974) 'A critique of political ecology', *New Left Review* 8: 3–32.

Escobar, A. (1995) *Encountering Development: the making and unmaking of the Third World*, Princeton University Press, Princeton, NJ.

Escobar, A. (1996) 'Poststructural political ecology', pp. 46–68 in R. Peet and M. Watts (eds) *Liberation Ecologies: environment, development, social movements*, Routledge, London.

Esteva, G. (1992) 'Development', pp. 6–25 in W. Sachs (ed.) *The Development Dictionary: a guide to knowledge as power*, Witwatersrand University Press, Johannesburg, and Zed Books, London.

European Commission (1997) *Addressing desertification: a review of EC policies, programmes, financial instruments and projects*, European Commission, Brussels.

Evans, D. (1992) *A History of Nature Conservation in Great Britain*, Routledge, London.

Evans, G.C. (1939) 'Ecological studies on the rainforest of southern Nigeria: II The atmospheric environmental conditions', *Journal of Ecology* 26: 436–82.

Fairhead, J. and Leach, M. (1995a) 'Local agro-ecological management and forest–savanna transitions: the case of Kissidougou, Guinea', pp. 163–70 in T. Binns (ed.) *People and Environment in Africa*, Wiley, Chichester.

Fairhead, J. and Leach, M. (1995b) 'False forest history, complicit social analysis: rethinking some West Africa environmental narratives', *World Development* 23: 1023–35.

Fairhead, J. and Leach, M. (1996) *Misreading the African Landscape: society and ecology in a forest savanna land*, Cambridge University Press, Cambridge.

Fairhead, J. and Leach, M. (1998) *Reframing Deforestation: global analysis and local realities*, Routledge, London.

FAO (1994a) 'Expert consultation on cotton pest problems and their control in the Near East', *FAO Plant Protection Bulletin* 42: 139–48.

FAO (1994b) *Mangrove Forest Management Guidelines*, FAO Forestry Paper 117, Food and Agriculture Organisation, Rome.

Farid, M.A. (1975) 'The Aswan High Dam development project', pp. 89–102 in N.F. Stanley and M.P. Alpers (eds) *Man-Made Lakes and Human Health*, Academic Press, London.

Farid, M.A. (1977) 'Irrigation and malaria in arid lands', pp. 413–19 in E.B. Worthington (ed.) *Arid Land Irrigation in Developing Countries: environmental problems and effects*, Pergamon Press, Oxford.

Farvar, M.T. and Milton, J.P. (eds) (1973) *The Careless Technology: ecology and international development*, Stacey, London.

Faulkner, O.T. and Mackie, J.R. (1933) *West African Agriculture*, Cambridge University Press, Cambridge.

Fearnside, P.M. (1980) 'The effects of cattle pasture on soil fertility in the Brazilian Amazon: consequences for beef production sustainability', *Tropical Ecology* 21: 125–37.

Fearnside, P.M. (1986) 'Spatial concentration of deforestation in the Brazilian Amazon', *Ambio* 15: 74–81.

Ferau, A.T. (1985) 'Environmental actors', pp. 43–67 in K.A. Dahlberg *et al.* (1985) *Environment and the Global Arena: actors, values, policies and futures*, Duke University Press, Durham, NC.

Fitsimmons, M. (1989) 'The matter of nature', *Antipode* 21(2): 106–20.

Fitter, R.S.R. and Scott, P. (1978) *The Penitent Butchers: the Fauna Preservation Society, 1903–1978*, Collins, London.

Flanders, L. (1997) 'The United Nations' Department for Policy Coordination and Sustainable Development (DPCSD)', *Global Environmental Change* 7: 391–4.

Flohn, H. and Nicholson, S.E. (1980) 'Climatic fluctuations in the arid belt of the "Old World" since the last glacial maximum: possible causes and future implications', *Palaeoecology of Africa* 12: 3–21.

Folke, C., Hamer, M., Costanza, R. and Jansson, A. (1994) 'Investing in natural capital – why what and how?', pp. 1–20 in A. Jansson, M. Hammer, C. Folke and R. Costanza (eds) *Investing in Natural Capital: the ecological economics approach to sustainability*, Island Press, Washington, DC.

Folland, C.K., Palmer, T.N. and Parker, D.E. (1986) 'Sahel rainfall and worldwide sea temperatures 1901–1985', *Nature* 320: 602–7.

Ford, J. (1971) *The Role of Trypanosomiasis in African Ecology: a study of the tsetse fly problem*, Clarendon Press, Oxford.

Forrester, J.W. (1971) *World Dynamics*, Wright-Allen Press, Cambridge, MA.

Fosberg, F.R. (1963) *Man's Place in the Island Ecosystem*, Bishop Museum Press, Honolulu.

Fox, J.A. (1998) 'When does reform policy influence practice?', pp. 303–44 in J.A. Fox and L.D. Brown (eds) *The Struggle for Accountability: the World Bank, NGOs and grassroots movements*, MIT Press, Cambridge, MA.

Fox, J.A. and Brown, L.D. (1998a) 'Introduction', pp. 1–47 in J.A. Fox and L.D. Brown (eds) *The Struggle for Accountability: the World Bank, NGOs and grassroots movements*, MIT Press, Cambridge, MA.

Fox, J.A. and Brown, L.D. (eds) (1998b) *The Struggle for Accountability: the World Bank, NGOs and grassroots movements*, MIT Press, Cambridge, MA.

Fox, W. (1984) 'Deep ecology: a new philosophy of our time?', *The Ecologist* 14: 194–200.

Fox, W. (1990) *Toward a Transpersonal Ecology: developing new foundations for environmentalism*, Shambhala, Boston.

Frank, A.G. (1980) 'North–South and East–West: Keynesian paradoxes in the Brandt Report', *Third World Quarterly* 2: 669–80.

Frank, A.G. (1981) *Crisis in the Third World*, Heinemann, London.

Frank, L. (1987) 'The development game', *Granta* 22: 231–43.

Franke, R.W. and Chasin, B.H. (1979) 'Peanuts, peasants and pastoralists: the social and economic background to ecological deterioration in Niger', *Journal of Peasant Studies* 8: 1–30.

Franke, R.W. and Chasin, B.H. (1980) *Seeds of Famine: ecological destruction and the development dilemma in the West African Sahel*, Allenheld and Osman, Montclair, NJ.

Fraser Darling, F. (1955) *West Highland Survey: an essay in human ecology*, Oxford University Press, Oxford.

Freeman, P.H. (1977) *Large Dams and the Environment: recommendations for development planning*, Report for the United Nations 1977 Water Conference, IIED, London.

Fresco, L.O. (1997) 'The 1996 World Food Summit', *Global Environmental Change* 7: 1–3.

Fresco, L.O. and Kroonenberg, S.B. (1992) 'Time and spatial scales in ecological sustainability', *Land Use Policy* July: 155–68.

Friberg, M. and Hettne, B. (1985) 'The greening of the world: towards a non-deterministic model of global processes', pp. 204–70 in H. Addo, S. Amin, G. Aseniero,

A.G. Frank, M. Friberg, F. Frobel, J. Heinrichs, B. Heltne, O. Kreye and H. Seki, *Development as Social Transformation: reflections on the global problematique*, Hodder and Stoughton, Sevenoaks, for the United Nations University.

Frobel, F., Heinrichs, J. and Kreye, O. (1985) 'The global crisis and developing countries', pp. 111–24 in H. Addo, S. Amin, G. Aseniero, A.G. Frank, M. Friberg, F. Frobel, J. Heinrichs, B. Heltne, O. Kreye and H. Seki, *Development as Social Transformation: reflections on the global problematique*, Hodder and Stoughton, Sevenoaks, for the United Nations University.

Frynas, J.G. (2000) *Oil in Nigeria: conflict and litigation between oil companies and village communities*, LIT, Müenster/Hamburg.

Furley, P.A. (1994) *The Forest Frontier: settlement and change*, Routledge, London.

Furon, R. (1947) *L'Érosion du sol*, Payot, Paris.

Gadgil, M. and Guha, R. (1995) *Ecology and Equity: the use and abuse of nature in contemporary India*, Routledge, London.

Galois, B. (1976) 'Ideology and the idea of nature: the case of Peter Kropotkin', *Antipode* 8: 1–16.

Galtung, J. (1984) 'Perspectives on environmental politics in overdeveloped and under-developed countries', pp. 9–21 in B. Glaeser (ed.) *Ecodevelopment: concepts, projects, strategies*, Pergamon Press, Oxford.

Gammelsrød, T. (1996) 'Effect of Zambezi management on the prawn fishery of the Sofala Bank', pp. 119–23 in M.C. Acreman and G.E. Hollis (eds) *Water Management and Wetlands in Sub-Saharan Africa*, IUCN, Gland, Switzerland.

Gash, J.H.C., Nobre, C.A., Roberts, J.M. and Victoria, R.L. (eds) (1976) *Amazonian Deforestation and Climate*, Wiley, Chichester.

Geijer, J.C.M.A., Svendsen, M. and Vermillion, D.L. (1996) *Transferring Irrigation Management Responsibility in Asia, Bangkok and Chiang Mai, 25–29 September 1995*, IIMI, Columbo, Sri Lanka. Short Report Series on Locally Managed Irrigation No. 13.

Geldof, B. (1986) *Is that It?*, Sidgwick and Jackson, London.

George, C.J. (1973) 'The role of the Aswan High Dam in changing the fisheries of the southeastern Mediterranean', pp. 159–78 in M.T. Farvar and J.P. Milton (eds) *The Careless Technology: ecology and international development*, Stacey, London.

Ghai, D. and Vivian, J.M. (eds) (1992a) *Grassroots Environmental Action: people's participation in sustainable development*, Routledge, London.

Ghai, D. and Vivian, J.M. (1992b) 'Introduction', pp. 1–19 in D. Ghai and J.M. Vivian *Grassroots Environmental Action: people's participation in sustainable development*, Routledge, London.

Ghatak, S. and Turner, R.K. (1978) 'Pesticide use in less developed countries: economic and environmental considerations', *Food Policy* 3: 136–46

Ghimire, K.B. and Pimbert, M.P. (eds) (1996) *Social Change and Conservation*, Earthscan, London.

Gibson, C.C. and Marks, S.A. (1995) 'Transforming rural hunters into conservationists: an assessment of community-based wildlife management programs in Africa', *World Development* 23: 941–57.

Gilbert, V.C. and Christy, E.J. (1981) 'The UNESCO Program on Man and the Biosphere (MAB)', pp. 701–20 in E.J. Kormondy and J.F. McCormick (eds) *Handbook of Contemporary Developments in World Ecology*, Greenwood Press, Westport, CT.

Glaeser, B. (1984a) *Ecodevelopment in Tanzania: an empirical contribution on needs, self-sufficiency and environmentally sound agriculture on peasant farms*, Mouton, Berlin.

Glaeser, B. (ed.) (1984b) *Ecodevelopment: concepts, projects, strategies*, Pergamon Press, Oxford.

Glaeser, B. (ed.) (1987) *The Green Revolution Revisited: critique and alternatives*, Allen and Unwin, Hemel Hempstead.

Glaeser, B. and Vyasulu, V. (1984) 'The obsolescence of ecodevelopment?', pp. 23–36 in B. Glaeser (ed.) *Ecodevelopment: concepts, projects, strategies*, Pergamon Press, Oxford.

Glantz, M. (1992) 'Global warming and environmental change in sub-Saharan Africa', *Global Environmental Change* September 183–205.

Gliessman, S.R. (1984) 'Resource management in traditional agroecosystems: southeast Mexico', pp. 191–201 in G.K. Douglass (ed.) *Agricultural Sustainability in a Changing World Order*, Westview, Boulder, CO.

Goldsmith, E. (1987) 'Open letter to Mr. Conable, President of the World Bank', *The Ecologist* 17: 58–61.

Goldsmith, E. and Hildyard, N. (1984) *Social and Environmental Impacts of Large Dams*, vol. 1, Ecologist Magazine, Wadebridge, Cornwall.

Goldsmith, E., Allen, R., Allaby, M., Davoll, J. and Lawrence, S. (1972) *Blueprint for Survival*, *The Ecologist* 2: 1–50 (also 1972, Penguin Books, Harmondsworth).

Golub, R. and Townsend, J. (1977) 'Malthus, multinationals and the Club of Rome', *Social Studies of Science* 7: 202–22.

Gómez-Pompa, A., Whitmore, T.C. and Hadley, M. (eds) (1991) *Rain Forest Regeneration and Management*, UNESCO, Paris.

Good, K. (1986) 'The reproduction of weakness in the state and agriculture: the Zambian experience', *African Affairs* 85: 239–65.

Goodland, R.J. (1975) 'The tropical origins of ecology: Eugen Warming's Jubilee', *Oikos* 26: 240–5.

Goodland, R.J. (1978) *Environmental Assessment of the Tucurui Hydroproject, Rio Tocatins, Amazonia, Brazil*, Eletronorte SA, Braszilia.

Goodland, R.J. (1980) 'Environmental rankings of Amazonian development projects in Brazil', *Environmental Conservation* 7: 9–26.

Goodland, R.J. (1984) *Environmental Requirements of the World Bank*, Environment and Science Unit, Projects Policy Department, World Bank, Washington, DC.

Goodland, R.J. (1990) 'Environment and development: progress of the World Bank', *Geographical Journal* 156: 149–57.

Goodland, R.J. and Ledec, G. (1984) *Neoclassical Economics and Principles of Sustainable Development*, World Bank Office of Environmental Affairs, Washington, DC.

Goodland, R.J.A., Daly, H.E. and El Serafy, S. (1993) 'The urgent need for rapid transition to global environmental sustainability', *Environmental Conservation* 20: 297–309.

Gordon, R.J. (1985) 'Conserving bushmen to extinction in Southern Africa', pp. 28–42 in *An End to Laughter? Tribal peoples and economic development*, Review No. 44, Survival International, London.

Gornitz, V. and NASA (1985) 'A survey of anthropogenic vegetation change in West Africa during the last century: climatic implications', *Climatic Change* 7: 285–326.

Gosling, L. (1979) 'Resettlement losses and compensation', pp. 119–31 in L. Gosling (ed.) *Population Resettlement in the Mekong River Basin*, Studies in Geography No. 10, University of Northern Carolina, Chapel Hill, NC.

Gottfried, R.R., Brockett, C.D. and Davis, W.C. (1994) 'Models of sustainable development and forest resource management in Costa Rica', *Ecological Economics* 9: 107–20.

Gouldner, L.H. and Kennedy, D. (1997) 'Valuing ecosystem services: philosophical bases and empirical methods', pp. 23–47 in G.C. Daily (ed.) *Nature's Services: societal dependence on natural ecosystems*, Island Press, Washington, DC.

Goulet, D. (1971) *The Cruel Choice: a new concept in the theory of development*, Athanaeum, London.

Goulet, D. (1992) 'Development: creator and destroyer of values', *World Development* 20: 467–75.

Goulet, D. (1995) *Development Ethics: a guide to theory and practice*, Zed Books, London.

Graber, L.H. (1976) *Wilderness as Sacred Space*, Association of American Geographers, Washington, DC.

Gradwohl, J. and Greenberg, R. (1988) *Saving the Tropical Forests*, Earthscan, London.

Graham, A. (1973) *The Gardeners of Eden*, Allen and Unwin, Hemel Hempstead.

Grainger, A. (1982) *Desertification: how people make deserts, how people can stop and why they don't*, Earthscan/IIED, London.

Grainger, A. (1995) 'The forest transition: an alternative approach', *Area* 27: 242–51.

Grainger, A. (1996) 'An evaluation of the FAO *Tropical Forest Resource Assessment 1990*', *Geographical Journal* 162: 73–9.

Green, K.M. (1983) 'Using Landsat to monitor tropical forest ecosystems', pp. 397–409 in S.L. Sutton, T.C. Whitmore and A.C. Chadwick (eds) *Tropical Rain Forest: ecology and management*, Blackwell, Oxford.

Greener, L. (1962) *High Dam over Nubia*, Cassell, London.

Gregory, D. (1980) 'The ideology of control: systems theory and geography', *Tijdschrift voor Economische en Sociale Geographie* 81: 327–42.

Grojean, R. (1991) *Sand Encroachment Control in Mauritania*, United Nations Sudan-Sahelian Office Technical Publication Series No. 5, New York.

Groombridge, B. (1992) *Global Biodiversity: status of the earth's living resources*, Chapman and Hall, London.

Grove, A.T. (1958) 'The ancient erg of Hausaland, and similar formations on the south side of the Sahara', *Geographical Journal* 124: 526–33.

Grove, A.T. (1973) 'A note on the remarkably low rainfall of the Sudan Zone in 1913', *Savanna* 2: 133–8.

Grove, A.T. (1977) 'Desertification', *Progress in Physical Geography* 1: 296–310.

Grove, A.T. (1981) 'The climate of the Sahara in the period of meteorological records', in J.A. Allan (ed.) *Sahara: ecological change and early economic history*, Menas Press, London.

Grove, A.T. and Warren, A. (1968) 'Quaternary landforms and climate on the south side of the Sahara', *Geographical Journal* 134: 194–208.

Grove, R.H. (1987) 'Early themes in African conservation: the Cape in the nineteenth century', pp. 21–40 in D.M. Anderson and R.H. Grove (eds) *Conservation in African people, policies and practice*, Cambridge University Press, Cambridge.

Grove, R.H. (1990a) 'The origins of environmentalism', *Nature* 345(6270): 11–14.

Grove, R.H. (1990b) 'Colonial conservation, ecological hegemony, and popular resistance: towards a global synthesis', pp. 15–50 in J.M. McKenzie (ed.) *Imperialism and the Natural World*, Manchester University Press, Manchester.

Grove, R.H. (1992) 'Origins of western environmentalism', *Scientific American* 267: 42–7.

Grove, R.H. (1995) *Green Imperialism: colonial expansion, tropical island Edens and the origins of environmentalism, 1600–1800*, Cambridge University Press, Cambridge.

Grove, R.H., Damodaran, V. and Sangwan, S. (eds) (1998) *Nature and the Orient: the environmental history of South and South East Asia*, Oxford University Press, Delhi.

Grubb, M., Koch, M., Thompson, K., Munson, A. and Sullivan, F. (eds) (1993) *The 'Earth Summit' Agreements: a guide and assessment*, Earthscan, London (for the Royal Institute of International Affairs, London).

Grübler, A. (1994) 'Industrialization as a historical phenomenon', pp. 43–68 in R. Socolow, C. Andrews, F. Berkhout and V. Thomas (eds) *Industrial Ecology and Global Change*, Cambridge University Press, Cambridge.

Guha, R. (1989) *The Unquiet Woods: ecological change and peasant resistance in the Himalaya*, Oxford University Press, Delhi.

Guha, R. and Martinez-Alier, J. (1997) *Varieties of Environmentalism: essays North and South*, Earthscan, London.

Gullison, R.E. and Losos, E.C. (1993) 'The role of foreign debt in deforestation in Latin America', *Conservation Biology* 7: 140–7.

Guy, P.R. (1981) 'River bank erosion in the mid-Zambezi Valley downstream of Lake Kariba', *Biological Conservation* 19: 199–212.

Gwynne, M.D. (1982) 'The Global Environment Monitoring System (GEMS) of UNEP', *Environmental Conservation* 9: 35–42.

Hadfield, P. (1993) 'Japanese aid may upset Cambodia's harvests', *New Scientist* 1864 (13 March): 5.

Haila, Y. and Levins, R. (1992) *Humanity and Nature: ecology, science and society*, Pluto Press, London.

Hailey, Lord (1938) *An African Survey*, Royal Institute for African Affairs and Oxford University Press, London.

Hajer, M.A. (1995) *The Politics of Environmental Discourse: ecological modernisation and the policy process*, Oxford University Press, Oxford.

Hajer, M.A. (1996) 'Ecological modernisation as cultural politics', pp. 246–68 in S. Lash, B. Szerzynski and B. Wynne (eds) *Risk, Environment and Modernity: towards a new ecology*, Sage, London.

Hall, P. (1980) *Great Planning Disasters*, Weidenfeld and Nicolson, London.

Hamnet, I. (1970) 'A social scientist among technicians', *IDS Bulletin* 3: 24–29.

Handlos, W.L. and Williams, G.J. (1984) *Development of the Kafue Flats: the last five years*, Kafue Basin Research Committee, University of Zambia, Lusaka.

Hanlon, J. (1996) *Peace without Profit: how the IMF blocks rebuilding in Mozambique*, James Currey/Heinemann for the International African Institute, London.

Hannah, L. (1992) *African People, African Parks: an evaluation of development initiatives as a means of improving protected area conservation in Africa*, USAID, Washington, DC.

Hannah, L., Lohse, D., Hutchinson, C., Carr, J.L. and Lankerani, A. (1994) 'A preliminary inventory of human disturbance of world ecosystems', *Ambio* 23, 246–50.

Haraway, D.J. (1991) *Simians, Cyborgs and Women: the reinvention of nature*, Free Association Books, London.

Hardin, G. (1968) 'The tragedy of the commons', *Science* 1628: 1243–8.

Hardin, G. (1974) 'Living on a lifeboat', *Bioscience* 24: 561–8.

Hardjono, J.M. (1977) *Transmigration in Indonesia*, Oxford University Press, Selangor, Malaysia.

Hardjono, J.M. (1983) 'Rural development in Indonesia: the "top down" approach', pp. 38–66 in D.A.M. Lea and D.P. Chaudhri (eds) *Rural Development and the State: contradictions and dilemmas in developing countries*, Methuen, London.

Hardoy, J.E., Mitlin, D. and Satterthwaite, D. (1992) *Environmental Problems in Third World Cities*, Earthscan, London.

Hare, F.K. (1984) 'Recent climatic experience in the arid and semi-arid lands', *Desertification Control Bulletin* 10: 15–20.

Harrell-Bond, B. (1985) 'Humanitarianism in a straightjacket', *African Affairs* 84: 3–14.

Harris, F.M.A. (1998) 'Farm-level assessment of the nutrient balance in northern Nigeria', *Agriculture, Ecosystems and Environment* 71: 201–14.

Harris, F.M.A. (1999) 'Nutrient management of smallholder farmers in a short-fallow farming system in north-east Nigeria', *Geographical Journal* 165: 275–85.

Harrison, P. (1987) *The Greening of Africa: breaking through in the battle for land and food*, Paladin, London.

Harrison, P. (1992) *The Third Revolution: population, environment and a sustainable world*, Penguin Books, Harmondsworth.

Harrison, T.H. (1933) 'The Oxford University Expedition to Sarawak, 1932', *Geographical Journal* 82: 385–410.

Harriss, J. and de Renzio, P. (1997) '"Missing link" or analytically missing? The concept of social capital. An introductory bibliographic essay', *Journal of International Development* 9: 919–37.

Harroy, J.-P. (1949) *Afrique: terre qui meurt: la dégradation des sols africains sous l'influence de la colonisation*, Marcel Hayez, Brussels.

Hart, D. (1980) *The Volta River Project: a case study in politics and technology*, Edinburgh University Press, Edinburgh.

Hart, K. (1982) *The Political Economy of West African Agriculture*, Cambridge University Press, Cambridge.

Harvey, D. (1974) 'Population, resources and the ideology of science', *Economic Geography* 50: 256–77.

Harvey, D. (1990) *The Condition of Postmodernity: an enquiry into the origins of cultural change*, Blackwell, Oxford.

Harvey, D. (1996) *Justice, Nature and the Geography of Difference*, Blackwell, Oxford.

Haub, C. (1999) 'Six billion – and counting', *People and the Planet* 8(1): 6–9.

Haugerud, A. (1986) 'An anthropologist in an African Research Institute: an informal essay', *Development Anthropology Network* 4: 4–9.

Hays, S.P. (1959) *Conservation and the Gospel of Efficiency: the progressive conservation movement, 1890–1920*, Harvard University Press, Cambridge, MA.

Hays, S.P. (1987) *Beauty, Health and Permanence: environmental politics in the United States, 1955–1985*, Cambridge University Press, Cambridge.

Hecht, J. (1996) 'Grand Canyon flood a roaring success', *New Scientist* 31 August: 8.

Hecht, S.B. (1980) 'Deforestation in the Amazon basin: magnitude, dynamics and soil resource effects', pp. 61–108 in V.H. Sutlive, N. Altshuler and M.D. Zamora (eds) *Where Have All The Flowers Gone? Deforestation in the Third World*, Studies in Third World Societies No. 13, College of William and Mary, Williamsburg, VA.

Hecht, S.B. (1984) 'Cattle ranching in Amazonia: political and ecological considerations', pp. 366–400 in M. Schmick and C.H. Wood (eds) *Frontier Expansion in Amazonia*, University of Florida Press, Gainesville.

Hecht, S.B. (1985) 'Environment, development and politics: capital accumulation and the livestock sector in Eastern Amazonia', *World Development* 13: 663–84.

Hecht, S. and Cockburn, A. (1989) *The Fate of the Forest: developers and defenders of the Amazon*, Verse, London.

Hellden, U. (1988) 'Desertification monitoring: is the desert encroaching?', *Desertification Control Bulletin* 17: 8–12.

Henderson-Sellers, A. (1994) 'Numerical modelling of global climates', pp. 99–124 in N. Roberts (ed.) *The Changing Global Environment*, Blackwell, Oxford.

Henderson-Sellers, A. and Gornitz, V. (1984) 'Possible climatic impacts of land cover transformations, with particular emphasis on Tropical deforestation', *Climatic Change* 6: 231–58.

Herlocker, D. (1979) *Vegetation of Southwestern Marsabit District Kenya*, Integrated Project on Arid Lands Technical Report No. D-1 (Man and the Biosphere Project 3: Impact of Human Land Use Practices on Grazing Lands), Integrated Project on Arid Lands, Nairobi.

Herrera, R., Jordan, C.F., Medina, E. and Klinge, H. (1981) 'How human activities disturb the nutrient cycle of a tropical rainforest in Amazonia', *Ambio* 10: 109–14.

Hewitt, K. (ed.) (1983) *Interpretations of Calamity*, Allen and Unwin, London.

Hickling, C.F. (1961) *Tropical Inland Fisheries*, Longman, London.

Higgins, G.M., Kassam, A.H. and Naiken, L. (1982) *Potential Population Supporting Capacities of Lands in the Developing World*, Food and Agriculture Organisation, Rome.

Hildermeier, M. (1979) 'Agrarian social protest, populism and economic development: some problems and results from recent studies', *Social History* 4: 319–32.

Hill, F. (1978) 'Experiments with a public sector peasantry: agricultural schemes and class formation in Africa', pp. 25–41 in A.K. Smith and C.E. Welch (eds) *Peasants in Africa*, African Studies Association, Boston, MA.

Hill, K.A. (1995) 'Conflicts over development and environmental values: the international ivory trade in Zimbabwe's historical context', *Environment and History* 1: 335–49.

Hill, M.A. and Press, A.J. (1994) 'Kakadu National Park: an Australian experience in comanagement', pp. 135–57 in D. Western, R.M. White and S.C. Strum (eds) *Natural Connections: perspectives in community-based conservation*, Island Press, Washington, DC.

Hill, P. (1986) *Development Economics on Trial: the anthropological case for the prosecution*, Cambridge University Press, Cambridge.

Hilton, T.E. and Kuwo-Tsri, J.Y. (1970) 'The impact of the Volta Scheme on the Lower Volta floodplains', *Journal of Tropical Geography* 30: 29–37.

Hingston, R.W.G. (1930) 'The Oxford University expedition to British Guiana', *Geographical Journal* 76: 1–24.

Hingston, R.W.G. (1931) 'Proposed British national parks for Africa', *Geographical Journal* 77: 401–28.

Hiraoka, M. and Yamamoto, S. (1980) 'Agricultural development in the Upper Amazon of Ecuador', *Geographical Review* 70: 423–45.

Hoben, A. (1995) 'Paradigms and politics: the cultural construction of environmental politics in Ethiopia', *World Development* 23: 1007–22.

Hobley, C.W. (1914) 'The alleged desiccation of East Africa', *Geographical Journal* 44: 467–77.

Hogg, R. (1983) 'Irrigation agriculture and pastoral development: a lesson from Kenya', *Development and Change* 14: 577–91.

Hogg, R. (1987a) 'Development in Kenya: drought, desertification and food security', *African Affairs* 86: 47–58.

Hogg, R. (1987b) 'Settlement, pastoralism and the commons: the ideology and practice of irrigation development in northern Kenya', pp. 293–306 in D.M. Anderson and R.H. Grove (eds) *Conservation in Africa: people, policies and practice*, Cambridge University Press, Cambridge.

Holden, C. (1987) 'World Bank launches new environmental policy', *Science* 236: 769.

Holden, C. (1988) 'The greening of the World Bank', *Science* 240: 1610–11.

Holdgate, M. (1996) *From Care to Action: making a sustainable world*, Earthscan, London.

Holdgate, M. (1999) *The Green Web: a union for world conservation*, Earthscan, London.

Holdgate, M.W., Kassas, M. and White, G.F. (1982) 'World environmental trends between 1972 and 1982', *Environmental Conservation* 9: 11–29.

Holland, A. and Roxbee Cox, J. (1992) 'The valuing of environmental goods: a modest proposal', pp. 12–24 in A. Coker and C. Richards (eds) *Valuing the Environment: economic approaches to environmental valuation*, Belhaven, London.

Hollis, G.E. (1990) 'Environmental impacts of development on wetlands in arid and semi-arid lands', *Hydrological Sciences Journal* 35: 411–28.

Hollis, G.E. (1996) Hydrological inputs to management policy for the Senegal River and its floodplain', pp. 155–184 in M.C. Acreman and G.E. Hollis *Water Management and Wetlands in Sub-Saharan Africa*, IUCN, Gland, Switzerland.

Hollis, G.E., Adams, W.M. and Aminu-Kano, M. (eds) (1994) *The Hadejia-Nguru Wetlands: environment, economy and sustainable use of a Sahelian floodplain wetland*, IUCN Wetlands Programme, Gland, Switzerland.

Holmberg, J., Thomson, K. and Timberlake, L. (1993) *Facing the Future: beyond the Earth Summit*, Earthscan/International Institute for Environment and Development, London.

Homewood, K. and Rodgers, W.A. (1984) 'Pastoralism and conservation', *Human Ecology* 12: 431–41.

Homewood, K. and Rodgers, W.A. (1987) 'Pastoralism, conservation and the overgrazing controversy', pp. 111–28 in D.M. Anderson and R.H. Grove (eds) *Conservation in Africa: people, policies and practice*, Cambridge University Press, Cambridge.

Homewood, K. and Rodgers, W.A. (1991) *Maasailand Ecology*, Cambridge University Press, Cambridge.

Hopper, W. (1988) *The World Bank's Challenge: balancing economic need with environmental protection*, Seventh Annual World Conservation Lecture, 3 March, World Wide Fund for Nature UK, Godalming.

Horberry, J.A.J. (1988) 'Fitting USAID to the environmental assessment provisions of NEPA', pp. 286–99 in P. Wathern (ed.) *Environmental Impact Assessment: theory and practice*, Unwin Hyman, London.

Horowitz, M.M. (1987) 'Destructive development', *Institute of Development Anthropology Network* 5: 1–2.

Horowitz, M.M. and Little, P.D. (1987) 'African pastoralism and poverty: some implications for drought and famine', pp. 59–82 in M. Glantz (ed.) *Drought and Hunger in Africa: denying famine a future*, Cambridge University Press, Cambridge.

Horowitz, M.M. and Salem-Murdock, M. (1991) 'Management of an African floodplain: a contribution to the anthropology of public policy', *Landscape & Urban Planning* 20: 215–21.

Houghton, J. (1997) *Global Warming: the definitive guide*, Cambridge University Press, Cambridge (2nd edition).

Houghton, J.T., Meira Filho, L.G., Callander, B.A., Harris, N., Kattenberg, A. and Maskell, K. (eds) (1995) *Climate Change 1995: the science of climate change (contribution of Working Group I to Second Assessment Report of the Intergovernmental Panel on Climate Change)*, Cambridge University Press, Cambridge.

Howard, G.W. and Williams, G.J. (eds) (1982) *The Consequences of Hydroelectric Power*

Development on the Kafue Flats, Kafue Basin Research Project, Lusaka (Proceedings of the National Seminar on Environment and Change, Lusaka).

Howard, P. (1978) *Weasel Words*, Hamish Hamilton, London.

Howarth, D. (1961) *The Shadow of the Dam*, Collins, London.

Howell, P., Lock, M. and Cobb, S. (1988) *The Jonglei Canal: impact and opportunity*, Cambridge University Press, Cambridge.

Howell, P.P. (1953) 'The Equatorial Nile Project and its effects in the Sudan', *Geographical Journal* 119: 33–48.

Hughes, F.M.R. (1983) 'Helping the World Conservation Strategy? Aid agencies on the Tana River in Kenya', *Area* 15: 177–83.

Hughes, F.M.R. (1984) 'A comment on the impact of development schemes on the flood-plain forests of the Tana River of Kenya', *Geographical Journal* 150: 230–44.

Hughes, F.M.R. (1987) 'Conflicting uses for forest resources in the lower Tana River Basin of Kenya', pp. 211–28 in D.M. Anderson and R.H. Grove (eds) *Conservation in Africa: people, policies and practice*, Cambridge University Press, Cambridge.

Hughes, F.M.R. (1997) 'Floodplain biogeomorphology', *Progress in Physical Geography* 21, 501–29.

Hughes, F.M.R. (1990) 'The influence of flooding regimes on forest distribution and composition in The Tana River Floodplain, Kenya', *Journal of Applied Ecology* 27: 475–91.

Huijsman, B. and Savenije, H. (1991) 'Making haste slowly', pp. 13–34 in H. Savenije and B. Huijsman (eds) *Making Haste Slowly: strengthening local environmental management in agricultural development*, Royal Tropical Institute, Amsterdam.

Hulme, D. and Murphree, M. (1999) 'Communities, wildlife and the "new conservation" in Africa', *Journal of International Development* 11: 277–86.

Hulme, M. (1987) 'Secular changes in wet season structure in central Sudan', *Journal of Arid Environments* 13: 31–46.

Hulme, M. (ed.) (1995) *Climate Change and Southern Africa: an exploration of some potential impacts and implications in the SADC region*, Climate Research Unit and World Wide Fund for Nature, Norwich.

Hulme, M. (1996) 'Climate change within the period of meteorological records', pp. 88–102 in W.M. Adams, A.S. Goudie and A.R. Orme (eds) *The Physical Geography of Africa*, Oxford University Press, Oxford.

Huq, S. (1994) 'Global industrialization: a developing country perspective', pp. 107–13 in R. Socolow, C. Andrews, F. Berkhout and V. Thomas (eds) *Industrial Ecology and Global Change*, Cambridge University Press, Cambridge.

Huxley, E. (1960) *A New Earth: an experiment in colonialism*, Chatto and Windus, London.

Huxley, J.L. (1930) *African View*, Chatto and Windus, London.

Huxley, J.L. (1977) *Memories II*, Harper and Row, New York.

Hviding, E. and Bayliss-Smith, T. (2000) *Islands of Rainforest: agroforestry, logging and ecotourism in Solomon Islands*, Ashgate, Aldershot.

Hyden, G. (1980) *Beyond Ujamaa in Tanzania: underdevelopment and an uncaptured peasantry*, Heinemann, London.

Hyndman, D. (1994) *Ancestral Rain Forests and the Mountain of Gold: indigenous people and mining in New Guinea*, Westview Press, Boulder, CO.

Hynes, H.B.N. (1970) *The Ecology of Running Waters*, University of Toronto Press, Toronto.

ICOLD (1980) *Dams and their Environment*, International Commission on Large Dams, Paris.

ICOLD (1981) *Dam Projects and Environmental Success*, International Commission on Large Dams, Paris.

Idso, S.B. (1977) 'A note on some recently proposed mechanisms of genesis of deserts', *Quarterly Journal of the Royal Meteorological Society* 103: 369–70.

Ikporukpo, C.O. (1983) 'Environmental deterioration and public policy in Nigeria', *Applied Geography* 3: 303–16.

Iliffe, J. (1995) *Africans: the history of a continent*, Cambridge University Press, Cambridge.

Illich, I. (1973) 'Outwitting the "developed" countries', pp. 357–68 in H. Bernstein (ed.) *Underdevelopment and Development: the Third World today*, Penguin, Harmondsworth.

Infield, M. and Adams, W.M. (1999) 'Institutional sustainability and community conservation: a case study from Uganda', *Journal of International Development* 11: 305–15.

International Institute for Sustainable Development (1997) 'Summary of the First Conference of the Parties to the Convention to Combat Desertification', *Desertification Control Bulletin* 31: 1–5.

Ite, U.E. (1996) 'Small farmers and forest loss in Cross River National Park, Nigeria', *Geographical Journal* 163: 47–56.

IUCN (1980) *The World Conservation Strategy*, International Union for Conservation of Nature and Natural Resources, United Nations Environment Programme, World Wildlife Fund, Geneva.

IUCN (1984a) *National Conservation Strategies: a framework for sustainable development*, IUCN, Geneva.

IUCN (1984b) *Towards Sustainable Development: a national conservation strategy for Zambia*, IUCN, Geneva.

IUCN (1991) *Caring for the Earth: a strategy for sustainability*, International Union for the Conservation of Nature, Gland, Switzerland.

Ives, J. and Messerli, B. (1989) *The Himalayan Dilemma: reconciling development and conservation*, Routledge, London.

Jacks, G.V. and Whyte, R.O. (1938) *The Rape of the Earth: a world survey of soil erosion*, Faber and Faber, London.

Jackson, C. (1993) 'Questioning synergism: win–win with women in population and environment policies', *Journal of International Development* 5: 651–68.

Jackson, C. (1994) 'Gender analysis and feminisms', pp. 113–49 in M.R. Redclift and T. Benton (eds) *Social Theory and the Global Environment*, Routledge, London.

Jackson, P.B.N. (1966) 'The establishment of fisheries in man-made lakes in the tropics', in R.H. Lowe-McConnell (ed.) *Man-Made Lakes*, Academic Press, London.

Jackson, R.D. and Idso, S.B. (1975) 'Surface albedo and desertification', *Science* 189: 1012–13.

Jacobs, M. (1991) *The Green Economy: environment, sustainable development and the politics of the future*, Pluto Press, London.

Jacobs, M. (1995) 'Sustainable development, capital substitution and economic humility: a response to Beckerman', *Environmental Values* 4: 57–68.

Jäger, J. and O'Riordan, T. (1996) 'The history of climate change science and politics', pp. 1–31 in T. O'Riordan and J. Jäger (eds) *Politics of Climate Change: a European perspective*, Routledge, London.

Jago, N. (1987) 'The return of the eighth plague', *New Scientist* 18 June: 47–51.

Jansson, A., Hammer, M., Folke, C. and Costanza, R. (eds) (1994) *Investing in Natural Capital: the ecological economics approach to sustainability*, Island Press, Washington, DC.

Jarosz, L. (1996) 'Defining deforestation in Madagascar', pp. 148–64 in R. Peet and M. Watts (eds) *Liberation Ecologies: environment, development, social movements*, Routledge, London.

Jepma, C.J. (1995) *Tropical Deforestation: a socio-economic approach*, Earthscan, London.

Jewell, P.A. (1980) 'Ecology and management of game animals and domestic livestock in African savannas', pp. 353–81 in D.R. Harris (ed.) *The Human Ecology of Savanna Environments*, Academic Press, London.

Jewitt, S. (1995) 'Europe's "others"? Forestry policy and practices in colonial and post-colonial India', *Environment and Planning D: Society and Space* 13: 67–90.

Jodha, N.S. (1980) 'Intercropping in traditional farming systems', *Journal of Development Studies* 16: 427–42.

Johns, A.D. (1985) 'Selective logging and wildlife conservation in tropical rain-forest: problems and recommendations', *Biological Conservation* 31: 355–76.

Johns, A.G. and Johns, B.G. (1995) 'Tropical forest primates and logging: long-term coexistence?', *Oryx* 29: 205–11.

Johnson, B. (1985) 'Chimera or opportunity? An environmental appraisal of the recently concluded International Tropical Timber Agreement', *Ambio* 14: 42–4.

Johnson, B. and Blake, R.O. (1980) *The Environment and Bilateral Aid Agencies*, IIED, Washington, DC.

Johnston, R.J. (1989) *Environmental Problems: nature, economy and state*, Belhaven, New York.

Jones, B. (1938) 'Desiccation and the West African Colonies', *Geographical Journal* 41: 401–23.

Jones, G.H. (1936) *The Earth Goddess: a study of native farming in the West African context*, Longman, Green, London.

Jones, S. (1996) 'Farming systems and nutrient flows: a case of degradation? *Geography* 81: 289–300.

Jordan, A. and Voisey, H. (1998) 'The "Rio Process": the politics and substantive outcomes of "Earth Summit II"', *Global Environmental Change* 8: 93–7.

Jubb, R.A. (1972) 'The J.G. Strydon Dam, Pongolo River, northern Zululand: the importance of floodplain pans below it', *Piscator* 86: 104–9.

Justice, C.O., Townshend, J.R.G., Holben, B.N. and Tucker, C.J. (1985) 'Analysis of the phenology of global vegetation using meteorological satellite data', *International Journal of Remote Sensing* 6: 1271–318.

Kahn, H. (1979) *World Economic Development: 1979 and beyond*, Croom Helm, London.

Kahn, H. and Wiener, A.J. (1967) *The Year 2000: a framework for speculation on the next 33 years*, Macmillan, London.

Kahn, J. and McDonald, J. (1994) 'International debt and deforestation', pp. 57–67 in K. Brown and D. Pearce (eds) *The Causes of Tropical Deforestation: the economic and statistical analysis of factors giving rise to the loss of the tropical forests*, UCL Press, London.

Karrar, G. (1984) 'The UN Plan of Action to Combat Desertification and the concommitent UNEP campaign', *Environmental Conservation* 11: 99–102.

Kartawinata, K., Adisoemarto, S., Riswar, S. and Vayda, A.P. (1980) 'The environmental consequences of the removal of the forest in Indonesia', pp. 191–214 in V.H. Sutlive, N. Altshuler and M.D. Zamora (eds) *Where Have All the Flowers Gone? Deforestation in the Third World*, Studies in Third World Societies No. 13, College of William and Mary, Williamsburg, VA.

Kartawinata, K., Adisoemarto, S., Riswa, S. and Vayda, A.P. (1981) 'The impact of man on a tropical forest in Indonesia', *Ambio* 10: 115–19.

Kassas, M. (1973) 'Impact of river control schemes on the shoreline of the Nile Delta', pp. 179–88 in M.T. Farvar and J.P. Milton (eds) *The Careless Technology: ecology and international development*, Stacey, London.

Katz, C. and Kirby, A. (1991) 'In the nature of things: the environment and everyday life', *Transactions of the Institute of British Geographers* 16: 259–71.

Kemf, E. (ed.) (1993) *The Law of the Mother: protecting indigenous peoples in protected areas*, Sierra Club Books, San Francisco.

Kennedy, V.W. (1988) 'Environmental impact assessment and bilateral development aid: an overview', pp. 272–82 in P. Wathern (ed.) *Environmental Impact Assessment: theory and practice*, Unwyn Hyman, London.

Khogali, M.M. (1982) 'The problem of siltation in Khasm el Girba Reservoir: its implications and suggested solutions', pp. 96–106 in H.G. Mensching (ed.) *Problems of the Management of Irrigated Land in Areas of Traditional and Modern Cultivation*, International Geographical Union Working Group on Resource Management in Drylands, Hamburg.

Kiessling, K.L. and Landberg, H. (eds) (1994) *Population, Economic Development and the Environment: the making of our common future*, Clarendon Press, Oxford.

Kimmage, K. (1991) 'Small scale irrigation initiatives in Nigeria: the problems of equity and sustainability', *Applied Geography* 11: 5–20.

Kimmage, K. and Adams, W.M. (1990) 'Small-scale farmer-managed irrigation in northern Nigeria', *Geoforum* 21(4): 435–43.

Kirchner, J.W., Ledec, G., Goodland, R.J.A and Drake, J.M. (1985) 'Carrying capacity, population growth and sustainable development', pp. 42–89 in D.J. Mahar (ed.) *Rapid Population Growth and Human Carrying Capacity*, World Bank Staff Working Paper 690, Washington, DC.

Kishk, M.A. (1986) 'Land degradation in the Nile Valley', *Ambio* 15: 226–30.

Kitching, G. (1982) *Development and Underdevelopment in Historical Perspective: populism, nationalism and industrialism*, Methuen, London.

Kleymeyer, C.D. (1994) 'Cultural traditions and community-based conservation', pp. 323–46 in D. Western, R.M. White and S.C. Strum (eds) *Natural Connections: perspectives in community-based conservation*, Island Press, Washington, DC.

Kloezen, W. and Slabbers, J. (1992) 'Turnover of irrigation systems: a role for engineers?', pp. 275–84 in G. Diemer and J.H. Slabbers (eds) *Irrigators and Engineers: essays in honour of Lucas Horst*, Thesis Publishers, Amsterdam.

Koh, T.T.-B. (1993) 'The Earth Summit's negotiating process: some comments on the art and science of negotiation', pp. v–xiii in N. Robinson (ed.) *Agenda 21: Earth's action plan*, IUCN Environmental Policy and Law Paper 27, Oceana Publications, New York.

Kolawole, A. (1987) 'Environmental change and the South Chad Irrigation Project (Nigeria)', *Journal of Arid Environments* 13: 169–76.

Kovda, V.A. (1977) 'Arid land irrigation and soil fertility: problems of salinity, alkalinity, compaction', pp. 211–35 in E.B. Worthington (ed.) *Arid Land Irrigation in Developing Countries: environmental problems and effects*, Pergamon Press, Oxford.

Kowal, J.M. and Adeoye, K.B. (1973) 'An assessment of aridity and the severity of the 1972 drought in northern Nigeria and neighbouring countries', *Savanna* 2: 145–58.

Kowal, J.M. and Kassam, A.H. (1978) *Agricultural Ecology of Savanna: a study of West Africa*, Clarendon Press, Oxford.

Kropotkin, P. (1972) *The Conquest of Bread*, ed. P. Avrich, Allen Lane, London (first published in English as articles in *Freedom* 1892–4, and as a book in 1906, London).

Kropotkin, P. (1974) *Fields, Factories and Workshops Tomorrow*, ed. C. Ward, Allen and Unwin, London (1st edition Hutchinson, London, 1899).

Lagler, K.F. (ed.) (1969) *Man-Made Lakes: planning and development*, Food and Agriculture Organisation, Rome.

Lagler, K.F. (1971) 'Ecological effects of hydrological dams', pp. 133–57 in D.A. Berkowitz and A.M. Squires (eds) *Power Generation and Environmental Change*, MIT Press, Cambridge, MA.

Lako, G.T. (1985) 'The impact of the Jonglei Scheme on the economy of the Dinka', *African Affairs* 84: 15–38.

Lamb, D. (1980) 'Some ecological consequences of logging rainforests on Papua New Guinea', pp. 55–64 in J.I. Furtado (ed.) *Tropical Ecology and Development*, International Society for Tropical Ecology, Kuala Lumpur (Proceedings of the Fifth International Symposium on Tropical Ecology).

Lamb, P.J. (1979) 'Some perspectives on climate and climatic dynamics', *Progress in Physical Geography* 3: 215–35.

Lamprey, H. (1988) 'Report on the desert encroachment reconnaissance in northern Sudan, October 21–November 10, 1875', *Desertification Control Bulletin* 17: 1–7 (reprinted).

Lane, C. (1992) 'The Barabaig pastoralists of Tanzania: sustainable land use in jeopardy', pp. 81–105 in D. Ghai and J.M. Vivian (eds) *Grassroots Environmental Action: people's participation in sustainable development*, Routledge, London.

Lanly, J.-P. (1982) *Tropical Forest Resources*, Forestry Paper No. 30, Food and Agriculture Organisation, Rome.

Lanning, G. and Mueller, M. (1979) *Africa Undermined: mining companies and the underdevelopment of Africa*, Penguin, Harmondsworth.

Lash, S. and Urry, J. (1994) *Economies of Signs and Space*, Sage, London.

Lawson, R.M. (1963) 'The economic organisation of the *Egeria* fishery on the River Volta', *Proceedings of the Malacological Society of London* 35: 273–87.

Leach, M. and Mearns, R. (eds) (1996) *The Lie of the Land: challenging received wisdom on the African environment*, James Currey/International African Institute, London.

Lee, E. (1981) 'Basic needs strategies: a frustrated response to development from below?', pp. 107–27 in W.B. Stöhr and D.R.F. Taylor (eds) *Development: from above or below?*, Wiley, Chichester.

Lee, N. (1983) 'Environmental impact assessment: a review', *Applied Geography* 3: 5–28.

Lélé, S.M. (1991) 'Sustainable development: a critical review', *World Development* 19: 607–21.

Lelek, A. and El-Zarka, S. (1973) 'Kainji Lake, Nigeria', pp. 655–60 in W.C. Ackerman, G.F. White and E.B. Worthington (eds) *Man-Made Lakes: their problems and environmental effects*, Geophysical Monograph 17, American Geophysical Union, Washington, DC.

Leopold, L.B., Clarke, F.E., Nanshaw, B.B. and Balsley, J.R. (1971) *A Procedure for Evaluating Environmental Impact*, US Geological Survey Circular 645, Washington, DC.

Lericollais, A. and Schmitz, J. (1984) '"La calebasse et la houe". Techniques et outils de décrue dans la vallée du Sénégal', *Cahiers ORSTOM série Sciences Humaines* 20: 127–59.

Lewis, M.W. (1992) *Green Delusions: an environmentalist critique of radical environmentalism*, Duke University Press, Durham, NC.

Lightfoot, R.P. (1978) 'The costs of resettling reservoir evacuees in NE Thailand', *Journal of Tropical Geography* 47: 63–74.

Lightfoot, R.P. (1979) 'Alternative resettlement strategies in Thailand: lessons from experience', pp. 28–38 in L. Gosling (ed.) *Population Resettlement in the Mekong River Basin*, Studies in Geography 10, University of North Carolina, Chapel Hill.

Lightfoot, R.P. (1981) 'Problems of resettlement in the development of river basins in Thailand', pp. 93–114 in S.K. Saha and C.J. Barrow (eds) *River Basin Planning: theory and practice*, Wiley, Chichester.

Lindblade, K.A., Carswell, G. and Tumuhairwe, J.K. (1998) 'Mitigating the relationship between population growth and land degradation: land use change and farm management in southwestern Uganda', *Ambio* 27: 565–71.

Lindemann, R.L. (1942) 'The trophic-dynamic aspect of ecology' *Ecology* 23: 399–418.

Lindsay, W.K. (1987) 'Integrating parks and pastoralists: some lessons from Amboseli', pp. 149–68 in D.M. Anderson and R.H. Grove (eds) *Conservation in Africa: people, policies and practice*, Cambridge University Press, Cambridge.

Lipton, M. (1977) *Why Poor People Stay Poor: a study of urban bias in world development*, Temple Smith, London.

Lipton, M. (1991) 'A note on poverty and sustainability', *IDS Bulletin* 22(4): 12–16.

Lipton, M. and Longhurst, R. (1989) *New Seeds and Poor People*, Johns Hopkins University Press, Baltimore.

Livingstone, D.N. (1995) 'The polity of nature: representation, virtue, strategy', *Ecumene* 2(4): 353–77.

Lockwood, J. (1986) 'The causes of drought with particular reference to the Sahel', *Progress in Physical Geography* 10: 111–19.

Loh, J., Randers, J., MacGillivray, A., Kapos, V., Jenkins, M., Groombridge, B., Cox, N. and Warren, B. (1999) *Living Planet Report*, World Wildlife Fund, Gland, Switzerland.

Lohani, B.N. and Thanh, N.C. (1980) 'Impacts of rural development and their assessment in southeastern Asia', *Environmental Conservation* 7: 213–18.

Lovejoy, T.E. Bierregaard, R.O., Rankin, J. and Schubart, H.O.R. (1983) 'Dynamics of forest fragments', pp. 377–84 in S.L. Sutton, T.C. Whitmore and A.C. Chadwick (eds) *Tropical Rain Forest: ecology and management*, Blackwell, Oxford.

Lövgren, L. (1997) 'Moratorium in Sweden: an account of the dams debate', pp. 21–30 in A.D. Usher (ed.) *Dams as Aid: a political anatomy of Nordic development thinking*, Routledge, London.

Low, D.A. and Lonsdale, J.M. (1976) 'Introduction: towards a new order 1945–1963', pp. 1–63 in D.A. Low and A. Smith (eds) *History of East Africa*, vol. 3, Clarendon Press, Oxford.

Low, N. and Gleeson, B. (1998) *Justice, Society and Nature: an exploration of political ecology*, Routledge, London.

Lowe, P.D. (1976) 'Amateurs and professionals: the institutional emergence of British plant ecology', *Journal of the Society for the Bibliography of Natural History* 7: 517–35.

Lowe, P.D. and Goyder, P. (1983) *Environmental Groups in Politics*, Allen and Unwin, Hemel Hempstead.

Lowe, P.D. and Warboys, M. (1975) 'Ecology and the end of ideology', *Antipode* 10: 12–21.

Lowe, P.D. and Warboys, M. (1976) 'The ecology of ecology', *Nature* 262: 432–33.

Lowe-McConnell, R.H. (ed.) (1966) *Man-Made Lakes*, Institute of Biology and Academic Press, London.

Lowe-McConnell, R.H. (1975) *Fish Communities of Tropical Freshwaters*, Longman, London.

Lowe-McConnell, R.H. (1985) 'The biology of the river systems with particular reference to the fish', pp. 101–40 in A.T. Grove (ed.) *The Niger and its Neighbours: environment, history and hydrobiology, human use and health hazards of the major West African rivers*, Balkema, Rotterdam.

Lugo, A.E., Parrotta, J.A. and Brown, S. (1993) 'Loss of species caused by tropical defor-estation and their recovery through management', *Ambio* 22: 106–69.

Lumsden, D.P. (1975) 'Towards a systems model of stress: feedback from an anthropolog-ical study of the impact of Ghana's Volta River Project', *Stress and Anxiety* 2: 191–227.

Mabbutt, J.A. (1984) 'A new global assessment of the status and trends of desertification', *Environmental Conservation* 11: 103–13.

Mabogunje, A.L. (ed.) (1973) *Kainji: a Nigerian man-made lake*, Kainji Lake Studies vol. 2: *Socio-economic conditions*, Nigerian Institute for Social and Economic Research, Ibadan.

McCormick, J. (1997) *Acid Earth*, Earthscan, London.

McCormick, J.S. (1986) 'The origins of the World Conservation Strategy', *Environmental Review* 10(2): 177–87.

McCormick, J.S. (1992) *The Global Environmental Movement: reclaiming paradise*, Belhaven, London (first published 1989, Indiana University Press, Bloomington).

McCully, P. (1996) *Silenced Rivers: the ecology and politics of large dams*, Zed Press, London.

Mace, R. (1991) 'Overgrazing overstated', *Nature* 349 (24 January): 280–1.

McGinn, A.P. (2000) 'POPs culture', *World Watch* 13(2): 26–36.

McHarg, I. (1969) *Design with Nature*, Natural History Press, Garden City, NY.

McIntosh, R.P. (1985) *The Background of Ecology: concept and theory.* Cambridge University Press, Cambridge.

MacKenzie, D. (1987) '"Thousands poisoned" by pesticide in Guyana', *New Scientist* 19 March: 18.

MacKenzie, J.M. (1987) 'Chivalry, social Darwinism and ritualised killing: the hunting ethos in Central Africa up to 1914', pp. 41–62 in D.M. Anderson and R.H. Grove (eds) *Conservation in Africa: people, policies and practice*, Cambridge University Press, Cambridge.

MacKenzie, J.M. (1989) *The empire of nature: hunting, conservation and British imperial-ism*, University of Manchester Press, Manchester.

McKibben, B. (1990) *The End of Nature*, Penguin Books, Harmondsworth.

McLean, R.C. (1919) 'Studies in the ecology of Tropical rain forest: with special reference to the forests of south Brazil', *Journal of Ecology* 7: 121–72.

McNeely, J.A. (1984) 'Introduction: protected areas are adapting to new realities', pp. 1–7 in J.A. McNeely and K.R. Miller (eds) *National Parks, Conservation and Development: the role of protected areas in sustaining society*, Smithsonian Institute Press, Washington, DC.

McNeely, J.A. (1993) 'Economic incentives for conserving biodiversity: lessons for Africa', *Ambio*, 22, 144–50.

McNeely, J.A. (1996) 'Partnerships for conservation: an introduction', in J.A. McNeely (ed.) *Expanding Partnerships in Conservation*, Island Press, Washington, DC.

McNeely, J.A. and Miller, K.R. (eds) (1984) *National Parks, Conservation and Development: the role of protected areas in sustaining society*, Smithsonian Institute Press, Washington, DC.

McNeely, J. and Pitt, D. (eds) (1987) *Culture and Conservation: the human dimension in environmental planning*, Croom Helm, London.

McRobie, G. (1981) *Small is Possible*, Jonathan Cape, London.

Maddox, J. (1972) *The Doomsday Syndrome*, Macmillan, London.

Madely, J. (1995) Feeding 8 billion', *People and the Planet* 4(4): 7–9.

Malanson, G.P. (1993) *Riparian Landscapes*, Cambridge University Press, Cambridge.

Manley, R.E. and Wright, E.P. (1996) 'The review of the Southern Okavango Integrated Water Development Project', pp. 213–24 in M.C. Acreman and G.E. Hollis (eds) *Water Management and Wetlands in Sub-Saharan Africa*, International Union for the Conservation of Nature, Gland, Switzerland.

Margalef, R. (1968) *Perspectives in Ecological Theory*, University of Chicago Press, Chicago.

Marsh, G.P. (1864) *Man and Nature; or, physical geography as modified by human action*, Scribners, New York, and Sampson Low, London (reprinted Harvard University Press, 1965).

Marshall, B.K. (1999) 'Globalisation, environmental degradation and Ulrich Beck's Risk Society', *Environmental Values* 8: 253–75.

Maser, C. (1990) *The Redesigned Forest*, Stoddart, Toronto.

Mather, A. (1992) 'The forest transition', *Area* 24: 367–79.

Matthieson, P. and Douthwaite, B. (1985) 'The impact of tsetse control campaigns on African wildlife', *Oryx* 19: 202–9.

May, B. (1981) *The Third World Calamity*, Routledge and Kegan Paul, London.

Maybury-Lewis, D. (1984) 'Demystifying the second conquest', pp. 127–34 in M. Schmick and C.H. Wood (eds) (1984) *Frontier Expansion in Amazonia*, University of Florida Press, Gainesville.

Meadows, D.H., Meadows, D.K., Randers, J. and Behrens, W.W. III (1972) *The Limits to Growth*, Universe Books, New York.

Melillo, J.M., Palm, C.A. and Myers, N. (1985) 'A comparison of two recent estimates of forest disturbance in tropical forests', *Environmental Conservation* 12: 37–40.

Merchant, C. (1980) *The Death of Nature: women, ecology and the scientific revolution*, Harper and Row, New York.

Metcalfe, S. (1994) 'The Zimbabwe Communal Areas Management Programme for Indigenous Resources (CAMPFIRE)', pp. 161–92 in D. Western, R.M. White and S.C. Strum (eds) *Natural Connections: perspectives in community-based conservation*, Island Press, Washington, DC.

Middleton, N.J. (1985) 'Effect of drought on dust production in the Sahel', *Nature* 316: 431–4.

Middleton, T. and Thomas, D.S.G. (eds) (1997) *World Atlas of Desertification*, United Nations Environment Programme, Nairobi (2nd edition).

Midgley, J., Hall, A., Hardiman, M. and Narine, D. (1986) *Community Development, Social Participation and the State*, Methuen, London.

Mies, M. (1986) *Patriarchy and Accumulation on a World Scale: women in the international division of labour*, Zed Books, London.

Miller, R.B. (1994) 'Interactions and collaboration in global change across the social and natural sciences', *Ambio* 23: 19–24.

Mishan, E. (1969) *The Costs of Economic Growth*, Penguin Books, Harmondsworth (first published Staples Press, 1967).

Mishan, E.J. (1977) *The Economic Growth Debate: an assessment*, Allen and Unwin, Hemel Hempstead.

Mock, J.F. and Bolton, P. (eds) (1993) *The ICID Environmental Check-list: to identify environmental effects of irrigation, drainage and flood-control projects*, published for the International Commission on Irrigation and Drainage by HR Wallingford, Wallingford, Oxfordshire.

el Moghraby, A.I. (1982) 'The Jonglei Canal: needed development or potential ecodisaster?', *Environmental Conservation* 9: 141–8.

el Moghraby, A.I. and el Sammani, M.O. (1985) 'On the environmental and socioeconomic impact of the Jonglei Canal Project, S. Sudan', *Environmental Conservation* 12: 41–8.

Mohun, J. and Sattaur, O. (1987) 'The drowning of a culture', *New Scientist* 15 January: 37–42.

Mol, A.P. (1996) 'Ecological modernisation and institutional reflexivity: environmental reform in a late modern age', *Environmental Politics* 5: 302–23.

Momsen, J. (1991) *Women and Development in the Third World*, Routledge, London.

Mooney, H.A. and Ehrlich, P.R. (1997) 'Ecosystem services: a fragmentary history', pp. 11–19 in G.C. Daily (ed.) *Nature's Services: societal dependence on natural ecosystems*, Island Press, Washington, DC.

Moore, D. and Sklar, L. (1998) 'Reforming the World Bank's lending for water: the process and outcome of developing a water resources management policy', pp. 345–90 in J.A. Fox and L.D. Brown (eds) *The Struggle for Accountability: the World Bank, NGOs and grassroots movements*, MIT Press, Cambridge, MA.

Moore, H. and Vaughan, M. (1994) *Cutting Down Trees: gender, nutrition and agricultural change in Zambia*, James Currey, London.

Moore, M. (1993) 'Good government? Introduction', *IDS Bulletin* 24(1): 1–6.

Moorhead, R. (1988) 'Access to resources in the Niger Inland Delta, Mali', pp. 27–39 in J. Seeley and W.M. Adams (eds) *Environmental Issues in African Development Planning*, Cambridge African Monographs No. 9, African Studies Centre, Cambridge.

Moran, E.F. (ed.) (1983) *The Dilemma of Amazonian Development*, Westview, Boulder, CO.

Morgan, A.E. (1971) *Dams and Other Disasters: a century of the Army Corps of engineers in civil works*, Porter Sergent, Boston, MA.

Morgan, W.B. and Solarz, J.A. (1994) 'Agricultural crisis in sub-Saharan Africa: development constrains and policy problems', *Geographical Journal* 160: 57–73.

Moris, J. (1987) 'Irrigation as a privileged solution in African development', *Development Policy Review* 5: 99–123.

Moris, J.R. and Thom, D.J. (1990) *Irrigation Development in Africa: lessons from experience*, Westview Press, Boulder, CO.

Mortimore, M. (1989) *Adapting to Drought: farmers, famines and desertification in West Africa*, Cambridge University Press, Cambridge.

Mortimore, M. (1993) 'Population growth and land degradation', *GeoJournal* 31: 15–21.

Mortimore, M. (1998) *Roots in the African Dust: sustaining the subSaharan drylands*, Cambridge University Press, Cambridge.

Mortimore, M. and Adams, W.M. (1999) *Working the Sahel: environment and society in northern Nigeria*, Routledge, London.

Mortimore, M.J. and Tiffen, M. (1995) 'Population and environment in time perspective: the Machakos story', pp. 69–89 in T. Binns (ed.) *People and Environment in Africa*, Wiley, Chichester.

Mulligan, P. (1999) 'Greenwash or blueprint? Rio Tinto in Madagascar', *IDS Bulletin* 30(3): 50–7.

Munasinghe, M. (1993a) 'Environmental issues and economic decisions in developing countries', *World Development* 21: 1729–48.

Munasinghe, M. (1993b) *Environmental Economics and Sustainable Development*, World Bank Environmental Paper 3, World Bank, Washington, DC.

Munasinghe, M. (1993c) 'Environmental economics and biodiversity management', *Ambio* 22: 126–35.

Munn, R.E. (ed.) (1979) *Environmental Impact Assessment: principles and procedures*, SCOPE Report 5, Wiley, Chichester.

Munro, D.A. (1978) 'The thirty years of IUCN', *Nature and Resources* 14(2): 14–18.

Munslow, B., Katere, V., Ferf, A. and O'Keefe, P. (1988) *The Fuelwood Trap: a study of the SADCC region*, Earthscan, London.

Murdia, R. (1982) 'Forest development and tribal welfare: analysis of some policy issues', pp. 31–41 in E.G. Hallsworth (ed.) *Socio-economic Effects and Constraints in Tropical Forest Management*, Wiley, Chichester.

Murombedzi, J.S. (1999) 'Devolution and stewardship in Zimbabwe's CAMPFIRE Programme', *Journal of International Development* 11: 287–93.

Murphree, M.W. (1994) 'The role of institutions in community-based conservation', pp. 403–427 in D. Western, R.M. White and S.C. Strum (eds) *Natural Connections: perspectives in community-based conservation*, Island Press, Washington, DC.

Murphy, D.F. and Bendell, J. (1997) *In the Company of Partners: business, environmental groups and sustainable development post-Rio*, The Policy Press, Bristol.

Murphy, R. (1994) *Rationality and Nature: a sociological inquiry into a changing relationship*, Westview Press, Boulder, CO.

Murray, S.P., Coleman, J.M., Roberts, H.H. and Salama, M. (1981) 'Accelerated currents and sediment transport of the Damieth Nile promontory', *Nature* 293: 51–4.

Murton, J. (1999) 'Population growth and poverty in Machakos District, Kenya', *Geographical Journal* 165: 37–46.

Myers, N. (1980) *Conversion of Tropical Moist Forests*, National Academy of Sciences, Washington, DC.

Myers, N. (1981) 'The hamburger connection: how Central America's forests become North America's hamburgers', *Ambio* 10: 3–8.

Myers, N. (1984) *The Primary Source: tropical forests and our future*, Norton, New York.

Myers, N. (ed.) (1985) *The Gaia Atlas of Planet Management*, Pan Books, London.

Myers, N. and Myers, D. (1982) 'From "duck pond" to the global commons: increasing awareness of the supranational nature of emerging environmental issues', *Ambio* 11: 195–201.

Naess, A. (1973) 'The shallow and the deep, long-range ecology movement: a summary', *Inquiry* 16: 95–100.

Narayanayaya, D.V. (1928) 'The aquatic weeds in Deccan irrigation canals', *Journal of Ecology* 16: 123–33.

Nash, R. (1973) *Wilderness and the American Mind*, Yale University Press, New Haven, CT.

National Research Council (US) (1992) *Restoration of Aquatic Ecosystems: science, technology and public policy*, National Academy Press, Washington, DC.

Nature (1948) 'Aspects of colonial development', *Nature* 162: 547–50.

Nature (1952) 'Development of backward countries', *Nature* 170: 677–80.

Nelson, J.G. (1987) 'National parks and protected areas, national conservation strategies and sustainable development', *Geoforum* 18: 291–320.

Nesmith, C. and Radcliffe, S.A. (1993) '(Re)mapping Mother Earth: a geographical perspective on environmental feminisms', *Environment and Planning D: Society and Space* 11: 379–94.

Neumann, R.P. (1998) *Imposing Wilderness: struggles over livelihood and nature preservation in Africa*, University of California Press, Berkeley.

New Left Review (1974) 'Themes', *New Left Review* 8: 1–2.

Ngantou, D. (1994) 'Rehabilitation of the Waza Logone floodplain, Cameroon', *IUCN Wetlands Programme Newsletter* 10 (November): 9–10.

Nicholaides, J.J. II, Bandy, D.E., Sanchez, P.A., Villachica, J.H., Contu, A.J. and Valverde, C.S. (1984) 'Continuous cropping potential in the Upper Amazon Basin', pp. 337–65 in M. Schmink and C.H. Wood (eds) *Frontier Expansion in Amazonia*, University of Florida Press, Gainesville.

Nichols, P. (1991) *Social Survey Methods*, Oxfam Books, Oxford.

Nicholson, E.M. (1970) *The Environmental Revolution: a guide for the new masters of the world*, Hodder and Stoughton, London.

Nicholson, E.M. (1975) 'Conservation', pp. 12–14 in E.B. Worthington (ed.) *The Evolution of the IBP*, Cambridge University Press, Cambridge.

Nicholson, S.E. (1978) 'Climatic variation in the Sahel and other African regions during the past five centuries', *Journal of Arid Environments* 1: 3–24.

Nicholson, S.E. (1980) 'Saharan climates in historic times', pp. 173–200 in M.A.J. Williams and H. Faure (eds) *The Sahara and the Nile: Quaternary environments and prehistoric occupation in northern Africa*, Balkema, Rotterdam.

Nicholson, S.E. (1988) 'Land surface atmosphere interaction: physical processes and surface changes and their impact', *Progress in Physical Geography* 12: 36–55.

Nicholson, S.E. (1996) Environmental change within the historical period', pp. 60–88 in W.M. Adams, A.S. Goudie and A.R. Orme (eds) *The Physical Geography of Africa*, Oxford University Press, Oxford.

North, D.C. (1990) *Institutions, Institutional Change and Economic Performance*, Cambridge University Press, Cambridge.

Norton, B. (1991) *Toward Unity among Environmentalists*, Oxford University Press, London.

Norton-Griffiths, M. (1979) 'The influence of grazing, browsing and fire on the vegetation dynamics of the Serengeti', pp. 310–52 in A.R.E. Sinclair and M. Norton-Griffiths (eds) *Serengeti: Dynamics of an Ecosystem*, University of Chicago Press, Chicago.

Nye, P. and Greenland, D. (1960) *The Soil under Shifting Cultivation*, Technical Report 51, Commonwealth Agricultural Bureau, London.

Nyerere, J.K. (1985) 'Africa and the debt crisis', *African Affairs* 84: 489–98.

Oates, J.F. (1995) 'The dangers of conservation by rural development: a case study from the forests of Nigeria', *Oryx* 29: 115–22.

Oates, J.F. (1999) *Myth and Reality in the Rain Forest: how conservation strategies are failing in West Africa*, University of California Press, Berkeley.

Obeng, L.E. (ed.) (1969) *International Symposium on Man-Made Lakes, Accra, Ghana* University Press, Accra.

Oculi, O. (1981) 'Planning the Bakolori irrigation project', *Food Policy* 6: 201–4.

OECD (1995) *The Economic Appraisal of Environmental Projects and Policies: a practical guide*, Organisation for Economic Cooperation and Development, Paris.

OECD (1996) *The Global Environmental Goods and Services Industry*, Organisation for Economic Cooperation and Development, Paris.

Oglesby, R.T., Carlson, C.A. and McCann, J.A. (1972) *River Ecology and Man*, Academic Press, New York.

Oldfield, S. (1988) 'Rare tropical timbers', International Union for the Conservation of Nature, Gland, Switzerland.

Olofin, E. (1984) 'Some effects of the Tiga Dam on valleyside erosion in downstream reaches of the River Kano', *Applied Geography* 4: 321–32.

Olsson, L. (1993) 'On the causes of famine: drought, desertification and market failure in the Sudan', *Ambio* 22: 395–403.

Olthof, W. (1994) 'Wildlife resources and local development: experiences from Zimbabwe's CAMPFIRE Programme', pp. 111–28 in J.P.M. van den Breemer, C.A. Drijver and L.B. Venema (eds) *Local Resource Management in Africa*, Wiley, Chichester.

Oman, C.P. and Wignarajah, G. (1991) *The Postwar Evolution of Development Thinking*, Macmillan, London, in association with the OECD Development Centre.

O'Riordan, T. (1976a) *Environmentalism*, Pion, London (1st edition).

O'Riordan, T. (1976b) 'Beyond environmental impact assessment', pp. 202–21 in T. O'Riordan and R. Hey (eds) *Environmental Impact Assessment*, Saxon House, Farnborough.

O'Riordan, T. (1981) *Environmentalism*, Pion, London (2nd edition).

O'Riordan, T. (1988) 'The politics of sustainability', pp. 29–50 in R.K. Turner (ed.) *Sustainable Environmental Management: principles and practice*, Westview Press, Boulder, CO.

O'Riordan, T. and Turner, R.K. (1983) *An Annotated Reader in Environmental Planning and Management*, Pergamon Press, Oxford.

Ormerod, W.E. (1986) 'A critical study of the policy of tsetse eradication', *Land Use Policy* 3: 85–99.

Osborn, F. (1948) *Our Plundered Planet*, Faber and Faber, London.

Osborn, F. (1954) *The Limits of the Earth,* Faber and Faber, London.

Osemiebo, G.J. (1988) 'Impacts of multiple forest land use on wildlife conservation in Bendel State, Nigeria', *Biological Conservation* 45: 209–20.

Osmaston, A.E. (1922) 'Notes on the forest communities of the Garhwal Himalaya', *Journal of Ecology* 10: 129–67.

Ostrom, E. (1990) *Governing the Commons: the evolution of institutions for collective action*, Cambridge University Press, Cambridge.

Otten, M (1986) 'Transmigrasi: from poverty to bare subsistence', *The Ecologist* 16: 57–67.

Otterman, J. (1974) 'Baring high-albedo soils by overgrazing: a hypothesised desertification mechanism', *Science* 166: 531–3.

Owens, S.E. and Cowell, R. (1994) 'Lost land and limits to growth: conceptual problems for sustainable land use change', *Land Use Policy* 11: 168–80.

Pahl-Wostl, C. (1995) *The Dynamic Nature of Ecosystems: chaos and order intertwined* Wiley, Chichester.

Palmer-Jones, R. (1984) 'Mismanaging the peasants: some origins of low productivity on irrigation schemes in northern Nigeria', in W.M. Adams and A.T. Grove (eds) *Irrigation in Tropical Africa: problems and problem-solving*, Cambridge African Monographs No. 3, University of Cambridge African Studies Centre.

Palo, M. (1994) 'Population and deforestation', pp. 42–56 in K. Brown and D. Pearce (eds) *The Causes of Tropical Deforestation: the economic and statistical analysis of factors giving rise to the loss of the tropical forests*, UCL Press, London.

Parayil, G. and Tong, F. (1998) 'Pasture-led to logging-led deforestation in the Brazilian Amazon', *Global Environmental Change* 8: 63–79.

Parikh, J., Babu, P.G. and Kavi Kumar, K.S. (1997) 'Climate change, North–South cooperation and collective decision-making post-Rio', *Journal of International Development* 9: 403–13.

Parnwell, M.J.G. and Bryant, R.L. (1996) 'Introduction', pp. 1–20 in M.J.G. Parnwell and R.L. Bryant (eds) *Environmental Change in South-East Asia: people, politics and sustainable development*, Routledge, London.

Parsons, H.L. (ed.) (1979) *Marx and Engels on Ecology*, Greenwood Press, Westport, CT.

Pasek, J. (1992) 'Obligations to future generations: a philosophical notes', *World Development* 20: 513–21.

Payne, A.J. (1986) *The Ecology of Tropical Lakes and Rivers*, Wiley, Chichester.

Pearce, D. (1988) 'The sustainable use of natural resources in developing countries', pp. 102–17 in R.K. Turner (ed.) *Sustainable Environmental Management*, Belhaven, London.

Pearce, D. (1995) *Blueprint 4: capturing global environmental value*, Earthscan, London.

Pearce, D. and Brown, K. (1994) 'Saving the world's tropical forests', pp. 2–26 in K. Brown and D. Pearce (eds) *The Causes of Tropical Deforestation: the economic and statistical analysis of factors giving rise to the loss of the tropical forests*, UCL Press, London.

Pearce, D., Markandya, A. and Barbier, E. (1989) *Blueprint for a Green Economy*, Earthscan, London.

Pearce, F. (1987) 'Pesticide deaths: the price of the Green Revolution', *New Scientist* 18 June: 30.

Pearce, F. (1991) 'North–South rift bars path to summit', *New Scientist* 22 November: 20–1.

Pearce, F. (1992) *The Dammed: rivers, dams and the coming world water crisis*, The Bodley Head, London.

Pearce, F. (1994) 'Are Sarawak's forests sustainable?', *New Scientist* 26 November: 28–32.

Pearce, F. (2000) 'The floodgates open', *New Scientist* 25 March: 16–17.

Pearl, R. (1927) 'The growth of populations', *Quarterly Review of Biology* 2: 537–43.

Peet, R. (1991) *Global Capitalism: theories of societal development*, Routledge, London.

Peet, R. and Watts, M. (1996a) 'Liberation ecology: development, sustainability, and environment in an age of market triumphalism', pp. 1–45 in R. Peet and M. Watts (eds) *Liberation Ecologies: environment, development, social movements*, Routledge, London.

Peet, R. and Watts, M. (eds) (1996b) *Liberation Ecologies: environment, development, social movements*, Routledge, London.

Pepper, D. (1984) *The Roots of Modern Environmentalism*, Croom Helm, London.

Pepper, D. (1993) *Eco-socialism: from deep ecology to social justice*, Routledge, London.

Pepper, D. (1996) *Modern Environmentalism: an introduction*, Routledge, London.

Perfect, J. (1980) 'The environmental impacts of DDT in a tropical agro-ecosystem', *Ambio* 9: 16–22.

Petr, T. (1975) 'On some factors associated with high fish catches in new African man-made lakes', *Archiv für Hydrobiologie* 75: 32–49.

Petr, T. (1979) 'Possible environmental impact on inland waters of two planned major engineering projects in Papua New Guinea', *Environmental Conservation* 6: 281–6.

Petts, G.E. (1996) 'Sustaining the ecological integrity of large floodplain rivers', pp. 535–51 in M.G. Anderson, D.E. Walling and P.D. Bates (eds) *Floodplain Processes*, Wiley, Chichester.

Petts, G.E. (1984) *Impounded Rivers: perspectives for ecological management*, Wiley, Chichester.

Phillips, J. (1931) 'Ecological investigations in South, Central and East Africa: outline of a progressive scheme', *Journal of Ecology* 14: 474–82.

Pickup, G. (1980) 'Hydrological and sediment modelling studies in the environmental impact assessment of a major tropical dam project', *Earth Surface Processes* 5: 61–75.

Pimbert, M. (1997) 'Issues emerging in implementing the Convention on Biological Diversity', *Journal of International Development* 9: 415–25.

Pimm, S. (1997) 'The value of everything', *Nature* 387 (15 May): 231–2.

Pitt, D.C. (1976) *The Social Dynamics of Development*, Pergamon Press, Oxford.

Plumwood, V. (1993) *Feminism and the Mastery of Nature*, Routledge, London.

Polet, G. and Thompson, J.R. (1996) 'Maintaining the floods: hydrological and institutional aspects of managing the Komadugu-Yobe River basin and its floodplain wetlands', pp. 73–90 in M.C. Acreman and G.E. Hollis (eds) *Water Management and Wetlands in Sub-Saharan Africa*, International Union for the Conservation of Nature, Gland, Switzerland.

Polunin, N. (1984) 'Genesis and progress of the World Campaign and Council for the Biosphere', *Environmental Conservation* 11: 293–8.

Poore, D. (1976) *Ecological Guidelines for Development in Tropical Rainforests*, International Union for the Conservation of Nature, Gland, Switzerland.

Poore, D. and Sayer, J. (1987) *The Management of Tropical Moist Forest Lands: ecological guidelines*, International Union for the Conservation of Nature, Gland, Switzerland.

Prance, G.T. (1991) 'Rates of loss of biological diversity', pp. 27–44 in I.F. Spellerberg, F.B. Goldsmith and M.G. Morris (eds) *The Scientific Management of Temperate Communities for Conservation*, Blackwell, Oxford.

Pratt, M.L. (1992) *Imperial Eyes: travel writing and transculturation*, Routledge, London.

Pretty, J.N. (1995) *Sustainable Agriculture: policies and practice for sustainability and self-reliance*, Earthscan, London.

Pretty, J.N. and Guijt, I. (1992) 'Primary environmental care: an alternative paradigm for development assistance', *Environment and Urbanisation* 4: 22–36.

Princen, T. (1994) 'The ivory trade ban: NGOs and international conservation', pp. 121–59 in T. Princen and M. Finger (eds) *Environmental NGOs in World Politics*, Routledge, London.

Princen, T. and Finger, M. (1994) *Environmental NGOs in World Politics: linking the local and the global*, Routledge, London.

Prior, J. (1993) *Pastoral Development Planning*, Oxfam, Oxford.

Prospero, J.M. and Nees, R.T. (1977) 'Dust concentrations in the atmosphere of the equatorial North Atlantic: possible relationship to the Sahelian drought', *Science* 196: 1196–8.

Putnam, R. (1993) *Making Democracy Work: civic traditions in modern Italy*, Princeton University Press, Princeton, NJ.

Putzel, J. (1997) 'Accounting for the "dark side" of social capital: reading Robert Putnam on democracy', *Journal of International Development* 9: 939–49.

Rajan, R. (1998) 'Imperial environmentalism or environmental imperialism? European forestry, colonial foresters and the agendas of forest management in British India, 1800–1900', pp. 3324–71 in R.H. Grove, V. Damodaran and S. Sangwan, (eds) *Nature and the Orient: the environmental history of South and South East Asia*, Oxford University Press, Delhi.

Rambo, A.T. (1982) 'Human ecology research on agroecosystems in SE Asia', *Singapore Journal of Tropical Geography* 3: 86–99.

Rangan, H. (1996) 'From Chipko to Uttaranchal: development, environment, and social protest in the Garhwal Himalaya', pp. 205–226 in R. Peet and M. Watts (eds)

Liberation Ecologies: environment, development and social movements, Routledge, London.

Rapp, A. (1987) 'Desertification', pp. 425–43 in K.J. Gregory and D.E. Walling (eds) *Human Activity and Environmental Process*, Wiley, Chichester.

Rasid, H. (1979) 'The effects of regime regulation by the Gardiner Dam on downstream geomorphic processes in the South Saskatchewan River', *Canadian Geographer* 23: 140–58.

Rawcliffe, P. (1998) *Environmental Pressure Groups in Transition*, Manchester University Press, Manchester.

Raynaut, C., with E. Grégoire, P. Janin, J. Koechlin and P.L. Delville (1998) *Societies and Nature in the Sahel*, Routledge, London.

Reardon, T. and Vosti, S.A. (1995) 'Links between rural poverty and the environment in developing countries: asset categories and investment poverty', *World Development* 23: 1495–506.

Redclift, M. (1984) *Development and the Environmental Crisis: red or green alternatives?*, Methuen, London.

Redclift, M. (1987) *Sustainable Development: exploring the contradictions*, Methuen, London.

Redclift, M. (1996) *Wasted: counting the costs of global consumption*, Earthscan, London.

Redclift, M. (1997) 'Development and global environmental change', *Journal of International Development* 9: 391–401.

Redclift, M.R. and Benton, T. (eds) (1994) *Social Theory and the Global Environment*, Routledge, London.

Reed, P. and Rothenberg, D. (eds) (1993) *Wisdom in the Open Air: the Norwegian roots of deep ecology*, University of Minnesota Press, Minneapolis.

Rees, W.A. (1978) 'The ecology of the Kafue Lechwe as affected by the Kafue Gorge hydroelectric scheme', *Journal of Applied Ecology* 15: 205–17.

Reeves, R.R. and Chaudhry, A.A. (1998) 'Status of the Indus River dolphin *Platanista minor*', *Oryx* 32: 35–44.

Reich, C. (1970) *The Greening of America*, Random House, New York.

Reining, C.C. (1966) *The Zande Scheme: an anthropological case study of economic development in Africa*, Northwestern University Press, Evanston, IL.

Repetto, R. (1987) 'Creating incentives for sustainable forest development', *Ambio* 16: 94–9.

Rich, B. (1994) *Mortgaging the Earth: the World Bank, environmental impoverishment and the crisis of development*, Earthscan, London.

Richards, M. (1997) *Missing a Moving Target? Colonist technology development on the Amazon frontier*, ODI Publications, London.

Richards, P. (1985) *Indigenous Agricultural Revolution: ecology and food production in West Africa*, Longman, London.

Richards, P. (1986) *Coping with Hunger: hazard and experiment in an African rice-farming system*, Allen and Unwin, London.

Richards, P.W. (1939) 'Ecological studies on the rain forest of southern Nigeria; 1. The structure and floristic composition of the primary forest', *Journal of Ecology* 26: 1–61.

Riddell, R. (1981) *Ecodevelopment*, Gower, Aldershot.

Roberts, N. (1998) *The Holocene: an environmental history*, Blackwell, Oxford (2nd edition).

Robin, L. (1997) 'Ecology: a science of empire?', pp. 63–75 in T. Griffiths and L. Robin, (eds) *Ecology and Empire: environmental history of settler societies*, Keele University Press, Keele.

Robinson, M. (1993) 'Governance, democracy and conditionality: NGOs and the New Policy Agenda', in A. Clayton (ed.) *Governance, Democracy and Conditionality: what role for NGOs?*, INTRAC, Oxford.

Robinson, N. (ed.) (1993) *Agenda 21: earth's action plan*, IUCN Environmental Policy and Law Paper 27, Oceana Publications, New York.

Robinson, R.E. (1971) *Developing the Third World: the experience of the 1960s*, Cambridge University Press, Cambridge.

Rocheleau, D., Steinberg, P.E. and Bengamin, P.A. (1995) 'Environment, development, crisis and crusade: Ukambani, Kenya, 1890–1990', *World Development* 23: 1037–51.

Rocheleau, D., Thomas-Slayter, B. and Wangari, E. (1996a) 'Gender and environment: a feminist political ecology perspective', pp. 3–23 in D. Rocheleau, B. Thomas-Slayter and E. Wangari (eds) *Feminist Political Ecology: global issues and local experiences*, Routledge, London.

Rocheleau, D., Thomas-Slayter, B. and Wangari, E. (eds) (1996b) *Feminist Political Ecology: global issues and local experiences*, Routledge, London.

Roddick, J. (1997) 'Earth Summit north and south: building a safe house in the winds of change', *Global Environmental Change* 2: 147–65.

Roder, W. (1994) *Human Adjustment to Kainji Reservoir in Nigeria: an assessment of the economic and environmental consequences of a major man-made lake in Africa*, University Press of America, Lanham, MD.

Rodhe, H., Cowling, E., Galbally, I., Galloway, J. and Herrera, R. (1988) 'Acidification and regional air pollution in the tropics', pp. 3–39 in H. Rodhe and R. Herrera (eds) *Acidification in Tropical Countries*, Wiley, Chichester.

Roe, E. (1991) 'Development narratives, or making the best of blueprint development', *World Development* 19: 287–300.

Roe, E. (1995) 'Except-Africa: postscript to a special section on development narratives', *World Development* 23: 1065–9.

Roggeri, H. (1985) *African Dams: impacts on the environment*, Environment Liaison Centre, Nairobi.

Röling, N.G. and Wagemakers, A.E. (1998) 'A new practice: facilitating sustainable agriculture', pp. 3–22 in N.G. Röling and A.E. Wagemakers (eds) *Facilitating Sustainable Agriculture: participatory learning and adaptive management in times of environmental uncertainty*, Cambridge University Press, Cambridge.

Rose, C.I. (1993) 'Beyond the struggle for proof: factors changing the environmental movement', *Environmental Values* 2: 285–98.

Rosenberg, D.M., Bodaly, R.A. and Usher, P.J. (1995) 'Environmental and social impacts of large-scale hydro-electric development: who is listening?', *Global Environmental Change* 5: 127–58.

Rosenweig, C. and Parry, M.L. (1994) 'Potential impact of climate change on world food supply', *Nature* 367 (13 January): 133–8.

Rostow, W.W. (1978) *The World Economy: history and prospect*, Macmillan, London.

Roszak, T. (1979) *Person/planet: the creative disintegration of industrial society*, Victor Gollancz, London.

Rowbotham, E.J. (1996) 'Legal obligations and uncertainties: the climate change convention', pp. 32–50 in T. O'Riordan and J. Jäger (eds) *Politics of Climate Change: a European perspective*, Routledge, London.

Rowell, A. (1996) *Green Backlash: global subversion of the environment movement*, Routledge, London.

Rowley, J. (1999) 'Beyond 6 billion', *People and the Planet* 8(1): 3.

Rubin, N. and Warren, W.M. (eds) (1968) *Dams in Africa: an interdisciplinary study of man-made lakes in Africa*, Frank Cass, London.

Rudd, J.W.M., Harris, R., Kelly, C.A. and Hecky, R.E. (1993) 'Are hydroelectric reservoirs significant sources of greenhouse gases?', *Ambio* 22: 246–8.

Rudé, G. (1980) *Ideology and Popular Protest*, Lawrence and Wishart, London.

Rudel, T. and Roper, J. (1997) 'The paths of rainforest destruction: cross-national patterns of tropical deforestation', *World Development* 25: 53–65.

Russell, C.S. (1975) 'Environment and development', *Biological Conservation* 7: 227–37.

Russell, E.J. (1954) *World Population and World Food Supplies,* Allen and Unwin, London.

Rydzewski, J.R. (1990) 'Irrigation: a viable development strategy?', *Geographical Journal* 150: 175–80.

Rzóska, J. (ed.) (1976) *The Nile, Biology of an Ancient River*, Junk, The Hague.

Sachs, I. (1979) 'Ecodevelopment: a definition', *Ambio* 8(2/3): 113.

Sachs, I. (1980) *Stratégies de l'écodéveloppement*, Les Éditions Ouvrières, Paris.

Sachs, W. (1992a) 'Introduction', pp. 1–5 in W. Sachs (ed.) *The Development Dictionary: a guide to knowledge as power*, Witwatersrand University Press, Johannesburg, and Zed Books, London.

Sach, W. (1992b) 'Environment', pp. 26–37 in W. Sachs (ed.) *The Development Dictionary: a guide to knowledge as power*, Witwatersrand University Press, Johannesburg, and Zed Books, London

Sagua, V.O. (1978) 'Flood control of the River Niger at Kainji Dam, Nigeria and its use during drought conditions, the period 1970–1976', pp. 127–40 in G.J. van Apeldoorn (ed.) *The Aftermath of the 1972–4 Drought in Nigeria*, Centre for Social and Economic Research, Ahmadu Bello University, Zaria, Nigeria.

Said, E. (1979) *Orientalism*, Pantheon, New York.

Salem-Murdock, M. and Horowitz, M.M. (1991) 'Monitoring development in the Senegal river basin', *Development Anthropology Network* 9(1): 8–15.

Salisbury, E.J. (1964) 'The origin and early years of the British Ecological Society', *Journal of Ecology* 52: 13–18.

Sandbach, F. (1978) 'Ecology and the "limits to growth" debate', *Antipode* 10: 22–32.

Sandbach, F. (1980) *Environment, Ideology and Policy*, Blackwell, Oxford.

Sandford, S. (1983) *Management of Pastoral Development in the Third World*, Wiley, Chichester.

Saulei, S.M. (1984) 'Natural regeneration following clear-fell logging operations in the Gogol Valley, Papua New Guinea', *Ambio* 13(5–6): 351–4.

Schmick, M. and Wood, C.H. (eds) (1984) *Frontier Expansion in Amazonia*. University of Florida Press, Gainesville.

Schoepf, B.G. (1984) 'Man and the biosphere in Zaire', pp. 269–90 in J. Barker (ed.) *The Politics of Agriculture in Tropical Africa*, Sage, Beverly Hills, CA.

Schove, D.J. (1977) 'African droughts in the spectrum of time', pp. 38–53 in D. Dalby, R.J. Harrison Church and F. Bezzaz (eds) *Drought in Africa 2,* International African Institute, London.

Schreckenberg, K. and Hadley, M. (1991) *Economic and Ecological Sustainability of Tropical Rain Forest Management*, MAB Digest 8, UNESCO, Paris.

Schumacher, E.F. (1973) *Small is Beautiful: economics as if people mattered*, Blond and Briggs, London (paperback edition published in 1974 by Abacus).

Schuurman, F.J. (ed.) (1993) *Beyond the Impasse: new directions in development theory*, Zed Press, London.

Scoones, I. (1991) 'Wetlands in drylands: key resources for agricultural and pastoral production in Africa', *Ambio* 20: 366–71.

Scoones, I. (1994) *Living with Uncertainty: new directions in pastoral development in Africa*, IT Publications, London.

Scoones, I. (1996) 'Range management science and policy: politics, polemics and pasture in southern Africa', pp. 34–53 in M. Leach and R. Mearns (eds) *The Lie of the Land: challenging received wisdom on the African environment*, James Currey/Heinemann, London.

Scott, J. (1985) *Weapons of the Weak: everyday forms of peasant resistance*, Yale University Press, New Haven, CT.

Scudder, T. (1975) 'Resettlement', pp. 453–71 in N.F. Stanley and M.P. Alpers (eds) *Man-Made Lakes and Human Health*, Academic Press, London.

Scudder, T. (1980) 'River basin development and local initiative in African savanna environments', pp. 383–405 in D.R. Harris (ed.) *Human Ecology in Savanna Environments*, Academic Press, London.

Scudder, T. (1988) *The African Experience with River Basin Development: achievements to date, the role of institutions and strategies for the future.* Institute of Development Anthropology and Clark University, Binghamton, NY.

Scudder, T. (1991a) 'A sociological framework for the analysis of new land settlements', pp. 148–67 in M. Cernea (ed.) *Putting People First: sociological variables in rural development*, Oxford University Press, Oxford.

Scudder, T. (1991b) 'The need and justification for maintaining transboundary flood regimes: the Africa case', *Natural Resources Journal* 31(1): 75–107.

Scudder, T. and Acreman, M.C. (1996) 'Water management for the conservation of the Kafue wetlands, Zambia and the practicalities of artificial flood releases', pp. 101–6 in M.C. Acreman and G.E. Hollis (eds) *Water Management and Wetlands in Sub-Saharan Africa*, IUCN, Gland, Switzerland.

Scudder, T. and Colson, E. (1982) 'From welfare to development: a conceptual framework for the analysis of dislocated people', pp. 267–87 in A. Hansen and A. Oliver-Smith (eds) *Involuntary Migration and Resettlement: the problems and responses of dislocated people*, Westview Press, Boulder, CO.

Scudder, T. and Habarad, J. (1991) 'Local responses to involuntary relocation and development in the Zambian portion of the Middle Zambezi Valley', pp. 178–205 in J.A. Mollett (ed.) *Migrants in Agricultural Development*, Macmillan, London.

Secrett, C. (1985) *Rainforest: protecting the planet's richest resource*, Friends of the Earth, London.

Seddon, G. (1984) 'Logging in the Gogol Valley, Papua New Guinea', *Ambio* 13: 345–50.

Seeger, A. (1982) 'Native Americans and the conservation of flora and fauna in Brazil', pp. 177–90 in E.G. Hallsworth (ed.) *Socio-economic Effects and Constraints in Tropical Forest Management*, Wiley, Chichester.

Seeley, J. and Adams, W.M. (eds) (1987) *Environmental Issues in African Development Planning*, Cambridge African Monograph No. 9, African Studies Centre, University of Cambridge, Cambridge.

Seers, D. (1977) 'The new meaning of development', *International Development Review* 3: 2–7.

Seers, D. (1980) 'North–South: muddling morality and mutuality', *Third World Quarterly* 2: 681–92.

Sen, A. (1981) *Poverty and Famines: an essay on entitlement and deprivation*, Oxford University Press, Oxford.

Sessions, G. (ed.) (1995) *Deep Ecology for the 21st Century: readings on the philosophy and practice of the new environmentalism*, Shambhala, Boston, MA.

Shackley, S. (1997) 'The Intergovernmental Panel on Climate Change: consensual knowledge and global politics', *Global Environmental Change* 7: 77–9.

Shalash, S. (1983) 'Degradation of the River Nile', Parts 1 and 2, *Water Power and Dam Construction* 35(7): 7–43 and 35(8): 56–8.

Shankar Raman, T.R., Rawath, G.S. and Johnsingh, A.J.T. (1998) 'Recovery of tropical forest avifauna in relation to vegetation succession following shifting cultivation in Mizoma, north-east India', *Journal of Applied Ecology* 35: 214–31.

Shantz, H.L. and Marbut, C.F. (1923) *The Vegetation and Soils of Africa*, American Geographical Society Research Series 13, American Geographical Society and National Research Council, New York.

Sharaf el Din, S.H. (1977) 'Effects of the Aswan High Dam on the Nile flood on the estuarine and coastal circulation pattern along the Mediterranean Egyptian coast', *Limnology and Oceanography* 22: 194–207.

Sheail, J. (1976) *Nature in Trust: the history of nature conservation in Great Britain*, Blackie, Glasgow.

Sheail, J. (1984) 'Nature reserves, national parks and post-war reconstruction in Britain, *Environmental Conservation* 11: 29–34.

Sheail, J. (1985) *Pesticides and Nature Conservation: the British experience, 1950–1975*, Clarendon Press, Oxford.

Sheail, J. (1987) *Seventy-five Years of Ecology: the British Ecological Society.* Blackwell Scientific, Oxford.

Sheail, J. (1996) 'From aspiration to implementation: the establishment of the first National Nature Reserves in Britain', *Landscape Research* 21: 37–54.

Shelton, N. (1985) 'Logging versus the natural habitat in the survival of tropical forests', *Ambio* 14(1): 39–42.

Sheppe, W.A. (1985) 'Effects of human activities on Zambia's Kafue Flats ecosystem', *Environmental Conservation* 12: 49–57.

Shiva, V. (1988) *Staying Alive: women, ecology and development*, Zed Books, London.

Shiva, V. (1997) *Biopiracy: the plunder of nature and knowledge*, South End Press, Boston, MA.

Silva, J.P.O. (1997) 'In defence of the Búobío River', pp. 153–70 in A.D. Usher (ed.) *Dams as Aid: a political anatomy of Nordic development thinking*, Routledge, London.

Simon, J.L. (1981) *The Ultimate Resource*, Princeton University Press, Princeton, NJ.

Simonsen, A.H. (1995) 'Where oil kills: the Ogoni story', *Indigenous Affairs* 4: 52–7.

Sinclair, A.R. and Fryxell, J.M. (1985) 'The Sahel of Africa: ecology of a disaster', *Canadian Journal of Zoology* 63: 987–94.

Singh, A. (1986) 'Change detection in the tropical forest environment of northeast India using Landsat', pp. 237–54 in M.J. Eden and J.T. Parry (eds) *Remote Sensing and Tropical Land Management*, Wiley, Chichester.

Singh, G. (1980) 'Ecodevelopment: the environmental movement's viewpoint, pp. 1349–55 in J. Furtado (ed.) *Tropical Ecology and Development*, International Society for Tropical Ecology, Kuala Lumpur.

Singh, J.S., Singh, S.P., Saxena, A.K. and Rawat, Y.S. (1984) 'India's Silent Valley and its threatened rainforest ecosystems', *Environmental Conservation* 11: 223–33.

Singh, S. (1997) *Taming the Waters: the political ecology of large dams in India*, Oxford University Press, Delhi.

Sioli, H. (1985) 'The effects of deforestation in Amazonia', *Geographical Journal* 151: 197–203.

Skillings, R.F. (1984) 'Economic development of the Brazilian Amazon: opportunities and constraints', *Geographical Journal* 150: 48–54.

Skinner, J.R. (1992) 'Conservation of the Inner Niger Delta in Mali: the interdependence of ecological and socio-economic research', pp. 41–7 in E. Maltby, P.J. Dugan and J.C. Lefeuve (eds) *Conservation and Development: the sustainable use of wetland resources*, IUCN, Gland, Switzerland.

Slater, D. (1993) 'The geopolitical imagination and the enframing of development theory', *Transactions of the Institute of British Geographers* N.S. 18: 419–37.

Small, G.L. (1971) *The Blue Whale*, Columbia University Press, New York.

Smith, F.M., May, R.M., Pellew, R., Johnson, T.H. and Walter, K.S. (1993) 'Estimating extinction rates', *Nature* 364 (5 August): 494–6.

Smith, N. (1984) *Uneven Development*, Blackwell, Oxford.

Soerianegara, I. (1982) 'Socio-economic aspects of forest resources management in Indonesia', pp. 73–85 in E.G. Hallsworth (ed.) *Socio-economic Effects and Constraints in Tropical Forest Management*, Wiley, Chichester.

Spaargaren, G. and Mol, A.P. (1991) 'Sociology, environment and modernity: ecological modernisation as a theory of social change', *Society and Natural Resources* 5: 323–44.

Spittler, G. (1979) 'Peasants and the state in Niger (West Africa)', *Journal of Peasant Studies* 7: 30–47.

Spivak, G.C. (1990) *The Post-colonial Critic: interviews, strategies, dialogues*, Routledge, London.

Stamp, L.D. (1925) 'The aerial survey of the Irrawaddy Delta forests (Burma)', *Journal of Ecology* 13: 262–76.

Stamp, L.D. (1938) 'Land utilisation and soil erosion in Nigeria', *Geographical Review* 28: 32–45.

Stamp, L.D. (1940) 'The southern margin of the Sahara: comments on some recent studies on the question of desiccation', *Geograpical Review* 30: 297–300.

Stamp, L.D. (1953) *Our Undeveloped World*, Faber and Faber, London.

Stebbing, E.P. (1935) 'The encroaching Sahara: the threat to the West African Colonies', *Geographical Journal* 85: 506–24.

Stein, R.E. and Johnson, B. (1979) *Banking on the Biosphere? Environmental procedures and practices of nine multilateral aid agencies*, Lexington Press, New York.

Steinhart, E. (1989) 'Hunters, poachers and gamekeepers: towards a social history of hunting in colonial Kenya', *Journal of African History* 30: 247–64.

Steward, T.A., Pickett, V., Parker, T. and Feidler, P.L. (1992) 'The new paradigm in ecology: implications for conservation biology above the species level', pp. 65–88 in P.L. Feidler and S.K. Jain (eds) *Conservation Biology: the theory and practice of nature conservation, preservation and management*, Chapman and Hall, London.

Stewart, F. (1985) *Planning to Meet Basic Needs*, Macmillan, London.

Stocking, M. and Perkin, S. (1992) 'Conservation-with-development: an application of the concept in the Usambara Mountains, Tanzania', *Transactions of the Institute of British Geographers* N.S. 17: 337–49.

Stoddart, D.R. (1970) 'Our environment', *Area* 2(1): 1–3.

Stoddart, D.R. (1986) *On Geography and its History*, Blackwell, Oxford.

Stöhr, W.B. (1981) 'Development from below: the bottom-up and periphery-inward development paradigm', pp. 39–72 in W.B. Stöhr and D.R.F. Taylor (eds) *Development: from above or below?*, Wiley, Chichester.

Stolper, W. (1966) *Planning without Facts*, Harvard University Press, Cambridge, MA.

Stott, P. and Sullivan, S. (eds) (2000) *Political Ecology: science, myth and power*, Arnold, London.

Strange, S. (1986) *Casino Capitalism*, Blackwell, Oxford.

Street, F.A. and Grove, A.T. (1976) 'Environmental and climatic implications of late Quaternary lake-level fluctuations in Africa', *Nature* 261: 385–90.

Street-Perrott, F.A. and Roberts, N. (1994) 'Past climates and future greenhouse warming', pp. 47–68 in N. Roberts (ed.) *The Changing Global Environment*, Blackwell, Oxford.

Suckcharoen, S., Nuorteva, P. and Hasanen, E. (1978) 'Alarming signs of mercury pollution in a freshwater area of Thailand', *Ambio* 7: 113–16.

Sugden, R. and Williams, A. (1978) *The Principles of Practical Cost–Benefit Analysis*, Oxford University Press, Oxford.

Sullivan, F. (1993) 'Forest Principles', pp. 159–67 in M. Grubb, M. Koch, K. Thompson, A. Munson and F. Sullivan (eds) *The 'Earth Summit' Agreements: a guide and assessment*, Earthscan, London, for the Royal Institute of International Affairs.

Sullivan, S. (1999) 'The impacts of people and livestock on topographically diverse open wood- and shrub-lands in arid north-west Namibia', *Global Ecology and Biogeography* 8: 257–77.

Sulton, I. (1970) 'Dams and the environment', *Geographical Review* 60: 128–9.

Survival International (1985) *An End to Laughter? Tribal peoples and economic development*, Review 44, Survival International, London.

Sutcliffe, R.B. (1972) *Industry and Underdevelopment*, Addison Wesley, London.

Sutcliffe, R.B. (1984) 'Industry and underdevelopment re-examined', *Journal of Development Studies* 21: 121–33.

Sutlive, V.H., Altshuler, N. and Zamora, M.D. (eds) (1980) *Where Have All the Flowers Gone? Deforestation in the Third World*, Studies in Third World Societies No. 13, College of William and Mary, Williamsburg, VA.

Sutlive, V.H., Altshuler, N. and Zamora, M.D. (eds) (1981) *Blowing in the Wind: deforestation and long-range implications*, Studies in Third World Societies No. 14, College of William and Mary, Williamsburg VA.

Sutton, K. (1977) 'Population resettlement: traumatic upheavals and the Algerian experience', *Journal of Modern African Studies* 15: 279–300.

Svedin, U. (1987) 'The IUCN conference on conservation and development in Ottawa', *Ambio* 16(1): 65.

Swanson, T. and Barbier, E.B. (1992) *Economics for the Wilds: wildlife, wildlands, diversity and development*, Earthscan, London.

Swift, J. (1982) 'The future of African hunter-gatherer and pastoral people in Africa', *Development and Change* 13: 159–81.

Swift, J. (1996) 'Desertification narratives; winners and losers', pp. 73–90 in M. Leach and R. Mearns (eds) *The Lie of the Land: challenging received wisdom on the African environment*, James Currey/Heinemann, London.

Swyngedouw, E. (1997) 'Power, nature, and the city: the conquest of water and the political ecology of urbanization in Guayaquil, Ecuador, 1880–1990', *Environment and Planning A* 29: 311–32.

Tahir, A.A. (1980) 'The Sudd as a wetland ecosystem and the Jonglei Canal Project', *Water Supply and Management* 4: 53–4.

Tansley, A.G. (1911) *Types of British Vegetation*, Cambridge University Press, Cambridge.

Tansley, A.G. (1935) 'The use and abuse of vegetational terms', *Ecology* 14(3): 284–307.

Tansley, A.G. (1939) *The British Islands and their Vegetation*, Cambridge University Press, Cambridge.

Taylor, R.W.D. and Harris, A.H. (1994) 'Control of the larger grain border, *Prostephanus truncatus*, in bagged maize by fumigation under gas-proof sheets', *FAO Plant Protection Bulletin* 42: 129–37.

Thangam, E.S. (1982) 'Problems of shifting cultivation in north-eastern India', pp. 53–63 in E.G. Hallsworth (ed.) *Socio-economic Effects and Constraints in Tropical Forest Management*, Wiley, Chichester.

Thomas, D.H.L. and Adams, W.M. (1997) 'Space, time and sustainability in the Hadejia-Jama'are wetlands and the Komodugu Yobe basin, Nigeria', *Transactions of the Institute of British Geographers* N.S. 22: 430–49.

Thomas, D.H.L. and Adams, W.M. (1999) 'Adapting to dams: agrarian change downstream of Tiga Dam, Northern Nigeria', *World Development* 27. 919–35.

Thomas, D.S.G. (1984) 'Ancient ergs of the former arid zones of Zimbabwe, Zambia and Angola', *Transactions of the Institute of British Geographers* N.S. 9: 75–88.

Thomas, D.S.G. (1993) 'Sandstorm in a teacup? Understanding desertification', *Geographical Journal* 159: 318–31.

Thomas, D.S.G. and Middleton, T. (1994) *Desertification: exploding the myth*, Wiley, Chichester.

Thomas, K. (1983) *Man and the Natural World: changing attitudies in England, 1500–1800*, Allen Lane, London (paperback Penguin, Harmondsworth, 1984).

Thomas, S. (1995) 'The next 1000 million people: do we have a choice?', pp. 187–206 in S. Morse and M. Stocking (eds) *People and Environment*, UCL Press, London.

Thomas, W.L. (ed.) (1956) *Man's Role in Changing the Face of the Earth*, University of Chicago Press, Chicago.

Thompson, M. and Warburton, M. (1988) 'Uncertainty on a Himalayan scale', pp. 1–53 in J. Ives and D.C. Pitt (eds) *Deforestation: social dynamics in watersheds and mountain ecosystems*, Routledge, London.

Throup, D.W. (1987) *Economic and Social Origins of Mau Mau*, James Currey, London.

Tiffen, M. and Mortimore, M. (1994) 'Malthus controverted: the role of capital and technology in growth and environmental recovery in Kenya', *World Development* 22: 997–1010.

Tiffen, M., Mortimore, M.J. and Gichugi, F. (1994) *More People, Less Erosion: environmental recovery in Kenya*, Wiley, Chichester.

Timberlake, L. (1985) *Africa in Crisis: the causes, the cures of environmental bankruptcy*, Earthscan, London.

Tisdell, C. (1988) 'Sustainable development: differing perspectives of ecologists and economists, and relevance to LDCs', *World Development* 16: 373–84.

Tolba, M.K. (1986) 'Desertification in Africa', *Land Use Policy* 3: 260 8.

Toulmin, C. (1993) *Combating Desertification: setting the agenda for a global convention*, International Institute for Environment and Development Dryland Networks Programme Paper 42, London.

Townsend, J.T. (1995) (with Arrevillaga, U., Bain, J., Cancino, S., Frenk, S.F., Pacheco, S. and Pérez, E.) *Women's Voices from the Rainforest*, Routledge, London.

Toye, J. (1987) *Dilemmas of Development: reflections on the counter-revolution in development theory and practice*, Blackwell, Oxford.

Toye, J. (1993) *Dilemmas of Development*, Blackwell, Oxford (2nd edition).

Trapnell, C.G. (1943) *Soils, Vegetation and Agriculture of North-Eastern Rhodesia*, Government Printer, Lusaka.

Trapnell, C.G. and Clothier, J.N. (1937) *The Soils, Vegetation and Agricultural Systems of North-Western Rhodesia*, Government Printer, Lusaka.

Trenbath, B.R. (1984) 'Decline of soil fertility and the collapse of shifting cultivation systems under intensification', pp. 279–92 in A.C. Chadwick and S.C. Sutton (eds) *Tropical Rain-forest: the Leeds symposium*, Leeds Philosophical Society, Leeds.

Trumper, E.V. and Holt, J. (1998) 'Modeling pest population resurgence due to recolonisation of fields following an insecticide application', *Journal of Applied Ecology* 35: 273–85.

Tucker, C.J., Holben, B.N. and Goff, T.E. (1984) 'Intensive forest clearing in Rondônia, Brazil, as detected by satellite remote sensing', *Remote Sensing of the Environment* 15: 255–61.

Tucker, C.J., Justice, C.O. and Prince, S.D. (1986) 'Monitoring the grasslands of the Sahel, 1984–1985', *International Journal of Remote Sensing* 7: 1571–83.

Tucker, C.J., Townshend, J.E.G. and Goff, T.E. (1985) 'African land cover classification using satellite data', *Science* 277: 369–75.

Tunstall, S.M. and Coker, A. (1992) 'Survey-based valuation methods', pp. 104–26 in A. Coker and C. Richards (eds) *Valuing the Environment: economic approaches to environmental valuation*, Belhaven, London.

Tuntamiroon, N. (1985) 'The environmental impact of industrialisation', *The Ecologist* 15(4): 161–4.

Turner, B. (1994) 'Small-scale irrigation in developing countries', *Land Use Policy* 11: 251–61.

Turner, B.L. II and Brush, S.B. (eds) (1987) *Comparative Farming Systems*, Guilford, Hove.

Turner, B.L. II and Meyer, W.B. (1994) 'Global land-use and land-cover change', pp. 3–10 in W.B. Meyer and B.L. Turner II (eds) *Changes in Land Use and Land Cover: a global perspective*, Cambridge University Press, Cambridge.

Turner, D.J. (1971) 'Dams in ecology', *Civil Engineering* 41: 76–80.

Turner, M. (1993) 'Overstocking the range: a critical analysis of the environmental science of Sahelian pastoralism', *Economic Geography* 69: 402–21.

Turner, M. and Hulme, D. (1997) *Governance, Administration and Development: making the state work*, Macmillan, London.

Turner, R.K. (ed.) (1988a) *Sustainable Environmental Management: principles and practice*, Westview Press, Boulder, CO.

Turner, R.K. (1988b) 'Sustainability, resource conservation and pollution control: an overview', pp. 17–25 in R.K. Turner (ed.) *Sustainable Environmental Management: principles and practice*, Westview Press, Boulder, CO.

Turner, R.K., Bateman, I. and Brooke, J.S. (1992) 'Valuing the benefits of coastal defence: a case study of the Aldeburgh sea-defence scheme', pp. 77–100 in A. Coker and C. Richards (eds) *Valuing the Environment: economic approaches to environmental valuation*, Belhaven, London.

Turton, D. (1987) 'The Mursi and National Park development in the lower Omo Valley', pp. 169–86 in D.M. Anderson and R.H. Grove (eds) *Conservation in Africa: people, policies and practice*, Cambridge University Press, Cambridge.

UNDP (1996) *Human Development Report 1996*, Oxford University Press, Oxford, for the United Nations Development Programme.

UNDP (1999) *Human Development Report 1999*, Oxford University Press, Oxford, for the United Nations Development Programme.

UNEP (1978) *Review of Areas: environment and development and environmental management*, Report No. 3, UNEP Nairobi.

UNEP (1995) *Global Biodiversity Assessment*, Cambridge University Press, Cambridge.

UNEP (2000) *Global Environment Outlook 2000*, Earthscan, London, for the United Nations Environment Programme.

UNESCO (1963) *A Review of the Natural Resources of the African Continent*, UNESCO, Paris.

UNESCO (1973) *Programme on Man and the Biosphere (MAB). Expert Panel on Project 8: Conservation of natural areas and of the genetic material they contain. Final Report*, MAB Report 12, UNESCO, Paris.

UNESCO (1976) *Effects on Man and his Environment of Major Engineering Works*, Man and the Biosphere Report Series 37, UNESCO, Paris.w

UNESCO (not dated) 'Man and the Biosphere Programme', pamphlet, UNESCO, Paris.

United Nations (ed.) (1977) *Desertification: its causes and consequences*, Pergamon Press, Oxford.

United Nations (1993) *The Global Partnership for Environment and Developent: a guide to Agenda 21*, Post Rio Edition, United Nations, New York.

United Nations Population Division (1998) *World Population Projections to 2150*, United Nations, New York.

Upton, C. and Bass, S. (1995) *The Forest Certification Handbook*, Earthscan, London.

Usher, A.D. (1997a) 'About this book, the contributors and what this book is not about', pp. 3–10 in A.D. Usher (ed.) *Dams as Aid: a political anatomy of Nordic development thinking*, Routledge, London.

Usher, A.D. (1997b) 'Kvaerner's game', pp. 133–52 in A.D. Usher (ed.) *Dams as Aid: a political anatomy of Nordic development thinking*, Routledge, London.

Usher, A.D. (1997c) 'The mechanism of pervasive appraisal optimism', pp. 59–75 in A.D. Usher (ed.) *Dams as Aid: a political anatomy of Nordic development thinking*, Routledge, London.

Usher, A.D. (ed.) (1997d) *Dams as Aid: a political anatomy of Nordic development thinking*, Routledge, London.

Usher, A.D. and Ryder, G. (1997) 'Vattenfal abroad: damming the Theun River', pp. 77–104 in A.D. Usher (ed.) *Dams as Aid: a political anatomy of Nordic development thinking*, Routledge, London.

Vainio-Mattila, A. (1987) *Domestic Fuel Economy*, Bura Fuelwood Project, Institute of Development Studies Report 13, University of Helsinki.

van Apeldoorn, G.J. (1980) *Perspectives on Drought and Famine in Nigeria*, Allen and Unwin, Hemel Hempstead.

van der Schalie, H. (1960) 'Egypt's new high dam: asset or liability?', *Biologist* 43: 63–70.

Van Pelt, M.J.F., Kuyvenhoven, A. and Nijkamp, P. (1990) 'Project appraisal and sustainability: methodological challenges', *Project Appraisal* 5: 139–58.

Vanclay, F. (1999) 'Social impact assessment', pp. 301–26 in J. Petts (ed.) *Handbook of Environmental Impact Assessment*, Blackwell Science, Oxford.

Vanclay, F. and Bronstein, D.A. (1995) *Environmental and Social Impact Assessment*, Wiley, Chichester.

Veldman, M. (1994) *Fantasy, the Bomb and the Greening of Britain: romantic protest, 1945–1980*, Cambridge University Press, Cambridge.

Verhoef, H. (1996) 'Health aspects of floodplain development', pp. 35–50 in M.C. Acreman and G.E. Hollis (1996) *Water Management and Wetlands in Sub-Saharan Africa*, International Union for the Conservation of Nature, Gland, Switzerland.

Verstraete, M.M. (1986) 'Defining desertification: a review', *Climatic Change* 9: 5–18.

References 431

Vickers, W.T. (1984) 'Indian policy in Amazonian Ecuador', pp. 8–32 in M. Schmick and C.H. Wood (eds) *Frontier Expansion in Amazonia*, University of Florida Press, Gainesville.

Vincent, L. (1992) 'Sustainable small-scale irrigation development: issues for farmers, governments and donors', *Water Resources Development* 6: 250–9.

Vincent, L. (1995) *Hill Irrigation: water and development in mountain agriculture*, London: Intermediate Technology Publications for the Overseas Development Institute.

Vitousek, P.M., Ehrlich, P.R., Ehrlich, A.H. and Matson, P.A. (1986) 'Human appropriation of the products of photosynthesis', *BioScience* 36: 368–73.

Vogt, W. (1949) *Road to Survival*, Victor Gollancz, London.

Waddington, C.H. (1975) 'The origin', pp. 4–11 in E.B. Worthington (ed.) *The Evolution of the IBP*, Cambridge University Press, Cambridge.

Wade, R. (1982) 'The system of administrative and political corruption: canal irrigation in south India', *Journal of Development Studies* 18: 287–328.

Wainwright, C. and Wehrmeyer, W. (1998) 'Success in integrating conservation and development? A study from Zambia', *World Development* 26: 933–44.

Wallace, T. (1980) 'Agricultural projects and land in northern Nigeria', *Review of African Political Economy* 17: 59–70.

Wallace, T. (1981) The Kano River Project, Nigeria: the impact of an irrigation scheme on productivity and welfare, pp. 281–305 in J. Heyer, P. Roberts and G. Williams (eds) *Rural Development in Tropical Africa*, Macmillan, London.

Walls, J. (1984) 'Summons to action', *Desertification Control Bulletin* 10: 5–14.

Walsh, J. (1985) 'Onchocerciasis: river blindness' , pp. 269–94 in A.T. Grove (ed.) *The Niger and its Neighbours: environment, history and hydrobiology, human use and health hazards of the major West African rivers*, Balkema, Rotterdam.

Walter, I. and Ugelow, J.L. (1979) 'Environmental problems in developing countries', *Ambio* 8(2–3): 102–9.

Ward, B. and Dubos, R. (1972) *Only One Earth: the care and maintenance of a small planet*, André Deutsch, London.

Ward, J.V. and Stanford, J.A. (eds) (1979) *The Ecology of Regulated Streams*, Plenum Press, New York.

Warren, A. (1993) 'Desertification as a global environmental issue', *GeoJournal* 31: 11–14.

Warren, A. (1996) "Desertification', pp. 342–55 in W.M. Adams, A.S. Goudie and A. Orme (eds) *The Physical Geography of Africa*, Oxford University Press, Oxford.

Warren, A. and Khogali, M. (1992) *Assessment of Desertification and Drought in the Sudano-Sahelian region, 1985–1991*, United Nations Sudano-Sahelian Office, New York.

Warren, A. and Maizels, J.K. (1977) 'Ecological change and desertification', pp. 169–261 in Secretariat of the United Nations Conference (eds) *Desertification: its causes and consequences*, Pergamon Press, Oxford.

Warren, B. (1980) *Imperialism: pioneer of capitalism*, New Left Books/Verso, London.

Warren, K.J. (1994) (ed.) *Ecological Feminism*, Routledge, London.

Washbourn, C. (1967) 'Lake levels and Quaternary climates in the eastern Rift Valley of Kenya', *Nature* 216: 672–3.

Waterbury, J. (1979) *Hydropolitics of the Nile Valley*, Syracuse University Press, New York.

Wathern, P. (ed.) (1988) *Environmental Impact Assessment: theory and practice*, Unwin Hyman, London.

Watson, C. (1986) 'Working at the World Bank', pp. 268–75 in T. Hayter and C. Watson (eds) *Aid: rhetoric and reality*, Pluto Press, London.

Watson, E.E., Adams, W.M. and Mutiso, S.K. (1997) 'Indigenous irrigation, agriculture and development, Marakwet, Kenya', *Geographical Journal* 164: 67–84.

Watts, M.J. (1983a) *Silent Violence: food, famine and peasantry in northern Nigeria*, University of California Press, Berkeley.

Watts, M.J. (1983b) 'On the poverty of theory: natural hazards research in context', pp. 231–62 in K. Hewitt (ed.) *Interpretations of Calamity from the Viewpoint of Human Ecology*, Allen and Unwin, London.

Watts, M.J. (1984) 'The demise of the moral economy: food and famine in the Sudano-Sahelian region in historical perspective', pp. 128–48 in E.P. Scott (ed.) *Life before the Drought*, Allen and Unwin, London.

Watts, M.J. (1987) 'Drought, environment and food security: some reflections on peasants, pastoralists and commoditisation in dryland East Africa', pp. 171–217 in M.H. Glantz (ed.) *Drought and Hunger in Africa: denying famine a future*, Cambridge University Press, Cambridge.

Watts, M.J. (1989) 'The agrarian question in Africa: debating the crisis', *Progress in Human Geography* 13: 1–14.

Watts, M.J. (1993) 'Living under contract: work, production problems and the manufacture of discontent in a peasant society', pp. 65–105 in A. Pred and M.K.J. Watts (eds) *Reworking Modernity: capitalism and symbolic discontent*, Rutgers University Press, New Brunswick, NJ.

Watts, M.J. (1995) 'A new deal in emotions: theory, practice and the crisis of development', pp. 44–62 in J. Crush (ed.) *Power of Development*, Routledge, London.

Watts, M.J. and Bohle, H.G. (1993) 'The space of vulnerability: the causal structure of hunger and famine', *Progress in Human Geography* 17: 43–67.

Weber, M. (1922) *Economy and Society: an outline of interpretive sociology*, eds G. Roth and C. Wittich, University of California Press, Berkeley (1978 edition)

Weiskel, T. (1988) 'Toward an archaeology of colonialism: elements of ecological transformation of the Ivory Coast', pp. 141–71 in D. Worster (ed.) *The Ends of the Earth: perspectives on modern environmental history*, Cambridge University Press, Cambridge.

Welcomme, R.L. (1979) *The Fisheries Ecology of Floodplain Rivers*, Longman, London.

Welford, R. and Starkey, R. (1996) *The Earthscan Reader in Business and the Environment*, Earthscan, London.

Wellings, P. (1983) 'Making a fast buck in Lesotho: capital leakage and the public accounts of Lesotho', *African Affairs* 82: 495–507.

Wells, M. and Brandon, K. (1992) *People and Parks: linking protected area management with Local Communities*, World Bank, Washington, DC.

Wesseling, J.W., Naah, E., Drijver, C.A. and Ngantou, D. (1996) 'Rehabilitation of the Logone floodplain, Cameroon, through hydrological management', pp. 185–98 in M.C. Acreman and G.E. Hollis (eds) *Water Management and Wetlands in Sub-Saharan Africa*, International Union for the Conservation of Nature, Gland, Switzerland.

Western, D. (1982) 'Amboseli National Park: enlisting landowners to conserve migratory wildlife', *Ambio* 11: 302–8.

Western, D. and Wright, R.M. (1994) 'The background to community-based conservation', pp. 1–14 in D. Western, R.M. White and S.C. Strum (eds) *Natural Connections: perspectives in community-based conservation*, Island Press, Washington, DC.

Westing, A.H. (1981) 'A world in balance', *Environmental Conservation* 8: 177–83.

White, E. (1969) The place of biological research in the development of the resources of man-made lakes', pp. 37–48 in E.L. Obeng (ed.) *Man-Made Lakes*, Ghana Universities Press, Accra.

White, S. (1978) 'Cedar and mahogany logging in eastern Peru', *Geographical Review* 68: 394–416.

White, S.C. (1996) 'Depoliticising the environment: the uses and abuses of participation', *Development in Practice* 6: 6–15.

Whitton, B.A. (1975) *River Ecology*, Blackwell, London.

Wiener, M.J. (1981) *English Culture and the Decline of the Industrial Spirit, 1850–1980*, Cambridge University Press, Cambridge.

Wiggins, S. (1995) Change in African farming systems between the mid-1970s and the mid-1980s', *Journal of International Development* 7: 807–48.

Williams, G. (1976) 'Taking the part of the peasants: rural development in Nigeria and Tanzania', pp. 131–54 in P. Gutkind and I. Wallerstein (eds) *The Political Economy of Contemporary Africa*, Sage, Beverly Hills, CA.

Williams, G. (1981) 'The World Bank and the peasant problem', pp. 16–51 in J. Heyer, P. Roberts and G. Williams (eds) *Rural Development in Tropical Africa*, Macmillan, London.

Williams, G.J. (1977) 'The Kafue Hydroelectric Scheme and its environmental setting', pp. 13–27 in G.J. Williams and G.W. Howard (eds) *Development and Ecology in the Lower Kafue Basin in the Nineteen Seventies*, Kafue Basin Research Committee, Lusaka.

Williams, G.J. and Howard, G.W. (eds) (1977) *Development and Ecology in the Lower Kafue Basin in the Nineteen Seventies*, Kafue Basin Research Committee, Lusaka.

Williams, M. (1989) 'Deforestation: past and present', *Progress in Human Geography* 13: 176–208.

Williams, M. (1994) Forests and tree cover', pp. 97–124 in W.B. Meyer and B.L. Turner II (eds) *Changes in Land Use and Land Cover: a global perspective*, Cambridge University Press, Cambridge.

Williams, M.A.J. (1985) 'Pleistocene aridity in tropical Africa, Australia and Asia', pp. 219–38 in I. Douglas and T. Spencer (eds) *Environmental Change and Tropical Geomorphology*, Allen and Unwin, London.

Williams, M.A.J. and Balling, R.C. Jr (1995) 'Interactions of desertification and climate: an overview', *Desertification Control Bulletin* 26: 8–16.

Wilson, A. (1992) *The Culture of Nature: North American landscape from Disney to the 'Exxon Valdez'*, Blackwell, Oxford.

Wilson, E.O. (1992) *The Diversity of Life*, Penguin Books, Harmondsworth.

Winid, B. (1981) 'Comments on the development of the Awash Valley, Ethiopia', pp. 147–65 in S.K. Saha and C.J. Barrow (eds) *River Basin Planning: theory and practice*, Wiley, Chichester.

Winterbottom, R. and Hazelwood, P.T. (1987) 'Agroforestry and sustainable development: making the connection', *Ambio* 16(2–3): 100–10.

Wissenburg, M. (1993) 'The idea of nature and the nature of distributive justice', pp. 3–20 in A. Dobson and P. Lucardie (eds) *The Politics of Nature: explorations in Green political theory*, Routledge, London.

Wood, A. (1950) *The Groundnut Affair*, Bodley Head, London.

Wood, W.B. (1990) 'Tropical deforestation: balancing regional development demands and global environmental concerns', *Global Environmental Change* 1: 23–41.

World Bank (1981) *Accelerated Development in Sub-Saharan Africa*, World Bank, Washington, DC.

World Bank (1984a) *Tribal Peoples and Economic Development: human ecologic considerations*, World Bank, Washington, DC.

World Bank (1984b) *World Development Report 1984*, Oxford University Press, Oxford and New York.

World Bank (1984c) *Environmental Policies and Procedures of the World Bank*, Office of Environmental and Health Affairs, World Bank, Washington, DC.

World Bank (1990) *Sub-Saharan Africa: from crisis to sustainable growth*, World Bank, Washington.

World Bank (1992) *World Development Report 1992: development and the environment*, Oxford University Press, Oxford, for the World Bank.

World Bank (1996) *From Plan to Market: World Development Report 1996*, Oxford University Press, Oxford, for the World Bank.

World Bank (2000) *Entering the 21st century: the World Development Report 1999/2000*, Oxford University Press, Oxford, for the World Bank.

World Commission on Dams (2000) *Dams and Development: a new framework for decision-making*, Earthscan, London.

World Commission on Environment and Development (1987) *Our Common Future*, Oxford University Press, Oxford.

Worster, D. (1985) *Nature's Economy: a history of ecological ideas*, Cambridge University Press, Cambridge.

Worster, D. (1993) *The Wealth of Nature: environmental history and the ecological imagination*, Oxford University Press, Oxford.

Worthington, E.B. (1932) *A Report on the Fisheries of Uganda Investigated by the Cambridge Expedition to the East African Lakes 1932–33*, Crown Agents for the Colonies, London.

Worthington, E.B. (1938) *Science in Africa: a review of scientific research relating to tropical and Southern Africa*, Royal Institute of International Affairs, London.

Worthington, E.B. (1958) *Science in the Development of Africa: a review of the contribution of physical and biological knowledge south of the Sahara*, Commission for Technical Cooperation in Africa South of the Sahara and the Scientific Council for Africa South of the Sahara.

Worthington, E.B. (ed.) (1975) *The Evolution of the IBP*, Cambridge University Press, Cambridge.

Worthington, E.B. (1977) *Arid Land Irrigation in Developing Countries: environmental problems and effects*, Pergamon Press, Oxford.

Worthington, E.B. (1982a) 'World Campaign for the Biosphere', *Environmental Conservation* 9: 93–100.

Worthington, E.B. (1982b) 'The twentieth anniversary of the IBP', *Environmental Conservation* 9: 70.

Worthington, E.B. (1983) *The Ecological Century: a personal appraisal*, Cambridge University Press, Cambridge.

WRI, UNEP, UNDP and the World Bank (1996) *World Resources 1996–97: a guide to the global environment*, Oxford University Press, Oxford.

Wright, S. and Nelson, S. (1995) *Power and Participatory Development: theory and practice*, IT Books, London.

WWF (1988) 'Debt-for-nature swap in the Philippines', *WWF News* July/August: 6.

WWF (1991) *Tropical Forest Conservation*, World Wide Fund for Nature Position Paper 7, Gland, Switzerland.

WWF (1996) *The WWF 1995 Group: the full story*, World Wide Fund for Nature UK, Godalming, Surrey.

Wynne, B. (1992) 'Uncertainty in environmental learning: reconceiving science and policy in the preventative paradigm', *Global Environmental Change* 2: 111–27.

Yapa, L. (1996) 'Improved seeds and constructed scarcity', pp. 69–85 in R. Peet and M. Watts (eds) *Liberation Ecologies: environment, development and social movements*, Routledge, London.

Young, A.M. (1986) 'Eco-enterprises: eco-tourism and farming of exotics in the tropics', *Ambio* 15: 361–3.

Yuquian, L. and Qishun, Z. (1981) 'Sediment regulation problems in Sanmenxia Reservoir', *Water Supply and Management* 5: 351–60.

Zerner, C. (1994) 'Transforming customary law and coastal management practices in the Makalu islands, Indonesia, 1870–1992', pp. 80–112 in D. Western, R.M. White and S.C. Strum (eds) *Natural Connections: perspectives in community-based conservation*, Island Press, Washington, DC.

Zhao, D. and Sun, B. (1986) 'Air pollution and acid rain in China', *Ambio* 15(1): 2–5.

Index